Methods and Algorithms for Molecular Docking-Based Drug Design and Discovery

Siavoush Dastmalchi
Tabriz University of Medical Sciences, Iran

Maryam Hamzeh-Mivehroud
Tabriz University of Medical Sciences, Iran

Babak Sokouti
Tabriz University of Medical Sciences, Iran

A volume in the Advances in Medical Technologies and Clinical Practice (AMTCP) Book Series

Medical Information Science
REFERENCE
An Imprint of IGI Global

Published in the United States of America by
Medical Information Science Reference (an imprint of IGI Global)
701 E. Chocolate Avenue
Hershey PA, USA 17033
Tel: 717-533-8845
Fax: 717-533-8661
E-mail: cust@igi-global.com
Web site: http://www.igi-global.com

Library of Congress Cataloging-in-Publication Data

Names: Dastmalchi, Siavoush, 1966- editor. | Hamzeh-Mivehroud, Maryam, 1981- editor. | Sokouti, Babak, 1978- editor.
Title: Methods and algorithms for molecular docking-based drug design and discovery / Siavoush Dastmalchi, Maryam Hamzeh-Mivehroud, and Babak Sokouti, editors.
Description: Hershey, PA : Medical Information Science Reference, [2016] | Includes bibliographical references and index.
Identifiers: LCCN 2016003940| ISBN 9781522501152 (hardcover) | ISBN 9781522501169 (ebook)
Subjects: LCSH: Pharmaceutical technology. | Drugs--Design--Computer simulation. | Algorithms.
Classification: LCC RM301.25 .M47 2016 | DDC 615.1/9--dc23 LC record available at http://lccn.loc.gov/2016003940

This book is published in the IGI Global book series Advances in Medical Technologies and Clinical Practice (AMTCP) (ISSN: 2327-9354; eISSN: 2327-9370)

British Cataloguing in Publication Data
A Cataloguing in Publication record for this book is available from the British Library.

All work contributed to this book is new, previously-unpublished material. The views expressed in this book are those of the authors, but not necessarily of the publisher.

For electronic access to this publication, please contact: eresources@igi-global.com.

Advances in Medical Technologies and Clinical Practice (AMTCP) Book Series

Srikanta Patnaik
SOA University, India
Priti Das
S.C.B. Medical College, India

ISSN: 2327-9354
EISSN: 2327-9370

Mission

Medical technological innovation continues to provide avenues of research for faster and safer diagnosis and treatments for patients. Practitioners must stay up to date with these latest advancements to provide the best care for nursing and clinical practices.

The **Advances in Medical Technologies and Clinical Practice (AMTCP) Book Series** brings together the most recent research on the latest technology used in areas of nursing informatics, clinical technology, biomedicine, diagnostic technologies, and more. Researchers, students, and practitioners in this field will benefit from this fundamental coverage on the use of technology in clinical practices.

Coverage

- Patient-Centered Care
- Nutrition
- Clinical Data Mining
- Nursing Informatics
- Biomedical Applications
- Neural Engineering
- E-Health
- Clinical Nutrition
- Clinical High-Performance Computing
- Medical Imaging

IGI Global is currently accepting manuscripts for publication within this series. To submit a proposal for a volume in this series, please contact our Acquisition Editors at Acquisitions@igi-global.com or visit: http://www.igi-global.com/publish/.

The Advances in Medical Technologies and Clinical Practice (AMTCP) Book Series (ISSN 2327-9354) is published by IGI Global, 701 E. Chocolate Avenue, Hershey, PA 17033-1240, USA, www.igi-global.com. This series is composed of titles available for purchase individually; each title is edited to be contextually exclusive from any other title within the series. For pricing and ordering information please visit http://www.igi-global.com/book-series/advances-medical-technologies-clinical-practice/73682. Postmaster: Send all address changes to above address.

Titles in this Series

For a list of additional titles in this series, please visit: www.igi-global.com

Classification and Clustering in Biomedical Signal Processing
Nilanjan Dey (Techno India College of Technology, India) and Amira Ashour (Tanta University, Egypt)
Medical Information Science Reference • copyright 2016 • 463pp • H/C (ISBN: 9781522501404) • US $225.00 (our price)

Virtual Reality Enhanced Robotic Systems for Disability Rehabilitation
Fei Hu (University of Alabama, USA) Jiang Lu (University of Houston -Clear Lake, USA) and Ting Zhang (University of Alabama, USA)
Medical Information Science Reference • copyright 2016 • 383pp • H/C (ISBN: 9781466697409) • US $210.00 (our price)

Software Innovations in Clinical Drug Development and Safety
Partha Chakraborty (Cognizant Technology Solutions, India) and Amit Nagal (GVK Bioscience, India)
Medical Information Science Reference • copyright 2016 • 305pp • H/C (ISBN: 9781466687264) • US $225.00 (our price)

Handbook of Research on Holistic Perspectives in Gamification for Clinical Practice
Daniel Novák (Czech Technical University in Prague, Czech Republic) Bengisu Tulu (Worcester Polytechnic Institute, USA) and Håvar Brendryen (Norwegian Center for Addiction Research, University of Oslo, Norway)
Medical Information Science Reference • copyright 2016 • 636pp • H/C (ISBN: 9781466695221) • US $340.00 (our price)

Optimizing Assistive Technologies for Aging Populations
Yosry S. Morsi (Swinburne University of Technology, Australia) Anupam Shukla (ABV - Indian Institute of Information Technology and Management Gwalior, India) and Chandra Prakash Rathore (Oracle India Private Limited, India)
Medical Information Science Reference • copyright 2016 • 494pp • H/C (ISBN: 9781466695306) • US $220.00 (our price)

Modern Concepts and Practices in Cardiothoracic Critical Care
Adam S. Evans (Icahn School of Medicine at Mount Sinai, USA) Gregory E. Kerr (Weill Cornell Medical College, USA) Insung Chung (Icahn School of Medicine at Mount Sinai, USA) and Robin Varghese (Icahn School of Medicine at Mount Sinai, USA)
Medical Information Science Reference • copyright 2015 • 1110pp • H/C (ISBN: 9781466686038) • US $415.00 (our price)

Table of Contents

Detailed Table of Contents

The current chapter introduces different aspects of molecular docking technique in order to give an overview to the readers about the topics which will be dealt with throughout this volume. Like many other fields of science, molecular docking studies has experienced a lagging period of slow and steady increase in terms of acquiring attention of scientific community as well as its frequency of application, followed by a pronounced era of exponential expansion in theory, methodology, areas of application and performance due to developments in related technologies such as computational resources and theoretical as well as experimental biophysical methods. In the following sections the evolution of molecular docking will be reviewed and its different components including methods, search algorithms, scoring functions, validation of the methods, and area of applications plus few case studies will be touched briefly.

Molecular docking is the prediction of conformational complementarity between ligand and receptor molecule. The process of docking integrates two schematic approaches namely sampling of ligand conformations and ranking of selected conformations based on scoring functions. The authors have discussed established methodologies for molecular docking and well-known tools implementing these methods. A brief account of different classes of scoring functions such as force field based, empirical, knowledge based, and descriptor based scoring functions is given along with the exemplary implementations of these scoring functions. By replacing test and trial based ligand screening with structure based virtual screening, molecular docking has helped in shortening the duration of novel drug discovery up to some extent. However, the developments made in the field of drug discovery are assisted by the advances in the techniques of molecular docking, but there is strong need of enrichment in the techniques, especially in scoring functions, to tackle the inbound problems of de novo drug discovery.

Molecular docking is a computational technique used to predict the binding orientation of a molecule while interacting with other molecule and finally quantify the inter-molecular interactions in terms of a binding score or binding affinity. In general, every docking algorithm comprises one or more "scoring function(s)" which is/are responsible for finding a precise binding pose as well as for predicting the binding affinity. In the last two decades, a significant development has been found in the field of scoring functions. In this chapter, the authors will discuss in detail about various types of scoring functions used in the docking experiments. This chapter will get the readers acquainted with different types of scoring functions available, their theoretical background, essential components, desired properties, and the important task performed by the scoring functions. The authors will also discuss the challenges faced by the scoring functions and their recent progress.

This chapter, composed of two parts, firstly provides molecular docking overview and secondly two molecular docking case studies are described. In overview, basic information about molecular docking are presented such as description of search algorithms and scoring functions applied in various docking programs. Brief description of methods utilized in some of the most popular docking programs also applied in our experimental work is provided. AutoDock, CDOCKER, GOLD, FlexX and FRED were used for docking studies of the DC-SIGN protein, while GOLD was further used for docking studies of cathepsin K protein. Methods and results of our studies with their contribution to science and medicine are presented. Content of the chapter is therefore appropriate for public of Medicinal and Organic Chemistry as an overview of docking studies, and also for Computational Chemists at the beginning of their work as the introduction to application of molecular docking programs.

A most challenging part in docking-based virtual screening is the scoring functions implemented in various docking programs in order to evaluate different poses of the ligands inside the binding cavity of the receptor. Precise and trustable measurement of ligand–protein affinity for Structure-Based Virtual Screening (SB-VS) is therefore, an outstanding problem in docking studies. Empirical post-docking filters can be helpful as a way to provide various types of structure-activity information. Different types of interaction have been presented between the ligands and the receptor so far. Based on the diversity and importance of PLIF methods, this chapter will focus on the comparison of different protocols. The advantages and disadvantages of all methods will be discussed explicitly in this chapter as well as future sights for further progress in this field. Different classifications approaches for the protein-ligand interaction fingerprints were also discussed in this chapter.

Molecular docking plays an important role in drug discovery research by facilitating target identification, target validation, virtual screening for lead identification and lead optimization. Depending upon the nature of the disease of interest, targets can be either protein or DNA while drugs are mostly organic small molecules. Different types of molecular docking techniques like protein-protein or protein-DNA or protein-small molecule or DNA-small molecule are employed for achieving the above mentioned objectives. This chapter provides a clear idea of the position of molecular docking in drug discovery with detailed discussion on different types of molecular docking based on the varieties of interacting partners. Subsequently the authors provide a detailed list of tools that can be used for docking in drug discovery and discus some examples of molecular docking in drug discovery before concluding with a remark on future areas of improvement in molecular docking related to drug discovery.

Protein-protein docking algorithms are powerful computational tools, capable of analyzing the protein-protein interactions at the atomic-level. In this chapter, we will review the theoretical concepts behind different protein-protein docking algorithms, highlighting their strengths as well as their limitations and pointing to important case studies for each method. The methods we intend to cover in this chapter include various search strategies and scoring techniques. This includes exhaustive global search, fast Fourier transform search, spherical Fourier transform-based search, direct search in Cartesian space, local shape feature matching, geometric hashing, genetic algorithm, randomized search, and Monte Carlo search. We will also discuss the different ways that have been used to incorporate protein flexibility within the docking procedure and some other future directions in this field, suggesting possible ways to improve the different methods.

Molecular docking is a key tool in structural biology and computer-assisted drug design. Molecular docking is a method which predicts the preferred orientation of a ligand when bound in an active site to form a stable complex. It is the most common method used as a structure-based drug design. Here, the authors intend to discuss the various types of docking methods and their development and applications in modern drug discovery. The important basic theories such as sampling algorithm and scoring functions have been discussed briefly. The performances of the different available docking software have also been discussed. This chapter also includes some application examples of docking studies in modern drug discovery such as targeted drug delivery using carbon nanotubes, docking of nucleic acids to find the binding modes and a comparative study between high-throughput screening and structure-based virtual screening.

Scoring function of protein-ligand interactions is used to recognize the "native" binding pose of a ligand on the protein and to predict the binding affinity, so that the active small molecules can be discriminated from the non-active ones. Scoring function is widely used in computationally molecular docking and structure-based drug discovery. The development and improvement of scoring functions have broad implications in pharmaceutical industry and academic research. During the past three decades, much progress have been made in methodology and accuracy for scoring functions, and many successful cases have be witnessed in virtual database screening. In this chapter, the authors introduced the basic types of scoring functions and their derivations, the commonly-used evaluation methods and benchmarks, as well as the underlying challenges and current solutions. Finally, the authors discussed the promising directions to improve and develop scoring functions for future molecular docking-based drug discovery.

Molecular docking has advanced to such an extent that one can rapidly and accurately identify pharmaceutically useful lead compounds. It is being used routinely to understand molecular interactions between enzyme and ligand molecules. Several computational approaches are combined with experimental work to investigate molecular mechanisms in detail at the atomic level. Molecular docking method is also useful to investigate proper orientation and interactions between receptor and ligand. In this chapter we have discussed protein-protein approach to understand interactions between enzyme and amyloid beta (Aβ) peptide. The Aβ peptide is a causative agent of Alzheimer's disease. The Aβ peptides can be cleaved specifically by several enzymes. Their interactions with Aβ peptide and specific enzyme can be investigated using molecular docking. Thus, the molecular information obtained from docking studies might be useful to design new therapeutic approaches in treatment of Alzheimer's as well as several other diseases.

Nowadays molecular docking has become an important methodology in CADD (Computer-Aided Drug Design)-assisted drug discovery process. It is an important computational tool widely used to predict binding mode, binding affinity and binding free energy of a protein-ligand complex. The important factors responsible for accurate results in docking studies are correct binding site prediction, use of suitable small-molecule databases, consistent docking pose, high dock score with good MD (Molecular Dynamics), clarity whether the compound is an inhibitor or agonist, etc. However, still there are several limitations which make it difficult to obtain accurate results from docking studies. In this chapter, the main focus is on recent advancements in various aspects of molecular docking such as ligand sampling, protein flexibility, scoring functions, fragment docking, post-processing, docking into homology models and protein-protein docking.

Docking, a molecular modelling method, has wide applications in identification and optimization in modern drug discovery. This chapter addresses the recent advances in the docking methodologies like fragment docking, covalent docking, inverse docking, post processing, hybrid techniques, homology modeling etc. and its protocol like searching and scoring functions. Advances in scoring functions for e.g. consensus scoring, quantum mechanics methods, clustering and entropy based methods, fingerprinting, etc. are used to overcome the limitations of the commonly used force-field, empirical and knowledge based scoring functions. It will cover crucial necessities and different algorithms of docking and scoring. Further different aspects like protein flexibility, ligand sampling and flexibility, and the performance of scoring function will be discussed.

Molecular docking was earlier considered to predict the binding affinity of the receptor and ligand molecules. With the progress in computational power and developing approaches, new horizons are now opening for accurate prediction of molecular binding affinity. In the current book chapter, recent strategies for Computer-Aided Drug Designing (CADD) including virtual screening and molecular docking, encompassing molecular dynamics simulations, and binding free energy calculation methods are discussed. Brief overview of different binding free energy methods MMPBSA, MMGBSA, LIE and TI have also been given along with the recent Relaxed Complex Scheme protocol.

Foreword

Here is my stamp of approval for this book. I give it unreservedly.

This book has expertly assembled a diverse set of researchers to cover all aspects of Molecular Docking. The book will find favour with all comers, starting with those new to the area, but the specifics also ensure important lessons for all. Part of the secret of the approach is to have significant coverage of both methodologies as well as application and case studies.

Dastmalchi, who most credibly leads the editors of this book, is with the Medicinal Chemistry Department at the School of Pharmacy, at the Tabriz University of Medical Sciences, Tabriz and also has been Director of the Biotechnology Research Centre at the University. He is co-located with his co-editors Hamzeh-Mivehroud and Sokouti. The Centre encompasses Biochemistry, Genetics, Molecular Biology, Structural Biology and Molecular Design. The Molecular Design remit of the Centre has spawned an active and well balanced interest in all aspects of pharmaceutical biotechnology and computational biology including Molecular Docking and bioinformatics, just to mention these specific aspects. Also, at least some of Dastmalchi and Hamzeh-Mivehroud's interests have been supported by interludes at the University of Sydney.

The analogies of the discovery of new drug activities with the approaches taken to searches for valuable land by the explorers and navigators of the western world of the 17 and 18th Century I attribute to Richard (Dick) Cramer. Cramer, one of the doyens of the area, started working on the relationship between shape and biological activity in 1975, on a programme that became Comparative Molecular Field Analysis, the technique more often going by the name "CoMFA" alone. Mostly the "finds" of the explorers and navigators of 3 and 4 centuries ago were to be of small islands which had great import, although there are also some sizeable islands out there. So the mission in drug discovery becomes something akin to seeking and developing the "islands of activity" in "chemistry space". Of course land mass is valuable for different reasons, as there are also different drug activities.

The game changers in computation are now fast becoming algorithms that are termed artificial intelligence or machine learning. So, an important component is to accumulate the data from which the algorithms can learn, but the underlying assumption is fast computing. In principle all the required processing power exists…in the last 10 years supercomputers have broken the petaflop (10 to the power 15 floating operations per second) barrier, but now over 80 supercomputers have a performance over 1 petaflop. Biological applications can increasingly take advantage not only of clusters of multi-core processors, but also GPGPUs (general-purpose graphical processing units) and highly parallel coprocessors. Research teams can use local servers, shared High Performance Computer (HPC) centres, or turn to private cloud providers to create their own virtual computing environment. This volume has within it the clues about the algorithms that will be providing information on drug activity.

In the context of this book, there is no better way to point out recognition of the role of closely related areas than to point to the award of the 2013 Nobel Prize for Chemistry to Martin Karplus, Michael Levitt, and Arieh Warshel "for the development of multiscale models for complex chemical systems". They have contributed to computational chemistry at many levels for many years, and for those unfamiliar with all their work, be assured that many of their contributions are captured in much existing software. By all means seek the individual published works of researchers, but also be reassured that the output of researchers the calibre of these Nobel Prize winners has undergone much scrutiny. Remember that in science, there should be no obligation to fully trust any dogma. Helping one's faith can be important, and this is often achieved by great communicators and their willingness to communicate. I have crossed paths with Michael Levitt, none of the times more notable than in 1984, at the start of my career in research, when he was briefly in-house at the University of Sydney as a guest of the Inorganic Chemistry Department. He gave a series of lectures- speaking about *Protein Structure and Dynamics* - to chemists, albeit primarily to structural chemists and biochemists and protein chemists. A willingness to communicate can easily kick off a career, but communication takes the form of both the spoken and written word. The written word is amply on display and accessible in this volume.

Computational fields have generally blossomed due to the advances in computer architecture, but so too have areas of endeavour using three-dimensional structure. While I might ask you to assume for the moment that methods in small molecule structure determination have improved, it is a simple matter to cite the growth of the protein database, itself a protein structure database, and therefore representing a significant source of information on target structures of interest… after its establishment in 1972 it hit 114,697 entries at the end of 2015. It is a more simple matter to cite the protein structure numbers as there is really only the one database for protein structure, and hopefully there is nothing like a simple statistic to be convincing. Be aware that the depositions can represent structures with a different ligand or mutant proteins, rather than solely a new protein structure, but they can also merely represent a protein solved in different conditions. Unsurprisingly over the period of the accrual of these database entries all methodological aspects of protein structure determination have developed, and as these entries themselves are models, this is also an important consideration relevant to achieving a true comprehension of our biological world.

This book is a "has it all" volume for those wanting entry into the field of Structure-based Drug Design, with coverage from the theoretical to practical, and also from the here-and-now to the possible in the near future. The pedigree is right for these editors to produce a volume on Molecular Docking-Based Drug Design and Discovery. While the techniques discussed in the book are crucial, so too is the fundamental knowledge of the Drug Discovery field, including the interplay with biology, medicinal and pharmaceutics. The exquisite requirements we want to design into a therapeutic dictate that multidisciplinary approaches are a necessity. Let there be no mistake: let's have strong computational input in all the disciplines that feed into our studies, but clearly positive clinical studies are the ultimate- the clinic is the endpoint which ever way you look at it, but there can and should be a leading role for computational effort and algorithms at every point along the path of discovery.

Let me end by ensuring I clearly commend this book to you. The editors are most ably led by my colleague and friend Siavoush Dastmalchi, and the result is a wonderfully comprehensive achievement. The principles and learnedness of Siavoush are very worthy- a not so veiled reference to the Siavoush, another Prince of Persia!

W. Bret Church
University of Sydney, Australia

Preface

In drug discovery pipeline, one can have a feeling much like a "finding a needle in the haystack". Undoubtedly, computational approaches have evolved as a pivotal component to rationalize the drug design and development processes. Molecular docking, a crucial tool in computer-assisted drug design, has secured its position in modern structure–based drug discovery. Molecular docking tries to predict predominant intermolecular interactions between ligands and the target macromolecules with known (or predicted) three-dimensional structures. Orientation and conformational geometry of the ligand in its docked position are determined through defined force fields. Information obtained from docking studies in conjunction with biophysical experiments can provide a framework for engineering bioactive molecules with controlled interaction profile toward the target of interest. *Methods and Algorithms for Molecular Docking-Based Drug Design and Discovery* is a book with total of 15 chapters. It aims to provide comprehensive coverage and in-depth description of the basic concepts of molecular docking with special emphasis on search algorithms, scoring functions, validation methods, and applications of the docking studies taking into account the recent advances in the field. The case studies mostly presented in chapters included in the second volume give the readers demonstrated examples of the application of docking methods on specific problems. These volumes are useful for researchers both in academia and industry as well as the students working in the field of theoretical and computational chemistry. The current volume contains fifteen chapters focusing on basic concepts, algorithms and techniques in molecular docking, although the methodological aspects were also touched in couple of chapters in second volume

The introductory chapter, chapter 1 titled "Molecular Docking at a Glance", is authored by Maryam Hamzeh-Mivehroud, Babak Sokouti and Siavoush Dastmalchi has tried to give a light overview on docking algorithms and studies. In this chapter, the overall view of a molecular docking approach by considering its different stages including different methods, search algorithms, scoring functions, validation of the methods, and their applications in some case studies is discussed. Moreover, some of the limitations faced in the docking studies are categorized.

The molecular docking has emerged to shorten the process of developing novel drugs in drug discovery field. AA T Naqvi and Md Imtaiyaz Hassan are the authors of Chapter 2 titled "Methods for Docking and Drug Designing: Docking Methods". The authors in this chapter have discussed various recent docking developments along with some useful known tools for docking studies. Different classification of scoring function has also been discussed with some recent examples. The need for more improvements of the developed techniques in terms of performance and hardware has also been emphasized.

"Scoring Functions in Docking Experiments" constitutes the chapter 3 of this book in which Pravin Ambure and Kunal Roy have contributed. The in-depth discussion on various types of scoring functions, and whether they are used in isolation or in combination in the docking studies for finding the relevant pose of the ligands on the receptors are the issues put forward in this chapter. Their theoretical background has also been given special consideration by discussing the challenges and limitations faced by these functions.

The authors Marjana Novič, Tjaša Tibaut, Marko Anderluh, Jure Borišek, and Tihomir Tomašič have presented Chapter 4; "The Comparison of Docking Search Algorithms and Scoring Functions: An Overview and Case Studies". The chapter is organized in two sections including the basic overview on molecular docking and case studies. Application of docking tools including AutoDock, FRED, CDOCKER, FlexX, and GOLD are demonstrated on two target systems, namely DC-SIGN and catepsin K proteins. The differences in the results of these methods in terms of scoring functions and search algorithms are also discussed.

Chapter 5 titled "Protein Ligand Interaction Fingerprints" is authored by Ali HajEbrahimi, Hamioreza Ghafouri, Mohsen Ranjbar, and Amirhossein Sakhteman. Scoring functions are regarded as the most important part of a docking-based virtual screening study. It has also been outlined that the affinity values measured for structure-based virtual screening require additional post-docking filters for achieving better results. Based on diversity of PLIF algorithms in the literature, comparison of the methods and their advantages and disadvantages, as well as their classifications considering the future trends are also discussed.

"Different Types of Molecular Docking Based on Variations of Interacting Molecules: Variations of Molecular Docking" have been contributed by Amit Das and Simantu Bhattacharya in Chapter 6. In this chapter, by considering the types of diseases, a detailed discussion on various docking methodologies including protein-protein, protein-DNA, protein-small molecule, and DNA-small molecule is covered. Also, the relevant tools and software are listed and their applications in terms of molecular docking in drug discovery with some useful examples are highlighted.

Chapter 7 "Protein-Protein Docking: Are We There Yet?" presented by Horia Jalily Hasani and Khaled Barakat demonstrates the current position of protein-protein docking studies at the atomic level. In this chapter, various methodologies used for search and scoring techniques along with their pros and cons, and their related case studies are discussed. Additionally, incorporation of the term "protein flexibility" in docking studies has received a considerable attention. Authors provided informative future directions for attaining more improvements on flexible protein docking.

In Chapter 8 titled "Protein-Ligand Docking Methodologies and Applications in Drug Discovery", the authors Sanchaita Rajkhowa and Ramesh C. Deka have provided diverse details on types of docking methods by considering the receptor's side chain flexibility as a major problem. In this chapter, an ensemble set of proteins is selected and local move Monte Carlo method (LMMC) is used for ligand sampling. In a situation that there is no known docking and scoring algorithm method covering all requirements for obtaining better results, and also the relationship between accuracy and computational cost is not fully understood, the authors have emphasized that still virtual screening is so fast and less expensive compared to experimental high throughput screening methods, which justifies its routine application in drug discovery projects.

Scoring functions are playing important roles in predicting the binding affinity by selecting the active molecules from non-active ones which have had many progresses in recent decades. Chapter 9 "Scoring Functions of Protein-Ligand Interactions" authored by Zhiqiang Yan and Jin Wang discusses the promising role of these functions in achieving good results in terms of selecting optimized binding affinity values in the pool of various binding poses of ligands. It has also been noted that several factors including the solvation, entropic effects, and binding specificity can be fused for further optimization of the results gained from the scoring functions to make a balance between accuracy and computational costs.

Kailas Dashrath Sonawne and Maruti Kayram Dhanavade describe identifying the enzyme-ligand interactions using the docking techniques in Chapter 10. This helps the researchers to get a brief mind map on the docking process and its algorithms by joining the experimental outcomes to understand the interactions between several enzymes and amyloid beta peptide and find out the cause of Alzheimer disease at the molecular level.

Chapter 11 by Vijay Kumar Srivastav, Vineet Singh, and Meena Tiwari, discusses the latest progresses achieved in various stages of molecular docking in terms of ligand sampling, scoring functions, and algorithms. Besides, two case studies based on inverse docking for identifying the anti-cancer targets and predicting the homology models for dopamine and CCR2 receptors have been demonstrated by considering their thermodynamic aspects.

Chapter 12 focuses on the recent advancements in docking methodologies. In this chapter Ashwani Kumar, Ruchika Goyal and Sandeep Jain have covered recent advances in the docking methodologies such as fragment docking, covalent docking and inverse docking. Moreover, the progress in scoring functions and searching methods are highlighted.

Chapter 13 titled "Current Trends in Docking Methodologies" contributed by Shubhandra Tripathi, Akhil Kumar, Amandeep Kaur Kahlon and Ashok Sharma describes the role of molecular docking in high throughput screening as well as virtual screening. In addition, molecular dynamics simulation as a tool for validation of docking results is explained. The importance of Relaxed Complex Scheme (RCS) as a new methodology considering receptor flexibility is also reviewed.

To do the molecular docking, availability of the three dimensional (3D) structure of the target protein is a priori knowledge. Homology modeling can serve as a valuable tool for the prediction of 3D structure for the proteins. This strategy is extensively used where there is no experimentally solved structure for the protein of interest. Chapter 14 authored by Akanksha Gupta, Pallavi Mohanty, and Sonika Bhatnagar gives a brief overview on homology modeling based protein structure prediction. Different steps involved in homology modeling are discussed.

In the last chapter of this volume, i.e. Chapter 15, Adriana Isvoran introduces the available online molecular docking resources along with many useful illustrative and practical examples.

In conclusion, the current volume, which consists of fifteen chapters, reviews various aspects of molecular docking technique and covers computational algorithms for searching and scoring the diverse pool of ligands poses along with many experimental and theoretical examples. Moreover, many useful online and stand alone software are introduced. We hope that the topics included in this volume will be of benefit for the researchers working in the field of molecular docking, however, by all means, we don't intend to give the impression that the book provides the full coverage of all possible topics.

Siavoush Dastmalchi
Tabriz University of Medical Sciences, Iran

Maryam Hamzeh-Mivehroud
Tabriz University of Medical Sciences, Iran

Babak Sokouti
Tabriz University of Medical Sciences, Iran
December 30, 2015

Chapter 1
Molecular Docking at a Glance

Maryam Hamzeh-Mivehroud
Tabriz University of Medical Sciences, Iran

Babak Sokouti
Tabriz University of Medical Sciences, Iran

Siavoush Dastmalchi
Tabriz University of Medical Sciences, Iran

ABSTRACT

The current chapter introduces different aspects of molecular docking technique in order to give an overview to the readers about the topics which will be dealt with throughout this volume. Like many other fields of science, molecular docking studies has experienced a lagging period of slow and steady increase in terms of acquiring attention of scientific community as well as its frequency of application, followed by a pronounced era of exponential expansion in theory, methodology, areas of application and performance due to developments in related technologies such as computational resources and theoretical as well as experimental biophysical methods. In the following sections the evolution of molecular docking will be reviewed and its different components including methods, search algorithms, scoring functions, validation of the methods, and area of applications plus few case studies will be touched briefly.

INTRODUCTION

Molecules are seldom found in isolation and all over the chemical universe they are involved in intermolecular interactions in homogenous or heterogeneous systems observed in almost all of the materials around us. The biological systems are no exceptions to this rule and the life with no question is dependent on these interactions. One of the important types of the interactions takes place between biological macromolecules known as receptors and the corresponding ligands. Understanding the molecular modes of such interactions is one of the major goals in comprehending the mechanism of actions of drug substances and drug design and discovery pipeline. Designing bioactive compounds with a controlled interaction profile toward a target of interest with minimum level of adverse effects is of great importance. In this context, molecular docking as a valuable tool in structure based drug design provides a framework to

DOI: 10.4018/978-1-5225-0115-2.ch001

engineer such molecules by visualizing structural requirements at the atomic level. Such useful information can significantly facilitate the rational drug design processes. The current chapter will introduce the molecular docking technique and its importance in computer-aided drug design. Moreover, different types of molecular docking methods, search algorithms as well as scoring functions will be presented briefly. Different examples of molecular docking applications will be also highlighted.

BACKGROUND

The inevitable need for developing new drugs originates from their inadequate efficiency and toxicity in the body. The history of drug discovery dates back to thousands of years ago wherein the ancient civilization used to employ herbal medicine as remedies for treatment of diseases. In spite of historical role of the serendipity factor throughout the traditional drug discovery, it is not regarded as a major component in modern drug discovery (circulation 2009). Drug development from inception to market requires an average time of 10-15 years and a cost of approximately a billion dollars (Campbell, Lamb, & Joseph-McCarthy, 2014; Ou-Yang et al., 2012). Being time consuming, costly, and risky, this process combines variety of tools to work more rationally in the process of introducing new drugs to the market. Late twentieth and early twenty- first century faced the substantial changes in drug design and discovery processes. The philosophy behind the rational drug design lies in logical reasoning before synthesizing any therapeutic agent for the biological evaluation. It provides a framework for prioritizing of potential lead compounds as early as possible, based on rational guidelines by employing the computational and experimental methods (Hung & Chen, 2014; Veselovsky & Ivanov, 2003). In this context, computer-aided drug design (CADD) is of great importance in shortening the development process in terms of cost and timeline. The CADD is mainly divided into two categories; ligand-based drug design (LBDD) and structure-based drug design (SBDD). In the case of LBDD, there is no information about the target of interest, and the chemical/biological features of a set of ligands are analyzed. Quantative-structure activity relationship (QSAR) techniques and pharmacophore-based design are the typical examples of LBDD strategy. If the 3D structural information of the target of interest is available, SBDD can also be utilized. In this strategy, the main focus is the design of novel ligands guided by investigating the ligand-target mutual interactions (Gani, 2007; Hung & Chen, 2014; Wilson & Lill, 2011). The starting point of SBDD approach is to identify the active site of the target molecule based on the known 3D structure of the bound natural ligand and /or drug molecules, mostly via X-ray crystallography or NMR experiments (Campbell et al., 2014; Mandal, Moudgil, & Mandal, 2009).The availability of the geometry for the target binding site may provide useful information for designing novel ligands or screening ligands from a large collection of chemical compounds based on structural and chemical complementarity features between target and ligand molecules (Gani, 2007; Mandal et al., 2009; Veselovsky & Ivanov, 2003).The candidates for synthesis and further biological assays are chosen based on the predicted binding mode and affinity as well as synthetic feasibility (Campbell et al., 2014). The most routinely used techniques in the area of SBDD are but not limited to molecular modeling, molecular dynamics, and docking techniques (Wilson & Lill, 2011).

The molecular docking is one the crucial components of the *in silico* studies which is heavily used in modern drug design and discovery efforts. Current molecular docking techniques have passed the infancy period and taking into account the tremendous advances in terms of novel computational algorithms as well as high performance computing hardware have led to its maturity to the state that is considered

Figure 1. Docking Web servers based on citation

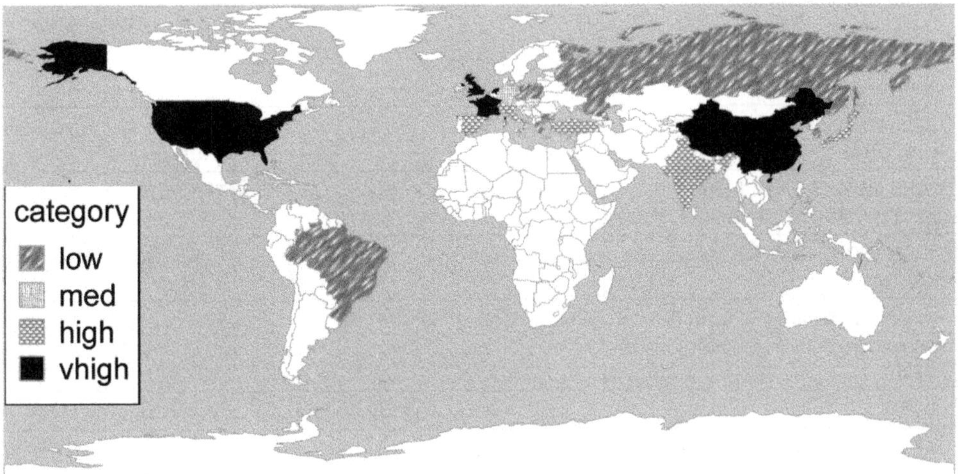

as an essential tool in rational drug design. In this method, the strength of intermolecular association is tried to be accurately predicted (Purisima & Hogues, 2012). A systematically driven statistics on the developed docking web servers all over the world and the fundamental principles of the molecular docking are described in the following sections.

WORLDWIDE USE OF MOLECULAR DOCKING

The numbers of docking web servers as a measure of popularity of docking calculations have been progressively increasing during the past decades. Figure 1 illustrates the distribution of the docking web servers based on main citation (original paper) all over the world during the period 2002-2015 which have also been listed in Table 1. In this list, docking programs are not included. Three top ranked cited web servers according to the country of origin are United States, United Kingdom and Netherlands, respectively.

To further appreciate the extent of penetration of docking studies in the scientific community, a simple query of PubMed for publications containing the terms "docking" and "ligand" was performed. The results shown in Figure 2 demonstrate that the number of docking publications from 1986 to 2014 has been increasing exponentially and it is estimated to reach about 700 by the end of 2015. This is just a naïve estimate and the exact statistics needs thorough inspection, however, it may provide a feeling for the widespread use of docking studies.

MOLECULAR DOCKING PROCESS

The therapeutical outcome observed for the drug molecules comes from a very fundamental phenomenon known as molecular recognition. Regulation of cellular signal transduction pathways and cellular processes are governed by variety of such recognition events (Sousa et al., 2013). These events occur at the atomic based level and form the foundation of ligand-receptor interactions. Binding a ligand to an

Table 1. A comprehensive list of existing molecular docking webservers and related websites

No.	Name	Type	Website Address
1	NPDock(Tuszynska, Magnus, Jonak, Dawson, & Bujnicki, 2015)	protein-nucleic acid	genesilico.pl/NPDock
2	GalaxyPepDock(H. Lee, Heo, Lee, & Seok, 2015)	protein-peptide	galaxy.seoklab.org/cgi-bin/submit.cgi?type=PEPDOCK
3	CABS-dock(Kurcinski, Jamroz, Blaszczyk, Kolinski, & Kmiecik, 2015)	protein-peptide	biocomp.chem.uw.edu.pl/CABSdock
4	DOCKSCORE(Malhotra, Mathew, & Sowdhamini, 2015)	protein-protein	caps.ncbs.res.in/dockscore
5	pyDockSAXS(Jimenez-Garcia, Pons, Svergun, Bernado, & Fernandez-Recio, 2015)	protein-protein	life.bsc.es/pid/pydocksaxs
6	webSDA(X. Yu et al., 2015)	protein/nucleic acid-protein/nucleic acid	mcm.h-its.org/webSDA
7	MTiOpenScreen(Labbe et al., 2015)	protein–ligand	bioserv.rpbs.univ-paris-diderot.fr/services/MTiOpenScreen
8	PSOVina(Ng, Fong, & Siu, 2015)	protein–ligand	cbbio.cis.umac.mo
9	14-3-3-Pred(Madeira et al., 2015)	phosphopeptides-protein	www.compbio.dundee.ac.uk/1433pred
10	AIDA(D. Xu, Jaroszewski, Li, & Godzik, 2014)	protein-protein	ffas.burnham.org/AIDA
11	ATTRACT(de Vries, Schindler, Chauvot de Beauchene, & Zacharias, 2015)	protein-protein	www.attract.ph.tum.de/services/ATTRACT/attract.html
12	CCharpPPI(Moal, Dapkunas, & Fernandez-Recio, 2015)	protein-protein	life.bsc.es/pid/ccharppi
13	CONSRANK(Chermak et al., 2015)	protein-protein/nucleic acid	https://www.molnac.unisa.it/BioTools/consrank
14	SEABED(Fenollosa et al., 2015)	protein–ligand	www.bsc.es/SEABED
15	DOCKovalent(London et al., 2014)	protein–ligand	covalent.docking.org
16	CHARMMing(Pevzner, Frugier, Schalk, Caflisch, & Woodcock, 2014)	protein–ligand	www.charmming.org/charmming
17	ClusPro(Mottarella et al., 2014)	protein-Heparin	cluspro.bu.edu
18	PEP-SiteFinder(Saladin et al., 2014)	protein-peptide	bioserv.rpbs.univ-paris-diderot.fr/PEP-SiteFinder
19	PLIC(Anand, Nagarajan, Mukherjee, & Chandra, 2014)	protein-ligand	proline.biochem.iisc.ernet.in/PLIC
20	GalaxySite(Heo, Shin, Lee, & Seok, 2014)	protein-ligand	galaxy.seoklab.org/cgi-bin/submit.cgi?type=SITE
21	DOCK/PIERR(Viswanath, Ravikant, & Elber, 2014)	protein-protein	clsb.ices.utexas.edu/web/dock.html
22	SwarmDock(Torchala & Bates, 2014; Torchala, Moal, Chaleil, Fernandez-Recio, & Bates, 2013)	protein-protein	bmm.cancerresearchuk.org/~SwarmDock
23	SPOT-Seq-RNA(Y. Yang, Zhao, Wang, & Zhou, 2014)	protein-RNA	sparks-lab.org/yueyang/server/SPOT-Struct-RNA
24	DINC(Dhanik, McMurray, & Kavraki, 2013)	protein-ligand	dinc.kavrakilab.org
25	ZDOCK (B. G. Pierce et al., 2014)	protein-protein	zdock.umassmed.edu

continued on following page

Table 1. Continued

No.	Name	Type	Website Address
26	Istar(H. Li, Leung, Ballester, & Wong, 2014)	protein-ligand	istar.cse.cuhk.edu.hk/idock
27	eSimDock(Brylinski, 2013)	protein-ligand	www.brylinski.org/esimdock
28	LigandRNA(Philips, Milanowska, Lach, & Bujnicki, 2013)	RNA-ligand	ligandrna.genesilico.pl
29	DockAFM(Chaves & Pellequer, 2013)	protein-protein/DNA	biodev.cea.fr/dockafm
30	GOMoDo(Sandal et al., 2013)	GPCR protein-ligand	molsim.sci.univr.it/gomodo
31	DockRank(Xue, Jordan, El-Manzalawy, Dobbs, & Honavar, 2014)	protein-protein	einstein.cs.iastate.edu/index.html
32	Score-MI(Jardin, Stefani, Eberhardt, Huber, & Sticht, 2013)	protein-protein	score-mi.biochem.uni-erlangen.de
33	proABC(Olimpieri, Chailyan, Tramontano, & Marcatili, 2013)	protein-ligand	circe.med.uniroma1.it/proABC
34	ROSIE(Lyskov et al., 2013)	protein-ligand	rosie.rosettacommons.org
35	CovalentDock Cloud(Ouyang, Zhou, Ge, Li, & Kwoh, 2013)	protein-ligand	docking.sce.ntu.edu.sg
36	pyDockWEB(Jimenez-Garcia, Pons, & Fernandez-Recio, 2013)	protein-protein	life.bsc.es/servlet/pydock/home
37	EpiDOCK(Atanasova, Patronov, Dimitrov, Flower, & Doytchinova, 2013)	MHC class II-ligand	epidock.ddg-pharmfac.net/ EpiDockPage.aspx
38	Sanjeevini(Jayaram et al., 2012)	protein/DNA-ligand	www.scfbio-iitd.res.in/sanjeevini/ sanjeevini.jsp
39	FINDSITE(comb)(H. Zhou & Skolnick, 2013)	protein-ligand	cssb.biology.gatech.edu/FINDSITE-COMB
40	ClusPro protein(Brenke et al., 2012; Comeau, Gatchell, Vajda, & Camacho, 2004a)	protein-protein	cluspro.bu.edu/home.php
41	Quantum.Ligand.Dock(Kantardjiev, 2012)	protein-ligand	87.116.85.141/LigandDock.html
42	idTarget(Wang, Chu, Chen, & Lin, 2012)	protein-ligand	idtarget.rcas.sinica.edu.tw
43	Pose & Rank(Fan et al., 2011)	protein-ligand	modbase.compbio.ucsf.edu/ poseandrank
44	BSP-SLIM(H. S. Lee & Zhang, 2012)	protein-ligand	zhanglab.ccmb.med.umich.edu/ BSP-SLIM
45	AADS(Singh, Biswas, & Jayaram, 2011)	protein-ligand	www.scfbio-iitd.res.in/dock/ ActiveSite_new.jsp
46	ProBiS(Konc, Konc, Penca, & Janezic, 2011)	protein-protein	probis.cmm.ki.si
47	KBDOCK(Ghoorah, Devignes, Smail-Tabbone, & Ritchie, 2011)	protein-protein	kbdock.loria.fr
48	GPU.proton.DOCK(Kantardjiev, 2011)	protein-protein docking	gpudock.orgchm.bas.bg
49	iScreen(Tsai, Chang, & Chen, 2011)	protein-ligand docking & virtual screening	iScreen.cmu.edu.tw
50	SDOCK(Zhang & Lai, 2011)	protein-protein docking	mdl.ipc.pku.edu.cn

continued on following page

Table 1. Continued

No.	Name	Type	Website Address
51	Albumin binding prediction(Zsila et al., 2011)	human serum albumin (HSA) interactions	albumin.althotas.com
52	DEPTH(Tan, Varadarajan, & Madhusudhan, 2011)	depth-based predictions of binding sites	mspc.bii.a-star.edu.sg/depth
53	PRUNE and PROBE(Mitra & Pal, 2011)	protein-protein docking	pallab.serc.iisc.ernet.in/prune & pallab.serc.iisc.ernet.in/probe
54	CPORT(de Vries & Bonvin, 2011)	protein-protein docking	haddock.chem.uu.nl/services/CPORT
55	Erebus(Shirvanyants, Alexandrova, & Dokholyan, 2011)	reliably locate rigid binding scaffolds by searching PDB	erebus.dokhlab.org
56	TPACM4(Mukherjee, Patra, Barua, & Jayaram, 2011)	modeling interactions of small molecules with biomolecular target	www.scfbio-iitd.res.in/software/drugdesign/charge.jsp
57	FILTREST3D (Gajda, Tuszynska, Kaczor, Bakulina, & Bujnicki, 2010)	scoring and ranking of models	filtrest3d.genesilico.pl
58	BSR(Brylinski, Lee, Zhou, & Skolnick, 2011)	Binding Site Refinement	cssb.biology.gatech.edu/bsr
59	IBIS(Shoemaker et al., 2012)	protein-protein, protein-small molecule, protein nucleic acids and protein-ion interaction	www.ncbi.nlm.nih.gov/Structure/ibis/ibis.cgi
60	Opal(Ren, Williams, Clementi, Krishnan, & Li, 2010)	Include AutoDock and AutoDockVina	ws.nbcr.ne
61	POLYVIEW-MM(Porollo & Meller, 2010)	Automated mapping of ligand binding sites	polyview.cchmc.org/conform.html
62	RosettaBackrub(Lauck, Smith, Friedland, Humphris, & Kortemme, 2010)	flexible backbone protein structure modeling and design	kortemmelab.ucsf.edu/backrub
63	HexServer(Macindoe, Mavridis, Venkatraman, Devignes, & Ritchie, 2010)	exhaustive GPU-accelerated FFT-based rigid-body docking predictions	hexserver.loria.fr
64	ConPlex (Choi et al., 2010)	analyze protein complex structures based on evolutionary conservation of protein-protein interface residues	sbi.postech.ac.kr/ConPlex
65	HADDOCK(de Vries, van Dijk, & Bonvin, 2010)	protein-protein, protein-nucleic acids and protein-ligand complexes	haddock.science.uu.nl/services/HADDOCK2.2
66	PharmMapper(X. Liu et al., 2010)	potential drug target identification using pharmacophore mapping approach	59.78.96.61/pharmmapper
67	AuPosSOM (Bouvier, Evrard-Todeschi, Girault, & Bertho, 2010)	Automatic clustering of docking poses in virtual screening process	www.aupossom.com
68	POCASA (J. Yu, Zhou, Tanaka, & Yao, 2010)	predict binding sites by detecting pockets and cavities of proteins with a rolling spher	altair.sci.hokudai.ac.jp/g6/Research/POCASA_e.html
69	DOCK Blaster(Irwin et al., 2009)	Free Virtual Screening Server	blaster.docking.org
70	LIGCRYST(Sondergaard, Garrett, Carstensen, Pollastri, & Nielsen, 2009)	detecting crystal contacts in protein-ligand complexes	enzyme.ucd.ie/LIGCRYS
71	FRODOCK(Garzon et al., 2009)	fast rotational protein-protein docking	sbg.cib.csic.es/Software/FRODOCK
72	CT(Bayden, Fornabaio, Scarsdale, & Kellogg, 2009)	computational titration algorithm for preparing models for virtual screening and molecular docking	hinttools.isbdd.vcu.edu/CT
73	RosettaAntibody(Sircar, Kim, & Gray, 2009)	rational judgment in choosing the best model for docking	antibody.graylab.jhu.edu

continued on following page

Table 1. Continued

No.	Name	Type	Website Address
74	@TOME-2(J. L. Pons & Labesse, 2009)	web server for comparative modeling of protein-ligand complexes	abcis.cbs.cnrs.fr/AT2
75	MRC(Rueda, Bottegoni, & Abagyan, 2009)	Generation of receptor conformation	abagyan.scripps.edu/MRC
76	SLITHER(P. H. Lee, Kuo, Chu, Liu, & Lin, 2009)	determining adjacent binding poses	bioinfo.mc.ntu.edu.tw/slither
77	DDNA 2.0(B. Xu, Yang, Liang, & Zhou, 2009)	protein-DNA docking	sparks.informatics.iupui.edu
78	RosettaDock (Lyskov & Gray, 2008)	protein-protein docking	rosettadock.graylab.jhu.edu
79	3D-Garden(Lesk & Sternberg, 2008)	protein-protein docking	www.sbg.bio.ic.ac.uk/3dgarden
80	ClusPro (Comeau et al., 2004a; Comeau, Gatchell, Vajda, & Camacho, 2004b)	protein-protein docking	cluspro.bu.edu/login.php
81	SKE-DOCK(Terashi et al., 2007)	protein-protein docking	www.pharm.kitasato-u.ac.jp/bmd/files/ske_dock.html
82	HingeProt(Emekli, Schneidman-Duhovny, Wolfson, Nussinov, & Haliloglu, 2008)	automated prediction of hinges in protein structures	bioinfo3d.cs.tau.ac.il/HingeProt
83	ROGER(Bernauer, Aze, Janin, & Poupon, 2007)	protein-protein docking scoring function	genomics.eu.org/spip/-Bioinformatics-tools
84	HADDOCK version 2.0(van Dijk & Bonvin, 2006)	Solvated docking	www.nmr.chem.uu.nl/haddock
85	KinDOCK (Martin, Catherinot, & Labesse, 2006)	analysis of ATP-binding sites of protein kinases	abcis.cbs.cnrs.fr/kindock
86	GRAMM-X(Tovchigrechko & Vakser, 2006)	protein-protein docking	vakser.bioinformatics.ku.edu/resources/gramm/grammx
87	TarFisDock(H. Li et al., 2006)	identifying drug targets with docking approach	www.dddc.ac.cn/tarfisdock
88	Scoring server(Camacho, Ma, & Champ, 2006)	scoring server for protein-protein interactions	structure.pitt.edu
89	Refining server(Lindahl & Delarue, 2005)	Refinement of docked protein-ligand and protein-DNA structures	lorentz.immstr.pasteur.fr
90	MEDock (Chang, Oyang, & Lin, 2005)	fficient prediction of ligand binding sites	bioinfo.mc.ntu.edu.tw/medock
91	Q-SiteFinder and Pocket-Finder(Laurie & Jackson, 2005)	prediction of protein-ligand binding sites	www.bioinformatics.leeds.ac.uk/pocketfinder
92	PPI_PRED(Bradford & Westhead, 2005)	prediction of protein-protein binding sites	www.bioinformatics.leeds.ac.uk/ppi_pred
93	ODA(Fernandez-Recio, Totrov, Skorodumov, & Abagyan, 2005)	predicting protein-protein interaction sites	www.molsoft.com/oda

specific target is determined by shape and feature complementarity factors (C. Sotriffer, Klebe, Stahl, & Bohm, 2003). In the case of pharmaceuticals, the interactions are mostly established through non-covalent interactions such as intermolecular van der Waals, hydrogen bonds, and electrostatic interactions (Brooijmans & Kuntz, 2003). Comprehensive study of such mutual ligand-receptor recognition originated from interactions is of great importance in structure-based drug design. One of the tools providing beneficial information on these interactions is molecular docking-based experiments.

Figure 2. An illustration of molecular docking publications during 1986-2015 (the red highlighted part in the plot demonstrates the estimated additional publications by the end of 2015)

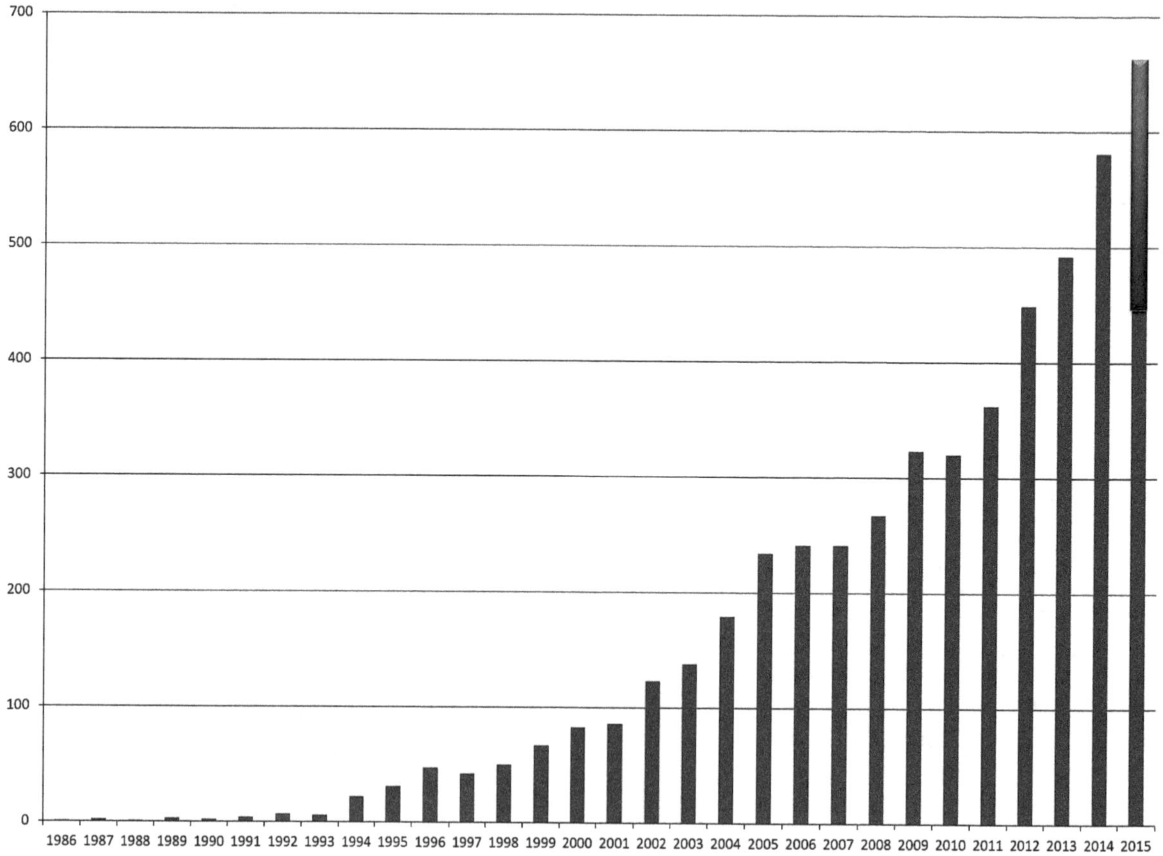

Molecular docking is a term used for prediction of location, conformation, and orientation (totally referred as pose) of a molecule to the binding site of a target biomacromolecule (Houston & Walkinshaw, 2013; Kitchen, Decornez, Furr, & Bajorath, 2004).The history of the first molecular docking experiments dates back to the mid 1970s when protein-protein docking was performed (Levinthal, Wodak, Kahn, & Dadivanian, 1975; Salemme, 1976). The early docking methods were based on the "lock-and-key" theory proposed by Emil Fischer in 1894 (Fischer, 1894; Nabuurs, Wagener, & de Vlieg, 2007). In this theory, the shape of the receptor active site and the corresponding ligand are exactly complementary and the ligand fits to the receptor like a key and lock. This theory was shifted to "induced-fit" concept in which the ligand induces the conformational changes to the receptor during the binding process (Gani, 2007).

Different purposes are followed in the molecular docking area. Those which are the most important include: i) predicting the ligand binding mode within the receptor active site; ii) identifying the novel ligands through virtual screening; and iii) predicting the compounds' binding affinities from a known active ligand (Kitchen et al., 2004; Kroemer, 2003; Leach, Shoichet, & Peishoff, 2006).

Any molecular docking experiment comprises several common key steps as listed below:

Figure 3. The workflow of molecular docking process

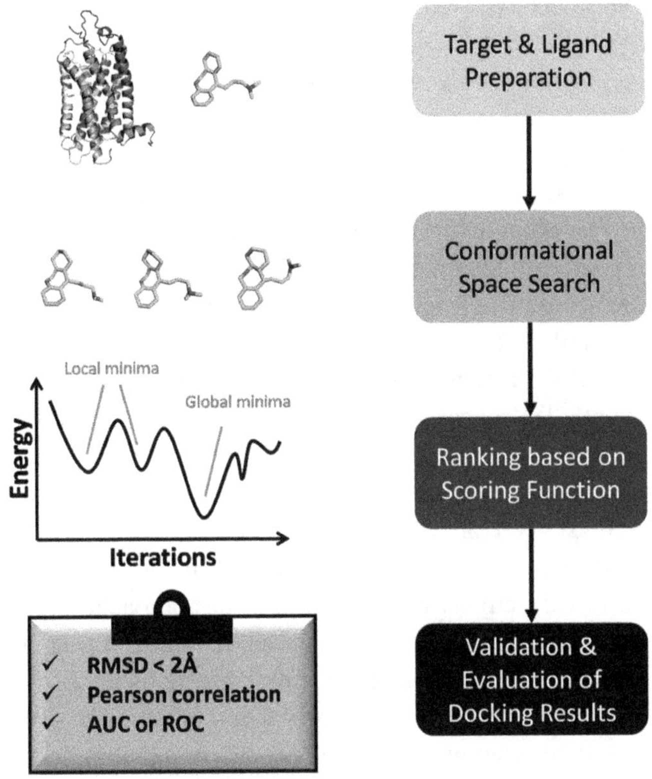

1. Target and ligand selection and representation,
2. Conformational space search,
3. Ranking of potential solution based on scoring functions, and
4. Validation and evaluation of docking results (G. M. Morris & Lim-Wilby, 2008).

Figure 3 shows a typical docking workflow. The following sections explain these steps with some more details.

MOLECULAR REPRESENTATION FOR MOLECULAR DOCKING

The two essential interplaying parts in any molecular docking are receptor and ligand molecules. The most widely used approaches for representation of the receptor relies on atomic, surface, and grid representations (Kitchen et al., 2004; C. Sotriffer et al., 2003). Atomic representation is generally used when the ranking is based on potential energy function. In some cases, surface based representation is employed where the rigid body approximation is considered. The grid representation is also applied based on potential energy grids in which the energy contribution of the receptor on the grid points (mainly electrostatic and van der Waals potentials) is calculated, then a map is generated for each atom or interaction types in order to provide the essential information for scoring during the ligand docking process (Goodford, 1985; C. Sotriffer et al., 2003).

In order to do a molecular docking experiment, the 3D-structure of the receptor which is used in the calculation is either an experimental (solved 3D structures of proteins deposited in the protein data bank) or a model structure predicted based on different prediction methodologies such as "comparative(homology and threading) modeling". In comparative method, the modeling is based on either a direct similarity between the sequences of protein of interest and that of a known structure called "template" whereas in threading or fold recognition method, the compatibility of the sequence of target protein with the structure of known folds (templates) in database is assessed (Kroemer, 2003). Three dimensional structure of protein can then be used for defining the binding site. If the structure of target along with its ligand is known (when the ligand co-crystallized with the target), the binding site is determined with certainty; otherwise, the binding site needs to be determined in several ways such as using tools to deduce protein function from structural homology or approaches derived from the inspection of physicochemical and geometrical characteristics of the receptor surface for locating the binding site. For example, cavities in the target protein, specially the biggest one, seem to be the most probable places to be defined as binding site. The catalytic site of the protein which is surrounded by the residues involved in catalysis can also be regarded as the binding site particularly if the inhibitors' docking is sought (G. M. Morris & Lim-Wilby, 2008; Veselovsky & Ivanov, 2003).

The other essential component of the molecular docking is ligand. Two strategies have been used for ligand handling: fragment-based methods and whole-molecule approaches. In both methods the binding site is considered to be known. In fragment-based method, a library of fragments is employed and the way the docking is completed proceeds by one of the following methods. In the first method, fragments are docked in a separate fashion followed by reconnection of them through searching the most favorable combination in terms of minimum energy and allowed degrees of freedom. In another method, the ligand is expanded increasingly during docking process using fragments referred to "anchor and grow". In whole-molecule approach, the entire molecule is docked as a whole (Brooijmans & Kuntz, 2003; G. M. Morris & Lim-Wilby, 2008; C. Sotriffer et al., 2003)(8, 9, 16). Different levels of approximation in terms of flexibility are used for whole molecule approaches which will be discussed in the following sections.

DOCKING CLASSIFICATION METHODS

Flexibility

In a molecular docking experiment, the flexibility is defined as six degrees of translational (along x-, y-, and z- axes) and rotational freedom. Moreover, each of ligand and receptor has its conformational degrees of freedom depending on torsion angles of each rotatable bond (Gani, 2007; Leach et al., 2006; G. M. Morris & Lim-Wilby, 2008). Since the search for all conformational space is a difficult task requiring spending much more time and computational cost, an approximation for reducing the dimensionality of the space is essential (Sousa, Fernandes, & Ramos, 2006).One of the most common classifications of molecular docking algorithms is based on extent of molecular flexibility. From the flexibility point of view, three levels of approximation in docking methods exist including: *i)* rigid docking *ii)* semi-flexible docking and *iii)* flexible docking. Earliest and simplest approaches of molecular docking were based on rigid-body approximation. In rigid docking the structural conformation of ligand and receptor are not changed during the docking procedure and both of them regarded as rigid solid bodies. In second approach i.e. semi-flexible docking, which is extensively used, the receptor is assumed as rigid structure while the conformational change will be occurred for the ligand molecule. In the case of flexible docking, conformational degrees of freedom along with translational and rotational degrees of freedom are also considered for both of ligand and receptor (Hung & Chen, 2014; Leach et al., 2006; Sousa et al., 2013). Accuracy and speed in molecular docking are two main limiting factors. Rigid-body treatment of two components (i.e. ligand and receptor) in molecular docking is faster than condition where the flexibility is included in the docking methodology (G. M. Morris & Lim-Wilby, 2008). This is due to smaller size of search space. Although high speed of a docking experiment is an advantage, the effect of different conformations of both ligand and receptor in unbound and bound states cannot be ignored. Upon binding of the ligand to the receptor (usually is a protein), significant conformational change such as rearrangement of the side chains and motion of loops and domains in the receptor structure can occur (see Koshland's induced-fit theory)(Lill, 2011). As a result, for an ideal system the flexibility of both ligand and receptor should be taken into account since there are several degrees of freedom in a real biological system. For this purpose, developing the algorithms capable of implementing these considerations is of great value. Soft-receptor modeling, side-chain flexibility, Monte Carlo (MC), molecular dynamics (MD), as well as ensemble grids docking are different categories employed in the existing methodologies (Huang & Zou, 2010a; Sousa et al., 2006). However, the receptor flexibility is still in its infancy and only limited motion for the receptor is considered in the currently used algorithms because of computing time and coverage of all conformational space. In order to enter the flexibility of the ligand, several algorithms have been developed consist of: systematic methods, random or stochastic methods, and simulation methods (Sousa et al., 2006). These methods and above mention methods will be discussed in the following sections.

Interacting Partners

Since the molecular bases of a biological response comes from the interactions formed between biological macromolecules such as enzymes, antibodies, nucleic acids and ligands which span a wide range in terms of size and molecular nature, the other classification of docking depends on considering these features of the interacting partners. Therefore, different types of docking from molecular perspective are: small molecule (e.g. drug or drug-like molecules)-protein, protein-protein, DNA/RNA-protein, and small molecule-DNA/RNA interactions (Ricci & Netz, 2009; Sousa et al., 2006; R. D. Taylor, Jewsbury, & Essex, 2002; Tuffery & Derreumaux, 2012). In most of the cases, small molecules constitute one of the participating elements in the docking experiments (Velec, Gohlke, & Klebe, 2005; Wang et al., 2012). High flexibility of small ligands makes them to be treated in small ligand-oriented docking schemes as flexible or semi-flexible. In addition, for small ligand-protein docking, the complementary contact interface is much smaller and less discriminating compared to the protein-protein docking. However, predicting three-dimensional binding pocket of the ligandis the main goal of small ligand-protein docking (Tuffery & Derreumaux, 2012). Protein-protein interactions are the source of many cellular functions and important in understanding the mechanism of signaling pathways. The most important issue in the field of protein-protein docking is presence of huge number of degrees of freedom making it impractical due to inability searching the entire conformational space. Because of this, rigid body handling of protein-protein docking is usually applied in many of the algorithms. Despite this fact, several attempts have been performed to tackle this problem (Baxter, Murray, Clark, Westhead, & Eldridge, 1998; Jerabek, Florian, Stiborova, & Martinek, 2014). Nucleic acids are another important type of targets for therapeutics like anti-bacterial, chemotherapeutic, and antiviral agents, and also is being recognized by many biological molecules such as proteins, the interaction of nucleic acids with protein or small molecules is of particular importance (Ricci & Netz, 2009). They are similar to proteins in terms of binding specificity (supplied by specific nucleic acid binding motifs) and conformational changes upon binding, but due to high charge density and unique helical geometry characteristics are distinct from proteins. Moreover, they exhibit no certain binding site and are highly solvent exposed. Studies showed that preferable binding site of protein to DNA is in its major groove whereas small ligands interact with minor groove of the DNA double helical structures (Ricci & Netz, 2009).

A BRIEF INTRODUCTION TO MOLECULAR DOCKING METHODS

Despite the various types of molecular docking protocols (i.e., protein-protein, protein-ligand, protein-RNA, protein-DNA, and etc) in which any mutual parts can be rigid or flexible, the main workflow includes searching for the correct pose with the most stable state (i.e., global minimum energy), and then evaluating the predicted conformations by relevant scoring functions for differentiating between true and all the other binding modes (Kitchen et al., 2004; Sousa et al., 2006). The computational cost of docking for placing two molecules near each other is exponentially high considering various degrees of freedoms including translational and rotational poses (Halperin, Ma, Wolfson, & Nussinov, 2002; Huang, 2014). This will be even more challenging when the flexibility of both molecules is considered which necessitates a need for exploring a huge conformational search space. As a rigorous method, one may exhaustively generate and explore all the possible binding modes considering their three translational and three rotational freedoms of the targets of interest, however this approach will be unfeasible considering the computational power of recently designed hardware. For an example provided by Taylor and his colleague who conducted a review on the protein-small molecule docking methods, by considering the 10 degree increments of angles and 6×10^{14} conformations, it will take about 2×10^{6} years for a computer with the capability of processing 10 conformations per second to complete the computations (R. D. Taylor et al., 2002). As a result, an essence is felt for providing a balance between the computational costs and the explored search space by developing efficient search algorithms as well as scoring functions.

Search Strategies

The search algorithms in the docking process can be mainly divided into two approaches. The first approach is to search for the potential solution candidates in the full solution space namely "exhaustive/systematic search" and the second approach is to search the solution space gradually or randomly using the intelligent or evolutionary methodologies (e.g., Monte Carlo (MC), genetic algorithm (GA), simulated annealing (SA), and Tabu search) which in turn only a part of the solution space is explored to reduce the computational costs (Halperin et al., 2002; Zsoldos, Reid, Simon, Sadjad, & Johnson, 2006). However, over a last decade, another field has been also emerged which is supposed to optimize the docking process further by considering it as a multi-objective problem (MOP) instead of only a one-objective problem (Nicolaou & Brown, 2013) due to the fact that drug discovery should be regarded as a challenging MOP with various pharmaceutical objectives (Schneider & Fechner, 2005). Considering the exhaustive search methodology, that samples all of the conformations with some/many unfavorable states, the optimization algorithm may not have a successful result in addition to the high cost of computational operations. For this purpose, in the early years of docking approaches, the algorithms were benefitted from the rigid bodies of both target and receptors to reduce the dimensionality (R. D. Taylor et al., 2002) to only six degrees of translational and rotational freedom which had its own shortcomings (R. D. Taylor et al., 2002). However, recently two subsets of direct search strategies in Cartesian space namely "local shape feature matching" and "Post-docking approach" have been discussed by Huang (Huang, 2014).

To reduce the highly computational cost of the exhaustive search (i.e., a standard search method) on exploring and evaluating the high number of binding orientations of complexes, *(i)* fast/spherical Fourier transform correlation and *(ii)* direct search in Cartesian space approaches have been introduced (Huang, 2014). In 1992, for speeding up the search process, a group of researchers proposed a mapping strategy from translational space to imaginary space for searching the target space using fast Fourier transform (FFT) calculations (Katchalski-Katzir et al., 1992). Some of the FFT-based docking programs with common principle behind their search strategies using a grid mapping structure are ZDOCK (Chen, Li, & Weng, 2003; Chen, Tong, Mintseris, Li, & Weng, 2003; Hwang, Vreven, Pierce, Hung, & Weng, 2010; B. Pierce, Tong, & Weng, 2005; B. G. Pierce, Hourai, & Weng, 2011; B. G. Pierce et al., 2014; Vreven, Pierce, Hwang, & Weng, 2009, 2013; Wiehe et al., 2005; Wiehe et al., 2007; Yi, Qiu, Cao, Wu, & Li, 2008; Zhang, Liu, & Zhou, 2005), SDOCK (Zhang & Lai, 2011), Cell-Dock (C. Pons et al., 2012), F2Dock (Bajaj, Chowdhury, & Siddavanahalli, 2011), PIPER (Kozakov, Brenke, Comeau, & Vajda, 2006), PIEDOCK (Viswanath et al., 2014), F(2)Dock2.0 protocol used for managing multiple docking jobs and visualizing the docked poses and interfaces (Chowdhury et al., 2013), and commercial ICM-Pro from Molsoft (Fernandez-Recio, Totrov, & Abagyan, 2003; Mendez, Leplae, De Maria, & Wodak, 2003; Mendez, Leplae, Lensink, & Wodak, 2005) from which ZDOCK has received much attention in docking research studies in recent decades. Although, the FFT-based search approach is applied over the 3D translational space in the above mentioned programs, however, it can also be applied over either the 3D rotational space (e.g., FRODOCK (Garzon et al., 2009) or both of the 3D translational and rotational spaces (e.g., HEX (Macindoe et al., 2010; Ritchie & Kemp, 2000; Ritchie, Kozakov, & Vajda, 2008; Ritchie & Venkatraman, 2010)). Furthermore, the ZDOCK has taken the advantage of the 3D convolutional library for accelerating the computational FFT calculations (B. G. Pierce et al., 2011) and recently, the graphics processing unit architecture and programming have been emerged for speeding up the parallel computations of FFT operations (Kantardjiev, 2011, 2012; Ritchie & Venkatraman, 2010; Ritchie, Venkatraman, & Mavridis, 2010; Shimoda, Suzuki, Ohue, Ishida, & Akiyama, 2015). Direct search in Cartesian space approach is known to be simpler but not more efficient than FFT-based methods in terms of computational costs while using the Boolean operators and heuristic rules with protein flexibility, however both of them use the Cartesian grid space as their mapping tool but the direct search looks for the shape matches between two proteins in the grid-based mapped system (Huang, 2014).

In the shape matching search approach, with its computational efficient advantage, the goal is to quickly position the target protein represented by molecular shapes (e.g., Connolly surface) in the binding site in its good shape complimentarily considering the six degrees of the freedom(Huang, 2014; Huang & Zou, 2010a) . And, if only a match is found and a binding orientation is generated, the transformation matrix of six degrees of freedom will be calculated which makes it more different from the exhaustive search algorithms (Huang, 2014). Examples of shape matching docking programs include DOCK (Kuntz, Blaney, Oatley, Langridge, & Ferrin, 1982), Surflex (Jain, 2003, 2007; Miteva, Lee, Montes, & Villoutreix, 2005; Spitzer & Jain, 2012), MDock(Huang & Zou, 2011), GAPDOCK (Gardiner, Willett, & Artymiuk, 2003), and FLOG (Miller, Kearsley, Underwood, & Sheridan, 1994).

Systematic search algorithms using exhaustive search approach is regarded as one of the time consuming and infeasible methods, therefore, fragmentation and database methods can also be included from which fragmentation based techniques are the most commonly used and continued to be a hot topic in docking studies (Sousa et al., 2006; Yuriev, Holien, & Ramsland, 2015). The fragment-based approach uses an incremental or stepwise algorithm to dock various molecular fragments covalently into the active site or divide them into two parts namely "core fragment" and "side chains" in which the core fragments (i.e., rigid part) are docked first and then the side chains (i.e., flexible parts) are added incrementally (Kitchen et al., 2004). Examples of fragment-based approaches include LUDI (Bohm, 1992a, 1992b), FlexX(Schellhammer & Rarey, 2004), DOCK (Kuntz et al., 1982), ADAM (Mizutani, Tomioka, & Itai, 1994), Hammerhead (Welch, Ruppert, & Jain, 1996), and eHiTs (Zsoldos et al., 2006). Database methods are used to minimize the computational costs of exhaustive search methodologies by taking advantage of pre-generated libraries of ligand conformations, for example, FLOG program produces 25 conformations for each molecule using the distance geometry (Miller et al., 1994; Sousa et al., 2006).

Unlike the previously mentioned search approaches, the randomized search algorithms which search at atomic level can be used either with or without molecular representations (i.e., grid or shape) to speed up the search computational process (Huang, 2014). The randomized search algorithms work based on a probability/fitness function that is optimized by using several heuristic algorithms such as Monte Carlo (MC) based on Metropolis criterion using the Boltzmann probability function (e.g., ICM (Abagyan, Totrov, & Kuznetsov, 2004), MCDOCK (M. Liu & Wang, 1999)), genetic algorithm (GA) (e.g.,GOLD(Jones, Willett, & Glen, 1995; Jones, Willett, Glen, Leach, & Taylor, 1997), AutoDock (Garrett M. Morris et al., 1999), FIPSDOCK (Y. Liu et al., 2013), MoDOCK (Gu, Yang, Kang, Wu, & Wang, 2015)), particle swarm ((Garcia-Godoy, Lopez-Camacho, Garcia-Nieto, & Aldana-Montes, 2015; Meier, Pippel, Brandt, Sippl, & Baldauf, 2010; Namasivayam & Gunther, 2007), Tabu search methods (e.g., PRO_LEADS (Baxter et al., 1998; Murray, Baxter, & Frenkel, 1999)) (Sousa et al., 2006), and combination of other algorithms such as GA or geometry search with Tabu search (Hou, Wang, Chen, & Xu, 1999; Tondel, Anderssen, & Drablos, 2006). The advantage of Tabu search is that the revisiting conformations are recorded as a list based on their RMSD values and only new conformations from the search space will be accepted and analyzed (Sousa et al., 2006).

Molecular dynamics (MD) is a computer simulation technique for modeling the complicated molecular systems at the atomic level which is categorized in the simulation search methods. In this approach, the solutions to Newton's equations of motions are calculated for finding the global minimum energy of a docked complex (R. D. Taylor et al., 2002). However, most of the times, the solutions are trapped in a local minimum of the energy surface (Brooijmans & Kuntz, 2003) and cannot cover the high-energy conformational barriers which counts as a main disadvantage for the MD technique (R. D. Taylor et al., 2002). As a result, the quality of the solutions will be dependent on the starting point of the molecular conformations. In some studies, the authors have overcome the shortcomings of MD simulation methods by utilizing a jumping method for not being trapped in local minima, or simulate parts of the docked system at different temperatures (Di Nola, Roccatano, & Berendsen, 1994) and taken advantage of MD performance in comparison to other methods (Jerabek et al., 2014; Pak & Wang, 2000). Another category of the simulation search methods is the energy minimization methodology which is a less widely applicable technique alone at its own (i.e., gradient, conjugate gradient, second derivatives, and least square methods (Sousa et al., 2006)), in contrast to the MD simulation method (Kitchen et al., 2004) and reaches the local minima solutions if not used with other evolutionary algorithms. However, there are some other new approaches that employ the well-known Monte Carlo method along with the energy minimization method for achieving the global minimum of the energy surface such as GlamDock (Tietze & Apostolakis, 2007) or using the Bezier splines based approach (Trosset & Scheraga, 1998). Most of the common docking algorithms such as QXP (McMartin & Bohacek, 1997), DARWIN (J. S. Taylor & Burnett, 2000), CRDOCK (Cortes Cabrera et al., 2012), and ADAM (Mizutani et al., 1994) use various types of energy minimization methods for their process.

The post-docking approach is a two stage protocol for refining the docking results and hence, it may not be directly categorized in the search algorithms. In this approach, by using a docking program all possible conformations are generated and in the next step, the top binding conformations of the first step are re-ranked based on other features such as biological information using relevant scoring methods (Huang, 2014). Some of the post-docking approaches which have also been assessed in Critical Assessment of PRedicted Interaction (CAPRI) experiments are RDOCK (L. Li, Chen, & Weng, 2003), PyDock (Cheng, Blundell, & Fernandez-Recio, 2007; C. Pons, Solernou, Perez-Cano, Grosdidier, & Fernandez-Recio, 2010), SmoothDock(Camacho & Gatchell, 2003), and MultiDock (Jackson, Gabb, & Sternberg, 1998).

Scoring Functions and Their Evaluations

During the above mentioned search algorithm process, a broad unmanageable range of binding modes of the conformations will be generated, however, generating of these conformations will not be sufficient and the correct conformation should be also recognizable (Sousa et al., 2006). For this purpose, an accurate and speedy tool for predicting the binding free energies (e.g., within 7-10 kJ/mol(Bissantz, Folkers, & Rognan, 2000)) known as "scoring function" is needed to discriminate between the correct (i.e., true positive) and the remaining false positive solutions (Halperin et al., 2002).On the other hand, a rigorous scoring function for calculating the energy values of all conformations is neither computationally cost effective nor with low complexity. So, by considering the speed and accuracy properties of scoring functions which is a main bottleneck in molecular docking studies (Kitchen et al., 2004), some pre-definitions, assumptions, and simplifications should be considered in order to reduce the complexity order in developing the scoring functions (Sousa et al., 2006). The various types of algorithms utilized in scoring functions can be divided into three main categories including force field, empirical, and knowledge-based scoring functions (Huang & Zou, 2010a).

Among various types of developed force field scoring function approaches (e.g., Tripos(Kramer, Rarey, & Lengauer, 1999) and AMBER (Weiner, Kollman, Nguyen, & Case, 1986) force fields), the summation value of two types of energies namely the "receptor-ligand interaction energy" and the "internal ligand energy" is denoted as a standard molecular mechanics force field (Kitchen et al., 2004). In most of the scoring approaches, the internal protein energy is simply ignored for reducing the calculation costs. However, the absence of solvation and entropic terms as well as not accurate treatment of long-range interactions involved in binding counts as one of the main shortcomings of standard force field scoring functions (Sousa et al., 2006). To overcome this limitation, recently, torsional entropy and protein-ligand hydrogen-bonding terms have been implemented in programs such as G-Score (Kramer et al., 1999), GOLD (Verdonk, Cole, Hartshorn, Murray, & Taylor, 2003) and AutoDock (G. M. Morris, Goodsell, Huey, & Olson, 1996; G. M. Morris et al., 2013; G. M. Morris et al., 2009), to increase the possibility of molecular recognition.

The second main types of scoring functions are the empirical functions which predict the affinity binding score based on a linear combination/or linear regression of multiple weighted energy terms (e.g., van der Waals, electrostatics, desolvation, hydrogen binding, entropy, and hydrophobicity (Huang,

Grinter, & Zou, 2010). However, the performance of the empirical scoring functions solely depends on the quality of the training set (i.e., the training set includes thousands of diverse protein-ligand complexes with known 3D structures) which fits the affinity binding scores to their corresponding features of the training set and the issues related many energy terms (Huang et al., 2010). Due to simple and linear structure of these functions, they are more efficient than force field scoring functions in terms of computational costs, speed, and accuracy (Huang et al., 2010; Vajda, Hall, & Kozakov, 2013). Some of the examples which are developed based on the empirical scoring functions are LUDI (Bohm, 1992b), MedusaScore(Yin, Biedermannova, Vondrasek, & Dokholyan, 2008), AIScore (Raub, Steffen, Kamper, & Marian, 2008), and LigScore (Krammer, Kirchhoff, Jiang, Venkatachalam, & Waldman, 2005).

The rationale behind the knowledge-based scoring functions is to define the potential of mean force (Zsoldos et al., 2006) by means of the statistical mechanics in the physics field which has been satisfied by employing the inverse Boltzmann method (Huang et al., 2010; McQuarrie, 1976). In this technique, the pairwise energy potentials are extracted in terms of rules from the experimental structures at the atomic level as well as their occurrence frequencies of atom pairs to calculate the scores of ligand binding poses (Huang et al., 2010; Zsoldos et al., 2006). Some of the common examples are DrugScore developed based on the Cambridge Structural Database (Gohlke & Klebe, 2002; Nisius & Gohlke, 2012; C. A. Sotriffer, Gohlke, & Klebe, 2002; Velec et al., 2005), SMoG/SMoG2001 (DeWitte & Shakhnovich, 1996), ITScore/SE (Huang & Zou, 2010b), PMF (Muegge & Martin, 1999) as well as QSAR studies for scoring the protein-ligand interactions (Ballester & Mitchell, 2010) –that are developed based on the knowledge-based scoring functions.

Moreover, recently, the performance of scoring functions have been improved using software or hardware based techniques such as consensus (combination of multiple scoring functions such as GFScore (Betzi, Suhre, Chetrit, Guerlesquin, & Morelli, 2006), PLS-DA (Avram, Pacureanu, Seclaman, Bora, & Kurunczi, 2011), VoteDock (Plewczynski, Lazniewski, von Grotthuss, Rychlewski, & Ginalski, 2011), CPORT (de Vries & Bonvin, 2011)), clustering and entropy-based (e.g., CyClus (Omori & Kitao, 2013)), machine learning based scoring functions (Ballester & Mitchell, 2010; Hsin, Ghosh, & Kitano, 2013; Khamis, Gomaa, & Ahmed, 2015; Kinnings et al., 2011), parallel computing on graphics processing units (GPUs) (e.g., Quantum.Ligand.Dock (Kantardjiev, 2012), GPU.proton.DOCK (Kantardjiev, 2011), and other GPU accelerated studies (Ritchie et al., 2010; Shimoda et al., 2015; Sukhwani & Herbordt, 2010; Wu, Hong, Takeda, & Guo, 2012)), and multi objective optimization programming (MOOP) (e.g., MoDock (Gu et al., 2015) and other MOOP approaches (H. Li et al., 2009; Nicolaou & Brown, 2013)).

For evaluating the performance of the scoring functions, there are three types of measurement criteria including *(i)* binding mode identification, *(ii)* binding affinity prediction, and *(iii)* virtual database screening (Huang et al., 2010). In the first criterion which is widely used in the (Huang, 2014) molecular docking studies, it is important to discriminate between the binding modes and decoys based on the predicted RMSD values (i.e., if the RMSD value is less than 2 Å, the prediction is regarded as successful and otherwise it is not). The second criterion is to measure the ability of the scoring function in predicting the binding affinity by the means of Pearson correlation between the predicted and actual experimental scores. The third approach is virtual screening to identify the potential hits in a large database. Moreover, the area under curve (AUC) and the area under the receiver operating characteristics (ROC) are also other common scoring function testing approaches used in virtual screening of drug design. If the performance of a scoring function is successfully validated by all of the three above mentioned techniques, the obtained results are highly reliable, however, in practice, it is very rare to find a scoring function to satisfy all of the three measurement criteria (Huang et al., 2010).

MOLECULAR DOCKING APPLICATIONS

Molecular docking is one of the foundations of the modern structure-based drug discovery. This technique tries to envisage the inter-molecular interactions in a virtual manner and has attracted enormous attention in drug discovery pipeline. Molecular docking can be utilized for many different purposes, but all may be distilled down to an *in silico* effort for theprediction of ligand binding mode. Having the binding mode predicted enables using the results for different aims. The binding affinity, as a very fundamental measure in ligand-target interaction, is one of the main properties, which can be dealt with using the predicted poses derived from the docking studies. For example, in a study by Politi and her coworkers, the binding affinity for a series of renin inhibitors was predicted through molecular docking study (Politi, Durdagi, Moutevelis-Minakakis, Kokotos, & Mavromoustakos, 2010). Yang et al. investigated the prediction of the binding affinities for small-molecule Smac (the second mitochondria-derived activator of caspase) mimetics toward the target using this technique (C. Y. Yang, Sun, Chen, Nikolovska-Coleska, & Wang, 2009).These are just few examples and obviously presenting a comprehensive list of such studies is impossible in this introductory chapter.

From qualitative point of view, there are numerous examples in which different conformations at different poses on the target molecule are generated and the energetically favored pose or poses are selected using a scoring function to identify the most probable interactions between the ligand and its target molecule (Brooijmans & Kuntz, 2003; Hamzeh-Mivehroud, Mahmoudpour, & Dastmalchi, 2012; Hamzeh-Mivehroud, Rahmani, Feizi, Dastmalchi, & Rashidi, 2014; Hamzeh-Mivehroud, Rahmani, Rashidi, Hosseinpour Feizi, & Dastmalchi, 2013).

Another important application of molecular docking is its significant contribution in virtual screening. Virtual screening is a multistep task in which molecular docking of a large data set of drug-like molecules (also called virtual chemical library) on drug targets is performed followed by post processing of the poses based on calculated binding affinities for selecting the top ranking ligands (Muegge & Enyedy, 2003; Schmidt, Bergner, & Schwede, 2014). This is performed to find molecules having *druglikeness* property capable of binding to a given target. If the candidate molecules (referred to as hits) or its modified relatives show good affinity to the target of interest in biochemical assays, they can be further evaluated as the potential "lead" compounds (Muegge & Enyedy, 2003). The lead optimization can be performed through structural modification in order to improve its biological activity, as well as its other properties collectively known as ADMET[1]. There are many studies in which molecular docking has served as a tool in virtual screening (Bleicher, Bohm, Muller, & Alanine, 2003; Brylinski, 2013; Elokely & Doerksen, 2013; Houston & Walkinshaw, 2013; Pirhadi, Shiri, & Ghasemi, 2013), however, the followings are just a few examples. Recently, Wei et al. have used an optimized hierarchical multistage virtual screening method targeting HIV-1 protease. Six compounds were selected for *in vitro* HIV-1 protease inhibitory assay and among them two molecules showed inhibitory activities against HIV-1 protease, which may be suitable for structural modification in drug development process (Wei et al., 2015). A similar work was conducted to find novel compounds against the signal transducer and activator of transcription 3 (STAT3) leading to identification of STX-0119 as STAT3 dimerization inhibitor. Anti-tumor activity of this compound by *in vivo* evaluation showed promising result in cancer treatment and can enter the structure optimization process for improved potency and ADMET properties (Matsuno et al., 2010). In an attempt to identify inhibitors of influenza virus PA endonuclease, docking based virtual screening method was used. Among the selected fifteen compounds as new chemical entities two of them had

IC_{50} values in micromolar range (Pala et al., 2015). For more comprehensive examples related to lead identification and optimization via molecular docking-based virtual screening see (Laurini et al., 2011; Nolan et al., 2011; Ou-Yang et al., 2012; Yakar & Akten, 2014).

Molecular docking experiments are also extensively utilized in 3D-QSAR analyses as a molecular alignment tool. In an investigation by Patel et al. inhibitory activity for benzimidazole derivatives as Hepatitis C Virus NS5B polymerase inhibitors were predicted using 3D-QSAR, where molecular docking was used for aligning the molecules (Patel, Patel, Kaushik-Basu, & Talele, 2008). In another study, the flexible docking of the nonnucleoside inhibitors of HIV-1 reverse transcriptase was performed in order to determine the active conformation of the studied ligands, and then the results were used in 3D-QSAR analysis (Z. Zhou & Madura, 2004). For more examples see (Harpsoe, Liljefors, & Balle, 2008; Tintori, Magnani, Schenone, & Botta, 2009; Tuccinardi et al., 2007; Vitorovic-Todorovic, Cvijetic, Juranic, & Drakulic, 2012).

The above mentioned case studies are few examples of molecular docking applications in structure-based drug design. However, as mentioned above, the citation is not comprehensive by any means.

FUTURE RESEARCH DIRECTIONS

Molecular docking, an on the move computational-experimental approach, is one of the most attention-received research fields in terms of quantity and the development of search algorithms as well as scoring functions. However, despite these advances there are still many challenges and limitations which need to be well addressed to improve the performance quality of the docking approaches in terms of accuracy and speed. To this end progress in the development of novel search strategies, scoring functions, and representation of protein flexibility as well as obtaining high accuracy with low computational costs by employing any software and hardware technologies, is needed simultaneously. Among these, the quality of the information in the search space is also a problem, as the more information does not always include the better information.

Among different types of search algorithms, the randomized search strategy may reserve the accuracy with high computational cost because the molecular representation is at the atomic level instead of grid based mapping approach which includes the complicated projection to grids and losing the accuracy of scoring functions. Some of the solutions which show increasing role in improving the docking results include (Huang, 2014): i) integration of scoring functions with sampling/searching processes; ii) Using additional information about binding by restricting the sampling space and template-based docking (i.e., these information can also be added in FFT/SFT based search algorithms); and iii) Considering the flexibility of target proteins at either receptor or ligand sides.

Moreover, the transition from considering the rigid to flexible bodies whether it is only the receptor or the ligand or both of them, will add more computational costs on all of the docking processes. Regarding the improvement of scoring functions, the intelligent machine learning approaches form an active and open area in molecular docking experiments to accelerate the docking process with high accuracy which also one may speed them up by utilizing several parallel and GPU based hardware improvement approaches.

In conclusion, a balance between accuracy and speed should be considered while developing new and high performance algorithms both in exploring the search space and in optimizing the selections by scoring functions.

CONCLUSION

Drug discovery looks like "finding a needle in a haystack" and is considered to be a time-consuming, tricky and complicated process. For the pharmaceutical companies the time and financial expenses are the main important criteria in the drug discovery and development projects with high failure rate. In modern drug discovery and development pipeline both of biophysical and computational approaches are employed side by side for discovering the bioactive compounds. Digitization of the biological systems has profound effect on shortening the way to identify therapeutic agent by reducing the number of compounds needed to be synthesized and biologically evaluated. In this sense, computational approaches have shed new light on drug design and discovery process and tremendous advances have been achieved during the past decades. Molecular docking as an invaluable tool in structure-based drug design has become a first line technique in early phase of hit and lead identification since the number of biological structures (as targets) in data banks are growing. Availability of the three dimensional structure for the target molecule responsible for the observed biological event, increases the possibility and accelerates the design of molecules, which can bind to the active site of the target and modulate its activity. Despite the fact that great progress and promising accomplishment has been made in this area, undoubtedly, the molecular docking and other related *in silico* prediction techniques cannot replace the information gathered by experimental investigation. We are far from the perfect and ideal molecular docking method and the issues dealing with fully flexible molecular docking in terms of computational cost and accuracy are still challenging.

REFERENCES

Abagyan, R., Totrov, M., & Kuznetsov, D. (2004). ICM—A new method for protein modeling and design: Applications to docking and structure prediction from the distorted native conformation. *Journal of Computational Chemistry*, *15*(5), 488–506. doi:10.1002/jcc.540150503

Anand, P., Nagarajan, D., Mukherjee, S., & Chandra, N. (2014). PLIC: Protein-ligand interaction clusters. *Database (Oxford)*, *2014*(0), bau029. doi:10.1093/database/bau029 PMID:24763918

Atanasova, M., Patronov, A., Dimitrov, I., Flower, D. R., & Doytchinova, I. (2013). EpiDOCK: A molecular docking-based tool for MHC class II binding prediction. *Protein Engineering, Design & Selection*, *26*(10), 631–634. doi:10.1093/protein/gzt018 PMID:23661105

Avram, S., Pacureanu, L. M., Seclaman, E., Bora, A., & Kurunczi, L. (2011). PLS-DA - Docking Optimized Combined Energetic Terms (PLSDA-DOCET) protocol: A brief evaluation. *Journal of Chemical Information and Modeling*, *51*(12), 3169–3179. doi:10.1021/ci2002268 PMID:22066983

Bajaj, C., Chowdhury, R., & Siddavanahalli, V. (2011). F2Dock: Fast Fourier protein-protein docking. *IEEE/ACM Transactions on Computational Biology and Bioinformatics*, *8*(1), 45–58. doi:10.1109/TCBB.2009.57 PMID:21071796

Ballester, P. J., & Mitchell, J. B. (2010). A machine learning approach to predicting protein-ligand binding affinity with applications to molecular docking. *Bioinformatics (Oxford, England)*, *26*(9), 1169–1175. doi:10.1093/bioinformatics/btq112 PMID:20236947

Baxter, C. A., Murray, C. W., Clark, D. E., Westhead, D. R., & Eldridge, M. D. (1998). Flexible docking using Tabu search and an empirical estimate of binding affinity. *Proteins*, *33*(3), 367–382. doi:10.1002/(SICI)1097-0134(19981115)33:3<367::AID-PROT6>3.0.CO;2-W PMID:9829696

Bayden, A. S., Fornabaio, M., Scarsdale, J. N., & Kellogg, G. E. (2009). Web application for studying the free energy of binding and protonation states of protein-ligand complexes based on HINT. *Journal of Computer-Aided Molecular Design*, *23*(9), 621–632. doi:10.1007/s10822-009-9270-7 PMID:19554265

Bernauer, J., Aze, J., Janin, J., & Poupon, A. (2007). A new protein-protein docking scoring function based on interface residue properties. *Bioinformatics (Oxford, England)*, *23*(5), 555–562. doi:10.1093/bioinformatics/btl654 PMID:17237048

Betzi, S., Suhre, K., Chetrit, B., Guerlesquin, F., & Morelli, X. (2006). GFscore: A general nonlinear consensus scoring function for high-throughput docking. *Journal of Chemical Information and Modeling*, *46*(4), 1704–1712. doi:10.1021/ci0600758 PMID:16859302

Bissantz, C., Folkers, G., & Rognan, D. (2000). Protein-based virtual screening of chemical databases. 1. Evaluation of different docking/scoring combinations. *Journal of Medicinal Chemistry*, *43*(25), 4759–4767. doi:10.1021/jm0010441 PMID:11123984

Bleicher, K. H., Bohm, H. J., Muller, K., & Alanine, A. I. (2003). Hit and lead generation: Beyond high-throughput screening. *Nature Reviews. Drug Discovery*, *2*(5), 369–378. doi:10.1038/nrd1086 PMID:12750740

Bohm, H. J. (1992a). The computer program LUDI: A new method for the de novo design of enzyme inhibitors. *Journal of Computer-Aided Molecular Design*, *6*(1), 61–78. doi:10.1007/BF00124387 PMID:1583540

Bohm, H. J. (1992b). LUDI: Rule-based automatic design of new substituents for enzyme inhibitor leads. *Journal of Computer-Aided Molecular Design*, *6*(6), 593–606. doi:10.1007/BF00126217 PMID:1291628

Bouvier, G., Evrard-Todeschi, N., Girault, J. P., & Bertho, G. (2010). Automatic clustering of docking poses in virtual screening process using self-organizing map. *Bioinformatics (Oxford, England)*, *26*(1), 53–60. doi:10.1093/bioinformatics/btp623 PMID:19910307

Bradford, J. R., & Westhead, D. R. (2005). Improved prediction of protein-protein binding sites using a support vector machines approach. *Bioinformatics (Oxford, England)*, *21*(8), 1487–1494. doi:10.1093/bioinformatics/bti242 PMID:15613384

Brenke, R., Hall, D. R., Chuang, G. Y., Comeau, S. R., Bohnuud, T., Beglov, D., & Kozakov, D. et al. (2012). Application of asymmetric statistical potentials to antibody-protein docking. *Bioinformatics (Oxford, England)*, *28*(20), 2608–2614. doi:10.1093/bioinformatics/bts493 PMID:23053206

Brooijmans, N., & Kuntz, I. D. (2003). Molecular recognition and docking algorithms. *Annual Review of Biophysics and Biomolecular Structure*, *32*(1), 335–373. doi:10.1146/annurev.biophys.32.110601.142532 PMID:12574069

Brylinski, M. (2013). Nonlinear scoring functions for similarity-based ligand docking and binding affinity prediction. *Journal of Chemical Information and Modeling*, *53*(11), 3097–3112. doi:10.1021/ci400510e PMID:24171431

Brylinski, M., Lee, S. Y., Zhou, H., & Skolnick, J. (2011). The utility of geometrical and chemical restraint information extracted from predicted ligand-binding sites in protein structure refinement. *Journal of Structural Biology*, *173*(3), 558–569. doi:10.1016/j.jsb.2010.09.009 PMID:20850544

Camacho, C. J., & Gatchell, D. W. (2003). Successful discrimination of protein interactions. *Proteins*, *52*(1), 92–97. doi:10.1002/prot.10394 PMID:12784373

Camacho, C. J., Ma, H., & Champ, P. C. (2006). Scoring a diverse set of high-quality docked conformations: A metascore based on electrostatic and desolvation interactions. *Proteins*, *63*(4), 868–877. doi:10.1002/prot.20932 PMID:16506242

Campbell, A. J., Lamb, M. L., & Joseph-McCarthy, D. (2014). Ensemble-based docking using biased molecular dynamics. *Journal of Chemical Information and Modeling*, *54*(7), 2127–2138. doi:10.1021/ci400729j PMID:24881672

Chang, D. T., Oyang, Y. J., & Lin, J. H. (2005). MEDock: a web server for efficient prediction of ligand binding sites based on a novel optimization algorithm. *Nucleic Acids Res*, *33*, W233-238.

Chaves, R. C., & Pellequer, J. L. (2013). DockAFM: Benchmarking protein structures by docking under AFM topographs. *Bioinformatics (Oxford, England)*, *29*(24), 3230–3231. doi:10.1093/bioinformatics/btt561 PMID:24078683

Chen, R., Li, L., & Weng, Z. (2003). ZDOCK: An initial-stage protein-docking algorithm. *Proteins*, *52*(1), 80–87. doi:10.1002/prot.10389 PMID:12784371

Chen, R., Tong, W., Mintseris, J., Li, L., & Weng, Z. (2003). ZDOCK predictions for the CAPRI challenge. *Proteins*, *52*(1), 68–73. doi:10.1002/prot.10388 PMID:12784369

Cheng, T. M., Blundell, T. L., & Fernandez-Recio, J. (2007). pyDock: Electrostatics and desolvation for effective scoring of rigid-body protein-protein docking. *Proteins*, *68*(2), 503–515. doi:10.1002/prot.21419 PMID:17444519

Chermak, E., Petta, A., Serra, L., Vangone, A., Scarano, V., Cavallo, L., & Oliva, R. (2015). CONSRANK: A server for the analysis, comparison and ranking of docking models based on inter-residue contacts. *Bioinformatics (Oxford, England)*, *31*(9), 1481–1483. doi:10.1093/bioinformatics/btu837 PMID:25535242

Choi, Y. S., Han, S. K., Kim, J., Yang, J. S., Jeon, J., Ryu, S. H., & Kim, S. (2010). ConPlex: a server for the evolutionary conservation analysis of protein complex structures. *Nucleic Acids Res*, *38*, W450-456. doi: 10.1093/nar/gkq328

Chowdhury, R., Rasheed, M., Keidel, D., Moussalem, M., Olson, A., Sanner, M., & Bajaj, C. (2013). Protein-protein docking with F(2)Dock 2.0 and GB-rerank. *PLoS ONE*, *8*(3), e51307. doi:10.1371/journal.pone.0051307 PMID:23483883

Comeau, S. R., Gatchell, D. W., Vajda, S., & Camacho, C. J. (2004a). ClusPro: a fully automated algorithm for protein-protein docking. *Nucleic Acids Res*, *32*, W96-99. doi: 10.1093/nar/gkh354

Comeau, S. R., Gatchell, D. W., Vajda, S., & Camacho, C. J. (2004b). ClusPro: An automated docking and discrimination method for the prediction of protein complexes. *Bioinformatics (Oxford, England)*, *20*(1), 45–50. doi:10.1093/bioinformatics/btg371 PMID:14693807

Cortes Cabrera, A., Klett, J., Dos Santos, H. G., Perona, A., Gil-Redondo, R., Francis, S. M., & Morreale, A. et al. (2012). CRDOCK: An ultrafast multipurpose protein-ligand docking tool. *Journal of Chemical Information and Modeling*, *52*(8), 2300–2309. doi:10.1021/ci300194a PMID:22764680

de Vries, S. J., & Bonvin, A. M. (2011). CPORT: A consensus interface predictor and its performance in prediction-driven docking with HADDOCK. *PLoS ONE*, *6*(3), e17695. doi:10.1371/journal.pone.0017695 PMID:21464987

de Vries, S. J., Schindler, C. E., Chauvot de Beauchene, I., & Zacharias, M. (2015). A web interface for easy flexible protein-protein docking with ATTRACT. *Biophysical Journal*, *108*(3), 462–465. doi:10.1016/j.bpj.2014.12.015 PMID:25650913

de Vries, S. J., van Dijk, M., & Bonvin, A. M. (2010). The HADDOCK web server for data-driven biomolecular docking. *Nature Protocols*, *5*(5), 883–897. doi:10.1038/nprot.2010.32 PMID:20431534

DeWitte, R. S., & Shakhnovich, E. I. (1996). SMoG: De Novo Design Method Based on Simple, Fast, and Accurate Free Energy Estimates. 1. Methodology and Supporting Evidence. *Journal of the American Chemical Society*, *118*(47), 11733–11744. doi:10.1021/ja960751u

Dhanik, A., McMurray, J. S., & Kavraki, L. E. (2013). DINC: A new AutoDock-based protocol for docking large ligands. *BMC Structural Biology*, *13*(Suppl 1), S11. doi:10.1186/1472-6807-13-S1-S11 PMID:24564952

Di Nola, A., Roccatano, D., & Berendsen, H. J. (1994). Molecular dynamics simulation of the docking of substrates to proteins. *Proteins*, *19*(3), 174–182. doi:10.1002/prot.340190303 PMID:7937732

Elokely, K. M., & Doerksen, R. J. (2013). Docking challenge: Protein sampling and molecular docking performance. *Journal of Chemical Information and Modeling*, *53*(8), 1934–1945. doi:10.1021/ci400040d PMID:23530568

Emekli, U., Schneidman-Duhovny, D., Wolfson, H. J., Nussinov, R., & Haliloglu, T. (2008). HingeProt: Automated prediction of hinges in protein structures. *Proteins*, *70*(4), 1219–1227. doi:10.1002/prot.21613 PMID:17847101

Fan, H., Schneidman-Duhovny, D., Irwin, J. J., Dong, G., Shoichet, B. K., & Sali, A. (2011). Statistical potential for modeling and ranking of protein-ligand interactions. *Journal of Chemical Information and Modeling*, *51*(12), 3078–3092. doi:10.1021/ci200377u PMID:22014038

Fenollosa, C., Oton, M., Andrio, P., Cortes, J., Orozco, M., & Goni, J. R. (2015). SEABED: Small molEcule activity scanner weB servicE baseD. *Bioinformatics (Oxford, England)*, *31*(5), 773–775. doi:10.1093/bioinformatics/btu709 PMID:25348211

Fernandez-Recio, J., Totrov, M., & Abagyan, R. (2003). ICM-DISCO docking by global energy optimization with fully flexible side-chains. *Proteins*, *52*(1), 113–117. doi:10.1002/prot.10383 PMID:12784376

Fernandez-Recio, J., Totrov, M., Skorodumov, C., & Abagyan, R. (2005). Optimal docking area: A new method for predicting protein-protein interaction sites. *Proteins*, *58*(1), 134–143. doi:10.1002/prot.20285 PMID:15495260

Fischer, E. (1894). Einfluss der Configuration auf die Wirkung der Enzyme. *Berichte der deutschen chemischen Gesellschaft*, *27*(3), 2985-2993.

Gajda, M. J., Tuszynska, I., Kaczor, M., Bakulina, A. Y., & Bujnicki, J. M. (2010). FILTREST3D: Discrimination of structural models using restraints from experimental data. *Bioinformatics (Oxford, England)*, *26*(23), 2986–2987. doi:10.1093/bioinformatics/btq582 PMID:20956242

Gani, O. A. (2007). Signposts of docking and scoring in drug design. *Chemical Biology & Drug Design*, *70*(4), 360–365. doi:10.1111/j.1747-0285.2007.00571.x PMID:17937781

Garcia-Godoy, M. J., Lopez-Camacho, E., Garcia-Nieto, J., & Aldana-Montes, A. J. (2015). Solving molecular docking problems with multi-objective metaheuristics. *Molecules (Basel, Switzerland)*, *20*(6), 10154–10183. doi:10.3390/molecules200610154 PMID:26042856

Gardiner, E. J., Willett, P., & Artymiuk, P. J. (2003). GAPDOCK: A Genetic Algorithm Approach to Protein Docking in CAPRI round 1. *Proteins*, *52*(1), 10–14. doi:10.1002/prot.10386 PMID:12784360

Garzon, J. I., Lopez-Blanco, J. R., Pons, C., Kovacs, J., Abagyan, R., Fernandez-Recio, J., & Chacon, P. (2009). FRODOCK: A new approach for fast rotational protein-protein docking. *Bioinformatics (Oxford, England)*, *25*(19), 2544–2551. doi:10.1093/bioinformatics/btp447 PMID:19620099

Ghoorah, A. W., Devignes, M. D., Smail-Tabbone, M., & Ritchie, D. W. (2011). Spatial clustering of protein binding sites for template based protein docking. *Bioinformatics (Oxford, England)*, *27*(20), 2820–2827. doi:10.1093/bioinformatics/btr493 PMID:21873637

Gohlke, H., & Klebe, G. (2002). DrugScore meets CoMFA: Adaptation of fields for molecular comparison (AFMoC) or how to tailor knowledge-based pair-potentials to a particular protein. *Journal of Medicinal Chemistry*, *45*(19), 4153–4170. doi:10.1021/jm020808p PMID:12213058

Goodford, P. J. (1985). A computational procedure for determining energetically favorable binding sites on biologically important macromolecules. *Journal of Medicinal Chemistry*, *28*(7), 849–857. doi:10.1021/jm00145a002 PMID:3892003

Gu, J., Yang, X., Kang, L., Wu, J., & Wang, X. (2015). MoDock: A multi-objective strategy improves the accuracy for molecular docking. *Algorithms for Molecular Biology; AMB*, *10*(1), 8. doi:10.1186/s13015-015-0034-8 PMID:25705248

Halperin, I., Ma, B., Wolfson, H., & Nussinov, R. (2002). Principles of docking: An overview of search algorithms and a guide to scoring functions. *Proteins*, *47*(4), 409–443. doi:10.1002/prot.10115 PMID:12001221

Hamzeh-Mivehroud, M., Mahmoudpour, A., & Dastmalchi, S. (2012). Identification of new peptide ligands for epidermal growth factor receptor using phage display and computationally modeling their mode of binding. *Chemical Biology & Drug Design*, *79*(3), 246–259. doi:10.1111/j.1747-0285.2011.01282.x PMID:22136656

Hamzeh-Mivehroud, M., Rahmani, S., Feizi, M. A., Dastmalchi, S., & Rashidi, M. R. (2014). In vitro and in silico studies to explore structural features of flavonoids for aldehyde oxidase inhibition. *Archiv der Pharmazie, 347*(10), 738–747. doi:10.1002/ardp.201400076 PMID:24956178

Hamzeh-Mivehroud, M., Rahmani, S., Rashidi, M. R., Hosseinpour Feizi, M. A., & Dastmalchi, S. (2013). Structure-based investigation of rat aldehyde oxidase inhibition by flavonoids. *Xenobiotica, 43*(8), 661–670. doi:10.3109/00498254.2012.755228 PMID:23282065

Harpsoe, K., Liljefors, T., & Balle, T. (2008). Prediction of the binding mode of biarylpropylsulfonamide allosteric AMPA receptor modulators based on docking, GRID molecular interaction fields and 3D-QSAR analysis. *Journal of Molecular Graphics & Modelling, 26*(5), 874–883. doi:10.1016/j.jmgm.2007.06.002 PMID:17644444

Heo, L., Shin, W. H., Lee, M. S., & Seok, C. (2014). GalaxySite: ligand-binding-site prediction by using molecular docking. *Nucleic Acids Res, 42*, W210-214.

Hou, T., Wang, J., Chen, L., & Xu, X. (1999). Automated docking of peptides and proteins by using a genetic algorithm combined with a tabu search. *Protein Engineering, 12*(8), 639–648. doi:10.1093/protein/12.8.639 PMID:10469824

Houston, D. R., & Walkinshaw, M. D. (2013). Consensus docking: Improving the reliability of docking in a virtual screening context. *Journal of Chemical Information and Modeling, 53*(2), 384–390. doi:10.1021/ci300399w PMID:23351099

Hsin, K. Y., Ghosh, S., & Kitano, H. (2013). Combining machine learning systems and multiple docking simulation packages to improve docking prediction reliability for network pharmacology. *PLoS ONE, 8*(12), e83922. doi:10.1371/journal.pone.0083922 PMID:24391846

Huang, S. Y. (2014). Search strategies and evaluation in protein-protein docking: Principles, advances and challenges. *Drug Discovery Today, 19*(8), 1081–1096. doi:10.1016/j.drudis.2014.02.005 PMID:24594385

Huang, S. Y., Grinter, S. Z., & Zou, X. (2010). Scoring functions and their evaluation methods for protein-ligand docking: Recent advances and future directions. *Physical Chemistry Chemical Physics, 12*(40), 12899–12908. doi:10.1039/c0cp00151a PMID:20730182

Huang, S. Y., & Zou, X. (2010a). Advances and challenges in protein-ligand docking. *International Journal of Molecular Sciences, 11*(8), 3016–3034. doi:10.3390/ijms11083016 PMID:21152288

Huang, S. Y., & Zou, X. (2010b). Inclusion of solvation and entropy in the knowledge-based scoring function for protein-ligand interactions. *Journal of Chemical Information and Modeling, 50*(2), 262–273. doi:10.1021/ci9002987 PMID:20088605

Huang, S. Y., & Zou, X. (2011). Construction and test of ligand decoy sets using MDock: Community structure-activity resource benchmarks for binding mode prediction. *Journal of Chemical Information and Modeling, 51*(9), 2107–2114. doi:10.1021/ci200080g PMID:21755952

Hung, C. L., & Chen, C. C. (2014). Computational approaches for drug discovery. *Drug Development Research, 75*(6), 412–418. doi:10.1002/ddr.21222 PMID:25195585

Hwang, H., Vreven, T., Pierce, B. G., Hung, J. H., & Weng, Z. (2010). Performance of ZDOCK and ZRANK in CAPRI rounds 13-19. *Proteins*, *78*(15), 3104–3110. doi:10.1002/prot.22764 PMID:20936681

Irwin, J. J., Shoichet, B. K., Mysinger, M. M., Huang, N., Colizzi, F., Wassam, P., & Cao, Y. (2009). Automated docking screens: A feasibility study. *Journal of Medicinal Chemistry*, *52*(18), 5712–5720. doi:10.1021/jm9006966 PMID:19719084

Jackson, R. M., Gabb, H. A., & Sternberg, M. J. (1998). Rapid refinement of protein interfaces incorporating solvation: Application to the docking problem. *Journal of Molecular Biology*, *276*(1), 265–285. doi:10.1006/jmbi.1997.1519 PMID:9514726

Jain, A. N. (2003). Surflex: Fully automatic flexible molecular docking using a molecular similarity-based search engine. *Journal of Medicinal Chemistry*, *46*(4), 499–511. doi:10.1021/jm020406h PMID:12570372

Jain, A. N. (2007). Surflex-Dock 2.1: Robust performance from ligand energetic modeling, ring flexibility, and knowledge-based search. *Journal of Computer-Aided Molecular Design*, *21*(5), 281–306. doi:10.1007/s10822-007-9114-2 PMID:17387436

Jardin, C., Stefani, A. G., Eberhardt, M., Huber, J. B., & Sticht, H. (2013). An information-theoretic classification of amino acids for the assessment of interfaces in protein-protein docking. *Journal of Molecular Modeling*, *19*(9), 3901–3910. doi:10.1007/s00894-013-1916-7 PMID:23828247

Jayaram, B., Singh, T., Mukherjee, G., Mathur, A., Shekhar, S., & Shekhar, V. (2012). Sanjeevini: A freely accessible web-server for target directed lead molecule discovery. *BMC Bioinformatics*, *13*(Suppl 17), S7. doi:10.1186/1471-2105-13-S17-S7 PMID:23282245

Jerabek, P., Florian, J., Stiborova, M., & Martinek, V. (2014). Flexible docking-based molecular dynamics/steered molecular dynamics calculations of protein-protein contacts in a complex of cytochrome P450 1A2 with cytochrome b5. *Biochemistry*, *53*(42), 6695–6705. doi:10.1021/bi500814t PMID:25313797

Jimenez-Garcia, B., Pons, C., & Fernandez-Recio, J. (2013). pyDockWEB: A web server for rigid-body protein-protein docking using electrostatics and desolvation scoring. *Bioinformatics (Oxford, England)*, *29*(13), 1698–1699. doi:10.1093/bioinformatics/btt262 PMID:23661696

Jimenez-Garcia, B., Pons, C., Svergun, D. I., Bernado, P., & Fernandez-Recio, J. (2015). pyDockSAXS: Protein-protein complex structure by SAXS and computational docking. *Nucleic Acids Research*, *43*(W1), W356-361. doi:10.1093/nar/gkv368 PMID:25897115

Jones, G., Willett, P., & Glen, R. C. (1995). Molecular recognition of receptor sites using a genetic algorithm with a description of desolvation. *Journal of Molecular Biology*, *245*(1), 43–53. doi:10.1016/S0022-2836(95)80037-9 PMID:7823319

Jones, G., Willett, P., Glen, R. C., Leach, A. R., & Taylor, R. (1997). Development and validation of a genetic algorithm for flexible docking. *Journal of Molecular Biology*, *267*(3), 727–748. doi:10.1006/jmbi.1996.0897 PMID:9126849

Kantardjiev, A. A. (2011). GPU.proton.DOCK: Genuine Protein Ultrafast proton equilibria consistent DOCKing. *Nucleic Acids Res*, *39*, W223-228. doi: 10.1093/nar/gkr412

Kantardjiev, A. A. (2012). Quantum.Ligand.Dock: protein-ligand docking with quantum entanglement refinement on a GPU system. *Nucleic Acids Res, 40*, W415-422. doi: 10.1093/nar/gks515

Katchalski-Katzir, E., Shariv, I., Eisenstein, M., Friesem, A. A., Aflalo, C., & Vakser, I. A. (1992). Molecular surface recognition: Determination of geometric fit between proteins and their ligands by correlation techniques. *Proceedings of the National Academy of Sciences of the United States of America, 89*(6), 2195–2199. doi:10.1073/pnas.89.6.2195 PMID:1549581

Khamis, M. A., Gomaa, W., & Ahmed, W. F. (2015). Machine learning in computational docking. *Artificial Intelligence in Medicine, 63*(3), 135–152. doi:10.1016/j.artmed.2015.02.002 PMID:25724101

Kinnings, S. L., Liu, N., Tonge, P. J., Jackson, R. M., Xie, L., & Bourne, P. E. (2011). A machine learning-based method to improve docking scoring functions and its application to drug repurposing. *Journal of Chemical Information and Modeling, 51*(2), 408–419. doi:10.1021/ci100369f PMID:21291174

Kitchen, D. B., Decornez, H., Furr, J. R., & Bajorath, J. (2004). Docking and scoring in virtual screening for drug discovery: Methods and applications. *Nature Reviews. Drug Discovery, 3*(11), 935–949. doi:10.1038/nrd1549 PMID:15520816

Konc, J., Konc, J. T., Penca, M., & Janezic, D. (2011). Binding-sites Prediction Assisting Protein-protein Docking. *Acta Chimica Slovenica, 58*(3), 396–401. PMID:24062097

Kozakov, D., Brenke, R., Comeau, S. R., & Vajda, S. (2006). PIPER: An FFT-based protein docking program with pairwise potentials. *Proteins, 65*(2), 392–406. doi:10.1002/prot.21117 PMID:16933295

Kramer, B., Rarey, M., & Lengauer, T. (1999). Evaluation of the FLEXX incremental construction algorithm for protein-ligand docking. *Proteins, 37*(2), 228–241. doi:10.1002/(SICI)1097-0134(19991101)37:2<228::AID-PROT8>3.0.CO;2-8 PMID:10584068

Krammer, A., Kirchhoff, P. D., Jiang, X., Venkatachalam, C. M., & Waldman, M. (2005). LigScore: A novel scoring function for predicting binding affinities. *Journal of Molecular Graphics & Modelling, 23*(5), 395–407. doi:10.1016/j.jmgm.2004.11.007 PMID:15781182

Kroemer, R. T. (2003). Molecular modelling probes: Docking and scoring. *Biochemical Society Transactions, 31*(Pt 5), 980–984. doi:10.1042/bst0310980 PMID:14505464

Kuntz, I. D., Blaney, J. M., Oatley, S. J., Langridge, R., & Ferrin, T. E. (1982). A geometric approach to macromolecule-ligand interactions. *Journal of Molecular Biology, 161*(2), 269–288. doi:10.1016/0022-2836(82)90153-X PMID:7154081

Kurcinski, M., Jamroz, M., Blaszczyk, M., Kolinski, A., & Kmiecik, S. (2015). CABS-dock web server for the flexible docking of peptides to proteins without prior knowledge of the binding site. *Nucleic Acids Research, 43*(W1), W419-424. doi:10.1093/nar/gkv456 PMID:25943545

Labbe, C. M., Rey, J., Lagorce, D., Vavrusa, M., Becot, J., Sperandio, O., & Miteva, M. A. et al. (2015). MTiOpenScreen: A web server for structure-based virtual screening. *Nucleic Acids Research, 43*(W1), W448-454. doi:10.1093/nar/gkv306 PMID:25855812

Lauck, F., Smith, C. A., Friedland, G. F., Humphris, E. L., & Kortemme, T. (2010). RosettaBackrub--a web server for flexible backbone protein structure modeling and design. *Nucleic Acids Res, 38*, W569-575. doi: 10.1093/nar/gkq369

Laurie, A. T., & Jackson, R. M. (2005). Q-SiteFinder: An energy-based method for the prediction of protein-ligand binding sites. *Bioinformatics (Oxford, England), 21*(9), 1908–1916. doi:10.1093/bioinformatics/bti315 PMID:15701681

Laurini, E., Col, V. D., Mamolo, M. G., Zampieri, D., Posocco, P., Fermeglia, M., & Pricl, S. et al. (2011). Homology Model and Docking-Based Virtual Screening for Ligands of the sigma1 Receptor. *ACS Med Chem Lett, 2*(11), 834–839. doi:10.1021/ml2001505 PMID:24900272

Leach, A. R., Shoichet, B. K., & Peishoff, C. E. (2006). Prediction of protein-ligand interactions. Docking and scoring: Successes and gaps. *Journal of Medicinal Chemistry, 49*(20), 5851–5855. doi:10.1021/jm060999m PMID:17004700

Lee, H., Heo, L., Lee, M. S., & Seok, C. (2015). GalaxyPepDock: A protein-peptide docking tool based on interaction similarity and energy optimization. *Nucleic Acids Research, 43*(W1), W431-435. doi:10.1093/nar/gkv495 PMID:25969449

Lee, H. S., & Zhang, Y. (2012). BSP-SLIM: A blind low-resolution ligand-protein docking approach using predicted protein structures. *Proteins, 80*(1), 93–110. doi:10.1002/prot.23165 PMID:21971880

Lee, P. H., Kuo, K. L., Chu, P. Y., Liu, E. M., & Lin, J. H. (2009). SLITHER: a web server for generating contiguous conformations of substrate molecules entering into deep active sites of proteins or migrating through channels in membrane transporters. *Nucleic Acids Res, 37*, W559-564. doi: 10.1093/nar/gkp359

Lesk, V. I., & Sternberg, M. J. (2008). 3D-Garden: A system for modelling protein-protein complexes based on conformational refinement of ensembles generated with the marching cubes algorithm. *Bioinformatics (Oxford, England), 24*(9), 1137–1144. doi:10.1093/bioinformatics/btn093 PMID:18326508

Levinthal, C., Wodak, S. J., Kahn, P., & Dadivanian, A. K. (1975). Hemoglobin interaction in sickle cell fibers. I: Theoretical approaches to the molecular contacts. *Proceedings of the National Academy of Sciences of the United States of America, 72*(4), 1330–1334. doi:10.1073/pnas.72.4.1330 PMID:1055409

Li, H., Gao, Z., Kang, L., Zhang, H., Yang, K., Yu, K., . . . Jiang, H. (2006). TarFisDock: a web server for identifying drug targets with docking approach. *Nucleic Acids Res, 34*, W219-224. doi: 10.1093/nar/gkl114

Li, H., Leung, K. S., Ballester, P. J., & Wong, M. H. (2014). istar: A web platform for large-scale protein-ligand docking. *PLoS ONE, 9*(1), e85678. doi:10.1371/journal.pone.0085678 PMID:24475049

Li, H., Zhang, H., Zheng, M., Luo, J., Kang, L., Liu, X., & Jiang, H. et al. (2009). An effective docking strategy for virtual screening based on multi-objective optimization algorithm. *BMC Bioinformatics, 10*(1), 58. doi:10.1186/1471-2105-10-58 PMID:19210777

Li, L., Chen, R., & Weng, Z. (2003). RDOCK: Refinement of rigid-body protein docking predictions. *Proteins, 53*(3), 693–707. doi:10.1002/prot.10460 PMID:14579360

Lill, M. A. (2011). Efficient incorporation of protein flexibility and dynamics into molecular docking simulations. *Biochemistry*, *50*(28), 6157–6169. doi:10.1021/bi2004558 PMID:21678954

Lindahl, E., & Delarue, M. (2005). Refinement of docked protein-ligand and protein-DNA structures using low frequency normal mode amplitude optimization. *Nucleic Acids Research*, *33*(14), 4496–4506. doi:10.1093/nar/gki730 PMID:16087736

Liu, M., & Wang, S. (1999). MCDOCK: A Monte Carlo simulation approach to the molecular docking problem. *Journal of Computer-Aided Molecular Design*, *13*(5), 435–451. doi:10.1023/A:1008005918983 PMID:10483527

Liu, X., Ouyang, S., Yu, B., Liu, Y., Huang, K., Gong, J., . . . Jiang, H. (2010). PharmMapper server: a web server for potential drug target identification using pharmacophore mapping approach. *Nucleic Acids Res, 38*, W609-614. doi: 10.1093/nar/gkq300

Liu, Y., Zhao, L., Li, W., Zhao, D., Song, M., & Yang, Y. (2013). FIPSDock: A new molecular docking technique driven by fully informed swarm optimization algorithm. *Journal of Computational Chemistry*, *34*(1), 67–75. doi:10.1002/jcc.23108 PMID:22961860

London, N., Miller, R. M., Krishnan, S., Uchida, K., Irwin, J. J., Eidam, O., & Taunton, J. et al. (2014). Covalent docking of large libraries for the discovery of chemical probes. *Nature Chemical Biology*, *10*(12), 1066–1072. doi:10.1038/nchembio.1666 PMID:25344815

Lyskov, S., Chou, F. C., Conchuir, S. O., Der, B. S., Drew, K., Kuroda, D., & Das, R. et al. (2013). Serverification of molecular modeling applications: The Rosetta Online Server that Includes Everyone (ROSIE). *PLoS ONE*, *8*(5), e63906. PMID:23717507

Lyskov, S., & Gray, J. J. (2008). The RosettaDock server for local protein-protein docking. *Nucleic Acids Res, 36*, W233-238. doi: 10.1093/nar/gkn216

Macindoe, G., Mavridis, L., Venkatraman, V., Devignes, M. D., & Ritchie, D. W. (2010). HexServer: an FFT-based protein docking server powered by graphics processors. *Nucleic Acids Res, 38*, W445-449. doi: 10.1093/nar/gkq311

Madeira, F., Tinti, M., Murugesan, G., Berrett, E., Stafford, M., Toth, R., & Barton, G. J. et al. (2015). 14-3-3-Pred: Improved methods to predict 14-3-3-binding phosphopeptides. *Bioinformatics (Oxford, England)*, *31*(14), 2276–2283. doi:10.1093/bioinformatics/btv133 PMID:25735772

Malhotra, S., Mathew, O. K., & Sowdhamini, R. (2015). DOCKSCORE: A webserver for ranking protein-protein docked poses. *BMC Bioinformatics*, *16*(1), 127. doi:10.1186/s12859-015-0572-6 PMID:25902779

Mandal, S., Moudgil, M., & Mandal, S. K. (2009). Rational drug design. *European Journal of Pharmacology*, *625*(1-3), 90–100. doi:10.1016/j.ejphar.2009.06.065 PMID:19835861

Martin, L., Catherinot, V., & Labesse, G. (2006). kinDOCK: a tool for comparative docking of protein kinase ligands. *Nucleic Acids Res, 34*, W325-329. doi: 10.1093/nar/gkl211

Matsuno, K., Masuda, Y., Uehara, Y., Sato, H., Muroya, A., Takahashi, O., & Asai, A. et al. (2010). Identification of a new series of STAT3 inhibitors by virtual screening. *ACS Med Chem Lett*, *1*(8), 371–375. doi:10.1021/ml1000273 PMID:24900220

McMartin, C., & Bohacek, R. S. (1997). QXP: Powerful, rapid computer algorithms for structure-based drug design. *Journal of Computer-Aided Molecular Design*, *11*(4), 333–344. doi:10.1023/A:1007907728892 PMID:9334900

McQuarrie, D. A. (1976). *StatisticalMechanics*. New York: Harper Collins Publishers.

Meier, R., Pippel, M., Brandt, F., Sippl, W., & Baldauf, C. (2010). ParaDockS: A framework for molecular docking with population-based metaheuristics. *Journal of Chemical Information and Modeling*, *50*(5), 879–889. doi:10.1021/ci900467x PMID:20415499

Mendez, R., Leplae, R., De Maria, L., & Wodak, S. J. (2003). Assessment of blind predictions of protein-protein interactions: Current status of docking methods. *Proteins*, *52*(1), 51–67. doi:10.1002/prot.10393 PMID:12784368

Mendez, R., Leplae, R., Lensink, M. F., & Wodak, S. J. (2005). Assessment of CAPRI predictions in rounds 3-5 shows progress in docking procedures. *Proteins*, *60*(2), 150–169. doi:10.1002/prot.20551 PMID:15981261

Miller, M. D., Kearsley, S. K., Underwood, D. J., & Sheridan, R. P. (1994). FLOG: A system to select 'quasi-flexible' ligands complementary to a receptor of known three-dimensional structure. *Journal of Computer-Aided Molecular Design*, *8*(2), 153–174. doi:10.1007/BF00119865 PMID:8064332

Miteva, M. A., Lee, W. H., Montes, M. O., & Villoutreix, B. O. (2005). Fast structure-based virtual ligand screening combining FRED, DOCK, and Surflex. *Journal of Medicinal Chemistry*, *48*(19), 6012–6022. doi:10.1021/jm050262h PMID:16162004

Mitra, P., & Pal, D. (2011). PRUNE and PROBE--two modular web services for protein-protein docking. *Nucleic Acids Res, 39*, W229-234. doi: 10.1093/nar/gkr317

Mizutani, M. Y., Tomioka, N., & Itai, A. (1994). Rational automatic search method for stable docking models of protein and ligand. *Journal of Molecular Biology*, *243*(2), 310–326. doi:10.1006/jmbi.1994.1656 PMID:7932757

Moal, I. H., Dapkunas, J., & Fernandez-Recio, J. (2015). Inferring the microscopic surface energy of protein-protein interfaces from mutation data. *Proteins*, *83*(4), 640–650. doi:10.1002/prot.24761 PMID:25586563

Morris, G. M., Goodsell, D. S., Halliday, R. S., Huey, R., Hart, W. E., Belew, R. K., & Olson, A. J. (1999). Automated docking using a Lamarckian genetic algorithm and an empirical binding free energy function. *Journal of Computational Chemistry*, *19*(14), 1639–1662. doi:10.1002/(SICI)1096-987X(19981115)19:14<1639::AID-JCC10>3.0.CO;2-B

Morris, G. M., Goodsell, D. S., Huey, R., & Olson, A. J. (1996). Distributed automated docking of flexible ligands to proteins: Parallel applications of AutoDock 2.4. *Journal of Computer-Aided Molecular Design*, *10*(4), 293–304. doi:10.1007/BF00124499 PMID:8877701

Morris, G. M., Green, L. G., Radic, Z., Taylor, P., Sharpless, K. B., Olson, A. J., & Grynszpan, F. (2013). Automated docking with protein flexibility in the design of femtomolar "click chemistry" inhibitors of acetylcholinesterase. *Journal of Chemical Information and Modeling*, *53*(4), 898–906. doi:10.1021/ci300545a PMID:23451944

Morris, G. M., Huey, R., Lindstrom, W., Sanner, M. F., Belew, R. K., Goodsell, D. S., & Olson, A. J. (2009). AutoDock4 and AutoDockTools4: Automated docking with selective receptor flexibility. *Journal of Computational Chemistry*, *30*(16), 2785–2791. doi:10.1002/jcc.21256 PMID:19399780

Morris, G. M., & Lim-Wilby, M. (2008). Molecular docking. In A. Kukol (Ed.), *Molecular Modeling of Proteins* (Vol. 443, pp. 365–382). Totowa: Humana Press. doi:10.1007/978-1-59745-177-2_19

Mottarella, S. E., Beglov, D., Beglova, N., Nugent, M. A., Kozakov, D., & Vajda, S. (2014). Docking server for the identification of heparin binding sites on proteins. *Journal of Chemical Information and Modeling*, *54*(7), 2068–2078. doi:10.1021/ci500115j PMID:24974889

Muegge, I., & Enyedy, I. (2003). Virtual screening. In D. J. Abraham (Ed.), *Burger's Medicinal Chemistry and Drug Discovery, Drug Discovery* (Vol. 1, pp. 243–279). New Jersey: John Wiley & Sons. doi:10.1002/0471266949.bmc006

Muegge, I., & Martin, Y. C. (1999). A general and fast scoring function for protein-ligand interactions: A simplified potential approach. *Journal of Medicinal Chemistry*, *42*(5), 791–804. doi:10.1021/jm980536j PMID:10072678

Mukherjee, G., Patra, N., Barua, P., & Jayaram, B. (2011). A fast empirical GAFF compatible partial atomic charge assignment scheme for modeling interactions of small molecules with biomolecular targets. *Journal of Computational Chemistry*, *32*(5), 893–907. doi:10.1002/jcc.21671 PMID:21341292

Murray, C. W., Baxter, C. A., & Frenkel, A. D. (1999). The sensitivity of the results of molecular docking to induced fit effects: Application to thrombin, thermolysin and neuraminidase. *Journal of Computer-Aided Molecular Design*, *13*(6), 547–562. doi:10.1023/A:1008015827877 PMID:10584214

Nabuurs, S. B., Wagener, M., & de Vlieg, J. (2007). A flexible approach to induced fit docking. *Journal of Medicinal Chemistry*, *50*(26), 6507–6518. doi:10.1021/jm070593p PMID:18031000

Namasivayam, V., & Gunther, R. (2007). pso@autodock: A fast flexible molecular docking program based on Swarm intelligence. *Chemical Biology & Drug Design*, *70*(6), 475–484. doi:10.1111/j.1747-0285.2007.00588.x PMID:17986206

Ng, M. C., Fong, S., & Siu, S. W. (2015). PSOVina: The hybrid particle swarm optimization algorithm for protein-ligand docking. *Journal of Bioinformatics and Computational Biology*, *13*(3), 1541007. doi:10.1142/S0219720015410073 PMID:25800162

Nicolaou, C. A., & Brown, N. (2013). Multi-objective optimization methods in drug design. *Drug Discovery Today. Technologies*, *10*(3), e427–e435. doi:10.1016/j.ddtec.2013.02.001 PMID:24050140

Nisius, B., & Gohlke, H. (2012). Alignment-independent comparison of binding sites based on DrugScore potential fields encoded by 3D Zernike descriptors. *Journal of Chemical Information and Modeling*, *52*(9), 2339–2347. doi:10.1021/ci300244y PMID:22881363

Nolan, T. L., Lapinsky, D. J., Talbot, J. N., Indarte, M., Liu, Y., Manepalli, S., & Surratt, C. K. et al. (2011). Identification of a novel selective serotonin reuptake inhibitor by coupling monoamine transporter-based virtual screening and rational molecular hybridization. *ACS Chemical Neuroscience*, *2*(9), 544–552. doi:10.1021/cn200044x PMID:21966587

Olimpieri, P. P., Chailyan, A., Tramontano, A., & Marcatili, P. (2013). Prediction of site-specific interactions in antibody-antigen complexes: The proABC method and server. *Bioinformatics (Oxford, England)*, *29*(18), 2285–2291. doi:10.1093/bioinformatics/btt369 PMID:23803466

Omori, S., & Kitao, A. (2013). CyClus: A fast, comprehensive cylindrical interface approximation clustering/reranking method for rigid-body protein-protein docking decoys. *Proteins*, *81*(6), 1005–1016. doi:10.1002/prot.24252 PMID:23344972

Ou-Yang, S. S., Lu, J. Y., Kong, X. Q., Liang, Z. J., Luo, C., & Jiang, H. (2012). Computational drug discovery. *Acta Pharmacologica Sinica*, *33*(9), 1131–1140. doi:10.1038/aps.2012.109 PMID:22922346

Ouyang, X., Zhou, S., Ge, Z., Li, R., & Kwoh, C. K. (2013). CovalentDock Cloud: a web server for automated covalent docking. *Nucleic Acids Res, 41*, W329-332.

Pak, Y., & Wang, S. (2000). Application of a Molecular Dynamics Simulation Method with a Generalized Effective Potential to the Flexible Molecular Docking Problems. *The Journal of Physical Chemistry B*, *104*(2), 354–359. doi:10.1021/jp993073h

Pala, N., Stevaert, A., Dallocchio, R., Dessi, A., Rogolino, D., Carcelli, M., . . . Naesens, L. (2015). Virtual screening and biological validation of novel influenza virus PA endonuclease inhibitors. *ACS Medicinal Chemistry Letters ACS Med. Chem. Lett.*

Patel, P. D., Patel, M. R., Kaushik-Basu, N., & Talele, T. T. (2008). 3D QSAR and molecular docking studies of benzimidazole derivatives as hepatitis C virus NS5B polymerase inhibitors. *Journal of Chemical Information and Modeling*, *48*(1), 42–55. doi:10.1021/ci700266z PMID:18076152

Pevzner, Y., Frugier, E., Schalk, V., Caflisch, A., & Woodcock, H. L. (2014). Fragment-based docking: Development of the CHARMMing Web user interface as a platform for computer-aided drug design. *Journal of Chemical Information and Modeling*, *54*(9), 2612–2620. doi:10.1021/ci500322k PMID:25151852

Philips, A., Milanowska, K., Lach, G., & Bujnicki, J. M. (2013). LigandRNA: Computational predictor of RNA-ligand interactions. *RNA (New York, N.Y.)*, *19*(12), 1605–1616. doi:10.1261/rna.039834.113 PMID:24145824

Pierce, B., Tong, W., & Weng, Z. (2005). M-ZDOCK: A grid-based approach for Cn symmetric multimer docking. *Bioinformatics (Oxford, England)*, *21*(8), 1472–1478. doi:10.1093/bioinformatics/bti229 PMID:15613396

Pierce, B. G., Hourai, Y., & Weng, Z. (2011). Accelerating protein docking in ZDOCK using an advanced 3D convolution library. *PLoS ONE*, *6*(9), e24657. doi:10.1371/journal.pone.0024657 PMID:21949741

Pierce, B. G., Wiehe, K., Hwang, H., Kim, B. H., Vreven, T., & Weng, Z. (2014). ZDOCK server: Interactive docking prediction of protein-protein complexes and symmetric multimers. *Bioinformatics (Oxford, England)*, *30*(12), 1771–1773. doi:10.1093/bioinformatics/btu097 PMID:24532726

Pirhadi, S., Shiri, F., & Ghasemi, J. B. (2013). Methods and applications of structure based pharmacophores in drug discovery. *Current Topics in Medicinal Chemistry*, *13*(9), 1036–1047. doi:10.2174/1568026611313090006 PMID:23651482

Plewczynski, D., Lazniewski, M., von Grotthuss, M., Rychlewski, L., & Ginalski, K. (2011). VoteDock: Consensus docking method for prediction of protein-ligand interactions. *Journal of Computational Chemistry*, *32*(4), 568–581. doi:10.1002/jcc.21642 PMID:20812324

Politi, A., Durdagi, S., Moutevelis-Minakakis, P., Kokotos, G., & Mavromoustakos, T. (2010). Development of accurate binding affinity predictions of novel renin inhibitors through molecular docking studies. *Journal of Molecular Graphics & Modelling*, *29*(3), 425–435. doi:10.1016/j.jmgm.2010.08.003 PMID:20855222

Pons, C., Jimenez-Gonzalez, D., Gonzalez-Alvarez, C., Servat, H., Cabrera-Benitez, D., Aguilar, X., & Fernandez-Recio, J. (2012). Cell-Dock: High-performance protein-protein docking. *Bioinformatics (Oxford, England)*, *28*(18), 2394–2396. doi:10.1093/bioinformatics/bts454 PMID:22815362

Pons, C., Solernou, A., Perez-Cano, L., Grosdidier, S., & Fernandez-Recio, J. (2010). Optimization of pyDock for the new CAPRI challenges: Docking of homology-based models, domain-domain assembly and protein-RNA binding. *Proteins*, *78*(15), 3182–3188. doi:10.1002/prot.22773 PMID:20602351

Pons, J. L., & Labesse, G. (2009). @TOME-2: a new pipeline for comparative modeling of protein-ligand complexes. *Nucleic Acids Res*, *37*, W485-491. doi: 10.1093/nar/gkp368

Porollo, A., & Meller, J. (2010). POLYVIEW-MM: web-based platform for animation and analysis of molecular simulations. *Nucleic Acids Res*, *38*, W662-666. doi: 10.1093/nar/gkq445

Purisima, E. O., & Hogues, H. (2012). Protein-ligand binding free energies from exhaustive docking. *The Journal of Physical Chemistry B*, *116*(23), 6872–6879. doi:10.1021/jp212646s PMID:22432509

Raub, S., Steffen, A., Kamper, A., & Marian, C. M. (2008). AIScore chemically diverse empirical scoring function employing quantum chemical binding energies of hydrogen-bonded complexes. *Journal of Chemical Information and Modeling*, *48*(7), 1492–1510. doi:10.1021/ci7004669 PMID:18597446

Ren, J., Williams, N., Clementi, L., Krishnan, S., & Li, W. W. (2010). Opal web services for biomedical applications. *Nucleic Acids Res*, *38*, W724-731. doi: 10.1093/nar/gkq503

Ricci, C. G., & Netz, P. A. (2009). Docking studies on DNA-ligand interactions: Building and application of a protocol to identify the binding mode. *Journal of Chemical Information and Modeling*, *49*(8), 1925–1935. doi:10.1021/ci9001537 PMID:19655805

Ritchie, D. W., & Kemp, G. J. (2000). Protein docking using spherical polar Fourier correlations. *Proteins*, *39*(2), 178–194. doi:10.1002/(SICI)1097-0134(20000501)39:2<178::AID-PROT8>3.0.CO;2-6 PMID:10737939

Ritchie, D. W., Kozakov, D., & Vajda, S. (2008). Accelerating and focusing protein-protein docking correlations using multi-dimensional rotational FFT generating functions. *Bioinformatics (Oxford, England)*, *24*(17), 1865–1873. doi:10.1093/bioinformatics/btn334 PMID:18591193

Ritchie, D. W., & Venkatraman, V. (2010). Ultra-fast FFT protein docking on graphics processors. *Bioinformatics (Oxford, England)*, *26*(19), 2398–2405. doi:10.1093/bioinformatics/btq444 PMID:20685958

Ritchie, D. W., Venkatraman, V., & Mavridis, L. (2010). Using graphics processors to accelerate protein docking calculations. *Studies in Health Technology and Informatics*, *159*, 146–155. PMID:20543434

Rueda, M., Bottegoni, G., & Abagyan, R. (2009). Consistent improvement of cross-docking results using binding site ensembles generated with elastic network normal modes. *Journal of Chemical Information and Modeling*, *49*(3), 716–725. doi:10.1021/ci8003732 PMID:19434904

Saladin, A., Rey, J., Thevenet, P., Zacharias, M., Moroy, G., & Tuffery, P. (2014). PEP-SiteFinder: a tool for the blind identification of peptide binding sites on protein surfaces. *Nucleic Acids Res*, *42*, W221-226.

Salemme, F. R. (1976). An hypothetical structure for an intermolecular electron transfer complex of cytochromes c and b5. *Journal of Molecular Biology*, *102*(3), 563–568. doi:10.1016/0022-2836(76)90334-X PMID:178879

Sandal, M., Duy, T. P., Cona, M., Zung, H., Carloni, P., Musiani, F., & Giorgetti, A. (2013). GOMoDo: A GPCRs online modeling and docking webserver. *PLoS ONE*, *8*(9), e74092. doi:10.1371/journal.pone.0074092 PMID:24058518

Schellhammer, I., & Rarey, M. (2004). FlexX-Scan: Fast, structure-based virtual screening. *Proteins*, *57*(3), 504–517. doi:10.1002/prot.20217 PMID:15382244

Schmidt, T., Bergner, A., & Schwede, T. (2014). Modelling three-dimensional protein structures for applications in drug design. *Drug Discovery Today*, *19*(7), 890–897. doi:10.1016/j.drudis.2013.10.027 PMID:24216321

Schneider, G., & Fechner, U. (2005). Computer-based de novo design of drug-like molecules. *Nature Reviews. Drug Discovery*, *4*(8), 649–663. doi:10.1038/nrd1799 PMID:16056391

Shimoda, T., Suzuki, S., Ohue, M., Ishida, T., & Akiyama, Y. (2015). Protein-protein docking on hardware accelerators: Comparison of GPU and MIC architectures. *BMC Systems Biology*, *9*(Suppl 1), S6. doi:10.1186/1752-0509-9-S1-S6 PMID:25707855

Shirvanyants, D., Alexandrova, A. N., & Dokholyan, N. V. (2011). Rigid substructure search. *Bioinformatics (Oxford, England)*, *27*(9), 1327–1329. doi:10.1093/bioinformatics/btr129 PMID:21460026

Shoemaker, B. A., Zhang, D., Tyagi, M., Thangudu, R. R., Fong, J. H., Marchler-Bauer, A., & Panchenko, A. R. et al. (2012). IBIS (Inferred Biomolecular Interaction Server) reports, predicts and integrates multiple types of conserved interactions for proteins. *Nucleic Acids Research*, *40*(Database issue), D834–D840. doi:10.1093/nar/gkr997 PMID:22102591

Singh, T., Biswas, D., & Jayaram, B. (2011). AADS--an automated active site identification, docking, and scoring protocol for protein targets based on physicochemical descriptors. *Journal of Chemical Information and Modeling*, *51*(10), 2515–2527. doi:10.1021/ci200193z PMID:21877713

Sircar, A., Kim, E. T., & Gray, J. J. (2009). RosettaAntibody: antibody variable region homology modeling server. *Nucleic Acids Res*, *37*, W474-479. doi: 10.1093/nar/gkp387

Sondergaard, C. R., Garrett, A. E., Carstensen, T., Pollastri, G., & Nielsen, J. E. (2009). Structural artifacts in protein-ligand X-ray structures: Implications for the development of docking scoring functions. *Journal of Medicinal Chemistry*, *52*(18), 5673–5684. doi:10.1021/jm8016464 PMID:19711919

Sotriffer, C., Klebe, G., Stahl, M., & Bohm, H.-J. (2003). Docking and scoring functions/Virtual screening. In D. J. Abraham (Ed.), *Burger's Medicinal Chemistry and Drug Discovery, Drug Discovery* (Vol. 1, pp. 281–416). John Wiley & Sons. doi:10.1002/0471266949.bmc007

Sotriffer, C. A., Gohlke, H., & Klebe, G. (2002). Docking into knowledge-based potential fields: A comparative evaluation of DrugScore. *Journal of Medicinal Chemistry*, *45*(10), 1967–1970. doi:10.1021/jm025507u PMID:11985464

Sousa, S. F., Fernandes, P. A., & Ramos, M. J. (2006). Protein-ligand docking: Current status and future challenges. *Proteins*, *65*(1), 15–26. doi:10.1002/prot.21082 PMID:16862531

Sousa, S. F., Ribeiro, A. J., Coimbra, J. T., Neves, R. P., Martins, S. A., Moorthy, N. S., & Ramos, M. J. et al. (2013). Protein-ligand docking in the new millennium--a retrospective of 10 years in the field. *Current Medicinal Chemistry*, *20*(18), 2296–2314. doi:10.2174/0929867311320180002 PMID:23531220

Spitzer, R., & Jain, A. N. (2012). Surflex-Dock: Docking benchmarks and real-world application. *Journal of Computer-Aided Molecular Design*, *26*(6), 687–699. doi:10.1007/s10822-011-9533-y PMID:22569590

Sukhwani, B., & Herbordt, M. C. (2010). FPGA acceleration of rigid-molecule docking codes. *IET Comput Digit Tech*, *4*(3), 184–195. doi:10.1049/iet-cdt.2009.0013 PMID:21857870

Tan, K. P., Varadarajan, R., & Madhusudhan, M. S. (2011). DEPTH: a web server to compute depth and predict small-molecule binding cavities in proteins. *Nucleic Acids Res*, *39*, W242-248. doi: 10.1093/nar/gkr356

Taylor, J. S., & Burnett, R. M. (2000). DARWIN: A program for docking flexible molecules. *Proteins*, *41*(2), 173–191. doi:10.1002/1097-0134(20001101)41:2<173::AID-PROT30>3.0.CO;2-3 PMID:10966571

Taylor, R. D., Jewsbury, P. J., & Essex, J. W. (2002). A review of protein-small molecule docking methods. *Journal of Computer-Aided Molecular Design*, *16*(3), 151–166. doi:10.1023/A:1020155510718 PMID:12363215

Terashi, G., Takeda-Shitaka, M., Kanou, K., Iwadate, M., Takaya, D., & Umeyama, H. (2007). The SKE-DOCK server and human teams based on a combined method of shape complementarity and free energy estimation. *Proteins*, *69*(4), 866–872. doi:10.1002/prot.21772 PMID:17853449

Tietze, S., & Apostolakis, J. (2007). GlamDock: Development and validation of a new docking tool on several thousand protein-ligand complexes. *Journal of Chemical Information and Modeling*, *47*(4), 1657–1672. doi:10.1021/ci7001236 PMID:17585857

Tintori, C., Magnani, M., Schenone, S., & Botta, M. (2009). Docking, 3D-QSAR studies and in silico ADME prediction on c-Src tyrosine kinase inhibitors. *European Journal of Medicinal Chemistry*, *44*(3), 990–1000. doi:10.1016/j.ejmech.2008.07.002 PMID:18722033

Tondel, K., Anderssen, E., & Drablos, F. (2006). Protein Alpha Shape (PAS) Dock: A new gaussian-based score function suitable for docking in homology modelled protein structures. *Journal of Computer-Aided Molecular Design*, *20*(3), 131–144. doi:10.1007/s10822-006-9041-7 PMID:16652207

Torchala, M., & Bates, P. A. (2014). Predicting the structure of protein-protein complexes using the SwarmDock Web Server. *Methods in Molecular Biology (Clifton, N.J.)*, *1137*, 181–197. doi:10.1007/978-1-4939-0366-5_13 PMID:24573482

Torchala, M., Moal, I. H., Chaleil, R. A., Fernandez-Recio, J., & Bates, P. A. (2013). SwarmDock: A server for flexible protein-protein docking. *Bioinformatics (Oxford, England)*, *29*(6), 807–809. doi:10.1093/bioinformatics/btt038 PMID:23343604

Tovchigrechko, A., & Vakser, I. A. (2006). GRAMM-X public web server for protein-protein docking. *Nucleic Acids Res*, *34*, W310-314. doi: 10.1093/nar/gkl206

Trosset, J. Y., & Scheraga, H. A. (1998). Reaching the global minimum in docking simulations: A Monte Carlo energy minimization approach using Bezier splines. *Proceedings of the National Academy of Sciences of the United States of America*, *95*(14), 8011–8015. doi:10.1073/pnas.95.14.8011 PMID:9653131

Tsai, T. Y., Chang, K. W., & Chen, C. Y. (2011). iScreen: World's first cloud-computing web server for virtual screening and de novo drug design based on TCM database@Taiwan. *Journal of Computer-Aided Molecular Design*, *25*(6), 525–531. doi:10.1007/s10822-011-9438-9 PMID:21647737

Tuccinardi, T., Nuti, E., Ortore, G., Supuran, C. T., Rossello, A., & Martinelli, A. (2007). Analysis of human carbonic anhydrase II: Docking reliability and receptor-based 3D-QSAR study. *Journal of Chemical Information and Modeling*, *47*(2), 515–525. doi:10.1021/ci600469w PMID:17295464

Tuffery, P., & Derreumaux, P. (2012). Flexibility and binding affinity in protein-ligand, protein-protein and multi-component protein interactions: Limitations of current computational approaches. *Journal of the Royal Society, Interface*, *9*(66), 20–33. PMID:21993006

Tuszynska, I., Magnus, M., Jonak, K., Dawson, W., & Bujnicki, J. M. (2015). NPDock: A web server for protein-nucleic acid docking. *Nucleic Acids Research*, *43*(W1), W425-430. doi:10.1093/nar/gkv493 PMID:25977296

Vajda, S., Hall, D. R., & Kozakov, D. (2013). Sampling and scoring: A marriage made in heaven. *Proteins*, *81*(11), 1874–1884. doi:10.1002/prot.24343 PMID:23775627

van Dijk, A. D., & Bonvin, A. M. (2006). Solvated docking: Introducing water into the modelling of biomolecular complexes. *Bioinformatics (Oxford, England)*, *22*(19), 2340–2347. doi:10.1093/bioinformatics/btl395 PMID:16899489

Velec, H. F., Gohlke, H., & Klebe, G. (2005). DrugScore(CSD)-knowledge-based scoring function derived from small molecule crystal data with superior recognition rate of near-native ligand poses and better affinity prediction. *Journal of Medicinal Chemistry*, *48*(20), 6296–6303. doi:10.1021/jm050436v PMID:16190756

Verdonk, M. L., Cole, J. C., Hartshorn, M. J., Murray, C. W., & Taylor, R. D. (2003). Improved protein-ligand docking using GOLD. *Proteins*, *52*(4), 609–623. doi:10.1002/prot.10465 PMID:12910460

Veselovsky, A. V., & Ivanov, A. S. (2003). Strategy of computer-aided drug design. *Current Drug Targets. Infectious Disorders*, *3*(1), 33–40. doi:10.2174/1568005033342145 PMID:12570731

Viswanath, S., Ravikant, D. V., & Elber, R. (2014). DOCK/PIERR: Web server for structure prediction of protein-protein complexes. *Methods in Molecular Biology (Clifton, N.J.)*, *1137*, 199–207. doi:10.1007/978-1-4939-0366-5_14 PMID:24573483

Vitorovic-Todorovic, M. D., Cvijetic, I. N., Juranic, I. O., & Drakulic, B. J. (2012). The 3D-QSAR study of 110 diverse, dual binding, acetylcholinesterase inhibitors based on alignment independent descriptors (GRIND-2). The effects of conformation on predictive power and interpretability of the models. *Journal of Molecular Graphics & Modelling*, *38*, 194–210. doi:10.1016/j.jmgm.2012.08.001 PMID:23073222

Vreven, T., Pierce, B. G., Hwang, H., & Weng, Z. (2009). Performance of ZDOCK in CAPRI rounds 20-26. *Proteomics*, *9*(12), 2175–2182. doi:10.1002/pmic.200800646 PMID:24123140

Vreven, T., Pierce, B. G., Hwang, H., & Weng, Z. (2013). Performance of ZDOCK in CAPRI rounds 20-26. *Proteins*, *81*(12), 2175–2182. doi:10.1002/prot.24432 PMID:24123140

Wang, J. C., Chu, P. Y., Chen, C. M., & Lin, J. H. (2012). idTarget: a web server for identifying protein targets of small chemical molecules with robust scoring functions and a divide-and-conquer docking approach. *Nucleic Acids Res*, *40*, W393-399.

Wei, Y., Li, J., Chen, Z., Wang, F., Huang, W., Hong, Z., & Lin, J. (2015). Multistage virtual screening and identification of novel HIV-1 protease inhibitors by integrating SVM, shape, pharmacophore and docking methods. *European Journal of Medicinal Chemistry*, *101*, 409–418. doi:10.1016/j.ejmech.2015.06.054 PMID:26185005

Weiner, S. J., Kollman, P. A., Nguyen, D. T., & Case, D. A. (1986). An all atom force field for simulations of proteins and nucleic acids. *Journal of Computational Chemistry*, *7*(2), 230–252. doi:10.1002/jcc.540070216

Welch, W., Ruppert, J., & Jain, A. N. (1996). Hammerhead: Fast, fully automated docking of flexible ligands to protein binding sites. *Chemistry & Biology*, *3*(6), 449–462. doi:10.1016/S1074-5521(96)90093-9 PMID:8807875

Wiehe, K., Pierce, B., Mintseris, J., Tong, W. W., Anderson, R., Chen, R., & Weng, Z. (2005). ZDOCK and RDOCK performance in CAPRI rounds 3, 4, and 5. *Proteins*, *60*(2), 207–213. doi:10.1002/prot.20559 PMID:15981263

Wiehe, K., Pierce, B., Tong, W. W., Hwang, H., Mintseris, J., & Weng, Z. (2007). The performance of ZDOCK and ZRANK in rounds 6-11 of CAPRI. *Proteins*, *69*(4), 719–725. doi:10.1002/prot.21747 PMID:17803212

Wilson, G. L., & Lill, M. A. (2011). Integrating structure-based and ligand-based approaches for computational drug design. *Future Med Chem*, *3*(6), 735–750. doi:10.4155/fmc.11.18 PMID:21554079

Wu, J., Hong, B., Takeda, T., & Guo, J. T. (2012). High performance transcription factor-DNA docking with GPU computing. *Proteome Science*, *10*(Suppl 1), S17. doi:10.1186/1477-5956-10-S1-S17 PMID:22759575

Xu, B., Yang, Y., Liang, H., & Zhou, Y. (2009). An all-atom knowledge-based energy function for protein-DNA threading, docking decoy discrimination, and prediction of transcription-factor binding profiles. *Proteins*, *76*(3), 718–730. doi:10.1002/prot.22384 PMID:19274740

Xu, D., Jaroszewski, L., Li, Z., & Godzik, A. (2014). AIDA: ab initio domain assembly server. *Nucleic Acids Res*, *42*, W308-313.

Xue, L. C., Jordan, R. A., El-Manzalawy, Y., Dobbs, D., & Honavar, V. (2014). DockRank: Ranking docked conformations using partner-specific sequence homology-based protein interface prediction. *Proteins*, *82*(2), 250–267. doi:10.1002/prot.24370 PMID:23873600

Yakar, R., & Akten, E. D. (2014). Discovery of high affinity ligands for beta2-adrenergic receptor through pharmacophore-based high-throughput virtual screening and docking. *Journal of Molecular Graphics & Modelling*, *53*, 148–160. doi:10.1016/j.jmgm.2014.07.007 PMID:25137647

Yang, C. Y., Sun, H., Chen, J., Nikolovska-Coleska, Z., & Wang, S. (2009). Importance of ligand reorganization free energy in protein-ligand binding-affinity prediction. *Journal of the American Chemical Society*, *131*(38), 13709–13721. doi:10.1021/ja9039373 PMID:19736924

Yang, Y., Zhao, H., Wang, J., & Zhou, Y. (2014). SPOT-Seq-RNA: Predicting protein-RNA complex structure and RNA-binding function by fold recognition and binding affinity prediction. *Methods in Molecular Biology (Clifton, N.J.)*, *1137*, 119–130. doi:10.1007/978-1-4939-0366-5_9 PMID:24573478

Yi, H., Qiu, S., Cao, Z., Wu, Y., & Li, W. (2008). Molecular basis of inhibitory peptide maurotoxin recognizing Kv1.2 channel explored by ZDOCK and molecular dynamic simulations. *Proteins*, *70*(3), 844–854. doi:10.1002/prot.21706 PMID:17729277

Yin, S., Biedermannova, L., Vondrasek, J., & Dokholyan, N. V. (2008). MedusaScore: An accurate force field-based scoring function for virtual drug screening. *Journal of Chemical Information and Modeling*, *48*(8), 1656–1662. doi:10.1021/ci8001167 PMID:18672869

Yu, J., Zhou, Y., Tanaka, I., & Yao, M. (2010). Roll: A new algorithm for the detection of protein pockets and cavities with a rolling probe sphere. *Bioinformatics (Oxford, England)*, *26*(1), 46–52. doi:10.1093/bioinformatics/btp599 PMID:19846440

Yu, X., Martinez, M., Gable, A. L., Fuller, J. C., Bruce, N. J., Richter, S., & Wade, R. C. (2015). webSDA: A web server to simulate macromolecular diffusional association. *Nucleic Acids Research*, *43*(W1), W220-224. doi:10.1093/nar/gkv335 PMID:25883142

Yuriev, E., Holien, J., & Ramsland, P. A. (2015). Improvements, trends, and new ideas in molecular docking: 2012-2013 in review. *Journal of Molecular Recognition*, *28*(10), 581–604. doi:10.1002/jmr.2471 PMID:25808539

Zhang, C., & Lai, L. (2011). SDOCK: A global protein-protein docking program using stepwise force-field potentials. *Journal of Computational Chemistry*, *32*(12), 2598–2612. doi:10.1002/jcc.21839 PMID:21618559

Zhang, C., Liu, S., & Zhou, Y. (2005). Docking prediction using biological information, ZDOCK sampling technique, and clustering guided by the DFIRE statistical energy function. *Proteins*, *60*(2), 314–318. doi:10.1002/prot.20576 PMID:15981255

Zhou, H., & Skolnick, J. (2013). FINDSITE(comb): A threading/structure-based, proteomic-scale virtual ligand screening approach. *Journal of Chemical Information and Modeling*, *53*(1), 230–240. doi:10.1021/ci300510n PMID:23240691

Zhou, Z., & Madura, J. D. (2004). CoMFA 3D-QSAR analysis of HIV-1 RT nonnucleoside inhibitors, TIBO derivatives based on docking conformation and alignment. *Journal of Chemical Information and Computer Sciences*, *44*(6), 2167–2178. doi:10.1021/ci049893v PMID:15554687

Zsila, F., Bikadi, Z., Malik, D., Hari, P., Pechan, I., Berces, A., & Hazai, E. (2011). Evaluation of drug-human serum albumin binding interactions with support vector machine aided online automated docking. *Bioinformatics (Oxford, England)*, *27*(13), 1806–1813. doi:10.1093/bioinformatics/btr284 PMID:21593135

Zsoldos, Z., Reid, D., Simon, A., Sadjad, B. S., & Johnson, A. P. (2006). eHiTS: An innovative approach to the docking and scoring function problems. *Current Protein & Peptide Science*, *7*(5), 421–435. doi:10.2174/138920306778559412 PMID:17073694

KEY TERMS AND DEFINITIONS

Induced-Fit Concept: Conformational changes of both the receptor and ligand induced during the binding process.

Ligand: Molecule able to bind to another molecule (usually a macromolecule).

Lock-and-Key Theory: Exact complementary fitting of the ligand to the receptor at the binding site in a manner similar to that of the key and lock required for the initiation of the response postulated by Emil Fischer in 1894.

Molecular Docking: Prediction of binding mode(s) of a ligand to the binding site of a target bio-macromolecule.

Pose: A candidate binding mode of ligand at the receptor binding site calculated based on the docking experiment.

Scoring Functions: Application of mathematical models used for the prediction of binding affinity between ligand and receptor within the molecular docking procedure.

Search Space: All possible orientations and conformations of a ligand as well as the target binding site (receptor) with respect to each other which can be explored during the docking calculation.

Structure-Based Drug Design: Designing and optimizing a ligand structure based on the known 3D-structure of its target molecule.

ENDNOTE

[1] ADMET is an abbreviation for "absorption, distribution, metabolism, excretion, and toxicity"

Chapter 2
Methods for Docking and Drug Designing

Ahmad Abu Turab Naqvi
Jamia Millia Islamia, India

Md. Imtaiyaz Hassan
Jamia Millia Islamia, India

ABSTRACT

Molecular docking is the prediction of conformational complementarity between ligand and receptor molecule. The process of docking integrates two schematic approaches namely sampling of ligand conformations and ranking of selected conformations based on scoring functions. The authors have discussed established methodologies for molecular docking and well-known tools implementing these methods. A brief account of different classes of scoring functions such as force field based, empirical, knowledge based, and descriptor based scoring functions is given along with the exemplary implementations of these scoring functions. By replacing test and trial based ligand screening with structure based virtual screening, molecular docking has helped in shortening the duration of novel drug discovery up to some extent. However, the developments made in the field of drug discovery are assisted by the advances in the techniques of molecular docking, but there is strong need of enrichment in the techniques, especially in scoring functions, to tackle the inbound problems of de novo drug discovery.

INTRODUCTION

All biological processes and molecular functions are carried out by diverse classes of macromolecules such as nucleic acids, proteins, carbohydrates, etc. Sometimes macromolecules perform certain activity on their own, but certain functions are carried out due to the intermolecular interactions in which the activity of a molecular species is induced by other molecules and vice versa. The concept of molecular docking is based on the phenomenon of intermolecular interactions between leading molecular species that play a key role in the working mechanisms of various biological processes and molecular functions. Basically, the event of docking upholds the binding of one molecule into the conformational space of another molecule leading to a specific activity or function. Docking can be seen between two proteins

DOI: 10.4018/978-1-5225-0115-2.ch002

having conformational complementariness such as protein- protein docking, or between one small molecule (usually called ligand) and a protein, DNA-protein docking, etc. In the context of computer-aided drug design, ligand- protein docking holds significant status. Therefore, in the upcoming paragraphs the authors will objectively discuss various aspects of ligand- protein docking in particular and molecular docking in general.

Several reviews and evaluation reports have been published in recent years covering important aspects of molecular docking, recent advances in methods of docking, evolution of search algorithms, developments in scoring functions, and advancements in docking tools (Ferreira, Dos Santos, Oliva, & Andricopulo, 2015; Kitchen, Decornez, Furr, & Bajorath, 2004; Meng, Zhang, Mezei, & Cui, 2011; G. M. Morris & Lim-Wilby, 2008; Oshiro, Kuntz, & Dixon, 1995; Verdonk, Cole, Hartshorn, Murray, & Taylor, 2003). In this chapter, the authors have tried to bring forward the objective study of small molecule-protein docking in the context of computer-aided drug design including brief idea of various preceding events of molecular docking such as target identification, lead optimization, and virtual screening. This chapter further elaborates the historical perspective of molecular docking, role of docking in computer aided drug design, an introduction to various methods and established tools for molecular docking and underlying scoring functions that assess the quality of conformational complementarity between ligand and receptor molecule. In this way, the flow of the chapter (Figure 1) goes with an introduction and historical background of molecular docking along the brief description of docking methodologies, docking tools, search algorithms and scoring function. In context of drug discovery, the authors have discussed the role of molecular docking in drug discovery by discussing some case studies. The authors also have discussed modern perspective of molecular modeling including the present day challenges of the area with respect to its involvement in drug discovery.

HISTORICAL PERSPECTIVE

Though, in *post genomic* era molecular docking has met revolutionary developments making it a more advanced concept aided by other fields of study such as computer science, physics, and statistics, etc., but the traces of the basic idea of docking way back to "Lock and Key Hypothesis" proposed by Emil Fischer in 1894 in the context of enzyme – substrate geometric interaction. During the development phases of molecular biology, several modifications have been proposed to understand the inbound complexity of docking. In 1958, Koshland came forward with the idea of "induced fit'' in his modification of "Lock and Key Hypothesis". He proposed that the conformational changes in the active site of the receptor are induced by the binding of the substrate (Koshland, 1958).

Before the period of late 1970s and 80s, molecular docking was studied more as a phenomenon of molecular interaction between the molecules having binding affinity. The idea of docking as a distinguished concept originates in the works of late 1970s. The term docking also has its origin in the studies conducted in late 70s (Wodak & Janin, 1978). The developments made in the 70s were more or less confined to interactions between a large molecule such as protein- protein docking (Wodak & Janin, 1978). More polished idea of docking of small molecules with proteins emerged in the early 1980s in the work of Kuntz et al., (Kuntz, Blaney, Oatley, Langridge, & Ferrin, 1982).

The field of molecule docking as a distinguished field of study has been developed parallel to the developments in computer science and technology. Molecular docking emerged significantly in the context of computer-aided drug design hand in hand with several techniques like target identification,

Figure 1. Flow chart showing covered topics of the chapter

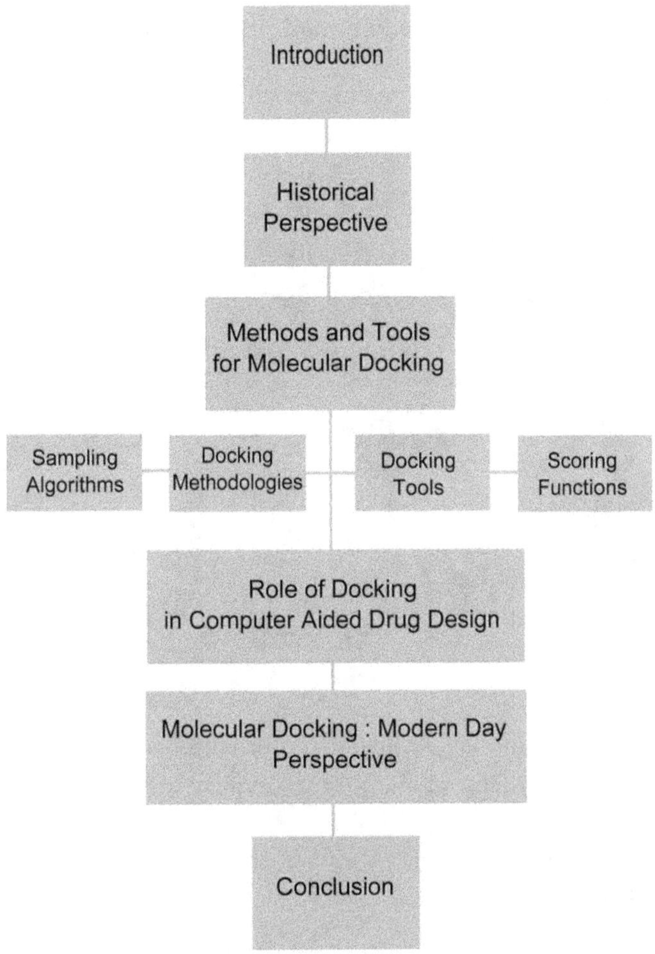

virtual screening or high throughput screening, molecular dynamics simulation, scoring functions, search algorithms, etc. (Kitchen, et al., 2004) (Meng, et al., 2011). Going through all the processes of development, ever since late 1970s, molecular docking has emerged as more developed field of study which integrates a huge variety of entities such drug discovery databases, ligand libraries, tools of molecular docking and various tools of evaluation.

METHODS AND TOOLS FOR MOLECULAR DOCKING

In this section, the authors will discuss some established methods of molecular docking with a brief introduction to some well developed tools based on these methodologies. Technically, docking is defined as the prediction of conformational complimentary between ligand and receptor molecule (proteins in the context of CADD) followed by the application of scoring function for the assessment of binding in terms of binding energy (G. M. Morris & Lim-Wilby, 2008). The whole process of docking integrates

Figure 2. Figure showing (A) three-dimensional structure of the ligand; (B) three-dimensional structure of the receptor; (C) the ligand is docked into the binding cavity of the receptor and the putative conformations are explored; (D) the most likely binding conformation and the corresponding intermolecular interactions are identified.
Figure was taken from the Ferreira et al. 2015.

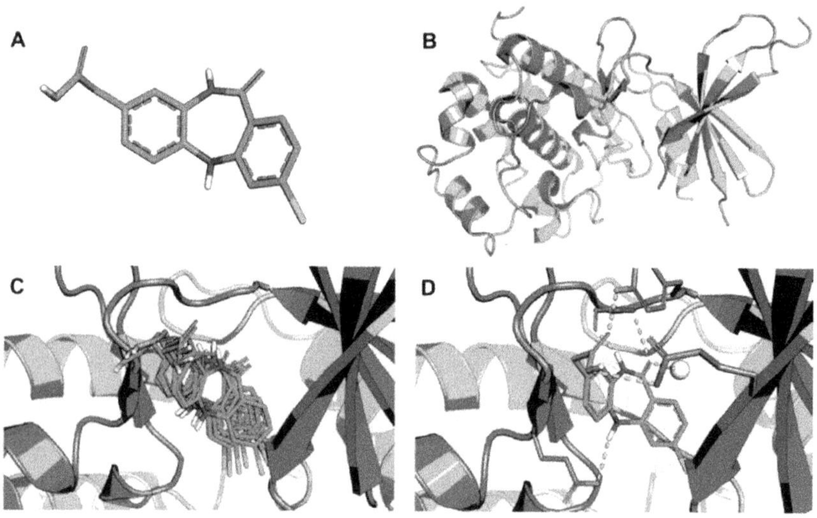

two schematic approaches i.e., sampling of ligand conformations and ranking of selected conformations based on scoring functions (Meng, et al., 2011). A typical example of outline of docking process is shown in Figure 2.

Sampling Algorithms

Sampling of ligand conformations in the active site of the receptor is done using various sampling algorithms. During the last decades, there is a revolutionary development in sampling algorithms. All of these sampling algorithms ranging from probabilistic to stochastic have their merits and demerits based on the computational cost and time efficiency. On the other hand, these algorithms are chosen on the priority basis depending upon the size of involved molecules. For example, molecular dynamics simulation has comparatively higher computational cost therefore it is usually preferred for small molecules.

Brooijmans and Kuntz (Brooijmans & Kuntz, 2003) in their review of sampling algorithms of molecular docking have proposed basic categories of searching algorithms based on the approach applied by these search algorithms. The basic categories of search algorithms are systematic, stochastic and simulation based (Kitchen, et al., 2004). In this way, incremental construction (Rarey, Kramer, Lengauer, & Klebe, 1996) and database search strategies are kept in systemic search categories. In incremental construction method, ligand is broken into fragments and the fragment showing higher complementarity to the active site conformation is docked inside the active site pocket as the 'anchor' then the remaining fragments of the ligand are docked in an incremental manner. Monte Carlo (Hart & Read, 1992) and Genetic algorithms (Jones, Willett, Glen, Leach, & Taylor, 1997; Garrett M. Morris et al., 1998) are the most established methods for stochastic searching of conformational space. Monte Carlo produces different conformations of ligand by inducing bond rotation and rigid body translation (Meng, et al., 2011).

Genetic algorithms are the second most used search algorithms in molecular docking. The concept of Genetic algorithms is based on Darwin's evolution theory (Oshiro, et al., 1995) . Genetic algorithms carry out ligand conformation search in term of genes, chromosomes and mutations. Energy based assessment of ligand structure is done and favoured conformation is taken to the next step based on the theory of 'survival of the fittest'. The ability of these methods to modify ligand conformation makes them more preferred to other search methods. Molecular dynamics simulation represents the simulation based methods of conformation search it is the most preferred approach in simulation ever since its development (Brooijmans & Kuntz, 2003).

Docking Methodologies

The differentiation between docking methodologies evolves on the basis of flexibility/rigidity of the ligand and/or receptor molecule. Thus, there are three basic classes of docking methods based the above-mentioned criteria:-

- **Rigid Ligand and Rigid Receptor Docking:** In this approach both ligand and receptor molecules are considered rigid. The relative rigidity is due to the lesser number of degrees of for rotational and translational space.
- **Flexible Ligand and Rigid Receptor Docking:** This is the most implemented docking methodology in almost all the leading docking software. The concept of flexible ligand rigid receptor docking is based on the 'induced fit' prototype of ligand- receptor interaction.
- **Flexible Ligand and Flexible Receptor Docking:** Ideally, docking in its traditional approach deals with ligand flexibility having rigid receptor in the frame. The idea of receptor flexibility is so far a challenging task in molecular docking (Lexa & Carlson, 2012). However, molecular dynamics simulation techniques have efficiency of handling the problem of receptor flexibility, but due to the high computation cost such approaches are less preferred.

Docking Tools

A vast range of docking tools has been developed by incorporating above mentioned search algorithms and docking methodologies so far such as AutoDock, GOLD, DOCK, Glide, FRED, FITTED and CDocker etc. Table 1 provides a brief description of some of the leading docking tools and web based docking servers including the information about implemented search algorithms and scoring functions.

Scoring Functions

In structure based virtual screening, the prediction of possible interaction between ligand and receptor in feasible conformational space is assessed in terms of binding energy. The elimination of 'false' ligand poses from the population of 'true' ones on the basis of minimal binding free energy is carried out using different classes of scoring functions. These scoring functions, on the basis of the fact that they consider for the estimation of binding energy, are divided into different classes such force field based scoring functions, empirical scoring functions, knowledge based scoring functions, scoring based on consensus (Kitchen, et al., 2004) and descriptor based scoring functions (Liu & Wang, 2015). Recent advances in molecular docking were mainly focused on strengthening the scoring strategies by including certain factors such as entropy effects, water molecules, and solvent effects were not considered in previous

Table 1. List of docking tools and web servers

S. No.	Software	Description	Platform
21.	Autodock 4.2.6	It is based on Monte Carlo search algorithm for flexible ligand search. The scoring functions used in AutoDock are force field based.	Desktop based
22.	DOCK 6.7	It uses incremental construction strategy for docking using AMBER molecular mechanics scoring function with implicit solvent	Desktop based
23.	GOLD	It is a Genetic algorithm based docking software. It uses force field based scoring functions for the evaluation of docking.	Desktop based
24.	Glide	It provides exhaustive search method for ligand screening.	Desktop based
25.	ICM	It is based on pseudo-Brownian sampling and local minimization. Provides ligand and protein flexibility.	Desktop based
26.	FlexX	It is an incremental construction based docking software. It comes with flexible ligand and flexible protein docking strategy.	Desktop based
27.	FITTED	It is a genetic algorithm based docking soft with more explicit features with explicit water molecule, side chain flexibility, displaceable water molecules etc.	Desktop based
28.	AutoDock Vina	It is Monte Carlo based. Provides ligand flexibility and side chain flexibility for receptor.	Desktop based
29.	FRED	It performs a systematic and non stochastic search for ligand conformations. It uses Chemgauss4 scoring function for the assessment of the search	Desktop based
30.	CDocker	It is a CHARMm based docking platform. It generates random ligand conformations using molecular dynamics.	Desktop based
31.	MOE suite 2014.9	Provides diverse range of modules used in molecular docking and conformations evaluation.	Desktop based
32.	ADAM	It is an Automated docking tool. Suitable for virtual high throughput screening.	Desktop based
33.	SwissDock	Suitable for predicting molecular interaction between small molecule and target.	Web based
34.	DockingServer	Online docking interface of docking.	Web based
35.	1-Click Docking	Online docking platform with WebGL/Javascript based molecule viewer of GLmol.	Web based
36.	Pardock	Provides all-atom energy based Monte Carlo, rigid protein ligand docking and parallel processing mode.	Web based
37.	PatchDock	It provides structure prediction of protein-protein and protein-small molecule complexes using shape complementarity principles.	Web based
38.	Dock blaster	Online server for structure based ligand discovery.	Web based
39.	PLATINUM	Calculates ligand- receptor interaction on the basis of Molecular Hydrophobicity Potential" (MHP).	Web based
40.	iScreen	Cloud computing web server for virtual screening and de novo drug design based on TCM database.	Web based

versions of these scoring functions. Computational time and cost were two main factors that are still to be addressed. The assessment of docking in the stochastic environment is needed to get better results, but the chances of increase in computational time and cost are the major obstacle in this approach. Parallel developments in computational techniques are paving the way for the elimination of this problem. In this section, the authors will discuss the aforementioned classes of scoring functions along with the docking tools that implement these scoring functions.

Force-Field Based Scoring Functions

Force field was conceptualized in the first half of the twentieth century. It remains the most used strategy in a variety of fields including molecular modeling through molecular dynamics simulation, energy minimization strategies, and molecular docking. Force field based scoring functions estimate the binding energy of ligand-receptor complex by considering non bonded interactions. These non bonded interactions include electrostatic interactions and van Waals interactions. The estimation of binding energy as a sum of different molecular mechanics terms is carried in using various functions. Electrostatic interactions are expressed in terms of coulombic formula whereas Lennard - Jones function is used for van der Waals interactions. Force field functions used by diverse docking tools are expressed in various parametric sets of molecular mechanics terms, such as DOCK (Allen et al., 2015) and AutoDock (G. M. Morris et al., 2009) use Amber force filed where as CDocker (Wu, Robertson, Brooks, & Vieth, 2003) uses CHARMm force field scoring functions. Advance versions of DOCK, AutoDock and GOLD have implemented some extensions in force field scoring functions such as hydrogen bonding terms, entropy and solvent factors (Meng, et al., 2011). Despite being widely used in a range of established docking tools, force field based scoring functions have certain limitations. The problem of high computational time leads to slow speed of estimations. To avoid slow speed, cut off distances are used for non bonded interactions. These cutoff distances affect the accuracy of estimation while considering long distance effects of non bonded interactions (Kitchen, et al., 2004; Meng, et al., 2011).

Empirical Scoring Functions

Empirical scoring functions estimate the binding energy of the ligand – receptor complex in terms of various energy terms such as hydrogen bonds, hydrophoboic effect, ionic interactions, binding entropy and rotatable bonds in the ligand (Bohm, 1994; Eldridge, Murray, Auton, Paolini, & Mee, 1997). Empirical scoring functions use the coefficients of the energy terms obtained by the regression analysis of the binding energies of known complex structures. LUDI (Bohm, 1992), Chemscore implemented in GOLD (Verdonk, et al., 2003), SCORE (Wang, Liu, Lai, & Tang, 1998) are the well-known examples of empirical scoring functions.

Knowledge Based Scoring Functions

Knowledge based scoring functions have certain advantages over other scoring functions due to the computational simplicity that they provide. The idea of knowledge based scoring functions is to use the statistical inferences from the binding energy potentials of known ligand – receptor complex structures hence the term knowledge based scoring functions. The frequent iteration of a certain pair wise atomic contact between ligand and receptor is considered energetically favoured. Thus, inferences for the query ligand-receptor complex are made based on these assumptions. SMoG (DeWitte & Shakhnovich, 1996), PMF (Muegge & Martin, 1999), and DrugScore (Gohlke, Hendlich, & Klebe, 2000) are the examples of knowledge based scoring functions.

Descriptor Based Scoring Functions

Jie Liu and Renxiao Wang (Liu & Wang, 2015) in a recent classification of scoring functions discussed this relatively new kind of scoring functions. The idea of descriptor based scoring functions is evolved from the quantitative structure - activity relationship (QSAR) analysis strategy. In the context of CADD, QSAR analysis is used for the evaluation of the interaction between the drug molecule and the target protein (receptor). QSAR equation is expressed using various molecular descriptors as parameters. These descriptors include surface or shape properties, hydrogen bonds, molecular weight, etc. RF-Score (Ballester & Mitchell, 2010), ID- Score (Li, Yang, Wang, Li, & Yang, 2013), and SF-Score (Zilian & Sotriffer, 2013) are the examples of descriptor based scoring functions.

Consensus Scoring

Consensus scoring has remained the preferred way of scoring in various techniques of computer based predictions. Scoring functions have certain limitations in assessing the quality of predictions. To tackle this situation, researchers have found the way of consensus scoring. Ligand-receptor complementarity is assessed using different scoring functions. The structure that gets better scores in maximum number of scoring strategies is accepted. CScore (Clark, Strizhev, Leonard, Blake, & Matthew, 2002) is an example of consensus based scoring functions. It uses DOCK, ChemScore, GOLD, and FlxX scoring functions.

ROLE OF DOCKING IN COMPUTER AIDED DRUG DESIGN

The principal objective of this chapter is to discuss molecular docking focusing its importance in computer aided drug design. During the last two decades, molecular docking has drastically revolutionized the field of drug discovery. Parallel developments in computer aided drug discovery having molecular docking incorporated as a principal phenomenon has made the process of drug discovery more time efficient. The integration of computer based techniques applied to preceding events of docking such as targeted identification, high throughput screening, lead optimization, etc., are now accomplished in less time, hence making the process less time-consuming (Gu, Zhang, & Yuan, 2014; Jorgensen, 2004). In order to understand the role of Molecular Docking and available docking programs, the authors are discussing here some case studies made during recent past in which molecular docking played essential role to bring forward the consequences favouring the possibility of drug discovery against certain diseases.

Design and Synthesis of Piperidine Derivatives as Novel Human Heat Shock Protein 70 Inhibitors for the Treatment of Drug-Resistant Tumors

Yanqun et al. (Zeng, Cao, Zhang, Li, & Zhong, 2015) have carried out designing and synthesis of piperidine derivatives as the inhibitors for human heat shock protein 70 (HSP70). HSP70s are significant due to their involvement in various essential biological processes such as maintenance of protein homeostasis, regulation of apoptosis and involvement in the binding of (Tumour necrosis factor (TNF) and TNF-related apoptosis inducing ligand (TRAIL) to the receptors. The study is mainly focused on the practicality of HSP70s in the treatment of cancers and drug-resistant tumours. HSP70s play a vital role in cancer resistance during the treatment of tumour cells. The phenomenon is backed by the over expression of HSP70s in cancer patients which also favours the spreading of cancerous infection to other cells falling in range of the cancer affected area. The study puts a significant stress on lapatinib-resistant breast cancer cells while performing the cell viability assays. Study suggests that compounds HSP70-36/37/40/43/46 are significant in inhibiting the growth of breast cancer cells. Developments made in the area were motivated by the unavailability of enough inhibiters against HSP70s.

The study is carried out with combination of *in silico* and *in vivo* techniques. In context, the authors will discuss the concerned part, i.e. *in silico* in detail. A set of tools and databases has been used for the purpose. The synthesis of novel inhibitors is based on N-terminal nucleotide binding domain of HSP70. Crystal structure of nucleotide binding domain in complex with adenosine diphosphate (ADP) is used for the screening. In this way, docking domain is detected using MVD4.0 software. Molergo Visual Docking (MVD) is an integrated docking platform. It provides high accuracy prediction of protein-ligan interaction. It is specifically equipped with the ability to predict potential binding sites in the target protein and also binding modes for the ligand. In this case, Cambridge Database and a patented library of 243,712 compounds are used for virtual fragment screening. The calculation of binding modes for the ligands in HSP70 ATPase domain has been performed using Discovery studio. Discovery Studio is an integrated program contained a diverse range of modules to perform various calculations and predictions used in molecular docking. Prediction of binding modes for all 67 compounds has been performed using Discovery Studio. Discovery studio is also equipped with modules to perform energy minimization and other energy related calculations. Energy minimization has also been performed using the modules available in Discovery Studio. To predict the drugability of the compounds ADMET screening was performed using the prescribed modules of Discovery Studio. In this study, an array of different programs for molecular docking, virtual fragment screening, and binding mode prediction has been used to maintain the credibility of the *in silico* approach. In recent time, an affinity for consensus based approach for computer based predictions has been developed in the researchers. It gives the researchers a way of comparing the results provided by different programs and then dragging consensus based consequences. Finally, Surflex-SYBYL6.9 was used for precise docking of the ligands. Novel derivatives of piperidine have good anticancer bioactivities. They also possess very low rate of inhibiting HSP70 ATPase activity. Therefore, newly derived inhibitors of HSP70 may provide ways of inhibiting the activity of HSP70 in cancer treatments.

Discovery of Akt Kinase Inhibitors through Structure-Based Virtual Screening and Their Evaluation as Potential Anticancer Agents

In recent years, several studies have been conducted for discovery and development of potential inhibitors against known target proteins that are involved in disease causing mechanisms. The structure based virtual screening and docking serves the purpose for successful development of such inhibiting compounds. Chuan et al. (Chuang et al., 2015) have conducted a similar study to discovery novel inhibitors against the activity of human Atk kinase using the structure based virtual screening. Akt is an essential regulator in the phosphatidylinositol 3-kinase PI3K/Akt signaling pathway therefore it is considered a potential drug target in cancer therapy. In this study, x-ray crystal structure of human Atk in complex with pyrrolopyrimidine inhibitor is used to perform virtual screening. Virtual screening has been performed using a dataset of 35,367 compounds that is derived from SPECS through Zinc database of chemical compounds. DOCK4.0 has been used for structure based virtual screening. ATP binding site of Atk is selected for screening. Consequently, a set of 48 compounds was selected that are showing inhibiting activity against Atk activity. Biological assessment of these 48 compounds for Atk kinase inhibition provided 26 final compounds that exhibited potential inhibition effect against Atk. Out of these compounds a48 (compounds were named as a1-a48) showed highest inhibition activity against Atk kinase. Finally, selected 26 test compounds were further taken for Cytotoxicity Evaluation on HCT-116 Cancer Cells and HEK-293 Normal Cells. Out of these 26 compounds 12 compounds expressed higher cytotoxicity against HCT-116 Cancer Cells. Eventually, compounds a46 and a48 were selected for further study on the basis of their IC_{50} values against colon cancer cells. Finally selected compounds show promising inhibition activity against Atk. Docking of both these compounds in the ATP binding site of Atk is performed using GOLD5.0 (Fig). GOLD performs genetic algorithm based flexible ligand docking simulation. On the basis of biological evaluation and docking studies, both compounds a46 and a48 are found potential inhibitors of Atk kinase activity that makes them potential anticancer agents.

Development of Effective Drug against H1N1 Pathogen

Mallipeddi et al., (Mallipeddi, Kumar, White, & Webb, 2014) in their review of recent developments in the field of computer aided drug discovery have discussed the applications of molecular docking in the development of anti-influenza drug. The pandemic outbreak of H1N1 influenza in 2009 has affected a majority of human population globally. Drugs effective against the H1N1 pathogen namely Zanamivir and oseltamivir are developed using the advanced techniques of molecular docking and virtual screening.

Development of Anti-HIV Drugs that Inhibit the C-C Chemokine Receptor Type 5 (CCR5) Activity

AIDS, ever since its first case detected in human has remained the one of the severest health issues for human race. Human immunodeficiency virus (HIV) causes AIDS in human beings. It enters the target cells through co receptors. In a review published, Planche & Cordeiro (Speck-Planche & Cordeiro, 2011) have highlighted the role of computer aided drug discovery methods in the development of anti-HIV drugs that inhibit the C-C chemokine receptor type 5 (CCR5) activity. CCR5 receptors are the most crucial co receptors that facilitate the process of HIV infection.

MOLECULAR DOCKING: MODERN DAY PERSPECTIVE

So far in this chapter, the authors have discussed molecular docking, its historical perspective, a brief introduction to various search algorithms, docking methods, docking programs and some case studies in which docking played role towards obtaining the desired results. However, with the advancement in Computation power that more strong computing devices, high performance servers, high power processing units, supercomputers, molecular docking is also undergone fruitful development in recent years, but there are still some challenges that this ever-developing field is facing. These challenges are evolved due to certain limitations and drawbacks of docking methods and programs. Though, on evolutionary level, docking programs have been developed effectively to throw out possible limitations, but the inability of most of the programs to contour the problems evolved due to the more stochastic behaviour of the molecular entities.

With the possible development of docking tools, in time, field of drug discovery has been revolutionized astronomically. The major concern for the development of docking programs has remained biomolecular flexibility specially flexibility of receptor protein. The other problem that has been managed, though not solved completely, is application of docking strategies for stochastic systems. In drug discovery process, the *in silico* part, i.e. computational predictions, target identification and final binding of ligand in receptor molecular hence identification of potential drug has been controlled up to some extent. The major problem is faced during laboratory experimentation of predicted compounds. This is still a challenge in the field of drug discovery for experimentalists.

Leading docking tools like, AutoDock, AutoDock Vina, GOLD, DOCK, and Glide, etc. have implemented advanced search algorithms and scoring functions to counter the problems of protein flexibility and stochasticity. However, the problem has been dealt up to some level; still there are practical instances where some of these programs fail to serve the purpose. Docking studies for highly stochastic and complex systems such as membrane proteins face various problems. To counter such kinds of difficulties, researchers have started using consensus approaches for molecular docking. For a problem, they use an array of docking and virtual screening programs in place of using single software for everything.

As the authors have discussed above, the major challenges in molecular docking are structural flexibility, stochasticity of the involved system and implementation of computational prediction at experimentation level with a promising precision that may pave the way of development of the drug. Most of the problems and limitations that docking programs face can be solved by advancing the search algorithms and scoring function. There is a need of more stochastic search algorithms such as genetic algorithm and implementation of molecular dynamics simulation with precise parameters. Currently, the parameters used in scoring functions and simulation are knowledge based. The deterministic parameters are less effective in making stochastic predictions. Hence, there is a need of developing the strategies to involve more and more stochastic parameters.

CONCLUSION

Molecular docking has become the principal weapon to conquest the problems of structure based virtual screening. It is in the way of revolutionary progress ever since its conceptualization. The parallel developments in computer technology have aided the progress of molecular docking in various manners. Though, docking has reached possible limits of enrichments up to some extent, but there are still some problems that are to be addressed for further developments in molecular docking. The intricate stochastic behaviour of the biological process is still a major challenge for the researchers. Furthermore, the interaction between ligand and receptor takes place in a more stochastic manner having so many factors affecting the phenomenon such as water molecules, entropy effects, solvent effects, etc. So, there is a strong need of enrichment of docking techniques with more focus on the strength of efficient scoring functions that can tackle the situation in a more realistic manner. There is also an undeniable inability of docking techniques and scoring functions to the problem of *de novo* drug design. Most of the scoring functions and search algorithms use the traditional approach of the parametric approximation on the basis of known experimental data. These approximations have their limits while applied in the process of *de novo* drug design. However, docking techniques become more powerful with efficient scoring functions, but further progress is strongly needed. Scoring functions with more and more stochastic parameters will be helpful in the betterment of *de novo* drug design processes.

REFERENCES

Allen, W. J., Balius, T. E., Mukherjee, S., Brozell, S. R., Moustakas, D. T., Lang, P. T., & Rizzo, R. C. et al. (2015). DOCK 6: Impact of new features and current docking performance. *Journal of Computational Chemistry*, *36*(15), 1132–1156. doi:10.1002/jcc.23905 PMID:25914306

Ballester, P. J., & Mitchell, J. B. (2010). A machine learning approach to predicting protein-ligand binding affinity with applications to molecular docking. *Bioinformatics (Oxford, England)*, *26*(9), 1169–1175. doi:10.1093/bioinformatics/btq112 PMID:20236947

Bohm, H. J. (1992). LUDI: Rule-based automatic design of new substituents for enzyme inhibitor leads. *Journal of Computer-Aided Molecular Design*, *6*(6), 593–606. doi:10.1007/BF00126217 PMID:1291628

Bohm, H. J. (1994). The development of a simple empirical scoring function to estimate the binding constant for a protein-ligand complex of known three-dimensional structure. *Journal of Computer-Aided Molecular Design*, *8*(3), 243–256. doi:10.1007/BF00126743 PMID:7964925

Brooijmans, N., & Kuntz, I. D. (2003). Molecular recognition and docking algorithms. *Annual Review of Biophysics and Biomolecular Structure*, *32*(1), 335–373. doi:10.1146/annurev.biophys.32.110601.142532 PMID:12574069

Chuang, C. H., Cheng, T. C., Leu, Y. L., Chuang, K. H., Tzou, S. C., & Chen, C. S. (2015). Discovery of Akt kinase inhibitors through structure-based virtual screening and their evaluation as potential anticancer agents. *International Journal of Molecular Sciences*, *16*(2), 3202–3212. doi:10.3390/ijms16023202 PMID:25648320

Clark, R. D., Strizhev, A., Leonard, J. M., Blake, J. F., & Matthew, J. B. (2002). Consensus scoring for ligand/protein interactions. *Journal of Molecular Graphics & Modelling*, *20*(4), 281–295. doi:10.1016/S1093-3263(01)00125-5 PMID:11858637

DeWitte, R. S., & Shakhnovich, E. I. (1996). SMoG: De Novo Design Method Based on Simple, Fast, and Accurate Free Energy Estimates. 1. Methodology and Supporting Evidence. *Journal of the American Chemical Society*, *118*(47), 11733–11744. doi:10.1021/ja960751u

Eldridge, M. D., Murray, C. W., Auton, T. R., Paolini, G. V., & Mee, R. P. (1997). Empirical scoring functions: I. The development of a fast empirical scoring function to estimate the binding affinity of ligands in receptor complexes. *Journal of Computer-Aided Molecular Design*, *11*(5), 425–445. doi:10.1023/A:1007996124545 PMID:9385547

Ferreira, L. G., Dos Santos, R. N., Oliva, G., & Andricopulo, A. D. (2015). Molecular docking and structure-based drug design strategies. *Molecules (Basel, Switzerland)*, *20*(7), 13384–13421. doi:10.3390/molecules200713384 PMID:26205061

Gohlke, H., Hendlich, M., & Klebe, G. (2000). Knowledge-based scoring function to predict protein-ligand interactions. *Journal of Molecular Biology*, *295*(2), 337–356. doi:10.1006/jmbi.1999.3371 PMID:10623530

Gu, W. G., Zhang, X., & Yuan, J. F. (2014). Anti-HIV drug development through computational methods. *The AAPS Journal*, *16*(4), 674–680. doi:10.1208/s12248-014-9604-9 PMID:24760437

Hart, T. N., & Read, R. J. (1992). A multiple-start Monte Carlo docking method. *Proteins*, *13*(3), 206–222. doi:10.1002/prot.340130304 PMID:1603810

Jones, G., Willett, P., Glen, R. C., Leach, A. R., & Taylor, R. (1997). Development and validation of a genetic algorithm for flexible docking. *Journal of Molecular Biology*, *267*(3), 727–748. doi:10.1006/jmbi.1996.0897 PMID:9126849

Jorgensen, W. L. (2004). The many roles of computation in drug discovery. *Science*, *303*(5665), 1813–1818. doi:10.1126/science.1096361 PMID:15031495

Kitchen, D. B., Decornez, H., Furr, J. R., & Bajorath, J. (2004). Docking and scoring in virtual screening for drug discovery: Methods and applications. *Nature Reviews. Drug Discovery*, *3*(11), 935–949. doi:10.1038/nrd1549 PMID:15520816

Koshland, D. E. (1958). Application of a Theory of Enzyme Specificity to Protein Synthesis. *Proceedings of the National Academy of Sciences of the United States of America*, *44*(2), 98–104. doi:10.1073/pnas.44.2.98 PMID:16590179

Kuntz, I. D., Blaney, J. M., Oatley, S. J., Langridge, R., & Ferrin, T. E. (1982). A geometric approach to macromolecule-ligand interactions. *Journal of Molecular Biology*, *161*(2), 269–288. doi:10.1016/0022-2836(82)90153-X PMID:7154081

Lexa, K. W., & Carlson, H. A. (2012). Protein flexibility in docking and surface mapping. *Quarterly Reviews of Biophysics*, *45*(3), 301–343. doi:10.1017/S0033583512000066 PMID:22569329

Li, G. B., Yang, L. L., Wang, W. J., Li, L. L., & Yang, S. Y. (2013). ID-Score: A new empirical scoring function based on a comprehensive set of descriptors related to protein-ligand interactions. *Journal of Chemical Information and Modeling*, *53*(3), 592–600. doi:10.1021/ci300493w PMID:23394072

Liu, J., & Wang, R. (2015). Classification of current scoring functions. *Journal of Chemical Information and Modeling*, *55*(3), 475–482. doi:10.1021/ci500731a PMID:25647463

Mallipeddi, P. L., Kumar, G., White, S. W., & Webb, T. R. (2014). Recent advances in computer-aided drug design as applied to anti-influenza drug discovery. *Current Topics in Medicinal Chemistry*, *14*(16), 1875–1889. doi:10.2174/1568026614666140929153812 PMID:25262796

Meng, X. Y., Zhang, H. X., Mezei, M., & Cui, M. (2011). Molecular docking: A powerful approach for structure-based drug discovery. *Current Computer-aided Drug Design*, *7*(2), 146–157. doi:10.2174/157340911795677602 PMID:21534921

Morris, G. M., Goodsell, D. S., Halliday, R. S., Huey, R., Hart, W. E., Belew, R. K., & Olson, A. J. (1998). Automated docking using a Lamarckian genetic algorithm and an empirical binding free energy function. *Journal of Computational Chemistry*, *19*(14), 1639–1662. doi:10.1002/(SICI)1096-987X(19981115)19:14<1639::AID-JCC10>3.0.CO;2-B

Morris, G. M., Huey, R., Lindstrom, W., Sanner, M. F., Belew, R. K., Goodsell, D. S., & Olson, A. J. (2009). AutoDock4 and AutoDockTools4: Automated docking with selective receptor flexibility. *Journal of Computational Chemistry*, *30*(16), 2785–2791. doi:10.1002/jcc.21256 PMID:19399780

Morris, G. M., & Lim-Wilby, M. (2008). Molecular docking. *Methods in Molecular Biology (Clifton, N.J.), 443*, 365–382. doi:10.1007/978-1-59745-177-2_19 PMID:18446297

Muegge, I., & Martin, Y. C. (1999). A general and fast scoring function for protein-ligand interactions: A simplified potential approach. *Journal of Medicinal Chemistry, 42*(5), 791–804. doi:10.1021/jm980536j PMID:10072678

Oshiro, C. M., Kuntz, I. D., & Dixon, J. S. (1995). Flexible ligand docking using a genetic algorithm. *Journal of Computer-Aided Molecular Design, 9*(2), 113–130. doi:10.1007/BF00124402 PMID:7608743

Rarey, M., Kramer, B., Lengauer, T., & Klebe, G. (1996). A fast flexible docking method using an incremental construction algorithm. *Journal of Molecular Biology, 261*(3), 470–489. doi:10.1006/jmbi.1996.0477 PMID:8780787

Speck-Planche, A., & Cordeiro, M. N. (2011). Current drug design of anti-HIV agents through the inhibition of C-C chemokine receptor type 5. *Current Computer-aided Drug Design, 7*(4), 238–248. doi:10.2174/157340911798260287 PMID:22050678

Verdonk, M. L., Cole, J. C., Hartshorn, M. J., Murray, C. W., & Taylor, R. D. (2003). Improved protein-ligand docking using GOLD. *Proteins, 52*(4), 609–623. doi:10.1002/prot.10465 PMID:12910460

Wang, R., Liu, L., Lai, L., & Tang, Y. (1998). SCORE: A New Empirical Method for Estimating the Binding Affinity of a Protein-Ligand Complex. *Molecular Modeling Annual, 4*(12), 379-394. doi: 10.1007/s008940050096

Wodak, S. J., & Janin, J. (1978). Computer analysis of protein-protein interaction. *Journal of Molecular Biology, 124*(2), 323–342. doi:10.1016/0022-2836(78)90302-9 PMID:712840

Wu, G., Robertson, D. H., Brooks, C. L. III, & Vieth, M. (2003). Detailed analysis of grid-based molecular docking: A case study of CDOCKER-A CHARMm-based MD docking algorithm. *Journal of Computational Chemistry, 24*(13), 1549–1562. doi:10.1002/jcc.10306 PMID:12925999

Zeng, Y., Cao, R., Zhang, T., Li, S., & Zhong, W. (2015). Design and synthesis of piperidine derivatives as novel human heat shock protein 70 inhibitors for the treatment of drug-resistant tumors. *European Journal of Medicinal Chemistry, 97*, 19–31. doi:10.1016/j.ejmech.2015.04.043 PMID:25935385

Zilian, D., & Sotriffer, C. A. (2013). SFCscore(RF): A random forest-based scoring function for improved affinity prediction of protein-ligand complexes. *Journal of Chemical Information and Modeling, 53*(8), 1923–1933. doi:10.1021/ci400120b PMID:23705795

KEY TERMS AND DEFINITIONS

Complementarity: Complimentary properties of ligand and receptor in docked complex structure.
Docking: Posing of Ligand molecule into the active site pocket of receptor.
Drug: Chemical compound having therapeutic properties against any disease.
Force Field: Function that expresses energy of a system as sum of non bonded interactions.

Lead Identification: Process of identifying suitable chemical compound showing the properties of drug.

Ligand: Small molecule that binds to the receptor.

Receptor: Molecule that provides conformational space for the binding of ligand.

Chapter 3
Scoring Functions in Docking Experiments

Pravin Ambure
Jadavpur University, India

Kunal Roy
Jadavpur University, India

ABSTRACT

Molecular docking is a computational technique used to predict the binding orientation of a molecule while interacting with other molecule and finally quantify the inter-molecular interactions in terms of a binding score or binding affinity. In general, every docking algorithm comprises one or more "scoring function(s)" which is/are responsible for finding a precise binding pose as well as for predicting the binding affinity. In the last two decades, a significant development has been found in the field of scoring functions. In this chapter, the authors will discuss in detail about various types of scoring functions used in the docking experiments. This chapter will get the readers acquainted with different types of scoring functions available, their theoretical background, essential components, desired properties, and the important task performed by the scoring functions. The authors will also discuss the challenges faced by the scoring functions and their recent progress.

INTRODUCTION

Molecular docking is a computational technique used to predict the binding orientation or native pose of a molecule while interacting with another molecule to form a stable complex. This technique is routinely carried out to study various kinds of inter-molecular interactions between ligand and protein, ligand and DNA, protein and protein, protein and DNA. In the pharmaceutical field, it is often used to find out the binding orientation of a lead molecule (ligand) to its target protein so as to understand the mechanism of action as well as to predict the binding affinity that plays a vital role in the rational drug designing (RDD). Each docking algorithm comprises one or more 'scoring function(s)', which plays the central role in exploring the correct binding orientation and understanding the outcome of the docking experiment. In docking, scoring function can be defined as a mathematical function that is used to compute

DOI: 10.4018/978-1-5225-0115-2.ch003

Figure 1. Hydrogen bonding
Snapshot taken in Discovery Studio software (DiscoveryStudio, 2007) for demonstration purpose only.

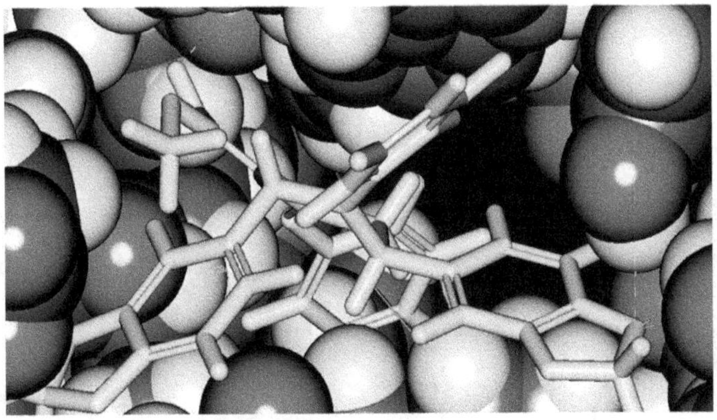

a representative score and to decide a binding pose, by evaluating the favorable (rewarding term) and unfavorable (penalty term) inter-molecular interactions found in a docked complex. The typical components of scoring functions are hydrogen bonding (illustrated in Figure 1), electrostatic interaction, van der Waals interaction (Figure 2), desolvation effect (Figure 3) and loss of torsional entropy upon binding (Figure 4).

BACKGROUND

The basic aim of a scoring function is to find a docking score that should somehow represent the experimentally-determined binding affinity of the docked ligand.

The scoring functions are intended to accomplish the following mentioned tasks:

1. To predict the absolute binding affinity between a ligand and a protein or at least the relative binding affinities among the multiple ligands and a protein.

Figure 2. van der Waal interactions
Snapshot taken in Discovery Studio software for demonstration purpose only.

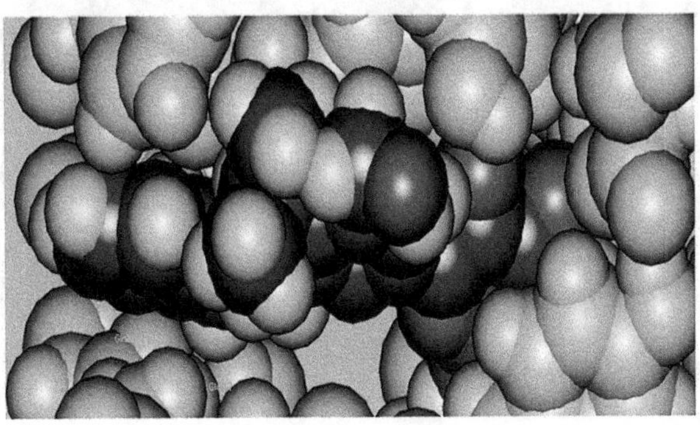

Figure 3. Desolvation effect
The figure was built using the snapshots taken in Discovery Studio software for demonstration purpose only.

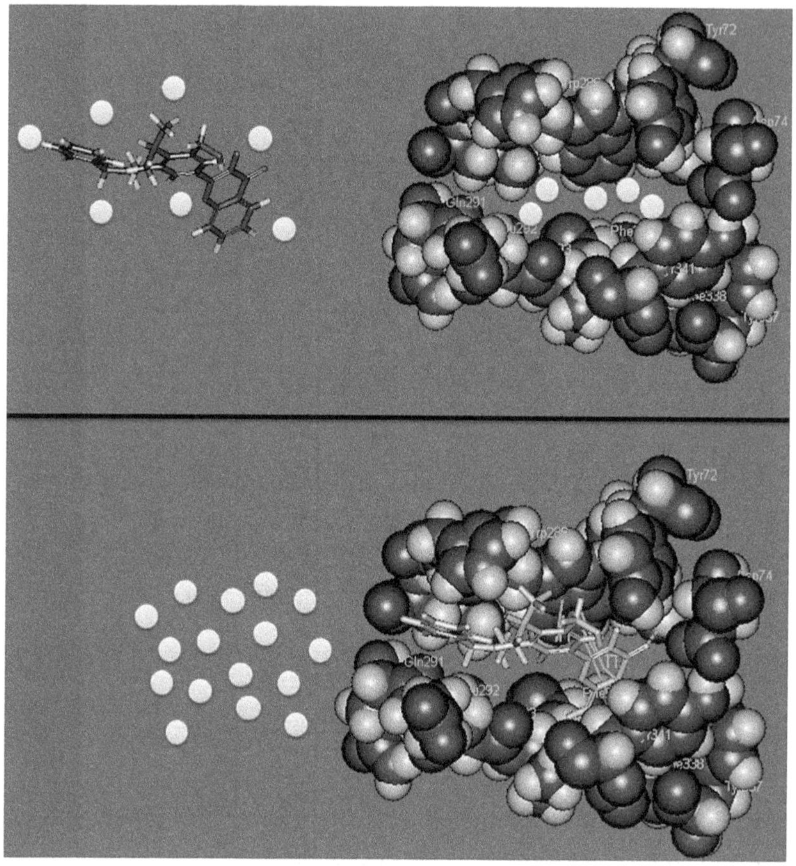

Figure 4. Loss of torsional entropy upon binding
The figure was built using a snapshot taken in Discovery Studio software for demonstration purpose only.

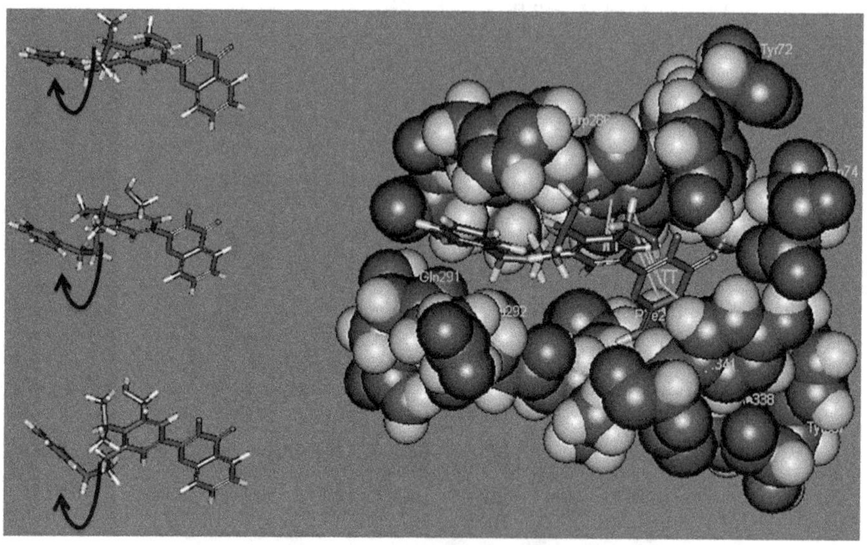

2. To decide the correct binding orientation (native pose) and position of a ligand in the active site of the protein.
3. To identify the potential lead compounds against a target protein by screening a huge ligand database.

The desired properties of a scoring function are as follows:

1. Ideally, the computed docking score should be significantly correlated with the experimental binding affinity of the ligand.
2. It should precisely distinguish between correctly and incorrectly docked structures.
3. The calculation of scoring function should be sufficiently fast, so that it can be used smoothly to screen huge databases (Huang, Grinter, & Zou, 2010).

Tuning of Scoring Functions

All docking algorithms comprise a key scoring function, but usually a single scoring function does not fulfill all the requisite tasks and hence one should be able to optimize a scoring function accordingly. The key scoring functions are often developed utilizing a large set of target proteins for having general applicability, but that does not necessarily mean that they are the best possible functions for any protein target. In such cases, if one has the experimental binding affinities against a specific target for a set of compounds and related native pose information from reported crystal structures or from a previous docking study, one can 'tune' the scoring function according to the target under study. Such tuned function is often called as '*targeted scoring function*'. One of the several ways to achieve this is *via* adjusting the existing regression coefficients to optimize the correlation between experimental and computed binding affinities or by adding new terms that are specific to the target. Another application of tuning is to modify the scoring function that has been originally developed for pose prediction into a scoring function that can now predict binding affinity (Kroemer, 2007). For example, docking success rates were found to be significantly better when a general scoring function was tuned to build targeted scoring function against cyclin–dependent kinase 2 (CDK2) target as reported in the literature (Mooij & Verdonk, 2005).

Classification of Scoring Functions

The scoring functions can be grouped into four different types (Huang, Grinter, & Zou, 2010; Liu & Wang, 2015) based on their theoretical background (see Figure 5):

1. Force field/Physics-based,
2. Empirical,
3. Knowledge-based potential, and
4. Descriptor-based.

Indeed, there are also hybridized scoring functions that are difficult to get classified into any one of the categories listed above, so the classification is often considered as rudimentary. For instance, one may easily point out that descriptor-based scoring functions are very similar to empirical scoring functions. But still both the categories are suggested to remain as separate (Liu & Wang, 2015). This will be more clearly understood as we will discuss each type of scoring functions in detail.

Figure 5. Classification of scoring functions used in docking experiments with some examples

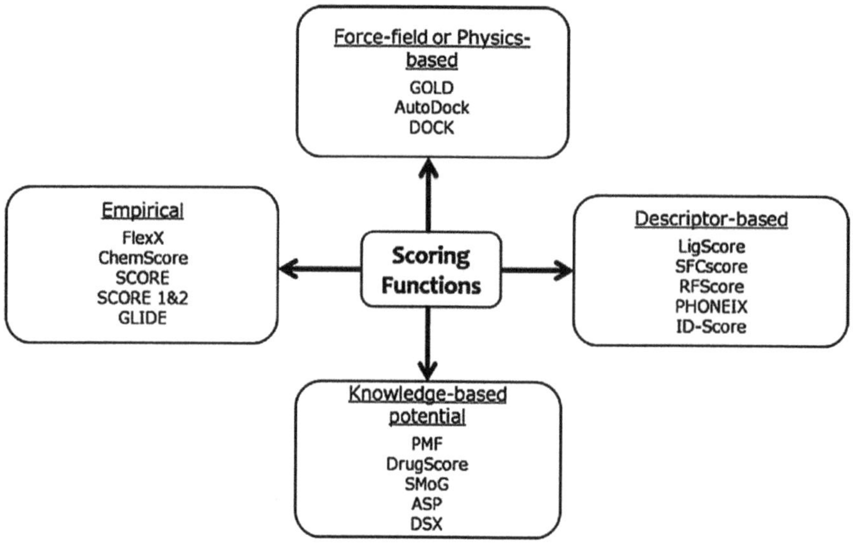

SCORING FUNCTIONS

Force Field Scoring Function/Physics-Based Methods

Force field (FF) scoring functions (*also termed as Physics-based methods*) are developed based on physical atomic interactions, such as van der Waals (vdW) interactions, electrostatic interactions, and bond stretching/bending/torsional forces. FFs are the set of functional forms and parameters used to calculate the potential energy (PE) of a system of atoms in molecular mechanics and molecular dynamics simulations. FF parameters of the energy function are usually derived from experimental data and quantum mechanical calculations based on the principles of physics. Thus a researcher often depends on the available force fields to find out the interactions between ligands and protein. Some examples of FFs are Merck molecular FF (MMFF94) (Halgren, 1996), Universal FF (UFF) (Rappe, Casewit, Colwell, Goddard III, & Skiff, 1992), AMBER FF (GAFF) (Wang, Wolf, Caldwell, Kollman, & Case, 2004), CHARMm FF (Momany & Rone, 1992)*etc.*

Theoretically, the most precise methods to estimate the ligand-protein binding enthalpy are based on quantum mechanical (QM) methods. Also, a quantum mechanics/ molecular mechanics (QM/MM) method that combines the advantages of the QM (*precise*) and MM (*speed*) was built to improve the prediction of native pose of ligands in protein by replacing the FF charges of ligands with QM/MM computed charges (Hensen, Hermann, Nam, Ma, Gao et al., 2004). Further, a semi-empirical QM based scoring function (Dobes, Fanfrlik, Rezac, Otyepka, & Hobza, 2011) was also developed to predict the binding affinities of diverse protein–ligand complexes, and also, to discriminate between native and decoy poses. However, at present such calculations are in fact computationally expensive to be utilized in high-throughput applications such as docking study of huge databases.

Linear Interaction Energy (LIE), the Linear Response Approximation (LRA) and Molecular Mechanics–Poisson-Boltzmann Surface Area/Generalized Born Surface Area (MM-PBSA/GBSA) methods were emerged in the 1990s. These methods utilize standard FFs to calculate potential energies and usually depend on molecular dynamics simulations in explicit solvent for configurational sampling. Such methods just evaluate the ligand and protein in their unbound and bound states in order to approximate the free energy change that occurred after ligand binding and hence are also called as end-point approximations. The methods discussed here are usually not employed as scoring functions since they calculate energies utilizing high computational power and thus cannot be utilized in high-throughput applications. But, such methods can be successfully used to compute precise binding energy of few selected protein–ligand complex structure (selected after initial screening). Further, these methods may also integrate empirical parameters to improve their performance (Liu & Wang, 2015).

The molecular mechanics (MM) methods are based on classical mechanics that allow computational studies to be carried out on large biomolecular systems containing more than 1,00,000 atoms, where both accuracy and efficiency can be achieved within certain approximations. Therefore, currently MM is the most practical means to study the interactions between ligands and protein in a physically realistic manner, with high-throughput applicability. Free energy calculations derived from MM simulations of protein– ligand complexes can account for ligand as well as protein flexibility and additionally the solvent effects can also be incorporated (Huang, Kalyanaraman, Bernacki, & Jacobson, 2006).

A general composition for FFs scoring function is as follows (Equation 1):

$$E = \sum_{bonds} K_b(b-b_0)^2 + \sum_{angles} K_\theta(\theta-\theta_0)^2 + \sum_{torsions} K_\varphi[1+\cos(n\varphi-\delta)] + \sum_{vdW} \varepsilon_{ij}\left[\left(\frac{R_{\min,ij}}{r_{ij}}\right)^{12} - 2\left(\frac{R_{\min,ij}}{r_{ij}}\right)^{6}\right] + \sum_{electrostatics} \frac{q_i q_j}{4\pi D r_{ij}} \tag{1}$$

where, the bonded interactions include bond, angle and dihedral terms (b, θ and ϕ), respectively, while, the non-bonded interactions include the vdW term represented by the Lennard-Jones potential, and electrostatic interactions, which are based on Coulomb's law (Huang, Kalyanaraman, Bernacki, & Jacobson, 2006).

The FF approach is challenging due to its complex nature. Important factors such as entropic changes after ligand binding, and solvation effects can significantly affect the precise computation of absolute/ relative binding affinities (Huang, Kalyanaraman, Bernacki, & Jacobson, 2006). Thus, current FF methods often need empirical scaling parameters to fit their outcome to experimental data.

Further, we will discuss some scoring functions in detail to understand the basic components, and the theoretical background behind the FF scoring functions.

Genetic Optimisation for Ligand Docking (GOLD)

In 1997, Genetic Optimisation for Ligand Docking (GOLD) was reported by Jones *et al.* as an automated ligand-protein docking program that employs genetic algorithm to study intermolecular interactions, while accounting for ligand conformational flexibility, partial protein flexibility, and ligand displacement of loosely bound water on binding. A fitness function (scoring function) was employed to rank the generated binding poses, and it was comprised of a term for hydrogen bonding (H-bonding), which took account of the water molecules displaced from both donor and acceptor during bond formation, a pair-wise dispersion potential to describe a significant contribution to the hydrophobic energy of binding, and a molecular mechanics term for computing the internal energy of the ligand.

Fitness Function or Scoring Function

The final fitness score can be computed from the sum of all the energy as follows (equation 2):

$$\text{Fitness\_score} = -\text{H\_Bond\_Energy} - (\text{Internal\_Energy} + \text{Complex\_Energy}) \quad (2)$$

where, 'H_Bond_Energy' represent H-bonding energies that are pre-calculated using model fragments and also account for the water-displacement. It is computed as the sum of all individual bond energies found from all combinations of ligand donor hydrogen atom and protein acceptor, and *vice versa*. The donor (d) and the acceptor (a) are considered in solution, and after formation of H-bond (da), water (w) is removed. The full bond energy, E_{pair}, is composed of four terms (equation 3):

$$E_{pair} = (E_{da} + E_{ww}) - (E_{dw} + E_{aw}) \quad (3)$$

Here, halogens like Cl, Br and F are not considered as acceptors and the associated fragments were removed from the set of fragments, but two metal ions (Mg and Zn) were added. For other metal ions like Fe, Mn and Ca, the empirical energies were used.

The E_{pair} between the donor and acceptor was then scaled by a weight (0 to 1). The weight is the product of two terms, a distance weight and an angle weight as shown below:

$$wt = distance\_wt \times angle\_wt \quad (4)$$

The distance weight is a function of '*d*' (the distance between the donor hydrogen fitting point and the acceptor). If $d < 0.25$Å then *distance_wt* is set to 1 and if $d >$ *max_distance,* then distance_*wt* is set to 0. Otherwise if *d is* in-between 0.25 and *max_distance*, in that case *d* is linearly rescaled to the interval (0, 1) and squared to compute *distance_wt*.

Further, if the acceptor has no lone-pair directional preference then *angle_wt* is set to 1.0. If the acceptor has H-bond directionality in the plane of the lone-pairs then *angle_wt* is a function of Θ (*the angle between the donor, donor hydrogen atom and the plane of the lone-pairs*). If Θ>60 then *angle_wt* is set to 0 and if Θ<20 then *angle_wt* is set to 1.0. Otherwise, when Θ is in-between 20 and 60, in that case it is linearly rescaled to (1, 0) and squared to compute *angle_wt*. But, if the acceptor has a directional preference along the lone-pair, then *angle_wt* is a function of Φ (*the largest H-bond angle between the donor hydrogen atom and any of the acceptors lone pairs*). If Φ>160, then *angle_wt* is 1.0 and if Φ<60, then *angle_wt* is set to 0. Otherwise, when Φ is between 160 and 60, in that case it is linearly rescaled to (1, 0) and squared to compute *angle_wt*.

Further, a pairwise energy (*Complex_Energy)* is obtained representing the steric energy of interaction between the protein and the ligand. Here, a 4-8 potential (softer than the Lennard-Jones 6-12 potential) with linear cut-off is used to determine the energy of interaction between the ligand and the protein. The 4-8 potential is defined as (equation 5):

$$E_{ij} = \frac{A}{d_{ij}^{8}} - \frac{B}{d_{ij}^{4}} \quad (5)$$

where, E_{ij} is the energy of interaction between two atoms '*i*' and '*j*'; d_{ij} is the distance between the two atoms. If the two atoms are involved in an H-bond, E_{ij} is set zero for the interaction between a donor hydrogen atom and an acceptor, while the distance between the donor and acceptor is scaled by a factor of 1.43. All pairwise interaction energies of ligand and protein atoms in close contact are summed to compute *Complex_Energy*. The van der Waals (vdW) potential used in GOLD is the much softer and favors close contacts between hydrophobic groups.

The term '*Internal_Energy*' is a sum of the ligand steric and torsional energies. The steric energy is computed using a 6-12 potential as given below (equation 6):

$$E_{ij} = \frac{C}{d_{ij}^{12}} - \frac{D}{d_{ij}^{6}} \tag{6}$$

The Tripos FF torsional potential is used to find torsional energies and is defined as follows (equation 7):

$$E_{ijkl} = \frac{1}{2} V_{ijkl} \left[1 + \frac{n_{ijkl}}{|\, n_{ijkl} \,|} \cos\left(|\, n_{ijkl} \,| . \omega_{ijkl} \right) \right] \tag{7}$$

where, E_{ijkl} is the torsional energy associated with four consecutively bonded atoms *i*, *j*, *k*, *l*; ω is a torsional angle; *n* is the periodicity; and *V* the barrier to rotation.

To account for covalently bound inhibitors, two energy terms are added to *Complex_Energy*: a) the first is a torsional term identical to the Tripos FF torsional potential (equation 7); b) angle bending term to ensure that the bond angle between the two bonds linking the ligand and the protein is correct. This second term is also taken from the Tripos FF and defined as follows (equation 8):

$$E_{ijk} = k_{ijk} \times (\omega - \theta_{ijk})^2 \tag{8}$$

where *i*, *j* and *k* are the three atoms forming the bond angle, ω is the measured angle and k_{ijk} and θ_{ijk} are parameters of the FF.

AutoDock 3.0

AutoDock is a molecular modeling software that is specially utilized to study ligand-protein interactions. The scoring function employed in the AutoDock version 3.0 (Morris, Goodsell, Halliday, Huey, Hart et al., 1998) is an semi-empirical FF function. It uses a *molecular mechanics* approach to evaluate enthalpic contributions such as dispersion/repulsion, electrostatics and H-bonding, and an *empirical* approach to evaluate the entropic contribution of changes in solvation and conformational mobility. The function was developed based on the linear regression technique and further validated for its ability to reproduce the known observed binding constants of ligand-protein complexes. Based on overall statistical results, a final scoring function was selected as follows (equation 9):

$$\Delta G = \Delta G_{vdW} \sum_{i,j} \left(\frac{A_{ij}}{r_{ij}^{12}} - \frac{B_{ij}}{r_{ij}^{6}} \right) + \Delta G_{hbond} \sum_{i,j} E(t) \left(\frac{C_{ij}}{r_{ij}^{12}} - \frac{D_{ij}}{r_{ij}^{10}} + E_{hbond} \right)$$
$$+ \Delta G_{elec} \sum_{i,j} \frac{q_i q_j}{\varepsilon(r_{ij}) r_{ij}} + \Delta G_{tor} N_{tor} + \Delta G_{sol} \sum_{t_C, j} S_i V_i e^{(-r_{ij}^2 / 2\sigma^2)} \quad (9)$$

Here, the first three terms are the typical molecular mechanics terms for dispersion/repulsion (*a Lennard-Jones 12-6 dispersion/repulsion term*), H-bonding (*a directional 12-10 H-bonding term, where E(t) is a directional weight based on the angle, t, between the probe and the target atom*), and electrostatics interactions (*Coulombic electrostatic potential*), respectively. The parameters '*A*' and '*B*' are taken from the AMBER FF. The parameters '*C*' and '*D*' are assigned to give a maximal well depth of 5 kcal/mol at 1.9 A° for O-H and N-H, and a depth of 1 kcal/ mol at 2.5 A° for S-H. Further, ΔG_{tor} (deformation term) represents the restriction of internal rotational bonds, global rotation and translation; N_{tor} is the number of sp^3 bonds in the ligand, and ΔG_{sol} represents the desolvation upon binding and the hydrophobic effect, which is well explained in the literature (Morris et al., 1998). The ΔG terms are weights/coefficients that are determined using linear regression analysis from a set of protein ligand complexes with known binding constants.

The summations are performed over all pairs of ligand atoms '*i*' and protein atoms '*j*' in addition to all pairs of atoms in the ligand that are separated by three or more bonds. The E_{hbond} is the estimated average energy of H-bonding of water with a polar atom, and the summation in the solvation term is performed over all pairs consisting of only carbon atoms in the ligand '*i*' and atoms of all types '*j*' in the protein. The intra-molecular interaction (internal energy) of the ligand is not included in the calculation of binding free energy, since it was found that adding the internal energy term does not improve the model; however, it is included in the total docked energy.

AutoDock 4.0

An improved semiempirical free energy FF was developed for AutoDock4.0 (Huey, Morris, Olson, & Goodsell, 2007) to address the significant limitations of the AutoDock3.0 FF. This FF is calibrated using a large set of diverse protein–ligand complexes and has two major achievements over AutoDock3.0 FF, i.e., a) an improved thermodynamic model of the binding process, which now includes intramolecular terms in the estimated free energy; b) it also includes a full desolvation model that includes terms for all atom types, including the favorable energy for desolvation of carbon atoms and the unfavorable energy for desolvation of polar and charged atoms.

The free energy of binding is computed as the difference between the energy of the ligand plus the protein in a separated unbound state and the energy of the ligand–protein complex. The FF scoring function includes six pair-wise evaluations (*V*) and an estimate of the conformational entropy lost upon binding (ΔS_{conf}):

$$\Delta G = (V_{bound}^{L-L} - V_{unbound}^{L-L}) + (V_{bound}^{P-P} - V_{unbound}^{P-P}) + (V_{bound}^{P-L} - V_{unbound}^{P-L} + \Delta S_{conf}) \quad (10)$$

Here in equation 10, L designates 'ligand' and P designates 'protein' in a protein–ligand complex. The first two terms are intra-molecular energies for the bound and unbound states of the ligand, and the following two terms are intra-molecular energies for the bound and unbound states of the protein. The change in intermolecular energy between the bound and unbound states is in the third parentheses. It is assumed that the two molecules are sufficiently distant from one another in the unbound state so that $V_{unbound}^{P-L}$ is zero. Also, here $V_{bound}^{P-P} - V_{unbound}^{P-P}$ is considered zero, since protein flexibility is not allowed; this means that bound state of the protein is identical with the protein unbound state, and therefore the difference in their intramolecular energy is zero.

The pair-wise atomic terms include dispersion/ repulsion, H-bonding, electrostatics, and desolvation similar to:

$$V = W_{vdW} \sum_{i,j} \left(\frac{A_{ij}}{r_{ij}^{12}} - \frac{B_{ij}}{r_{ij}^{6}} \right) + W_{hbond} \sum_{i,j} E(t) \left(\frac{C_{ij}}{r_{ij}^{12}} - \frac{D_{ij}}{r_{ij}^{10}} \right) + W_{elec} \sum_{i,j} \frac{q_i q_j}{\varepsilon(r_{ij}) r_{ij}} + W_{sol} \sum_{i,j} (S_i V_j + S_j V_i) e^{(-r_{ij}^2/2\sigma^2)} \qquad (11)$$

All the terms in Equation 11 are similar to the AutoDock3.0. The weighting constants W are optimized to calibrate the empirical free energy based on a set of experimentally characterized complexes. The term for the loss of torsional entropy upon binding (ΔS_{conf}) is directly proportional to the number of rotatable bonds in the molecule (N_{tors}) (see Equation 12):

$$\Delta S_{conf} = W_{conf} N_{tors} \qquad (12)$$

The number of rotatable bonds includes all torsional degrees of freedom, including rotation of polar hydrogen atoms on hydroxyl groups.

Empirical Scoring Function

In simple words, the empirical scoring functions compute the binding affinity of a protein-ligand complex on the basis of a set of weighted energy terms (also termed as *coefficients*) and have the following form:

$$\Delta G = \sum_i W_i \Delta G_i \qquad (13)$$

In Equation 13, ΔG_i may represent different energy terms such as H-bond, vdW energy, electrostatics, desolvation, entropy, and hydrophobicity. The respective coefficients W_i are determined by fitting the known binding affinity values of a set of protein–ligand complexes (*i.e.* training set) with available three-dimensional (3D) structures information. The coefficients are assigned using linear regression like multiple linear regression (MLR) or non-linear regression like support vector machine (SVM), methods that are used to fit the experimentally determined affinity values with the computed binding score. Compared to the FF scoring functions, the empirical scoring functions are faster in calculating the binding score due to their simple energy terms, but may have limited applicability domain, which is based on the number and diversity of protein-ligand complexes present in the training set.

Here, we will discuss some empirical scoring functions to understand how they are derived, and different components/terms present in each function.

FlexX

FlexX is an empirical scoring function that was developed using a set of 19 protein–ligand complexes whose 3D structures were known (Rarey, Kramer, Lengauer, & Klebe, 1996). This scoring function estimates the free binding energy (ΔG) of the protein–ligand complex, which is utilized in the ranking of generated docking pose. The model was derived from the work of Bohm (SCORE1 scoring function) (Bohm, 1994). The FlexX scoring function (equation 14) is as follows:

$$\begin{aligned}\Delta G = \Delta G_0 + \Delta G_{rot} \times N_{rot} + \Delta G_{hb} \sum_{neutralH-bonds} f(\Delta R, \Delta\alpha) + \Delta G_{io} \sum_{IonicInt.} f(\Delta R, \Delta\alpha) \\ + \Delta G_{aro} \sum_{Aro.Int.} f\left(\Delta R, \Delta\alpha\right) + \Delta G_{lipo} \sum_{Lipo.Cont.} f^*\left(\Delta R\right)\end{aligned} \quad (14)$$

Here, $f(\Delta R, \Delta\alpha)$ is a scaling function penalizing deviations from the ideal geometry and N_{rot} is the number of free rotatable bonds that are immobilized in the complex. The terms ΔG_{rot}, ΔG_{hb}, ΔG_{io}, and ΔG_0 are adjustable parameters. ΔG_{aro} accounts for the interactions of aromatic groups and is equal to -0.7 kJ/mol. ΔG_{lipo} is a modification of Bohm's lipophilic contact energy. In the Bohm's function, this energetic contribution is intended to be proportional to the lipophilic contact area estimated with a grid method, but here it is computed as a sum over all pair-wise atom-atom contacts. In equation 14, the function $f^*(\Delta R)$ accounts for the contacts that are within an ideal distance and penalizes the forbidden close contacts. Hence, the following criteria are utilized:

$$f^*(\Delta R) = \begin{cases} 0 & \Delta R > 0.6\dot{A} \\ 1 - \dfrac{\Delta R - 0.2}{0.4} & 0.2\dot{A} < \Delta R \leq 0.6\dot{A} \\ 1 & -0.2\dot{A} < \Delta R \leq 0.2\dot{A} \\ 1 - \dfrac{-\Delta R - 0.2}{0.4} & -0.6\dot{A} < \Delta R \leq -0.2\dot{A} \\ \dfrac{\Delta R + 0.6}{0.2} & \Delta R \leq -0.6\dot{A} \end{cases}$$

Here, R is the distance between the atom centers, R_0 is its ideal value assumed to be the sum of both vdWs radii, each increased by 0.3A° and $\Delta R = R - R_0$.

ChemScore

ChemScore (Eldridge, Murray, Auton, Paolini, & Mee, 1997) is a simple empirical scoring function, designed to estimate the free energy of binding for a protein–ligand complex. The function uses simple contact terms to estimate a simple explicit form for H-bonds, metal–ligand binding and lipophilic contributions, and a term which penalizes flexibility. The coefficients/weights of each term are obtained using a linear regression analysis based on 82 ligand–receptor complexes with known binding affinity values. The ChemScore scoring function (Equation 15) can be written as:

$$\begin{aligned}\Delta G_{binding} &= \Delta G_0 + \Delta G_{hbond} \sum_{iL} g_1(\Delta r) g_2(\Delta \alpha) + \Delta G_{metal} \sum_{aM} f(r_{aM}) \\ &+ \Delta G_{lipo} \sum_{lL} f(r_{lL}) + \Delta G_{rot} H_{rot}\end{aligned} \tag{15}$$

The ΔG coefficients were determined using MLR analysis. The H-bond term, $\sum_{iL} g_1 g_2$ is calculated for all complementary possibilities of H-bonds between ligand atoms '*i*' and receptor atoms '*I*'. The functions g_1 and g_2 are same as used by Böhm (Bohm, 1994):

$$g_1(\Delta r) = \begin{cases} 1 & if\, \Delta r \le 0.25\text{Å} \\ 1-(\Delta r - 0.25)/0.4 & if\, 0.25\text{Å} < \Delta r \le 0.65\text{Å} \\ 0 & if\, \Delta r > 0.65\text{Å} \end{cases}$$

and

$$g_2(\Delta \alpha) = \begin{cases} 1 & if\, \Delta \alpha \le 30^\circ \\ 1-(\Delta \alpha - 30)/50 & if\, 30^\circ < \Delta \alpha \le 80^\circ \\ 0 & if\, \Delta \alpha > 80^\circ \end{cases}$$

Here, Δr is the deviation of the H···O/N H-bond length from 1.85 Å and $\Delta \alpha$ is the deviation of the H-bond angle N/O-H···O/N from its ideal value of 180°. Solvent or water molecules are scored if in contact with the receptor, and those having more than one H-bond to the receptor are treated as a part of the receptor, while contacts between the ligand and receptor are evaluated. Further, no distinction is made between ionic and non-ionic H-bonds and hence, both types are assigned equal weights.

The metal term, $\sum_{aM} f(r_{aM})$, is computed for all acceptor and acceptor/donor atoms '*a*' in the ligand and any metal atoms '*M*' in the receptor. The function *f(r)* is a simple contact term. r_{aM} is the distance between ligand and receptor atoms of the appropriate atom type.

The lipophilic term, $\sum_{lL} f(r_{lL})$, is computed for all lipophilic ligand atoms '*l*' and all lipophilic receptor atoms '*L*'.

The final term first finds rotatable bonds that are immobilized in the complex (frozen rotatable bonds). Here, the rotatable bonds are defined as any sp^3-sp^3 bond and any sp^2-sp^3 bond, excluding terminal bonds such as in CH_3, CF_3, NH_2 or NH_3 groups (whose rotation does not produce any new conformation of heavy atoms). Also, bonds are considered frozen if atoms on both sides of the rotatable bond are in contact with the receptor, so that the distance between any two heavy atoms is less than the sum of the relevant vdW radii plus 0.5 Å. The flexibility penalty (entropy penalty; Equation 16) for molecules comprising frozen rotatable bonds is defined using the following function:

$$H_{rot} = 1 + (1 - 1 / N_{rot}) \sum_{r} (P_{nl}(r) + P_{nl}^{'}(r)) / 2 \tag{16}$$

Here, N_{rot} is the number of frozen rotatable bonds; the summation is over frozen rotatable bonds; $P_{nl}(r)$ and $P_{nl}^{'}(r)$ are the percentages of non-lipophilic heavy atoms on either side of the rotatable bond.

SCORE2

SCORE2 is a empirical scoring function (Bohm, 1998) that was developed using a dataset of 82 protein–ligand complexes of known 3D structure with binding constant (K_i). This scoring function is based on the following parameters: the number and geometry of H-bonds and ionic interactions between ligand and protein, size of the lipophilic contact surface, ligand flexibility, the electrostatic potential in the active site, water molecules in the active site, cavities along the protein-ligand interface and aromatic rings interactions. This scoring function also distinguishes between buried and solvent accessible hydrogen bonds. SCORE2 (Equation 17) is a modified version of SCORE1 by Bohm (Bohm, 1994) and is defined as follows:

$$\Delta G_{binding} = \Delta G_0 + \Delta G_{polar} + \Delta G_{apolar} + \Delta G_{solv} + \Delta G_{flexi} \tag{17}$$

ΔG_{polar} and ΔG_{apolar} represent the polar and apolar interactions, ΔG_{solv} accounts for desolvation effects and ΔG_{flexi} treats the ligand flexibility. ΔG_0 does not directly depend on any specific interactions with the protein and may be considered as a reduction of binding energy due to overall loss of translational and rotational entropy of the ligand. The ΔG_{flexi} term (Equation 18) is same as defined in SCORE1:

$$\Delta G_{flexi} = \Delta G_{rot} NROT \tag{18}$$

Here, ΔG_{rot} describes the loss of binding energy due to freezing of internal degrees of freedom in the ligand. The number of rotatable bonds 'NROT' is taken as the number of acyclic sp^3-sp^3 and sp^3-sp^2 bonds. Rotations of terminal -CH_3 or -NH_3 groups are not taken into account.

Polar Interactions

The polar contributions to $\Delta G_{binding}$ (Equation 19) are defined as follows:

$$\Delta G_{polar} = \Delta G_{hb} \sum_{h-bonds} f(\Delta R, \Delta\alpha) \times f(N_{neighb}) \times f_{pcs} + \Delta G_{ionic} \sum_{ionicInt.} f(\Delta R, \Delta\alpha) \times f(N_{neighb}) \times f_{pcs} + \Delta G_{esrep} N_{repulsivecontacts} \tag{19}$$

where,

$$f(\Delta R, \Delta\alpha) = f1(\Delta R) f2(\Delta\alpha)$$

$$f1(\Delta R) = \begin{cases} 1 & if \Delta R \le TOL \\ 1-(\Delta R - TOL)/0.4 & if \Delta R \le 0.4 + TOL \\ 0 & if \Delta R > 0.4 + TOL \end{cases}$$

$$f2(\Delta\alpha) = \begin{cases} 1 & if \Delta\alpha < 30^\circ \\ 1-(\Delta\alpha - 30)/50 & if \Delta\alpha \le 80^\circ \\ 0 & if \Delta\alpha > 80^\circ \end{cases}$$

TOL= tolerated deviation in the H-bond length, i.e., equal to 0.25 Å

Here, ΔG_{hb} describes the contribution from an ideal H-bond and accounts for all H-bonds (those having a neutral atom on at least one side of the bond). ΔG_{ionic} represents the contribution from an unperturbed ionic interaction. The geometric dependency is assumed same for both charged and uncharged interactions. $f1(\Delta R, \Delta\alpha)$ is a penalty function which accounts for large deviations of the H-bond geometry from ideal values. ΔR is the deviation of the H···O/N,H-bond length from the ideal value (1.9 Å). $\Delta\alpha$ is the deviation of the H-bond angle N=O-H···O=N from its idealized value (180°). The function tolerates little deviations in the ideal H-bond geometry without penalizing the score, i.e., ≤0.25 Å in the length and ≤30° in the H-bond angle. But it does not include a term accounting for distortions in the H-bond angle C=O···O/N. For interactions with metal atoms, $f2(\Delta\alpha)$ is set to 1 and the ideal distance metal···O/N is set to 2.1 Å.

Apolar Interactions

The apolar interactions (Equation 20) are described by two terms:

$$\Delta G_{apolar} = \Delta G_{lipo} \left| A_{lipo} \right| + \Delta G_{aro} \sum_{aro-\text{int.}} f(R, \Theta) \quad (20)$$

ΔG_{lipo} represents the contribution from lipophilic interactions. Here, the lipophilic interaction is assumed to be proportional to the protein–ligand contact surface area. In this function, the following atomic radii are used for both the ligand and the protein atoms, C: 1.8 Å, H: 1.2 Å, O,N: 1.6 Å, F: 1.5 Å, Cl: 1.8 Å, Br: 1.9 Å, S: 2.0 Å. All grid points inside an atomic sphere with the mentioned radius plus 0.4 Å are marked as occupied by the ligand. The same increased radii are utilized to mark the cubes that are adjacent to the ligand and overlap with the protein. ΔG_{aro} term describes interactions between aromatic rings. Here, any two aromatic rings are assumed to be interacting if the distance between any heavy atoms of the rings is smaller than 4.5 Å.

Desolvation Effects

In this scoring function, desolvation effect is only accounted for those water molecules that coincide with the occupied volume of the bound ligand. Here the desolvation effect is defined using a single term, *i.e.*, $\Delta G_{\text{lipowater}}$, as shown below (Equation 21):

$$\Delta G_{solv} = \Delta G_{lipowater} \sum unbound\_water\_molecules \quad (21)$$

Here, $\Delta G_{\text{lipowater}}$ is a volume dependent term. Further, the number of unbound water molecules for a protein cavity (*in lipophilic protein surface*) is computed.

SCORE

SCORE is another empirical scoring function developed using 170 protein-ligand complex structures taken form Protein Data Bank (PDB)(Wang, Liu, Lai, & Tang, 1998). According to this function, the binding affinity of a ligand is a sum of the contribution of individual ligand atoms. Each ligand atom gets a score termed as 'atomic binding score', which indicates its role in the binding process.

In this function, the free energy change in the protein-ligand binding process is defined using various basic components and is as follows (see Equation 22):

$$pK_d = K_0 + K_{vdw} + K_{metal} + K_{hbond} + K_{desolvation} + K_{deformation} \quad (22)$$

Here, pK_d is the negative logarithms of dissociation equilibrium constants; K_{vdw} is the contribution of vdW interaction between the protein and bound ligand; K_{metal} is the contribution of metal-ligand bonding, K_{hbond} is the contribution of H-bonding, $K_{\text{desolvation}}$ is the contribution of desolvation effect, and $K_{\text{deformation}}$ is the contribution of deformation. K_0 is the regression constant that may perhaps explain the translational and rotational entropy loss upon the binding process.

vdW Interaction

The term 'vdW interaction' is simply defined as a pair-wise counting of vdW bumps between the protein and the ligand (Equation 23):

$$K_{vdw} = \sum_i \sum_j VB(d_{ij}) \quad (23)$$

where, $VB(d_{ij}) = 1, d_{ij} < r_i + r_j - 0.60\text{Å}$

$= 0, d_{ij} \geq r_i + r_j - 0.60\text{Å}$

Here, r_i is the vdW radius of ligand atom '*i*' and r_j is the vdW radius of protein atom '*j*'; d_{ij} is the distance between atom *i* and *j*.

Metal-Ligand Bonding

In some protein molecules, metal cations are present in their active sites such as Mn^{2+}, Mg^{2+}, Zn^{2+}, and Ca^{2+} that may form coordinate bond with the ligand atom, and this bonding in turn helps in attaining more stability to the complex. In the SCORE algorithm, the metal cation (in active site) is considered as the protein component and the metal-ligand bond is distinguished from H-bond. The bond length of a metal·O/N bond is set to 2.0Å and a distance function is used to account for the deviation from the ideal value, as shown below:

$$MB(d) = \begin{cases} 1.0 & d < 2.0\text{Å} \\ 3.0 - d & 2.0\text{Å} \leq d < 3.0\text{Å} \\ 0 & 3.0\text{Å} \leq d \end{cases}$$

where, d is the metal···O/N bond length. The cutoff was set to 3.0Å.

The term 'metal-ligand bonding' is the sum over all the metal-O/N bonds, as given below (equation 24):

$$K_{metal} = \sum_i \sum_j MB(d_{ij}) \tag{24}$$

where, d_{ij} *is* the distance between ligand atom '*i*' and metal '*j*'.

H-Bonding

Generally, an H-bond donor is defined as a nitrogen or oxygen atom with attached hydrogen, while an H-bond acceptor is defined as nitrogen, oxygen, or fluorine atom with at least one vacant valence. Here, all the ligand and protein atoms are labeled as either donor (D), acceptor (A), donor/acceptor (DA), or none (N). The geometry of an H-bond is described using two parameters, namely, the bond length, (*i.e.* the distance between D and A) and the bond angle (*i.e.* the angle among D-H···A). The H-bond is present when the bond length is shorter than the sum of vdW radii of D and A. In case of the bond angle criteria, the function uses two angles involving only heavy atoms, since this scoring function avoids use of explicit hydrogen atoms. The two angles are computed are among X-D···A and D···A-X, where X stands for the adjacent heavy atom and if there are more than one adjacent heavy atom then their geometric centre is considered. The angle cutoff value set to define an H-bond is that either of these two angles should not be lower than 70 degrees.

In other empirical methods like SCORE1, FlexX, ChemScore, and SCORE2, the H-bonding strength is estimated typically by using a linear distance function. But in SCORE, a step function is used instead, and is defined as follows:

$SHB\ (d) = 1$, if $d<d_0$ - 0.60Å

= 0 otherwise

$MHB(d) = 1$, if d_0 - 0.60Å $\leq d < d_0$ – 0.30Å

= 0 otherwise

$WHB(d) = 1$, if d_0 – 0.30Å $\leq d < d_0$

= 0 otherwise

where '*d*' represents the distance between D and A; d_0 is the sum of vdW radii of D and A. SHB, MHB, and WHB are indicators for strong, moderate, and weak H-bond, respectively. Further, water-involved H-bond are also described separately using additional three indicators defined with the same step function (mentioned above) and designated as SWH, MWH and WWH to represent strong, moderate, and weak water-involved H-bond, respectively. The H-bonding term is defined as follows (Equation 25):

$$K_{hbond} = K_{SHB} + K_{MHB} + K_{WHB} + K_{SWH} + K_{MWH} + K_{WWH} = \sum_i \sum_j SHB(d_{ij}) + \sum_i \sum_j MHB(d_{ij}) + \sum_i \sum_j WHB(d_{ij}) + \sum_i \sum_j SWH(d_{ij}) + \sum_i \sum_j MWH(d_{ij}) + \sum_i \sum_j WWH(d_{ij}) \quad (25)$$

Desolvation Effect

Here, the desolvation effect is measured as summation of atomic contributions, where each atom (based on its atom type) is assigned a quantitative scale (*logP*) to represent its hydrophobicity. In this algorithm, an atom is considered as hydrophobic when its hydrophobic scale is larger than 0.20 units. Also, the environment of a ligand atom is defined as the assembly of all the neighboring protein atoms within 5Å and the hydrophobicity of the environment is expressed by the sum of the hydrophobic scales of all the protein atoms forming the environment. Therefore, if the sum is positive, the ligand atom is believed to be in a hydrophobic environment, otherwise it is in a hydrophilic environment. Hence, four situations were defined: a hydrophilic ligand atom in a hydrophilic environment, a hydrophilic ligand atom in a hydrophobic environment, a hydrophobic ligand atom in a hydrophilic environment, and a hydrophobic ligand atom in a hydrophobic environment. In the first three situations, the desolvation during binding is almost compensated by the interaction with a hydrophilic counterpart. Therefore, a significant change is not expected in the overall binding free energy. But, in the last situation, where such compensation is not possible, it can lead to favorable contribution to the protein-ligand binding process.

The term for 'desolvation effect' is a sum over all hydrophobic matching's between the protein and its ligand, and is defined as follows (Equation 26):

$$K_{HM} = \sum_i F_i \times HM_i \quad (26)$$

Here, HM_i is an indicator of hydrophobic matching (it is set to 1 if ligand atom '*i*' is hydrophobic and placed in a hydrophobic environment; otherwise it is set to 0). F_i is the atomic hydrophobic scale of ligand atom '*i*' and is used as a weight factor.

Deformation Effect

Deformation refers to the conformational changes during the binding process. It causes both adverse entropic changes (due to the freezing of internal rotations of both the protein and its ligand) and enthalpic change (due to the strain energy exerted during binding). Here, the number of rotatable bonds (or rotors) is utilized to represent both the entropic and enthalpic change in deformation. Hence, the term for the deformation effect is the number of rotors in the ligand. A rotor is split into half and assigned onto the two atoms involved, and the term is written as:

$$K_{RT} = \sum_{i} 0.5 \times RT_i \quad (27)$$

In Equation 27, RT_i is the number of rotors in which ligand atom '*i*' is involved. Rotatable bonds or rotors are defined as acyclic sp^3-sp^3 and sp^3-sp^2 single bond. Terminal rotors like $-CH_3$, $-NH_2$, or -OH, are not taken into account. The flexibility of cyclic portions of the ligand is also ignored.

After summarizing all the parameters, the SCORE scoring function can be written as (Equation 28):

$$pK_d = K_0 + c_1 \times K_{VB} + c_2 \times K_{MB} + c_3 \times K_{SHB} + c_4 \times K_{MHB} + c_5 \times K_{WHB} + c_6 \times K_{SWH} + c_7 \times K_{MWH} + c_8 \times K_{WWH} + c_9 \times K_{HM} + c_{10} \times K_{RT} \quad (28)$$

Fresno

A empirical scoring function 'Fresno' (Rognan, Lauemoller, Holm, Buus, & Tschinke, 1999) was developed to predict the binding free energy of peptides to class I major histocompatibility (MHC) proteins. It differs from previous scoring functions mainly by the explicit treatment of ligand desolvation and of unfavorable protein-ligand contacts. The Fresno scoring function was separately calibrated for two different training sets: (1) 5 HLA-A*0201-peptide structures, which were determined by X-ray crystallography, and (b) 37 H-2K^k-peptide structures that were developed using knowledge-based homology modeling.

- **Binding Score to HLA-A2 Alleles:** This scoring function was developed using ChemScore-based default coefficients and parameters (H-bonding, lipophilic, and rotational terms), with better correlation. Also, the ChemScore scoring function was recalibrated by adding two terms, *i.e.*, buried polar score and desolvation score. The final scoring function (equation 29) is as follows:

$$\Delta G_{bind} = -33.614 - 0.014HB - 0.076LIPO + 0.017ROT + 0.021BP + 0.026DESOLV \quad (29)$$

Here, HB is H-bonding score, LIPO is lipophilic score, ROT is rotational score, BP is buried polar score, and DESOLV is desolvation score.

- **Binding to the H-2Kk Protein:** This scoring function was developed using a training set of 37 related H-2Kk-binding octapeptides. All 3D-structures of 37 peptides were built using homology modeling, and the function was developed using MLR. The reported best model (Equation 30) is as follows:

$$\Delta G_{bind} = -35.596 - 5.44HB + 0.344ROT + 0.063BP + 0.021DESOLV \quad (30)$$

Glide

Glide scoring function (Friesner, Banks, Murphy, Halgren, Klicic et al., 2004) is a modified version of ChemScore scoring function (Eldridge et al., 1997) that is utilized to predict binding affinity (GlideScore) and for ranking the ligands in database screens. However, for selection of best docked pose, a model energy function is utilized that combines GlideScore, ligand- receptor molecular mechanics interaction energy, and the ligand strain energy. In a reported study (Friesner et al., 2004), the docking accuracy was tested by re-docking the ligands in 282 co-crystallized PDB complexes, and Glide was found to be more accurate than GOLD, FlexX and Surflex scoring functions.

Glide Scoring Functions

Glide 2.5 employs two forms of GlideScore (i) GlideScore Standard-Precision (SP) (ii) GlideScore Extra-Precision(XP) (Friesner, Murphy, Repasky, Frye, Greenwood et al., 2006). GlideScore SP is a softer function that can identify ligands with a reasonable propensity to bind. This function hence attempts to minimize false negatives and is appropriate for huge database screening. On the other hand, GlideScore XP is a harder function that enforces severe penalties for poses that violate established physical-chemistry principles. This function hence attempts in minimizing false positives and are preferably useful in lead optimization.

GlideScore SP (Equation 31) is defined as follows:

$$\begin{aligned}\Delta G_{bind} &= C_{lipo-lipo}\sum f(r_{lr}) + C_{hbond-neut-neut}\sum g(\Delta r)h(\Delta\alpha) \\ &+C_{hbond-neut-charged}\sum g(\Delta r)h(\Delta\alpha) + C_{hbond-charged-charged}\sum g(\Delta r)h(\Delta\alpha) \\ &+C_{\max-metal-ion}\sum f(r_{lm}) + C_{rotb}H_{rotb} + C_{polar-phob}V_{polar-phob} + C_{coul}E_{coul} \\ &+C_{vdW}E_{vdW} + solvationterms\end{aligned} \quad (31)$$

Here, the lipophilic-lipophilic term and H-bonding term is same as defined in ChemScore. But the H-bonding term is separated into different weights that depend on whether the donor and acceptor are both neutral, one is neutral and the other is charged, or both are charged. The metal-ligand interaction term also uses the same functional form as in ChemScore but varies in three principal ways: a) this term considers only interactions with anionic acceptor atoms; b) it counts just the single best interaction when two or more metal ligations are found; and c) the net charge on the metal ion is assessed in the unligated apo-protein. If the net charge is positive, the preference for an anionic ligand is incorporated into the scoring function. Else, if net charge is neutral, the preference is suppressed. The seventh term in the function rewards situations in which a polar but non-H-bonding atom is found in a hydrophobic region.

The major components are the contributions from the Coulomb and vdW interaction energies between the ligand and the receptor. Another major component is the introduction of a solvation model and incorporation of the solvation effects, Glide docks explicit waters into the active site for each energetically competitive ligand pose and utilize empirical scoring terms that assess the exposure of various groups to the explicit waters.

Glide 4.0 XP scoring function (Friesner et al., 2006) as discussed before is the harder function that is developed using a validation set of 198 complexes. In this function, besides water desolvation energy terms, the protein-ligand structural motifs that lead to enhanced binding affinity are also included, *i.e.*, (1) hydrophobic enclosure where groups of lipophilic ligand atoms are enclosed complementary to lipophilic protein atoms, (2) neutral-neutral single or correlated H-bonds in a hydrophobically enclosed environment, and (3) five categories of charged-charged H-bonds.

Glide XP Scoring Function

The XP Glide scoring function is computed as shown in Equation 32. Further, Equation 33 shows the principal terms that favor binding and Equation 34 shows those penalty-terms that hinder binding.

$$XPGlideScore = E_{coul} + E_{vdW} + E_{bind} + E_{penalty} \quad (32)$$

$$E_{bind} = E_{hyd\_enclosure} + E_{hb\_nn\_motif} + E_{hb\_cc\_motif} + E_{PI} + E_{hb\_pair} + E_{phobic\_pair} \quad (33)$$

where, $E_{hyd\_enclosure}$ represents hydrophobic enclosure; $E_{hb\_nn\_motif}$ represents the special neutral-neutral H-bond motifs; $E_{hb\_cc\_motif}$ represents special charged-charged H-bond motifs, E_{PI} represents *pi*-stacking and *pi*-cation interactions; and E_{phobic_pair} represents hydrophobic atom-atom pair energy term. E_{hb_pair} term is same as defined in ChemScore scoring function.

$$E_{penalty} = E_{desolv} + E_{ligand\_strain} \quad (34)$$

where, E_{desolv} represents desolvation penalties, and E_{ligand_strain} represents contact penalties (penalizing strain energy)

KNOWLEDGE-BASED SCORING FUNCTION

As the number of experimentally determined protein-ligand complexes (3D databases) grows rapidly, it makes more feasible to use this structural information to develop an energy function for studying the protein-ligand interactions. To get a reliable estimate, one might dissect the protein-ligand binding free energy into all its physically meaningful contributions (e.g., electrostatics, hydrophobic interactions, desolvation, and entropy), but evaluating all the energy terms explicitly will be a difficult task. One can avoid this complex task by deriving simplified potentials (from known structural data) to directly estimate the total protein-ligand binding free energy and thus, treat all the significant contributions implicitly. Knowledge-based scoring functions derive such energy potentials from the structural information present in experimentally determined atomic structures. There are two basic assumptions that allow us to derive energy functions using concepts of statistical mechanics that are as follows:

1. It is assumed that the protein-ligand complex is in a state of thermodynamic equilibrium and that this state retains the global minimum of the free energy of the complex.
2. The distribution of the molecules in their microscopic states obeys Boltzmann's law, which means that there is a relation between the PE of the system and a probability to find a molecule or a complex of molecules in a certain microscopic state (Muegge & Martin, 1999).

Knowledge-based potentials utilize statistics collected from the 3D structures of protein-ligand complexes to extract rules based on atomic interactions. It is assumed that those molecular interactions (pair-wise) that are close to the frequency maxima of the interactions in the database are favorable to the binding process and therefore presence of such interactions enhances the overall binding affinity, whereas interactions that have been found to be in low frequency in the database are likely to destabilize binding and hence decreases the affinity. The occurrence frequency of a pair-wise contact is assumed to be a measure of its energetic contribution to protein-ligand binding. As suggested in the literature (Liu & Wang, 2015), it is better to term this scoring function as "*Knowledge based potentials*" than "*Knowledge-based scoring function*" because later term is ambiguous in meaning, since all the scoring functions are in fact more or less "knowledge-based".

Further, these scoring function sums pair-wise statistical potentials between protein and ligand as binding score (equation 35) and are generally represented as:

$$BindingScore = \sum_{i}^{lig} \sum_{j}^{prot} \omega_{ij}(r) \tag{35}$$

Pair-wise potentials (ω_{ij}) are directly calculated from the occurrence frequency of atom pairs in a database using the inverse Boltzmann formula (equation 36).

$$\omega_{ij}(r) = -k_B T \ln[g_{ij}(r)] = -k_B T \ln\left[\frac{\rho_{ij}(r)}{\rho_{ij}^{*}}\right] \tag{36}$$

where k_B is the Boltzmann constant, T is the absolute temperature of the system, $\rho(r)$ is the number density of the protein–ligand atom pair at distance r, and ρ^* is the pair density in a reference state where the inter-atomic interactions are either considered zero or they comprise non-specific interactions that are common to all types of atom. In the theory of liquids, free energies are calculated using radial distribution function (RDF), $g(r)$, corresponding to the fraction $(\rho(r))/(\rho^*)$.

The standard approach to derive the desired pair-wise potentials is to employ a large set of protein-ligand complex structures from databases as the training set, often called as "knowledge base". Protein atoms as well as ligand atoms are classified into a number of degenerated atoms types. Then, distance-dependent potentials for each possible atom pair are derived from the occurrence frequency of this atom pair observed in the training set using the above formula.

Due to their pair-wise characteristics, knowledge based scoring function are much more efficient compared to the FF or physics-based methods that often require computationally expensive treatment of solvent. Also, unlike empirical scoring functions, knowledge-based potentials are derived through statistical analysis of pure structural information without any need of experimental binding affinity values.

Now we will discuss some knowledge-based potentials in details that are reported in the literature in the last two decades.

Potentials of Mean Force (PMF)

In 1999, Potentials of Mean Force (PMF or PMF99) was developed using a knowledge-based approach that exploited the structural information of 697 known protein-ligand complexes extracted from the Brookhaven PDB and then this information was converted into distance-dependent Helmholtz free interaction energies of protein-ligand atom pairs (Muegge & Martin, 1999). About 16 protein atom types and 34 ligand atom types were employed. All the ligand atoms with minimum distance greater than 5Å to any protein heavy atom were discarded (to avoid noise to the potentials). A cutoff for resolution of crystal structure was set to ≤2.5Å. The predicted PMF score using PMF function showed a standard deviation (SD) from experimental binding affinities of 1.8 log K_i units and an R^2 value of 0.61.

The idea of deriving appropriate PMF was earlier used by Sippl and co-workers (Sippl, Ortner, Jaritz, Lackner, & Flockner, 1996) in protein folding studies for characterizing arbitrary atom pair interactions of protein atoms and, the PMF scoring function is derived from this Sippl's methodology with some essential changes.

The protein-ligand interaction free energy between protein atoms of type '*i*' and a ligand atoms of type '*j*' can be written as shown in equation 37:

$$A_{ij}(r) = -k_B T \ln \left[f^{r}_{Vol\_corr}(r) \frac{\rho^{ij}_{seg}(r)}{\rho^{ij}_{bulk}} \right] \tag{37}$$

Here, k_B is the Boltzmann constant, T is the absolute temperature, and r is the atom pair distance. $f^{r}_{Vol\_corr}(r)$ represent the ligand volume correction factor, $\rho^{ij}_{seg}(r)$ is the number density of pairs of type '*ij*' in a protein database that exist in a specific radius range that is indicated by "*seg*", and ρ^{ij}_{bulk} represents the distribution of '*i*' and '*j*' when there is no interaction exist between '*i*' and '*j*'. The resulting value of $\rho^{ij}_{seg}(r) / \rho^{ij}_{bulk}$ in equation 37 represents the pair correlation function (or RDF) of a protein atom of type '*i*' paired with a ligand atom of type '*j*' in a structural database of protein-ligand complexes. The solvation effect is also captured in the ratio of protein to solvent volume occupied in a sphere of 12 Å radius around a given ligand atom '*j*'.

The PMF scoring function is defined as the sum over all inter-atom interactions ('*kl*') of the protein-ligand complex and can be written as:

$$PMF\_score = \sum_{\substack{kl \\ r<r^{ij}_{cut-off}}} A_{ij}(r) \tag{38}$$

Here in Equation 38, $r^{ij}_{cut-off}$ is the cutoff radius for the atom type pair '*ij*'. For scoring all non-carbon protein-ligand interactions 9 Å cut-off was set, while cutoff of 6 Å was set for carbon-carbon atom type interactions.

In 2006, the PMF scoring function (Muegge & Martin, 1999) was further updated to "PMF04 scoring function" (Muegge, 2006) that is developed using 7152 protein-ligand complexes obtained from the PDB. That makes around 10-fold increase in knowledge base size compared to the knowledge base of original PMF99 score (697 complexes). The distance cut-off (*i.e.* 5Å) between ligand atom and protein atom considered for computation as well as protein resolution cut-off (≤2.5Å) is same as in PMF99. Two additional potentials were added in PMF04, *i.e.*, potential for metal ions and halogens. The following changes to the atom types were made in PMF04, compared to the PMF99:

1. Histidine nitrogen atoms '*ND1*' and '*NE2*' were assigned atom types '*ND*' (nitrogen as H-bond donor) and '*NA*' (nitrogen as H-bond acceptor), respectively. Also, for standard histidine residue name, the derived PMF04 potential assumes the same protonation pattern (*HIS*) for all complexes.
2. A new general metal ion atom type '*ME*' was introduced that combines Zn, Ca, K, Mg, Mn, and Fe ions that are recorded as individual residues in the PDB. Metal ions that are part of larger prosthetic groups, for e.g., iron in heme, were not included. Such metals were treated as part of the ligand.
3. For ligands, the '*OR*' atom type (oxygen in a ring structure) was merged into the '*OE*' atom type (oxygen in ether bond) and a new atom type '*SO*' was assigned for sulfur bonded to more than two atoms or other than C and H to capture sulfone and sulfoxide functional groups. Specifically, the distances are binned in 0.2 Å increments. The PMF cut-off for the reference state is 12 Å. The cutoff for scoring is chosen to be 6 Å for carbon-carbon interactions and 9 Å for interactions involving heteroatoms.

DrugScore

DrugScore (Gohlke, Hendlich, & Klebe, 2000) or DrugScore^{PDB} was developed using the structural information extracted from crystallographically determined protein-ligand complexes present in PDB and the extracted information was converted into distance dependent pair-preferences and solvent-accessible surface (SAS) dependent singlet preferences for protein and ligand atoms. The DrugScore was found to be superior as compared to FlexX (Rarey, Kramer, Lengauer, & Klebe, 1996) and DOCK scoring function in the reported study (Gohlke, Hendlich, & Klebe, 2000). DrugScore considers implicit solvation and entropic contributions, and also shows correct geometry of directional interactions.

Distance-Dependent Pair-Potentials

The DrugScore scoring function also followed the formalism developed by Sippl (Sippl et al., 1996).

$$\Delta W_{i,j}(r) = W_{i,j}(r) - W(r) = -\ln \frac{g_{i,j}(r)}{g(r)} \tag{39}$$

$$g(r) = \frac{\sum_i \sum_j g_{i,j}(r)}{i * j} \tag{40}$$

The net statistical preferences $\Delta W_{i,j}$ (Equation 39) are obtained by comparing the mean statistical preferences of the subsystems *i*, *j* ($W_{i,j}$) to the reference system (*W*). $g_{i,j}(r)$ is the normalized radial pair distribution function for atoms of types '*i*' and '*j*', separated by a distance in the interval of '*r*' and '*r* + *dr*' *(radius r, thickness dr)*; *g*(*r*) is the normalized mean radial pair distribution function for a distance between any two atoms in the range of *r* and *r* +*dr*. Here, *g*(*r*) (Equation 40) corresponds to the reference state and incorporates all non-specific information that is common to all atom pairs present in an environment typical for proteins. Also, here, $g_{i,j}(r)$ (Equation 41) was compiled from occurrence frequencies *N* of atom pairs with types '*i*' and '*j*' as given below:

$$g_{i,j}(r) = \frac{N_{i,j}(r) / 4\pi r^2}{\sum_r (N_{i,j}(r) / 4\pi r^2)} \tag{41}$$

Here, distances $1.0 \leq r \leq 6.0$ Å for computation were considered and scaling to $4\pi r^2$ accounts for the volume of the spherical shell of the radius *r* and the thickness *dr*.

Non-Polar Surface-Dependent Singlet-Potential

Gohlke *et al.* derived a knowledge-based one-body potential scaled to the size of the solvent-accessible surface (SAS) of the protein and the ligand that becomes buried upon complex formation. The potential was defined as follows (Equation 42):

$$\Delta W_i(SAS, SAS_0) = W_i(SAS) - W_i(SAS_0) = -\ln \frac{g_i(SAS)}{g_i(SAS_0)} \tag{42}$$

Here, g_i is the normalized distribution function of the surface area of an atom '*i*' (considering ligand and protein atoms individually) in the buried state (*SAS*) in comparison to the solvated state (SAS_0). It is computed using an approximate cube algorithm, where it is assumed that any polar portion of the SAS that becomes buried in the complex in a polar environment is considered "solvent-accessible''. This is based on the rationale that changes in the Gibbs free energy is only expected if polar molecular portions are moved from a polar solvent to a non-polar protein environment. Hence, polar portions that are transferred from the polar solvent to a polar binding site environment are neglected. Here, $g_i(SAS_0)$ is not an averaged distribution function over all atom types but refers to the atom of type '*i*' only. Thus, ΔW_i reveals the contribution arising from differences in the solvent- accessible surface between the protein-bound and fully solvated state.

Finally, the total preference ΔW (see Equation 43) for defining a particular binding geometry can be approximated by summing all individual contributions (*i.e.* of k_i ligand atoms of type '*i*' and l_j protein atoms of type '*j*'):

$$\Delta W = \gamma \sum_{k_i} \sum_{l_j} \Delta W_{i,j}(r) + (1 - \gamma) \times \left[\sum_{k_i} \Delta W_i(SAS, SAS_0) + \sum_{l_j} \Delta W_j(SAS, SAS_0) \right] \tag{43}$$

where, γ is an adjustable parameter, optimized empirically to be 0.5.

DrugScore^{CSD}

In 2005, DrugScore^{CSD} (Velec, Gohlke, & Klebe, 2005) was reported to be an improved version of DrugScore^{PDB} (Gohlke, Hendlich, & Klebe, 2000). It was derived using Cambridge Structural Database (CSD)-based potentials. The study (Velec, Gohlke, & Klebe, 2005) showed that DrugScore^{CSD} has superior performance in recognition of near-native ligand binding modes and the mutual ranking with respect to experimentally determined binding affinities, as compared to DrugScore^{PDB}. However, the methodology for deriving pair potential was adapted from the DrugScore^{PDB} methodology.

DrugScore eXtended (DSX)

In 2011, based on the statistical formalism of DrugScore, however extended by more specialized set of atom types, a new knowledge-based scoring function DrugScore eXtended (DSX) (Neudert & Klebe, 2011) was developed employing both PDB (*37,067 X-ray structures with a resolution* $\leq$ *2.4 Å*) and CSD (*3,45,726 structures*). This scoring function comprised distance-dependent pair potentials, novel torsion angle potentials, and newly defined solvent accessible surface-dependent potentials. The torsion angle potentials were derived from the CSD, and the solvent accessible surface-area ratio (SAS-ratio or SR) potentials were derived from the PDB. The total DSX-score for a given protein-ligand complex was defined as shown in Equation 44:

$$score_{\mathrm{total}} = w_{\mathrm{p}} score_{\mathrm{pair}} + w_{\mathrm{t}} score_{\mathrm{tors}} + w_{\mathrm{s}} score_{\mathrm{SR}} \quad (44)$$

Here, the '*pair*' stands for pair potential, '*tors*' for torsion potential, '*SR*' is the SAS-ratio for a protein or ligand atom, and the $w_{\mathrm{p/t/s}}$ are the weighting factors employed.

SMall Molecule Growth (SMoG96 and SMoG2001)

SMoG96

SMall Molecule Growth (SMoG or SMoG96) is also based on a knowledge-based approach. The pairwise approximation is computed as the sum over all atom-atom contacts between protein and ligand atoms and is defined as (Equation 45):

$$F = \sum_{p} \sum_{l} F(\sigma_p, \sigma_1) \Delta(p, l) \quad (45)$$

In this Equation 45, p implies a protein atom of type σ_p, l implies a ligand atom of type σ_l, $\Delta\ (p,l)$ is the characteristic function of the contact (1, if atoms p and l are in contact and 0, otherwise), and $F(\sigma_p,\sigma_l)$ value represents the potential of mean force (PMF) of interaction between two atoms of specified atom types.

In the SMoG96 function (DeWitte & Shakhnovich, 1996), the two atoms are considered in contact if the distance between them is less than 5 Å. This distance is equivalent to the sum of radii of first coordination shells of water around two interacting atoms. When the atoms interact to form a contact, the coordination shells breaks, and the water molecules are transferred into the free solvent that lead to the net gain of solvent entropy. Thus, in this function the contribution of the solvent entropy is taken implicitly.

The $F(\sigma_p,\sigma_l)$ values in SMOG96 function (Equation 46) were calculated as follows:

$$F\left(\sigma_p,\sigma_l\right) = -\ln\left[\frac{p(\sigma_p,\sigma_l)}{p^{ref}}\right] \tag{46}$$

Here, $p(\sigma_p,\sigma_l)$ represents the measure of frequency of the contacts between atom types σ_p and σ_l in the knowledge database and p^{ref} is the probability of those contacts in the hypothetical reference state. The $p(\sigma_p,\sigma_l)$ values were computed using Equation 47 as follows:

$$p\left(\sigma_p,\sigma_l\right) = \frac{N\left(\sigma_p,\sigma_l\right)}{N(\sigma_p)^{1/2}N(\sigma_l)^{1/2}} \tag{47}$$

where, $N(\sigma_p,\sigma_l)$ is the number of (σ_p,σ_l) contacts calculated from the database, $N(\sigma_l)$ is the number of ligand atoms of type σ_l that forms at least one contact with any protein atom, and $N(\sigma_p)$ is the number of protein atoms of type σ_p that forms at least one contact with any ligand atom. Here, the reference state is described as a complex of randomly connected protein atoms and randomly connected ligand atoms that do not interact, *i.e.*, $F(\sigma_p,\sigma_l)=0$ for all (σ_p,σ_l) pairs. The simple approximation $p^{ref} = (p(\sigma_p,\sigma_l))_{(\sigma p,\sigma l)}$ was selected for computing the probabilities of the reference state.

SMoG2001

SMoG2001 is the updated and newer version of SMoG scoring function that was obtained using the set of 725 protein-ligand complexes (Ishchenko & Shakhnovich, 2002). In this version, the reference state was redefined, which ensures proper normalization of contact probabilities, *i.e.*, the sum of all values over atom types is equal to 1 and further introduced two distance intervals ("bins") over which the contact statistics are computed. This leads to the following formulation of the scoring function (equaTion 48):

$$F = \sum_r \sum_p \sum_l F\left(r,\sigma_p,\sigma_l\right)\Delta(r,p,l) \tag{48}$$

Here, for each distance interval, '*r*' is the outer radius of the bin and two atoms are considered to be in contact, if the distance between them is inside the bin $(r, r\text{-}\Delta r)$.

In the SMoG2001 function, $F(r, \sigma_p,\sigma_l)$ values of PMF are computed as a logarithm of the ratio of probabilities of contact formed in the complex and in the reference state (Equation 49):

$$F(r,\sigma_p,\sigma_l) = -\ln\left[\frac{N(r,\sigma_p,\sigma_l)}{C(r)\times\Omega(r,\sigma_p,\sigma_l)\times\sum_{\sigma_p}\sum_{\sigma_l}N(r,\sigma_p,\sigma_l)}\right] \tag{49}$$

Here, $N(r, \sigma_p, \sigma_l)$ is the number of contacts between atoms of specified atom types computed from the knowledge database, and $C(r) \times \Omega(r, \sigma_p, \sigma_l)$ is the probability of the contacts in the reference state for a given distance interval. $C(r)$ is a normalization constant, and $\Omega(r, \sigma_p, \sigma_l)$ is the term associated with the composition of the database, *i.e.*, number of atoms of the σ_l and σ_p types participating in the protein-ligand interaction.

To derive the expression for $C(r)$, the average of the Equation 49 is accounted over all atom types resulting in Equation 50:

$$\ln C(r) = \left\langle \ln \frac{N(r, \sigma_p, \sigma_l)}{\Omega(r, \sigma_p, \sigma_l) \times \sum_{\sigma_p} \sum_{\sigma_l} N(r, \sigma_p, \sigma_l)} \right\rangle_{(\sigma_p, \sigma_l)} + F_0(r) \quad (50)$$

where, $F_0(r) = (F(r, \sigma_p, \sigma_l))_{(\sigma p, \sigma l)}$ is a free parameter equal to an average free energy for a given distance interval. $\Omega(r, \sigma_p, \sigma_l)$ is described in the first approximation as:

$$\Omega(r, \sigma_p, \sigma_l) = N(r, \sigma_p)^{\alpha} N(r, \sigma_l)^{\beta} \quad (51)$$

In Equation 51, $N(r, \sigma_p)$ is the number of protein atoms of type (σ_p) in the interval $(r, r\text{-}\Delta r)$ that form at least one contact with any ligand atom, $N(r, \sigma_l)$ is the number of ligand atoms of type σ_l in the interval $(r, r\text{-}\Delta r)$ that form at least one contact with any protein atom, and α and β are free parameters.

The derivation of SMoG2001 scoring function is as follows: 1) first, the $N(r, \sigma_p, \sigma_l)$, $N(r, \sigma_p)$, and $N(r, \sigma_l)$ values are computed from the PDB of protein-ligand complexes that comprise the training set; 2) Then, first approximation values are chosen for the four free parameters, namely, α, β, $F_0(r)$, and r; 3) using these quantities and Equations 49-51, the $F(r, \sigma_p, \sigma_l)$ values are derived; 4) the scoring function is utilized to calculate the binding scores of the complexes for which the experimental binding constants are known (testing set); 5) the correlation coefficient between predicted and experimental binding free energies of the testing set complexes are employed as the measure of the performance of the scoring function.

Astex Statistical Potential (ASP)

To correct the deficiency in DrugScore and PMF reference states, the ASP (Mooij & Verdonk, 2005) scoring function was developed. It was suggested that essential information on the environment of protein atoms around ligand binding sites was missing from both the DrugScore and PMF reference states. The environment of protein atoms around ligand binding sites is derived by computing the available volume in spherical shells around each protein atom, which is within R_{max} of any ligand atom. Such a protein volume correction captures the differences in available volumes around protein atom types, which is similar to the ligand volume correction used in PMF does for ligand atoms.

The ASP potential (Mooij & Verdonk, 2005) incorporates available volume corrections for protein along with the ligand atoms. Both volume corrections are computed utilizing a grid, where the points are pointed out as protein, ligand, or solvent. Here, the co-factors are considered as the part of the protein.

The fraction of available volume for a protein atom type 'i' is calculated as the fraction of non-protein grid points at distance 'r', which is averaged over all protein atoms of type 'i' that are within 8.0Å of a ligand atom, in all complexes in the knowledge database (see Equation 52).

$$f_p(i,r) = \left\langle \left(gridpts_i^{total}(r) - gridpts_i^{protein}(r) \right) / \; gridpts_i^{total}(r) \right\rangle \tag{52}$$

The ligand volume correction (Equation 53) is computed similarly, averaging over all the ligand atoms of type 'j' in all the complexes:

$$f_l(j,r) = \left\langle \left(gridpts_j^{total}(r) - gridpts_j^{ligand}(r) \right) / \; gridpts_j^{total}(r) \right\rangle \tag{53}$$

In the ASP scoring function, the expected number of contacts for a given atom pair 'i' and 'j' was defined as (Equation 54):

$$n_{\exp}^{ASP}(i,j,r) = \left\langle \frac{n_{obs}(i,j,r')}{f_p(i,r')f_l(j,r')4\pi r'^2 \Delta r} \right\rangle_{r'=6.0}^{r'=8.0} \cdot f_p(i,r) \cdot f_l(j,r) \cdot 4\pi r^2 \Delta r \tag{54}$$

The expected number of contacts is the product of an average contact density with the doubly corrected volume of a sphere at distance 'r'. The average contact density is considered the average between 6.0 and 8.0 A of the corrected RDF, and not the overall average for the sphere with radius 'R_{max}'. At this long distance, atoms are believed to make no specific interactions and hence can be appropriately used to determine the reference contact density.

ITScore-PR

ITScore-PR (Huang & Zou, 2014) is a knowledge-based scoring function for protein-ribonucleic acid (RNA) binding mode prediction utilizing a statistical mechanics-based iterative method. For deriving potentials, a training set was constructed that comprise 110 diverse protein-RNA complex radiographic crystal structures with resolution <3.5A° and contain at least one protein and one RNA chain but no DNA chains.

Statistical Mechanics-Based Iterative Method to Derive the Interaction Potentials

In this method, the basic scheme is to optimize the inter-atomic pair potentials gradually through iterations by comparing the experimentally observed pair distribution functions (PDFs) of the native crystal structures present in the training set and the predicted PDFs of the protein-RNA complexes. The method is defined as follows using an iterative formula:

$$u_{ij}^{(n+1)}(r) = u_{ij}^{(n)}(r) + \Delta u_{ij}^{(n)}(r), \tag{55}$$

$$\Delta u_{ij}^{(n)}(r) = \frac{1}{2} k_B T \left[g_{ij}^{(n)}(r) - g_{ij}^{obs}(r) \right]$$

where, 'n' designate the iterative step, 'i' and 'j' represent the types of a pair of atoms in the protein and the RNA, respectively. $g_{ij}^{obs}(r)$ designate the PDF for atom pair 'ij' computed from the experimentally observed protein-RNA complex structures in the training set:

$$g_{ij}^{obs}(r) = \frac{1}{K} \sum_{k=1}^{K} g_{ij}^{k*}(r)$$

Here, 'K' is the total number of the protein–RNA complexes in the training set and $g_{ij}^{k*}(r)$ is the PDF of the kth native complex structure. $g_{ij}^{(n)}(r)$ is the PDF computed from the ensemble of the binding modes based on the binding score-dependent Boltzmann probabilities P_k^l that are predicted with the trial potentials $\{u_{ij}^{(n)}(r)\}$ at the nth step.

$$g_{ij}^{(n)}(r) = \frac{1}{K} \sum_{k=1}^{K} \sum_{l=0}^{L} P_k^l g_{ij}^{kl}(r)$$

where, $g_{ij}^{kl}(r)$ is the PDF for atom pair 'ij' observed in the lth binding state of the kth protein–RNA complex. $\{u_{ij}^{(n+1)}(r)\}$ are the improved potentials from $\{u_{ij}^{(n)}(r)\}$ after the correction and are used in the next iterative step. $k_B T$ is set to unit 1 in the iterations.

For a given set of initial potentials $\{u_{ij}^{(0)}(r)\}$,

$$u_{ij}^{(0)}(r) = \begin{cases} w_{ij}(r) \\ \dfrac{v_{ij}(r) e^{-v_{ij}(r)} + w_{ij}(r) e^{-w_{ij}(r)}}{e^{-v_{ij}(r)} + e^{-w_{ij}(r)}} \end{cases}$$

where, $v_{ij}(r)$ is the 6-12 vdW potential and $w_{ij}(r)$ is the potential of mean force and is defined as follows:

$$w_{ij}(r) \equiv -k_B T \ln g_{ij}^{obs}(r)$$

The first iterative step will lead to an improved set of pair potentials $\{u_{ij}^{(1)}(r)\}$ (using Equation (55)) followed by $\{u_{ij}^{(2)}(r)\}$, $\{u_{ij}^{(3)}(r)\}$ and so on in successive steps. The iteration was continued until the potentials discriminate all the native binding structures from the decoys.

In this way, the pairwise potentials for protein–RNA interactions were extracted for the protein–RNA complexes present in the training set. The 20 protein atom types and 12 RNA atom types were employed, which should have resulted in 240 different pairs. However, only the atom type pairs with >1000 occurrences were retained, which resulted in 206 pairs of interaction potentials. The scoring function was evaluated on its ability of identifying near-native structures on four diverse test sets, and compared with ten other scoring methods. Overall, ITScore-PR showed a success rate as high as 86.1% for bound docking and 24% for unbound docking if the top one prediction was considered.

DESCRIPTOR-BASED SCORING FUNCTION

Recently, a separate class of descriptor-based scoring functions was suggested. These functions were previously considered as empirical scoring functions (Liu & Wang, 2015). Quantitative structure-activity relationship (QSAR) technique has always been extensively employed for modeling of various biological and physicochemical properties. In QSAR, the descriptors are derived from the symbolic representation of the molecules and generally encode chemical or topological information, which can be utilized to study protein-ligand interaction patterns. The descriptors like conventional ligand-based descriptors *(molecular weight, number of rotatable single bonds, number of H-bond donor/acceptor),* structural interaction fingerprints between protein and ligand, geometrical descriptors (*surface or shape or volume properties*), specific interactions descriptors (*electrostatic interactions, hydrogen bonds, or aromatic stacking*) can be used in development of docking scoring functions. Further, linear regression methods like MLR or machine learning techniques like random forest, Bayesian classifiers, neural network, and support vector machines can be employed to develop statistical models for computing binding scores.

As we have mentioned earlier that the descriptor based scoring functions are actually similar to the empirical scoring functions. But there are sufficient differences between them requiring us to mention these scoring functions as two separate classes: a) empirical scoring functions always have a linear functional form, while descriptor-based can be linear or non-linear based on the regression or machine-learning technique employed; b) a descriptor-based scoring function usually consists of a larger number of descriptors as compared to an empirical scoring function; c) the individual terms in an empirical scoring function generally have interpretable physical meanings, while in the case of a descriptor-based method, the rationale for selecting a certain combination of descriptors is often fuzzy especially in machine learning techniques that behaves like a black box (Liu & Wang, 2015). The followings are some descriptor-based scoring functions reported in the literature.

LigScore

Two descriptor-based empirical scoring functions, LigScore1 and LigScore2, were reported (Krammer, Kirchhoff, Jiang, Venkatachalam, & Waldman, 2005) to predict the binding affinity of ligand molecules with their target protein. The LigScore scoring functions comprise 3 distinct terms that explain the vdWs interaction, the attractive polar interaction between the ligand and protein, and the desolvation effect (penalty). The linear regression approach was employed on a data set of 50 and 118 protein–ligand complexes in order to obtain two functions, LigScore1 and LigScore2, respectively, using these three terms.

LigScore1 was developed utilizing good resolution X-ray structures, and the experimental binding affinities range employed were from 1.49 to 13.4 (pK_i). The vdW energy term $\langle E_{vdW} \rangle$ was derived based on genetic function approximation (GFA) analysis. Other 23 descriptors were chosen to incorporate energy terms that stand for electrostatics and solvation effects. The final functional form of LigScore1 (Equation 56) comprises 3 descriptors and is as follows:

$$pK_i = C - \beta_1 \langle E_{vdW} \rangle + \beta_2 Cpos\_tot - \beta_3 TotPol^2 \quad (56)$$

where, $\langle\ \rangle$ indicates the summation of all vdW interactions between the ligand and protein atoms, '*Cpos_tot*' is the total surface area of the ligand that is involved in attractive polar interactions with the protein, and '$TotPol^2$' is equal to ($Cpos\_tot^2 + Cneg\_tot^2$), where '*Cneg_tot*' is the total surface area of the ligand that is involved in repulsive polar interactions with the protein. The '*Cpos_tot*' and '$TotPol^2$' values are always positive. Hence, the former term characterizes the favorable protein–ligand electrostatics, while the latter term represents a desolvation penalty. The overall fit of LigScore1 with experimental pK_i's showed a correlation coefficient, (r^2) of 0.79 with a standard deviation of 0.70.

However, the '$TotPol^2$' term in LigScore1 does not account for the desolvation of all protein atoms in the complex, since it does not consider various contacts of polar protein atoms with non-polar ligand atoms that replaces buried waters. A reasonable descriptor for protein desolvation should account for the entire desolvated polar protein surface and also a descriptor accounting for ligand desolvation that occurs when a polar portion of the ligand binds to a lipophilic protein binding site. Hence, following descriptors (equations 57 and 58) were defined as a replacement for TotPol2 in LigScore1:

$$SolvPlty\_lig^2 = (Bury\_tot\_lig - Bury\_lip\_lig)^2 \quad (57)$$

$$SolvPlty\_prot^2 = (Cpos\_tot + Cneg\_tot + Bury\_lip\_lig - Clip\_lip)^2 \quad (58)$$

The first term '$SolvPlty\_lig^2$' designates the overall polar ligand surface that is buried inside the protein-binding site, while the second term '$SolvPlty\_prot^2$' indirectly accounts for the polar protein surface when bound to the ligand molecule. Using these two descriptors resulted in a LigScore2 equation (developed employing GFA analysis; equation 59):

$$pK_i = C - \beta_1 \langle E_{vdW} \rangle + \beta_2 Cpos\_tot - \beta_3 (SolvPlty\_lig^2 + SolvPlty\_prot^2) \quad (59)$$

For LigScore2, the final averaged correlation coefficient, r^2, equaled 0.73 with a standard deviation value of 1.08, while the averaged predictive correlation coefficient, q^2, was 0.71 with a standard deviation of the error of prediction, s_{press}, of 1.11.

SFCscore

SFCscore scoring function was developed using 855 complexes obtained from AffinDB database and industrial data collected from Scoring Function Consortium (SFC) (Sotriffer, Sanschagrin, Matter, & Klebe, 2008). The experimental binding affinity values (K_i, K_d, IC_{50}) were related to the standard concentration (1 mol/L) and converted to their negative logarithms (pK_i, pK_d, pIC_{50}).

The descriptors utilized to develop SFCscore comprise ligand-related information and protein-ligand interaction information. The ligand-related descriptors include the molecular weight, the number of ligand atoms in terms of the Sybyl atom types, the number, type, and environment of rotatable bonds and ligand surface data (total, polar, hydrophobic, and aromatic ligand surface area). Descriptors related to the protein-ligand interaction in the complex represent a list of H-bonds, ring interactions, *i.e.*, interactions between two aromatic or heterocyclic rings, as well as between cationic centers and pi-ring systems, a matrix of atomic contacts between ligand and protein, listed by atom type and binned into 0.1 Å bins for distances up to 6 Å, the atomic hydrophobicity difference, the percentage of buried carbon atoms of the ligand and a series of surface measures: buried and exposed surface of the ligand, as well as size and type of the protein-ligand contact surfaces. Altogether, around 66 descriptors were employed to develop the scoring function.

Further regression analysis that includes stepwise multiple linear regression (S-MLR) and systematic searches, along with partial least squares (PLS) analysis was performed to select the best models with the most suitable descriptors. After validation, eight functions were selected with q^2 values up to 0.72 and standard errors (s_{press}) below 1.15 (Sotriffer, Sanschagrin, Matter, & Klebe, 2008).

In 2013, a new version of SFCscore (i.e. SFCscore^{RF}) was reported based on descriptors employed in original SFCscore scoring functions (Zilian & Sotriffer, 2013). The new version is derived utilizing a machine learning approach, namely, random forest. Here, the PDBbind database and the CSAR–NRC 2010 set of complexes served as a data set for development of this scoring function. Further on evaluation of SFCscore^{RF}, it was found that this function outperforms the original SFCscore scoring functions based on the values of Pearson's correlation coefficient (R_p), Spearman's correlation coefficient (ρ), Kendall's tau (τ), median of residuals, root mean square error (RMSE). However, the interpretation of a random forest model (SFCscore^{RF}) with respect to descriptor contributions is not as straightforward as compared to the MLR or PLS functions (original SFCscore functions).

RF-Score

In 2010, Ballester and Mitchell published RF-Score (Ballester & Mitchell, 2010), a random forest model based on sums of protein-ligand element pair counts in a radius of 12 Å around each ligand atom. Random Forest (RF) was used to implicitly capture binding effects that are hard to model explicitly. This model was developed and validated utilizing 1300 protein-ligand complexes present in the PDBbind07 database (2007 version). On the validation set that has been selected in a kind of space-filling manner, the model reaches $R^2 = 0.60$ and root mean square error ($RMSE$) = 1.58.

AutoDock Vina

The nature of the scoring function employed in AutoDock Vina (Trott & Olson, 2010) is more associated to "machine learning" than a direct physics-based method, when compared with the scoring function employed in other AutoDock version (3.0 or 4.0) discussed before. The functional form of the conformation-dependent part of the AutoDock Vina scoring function (equation 60) is as follows:

$$c = \sum_{i<j} f_{t_i t_j}(r_{ij}) \tag{60}$$

Here, the summation is over all the atoms pairs that can move relative to each other, but usually excluding atoms that are separated by three consecutive covalent bonds. Here, each atom '*i*' is assigned a type t_i, and a symmetric set of interaction functions $f_{t_i t_j}$ is defined for inter-atomic distance r_{ij}. The '*c*' value is a sum of intermolecular and intramolecular contributions. An optimization algorithm is employed to find the global minimum of '*c*' and other low-scoring conformations that are ranked accordingly. The predicted free energy of binding is then calculated from the intermolecular part of the lowest-scoring conformation. This scoring function combines the benefits of knowledge-based potentials as well as empirical scoring functions. It gains the empirical information from the experimental-affinity measurements and the conformational preferences of the receptor-ligand complexes.

The interaction functions $f_{t_i t_j}$ are defined relative to the surface distance (d_{ij}):

$$f_{t_i t_j}(r_{ij}) \equiv h_{t_i t_j}(d_{ij})$$

where, $d_{ij} = r_{ij} - R_{t(i)} - R_{t(j)}$ and R_t is the vdWs radius of atom type '*t*'.

$h_{t_i t_j}$ is a weighted sum of steric interactions (includes gauss1, gauss2 and repulsion terms), hydrophobic interactions and, H-bonding (all the respective terms and their weights are shown in Table 1).

The steric terms are defined as follows:

$$gauss_1(d) = e^{-(d/0.5\text{A})^2}$$

$$gauss_2(d) = e^{-((d-3\text{A})/2\text{A})^2}$$

$$repulsion(d) = \begin{cases} d^2, if d < 0 \\ 0, if d \geq 0 \end{cases}$$

The hydrophobic term is equal to 1, when $d < 0.5$Å and is equal to 0, when $d > 1.5$Å, and is linearly interpolated in between. The H-bonding term is equal to 1, when $d < -0.7$Å and is equal to 0, when $d > 0$, and is linearly interpolated in between. Metals are treated as H-bond donors. Further, all the interaction functions $f_{t_i t_j}$ are cut off at $r_{ij} = 8$Å.

Table 1. AutoDock Vina scoring function: weights and terms

Terms	Weights
gauss1	−0.0356
gauss2	−0.00516
Repulsion	0.840
Hydrophobic	−0.0351
H-bonding	−0.587
Nrot	0.0585

The conformation-independent function '*g*' is defined as follows:

$$g(c_{inter}) = \frac{c_{inter}}{1 + wN_{rot}}$$

where, N_{rot} = number of active rotatable bonds between heavy atoms in the ligand;

'*w*' = associated weight

Global Free Energy Scoring Functions Based on Distance-Dependent Atom-Type Pair Descriptors

Global free energy scoring functions (Kramer & Gedeck, 2011) based on distance-dependent atom-type pair descriptors were trained and evaluated on 1387 complexes taken from the PDBbind09 (2009 version) database. Distance-binned Crippen-*like* atom type pair counts were used as descriptors to build the scoring functions. The chemometric methods used to derive the functions were MLR and PLS. The best scoring function had $R^2 = 0.48$, $RMSE = 1.44$, and mean unsigned error (MUE) = 1.14. Each scoring function on an average consisted of 12 descriptors, and the most significant descriptors were found to be *logS*, *logP*, and molecular weight of ligands.

PHOENIX

PHOENIX (Tang & Marshall, 2011), a scoring function developed utilizing the vast data available in the PDB to improve binding affinity predictions and employing biophysical methods such as isothermal titration calorimetry to measure thermodynamic parameters that contribute to binding free energy (ΔG). Current empirical scoring functions employ descriptors that take into account the changes of enthalpy (ΔH) in binding, but to describe changes in entropic forces ($T\Delta S$), elementary methods such as the number of rotamers on a ligand, calculated partition coefficient (XlogP), and complementary hydrophobic surface area estimation, are used. Most of the scoring function can predict enthalpic contributions accurately, but fail to predict entropic contributions, which can limit their accuracy in predicting binding free energy. Thus, in the PHOENIX scoring function, additional terms were included to describe the shape and volume of both the ligand and protein active site. These terms contribute to the entropic changes by implicitly capturing the desolvation effect and ligand expulsion of solvent from the active site. Volume-based descriptors may also capture the rotational and translation entropy contributing to the configurational entropy of the system.

PHOENIX scoring function was developed using 112 protein-ligand complexes (82 in the training set and 30 in the test set). The final functions were derived using PLS to assign coefficients to each of the terms to derive the master equation to calculate change of enthalpy (ΔH) and change of entropy (TΔS), and change of binding free energy (ΔG= ΔH- TΔS). A set of 42 descriptors were included in the ΔH, TΔS, and ΔG (ΔH-TΔS) models: 34 derived from molecular mechanics calculations, various surface area terms, H-bond donors and acceptor count from VALIDATE, 1 to estimate the ligand partition coefficient (XlogP) and 7 shape- and volume-based descriptors from FPOCKET to better capture entropic contributions. The model has a predictive r^2 (r^2_{pred}) of 0.55 and a standard error (SE) of 1.34 kcal/mol. External validation resulted in a Pearson correlation coefficient (R_p) of 0.575 and a mean error (ME) of 1.41 pK_d.

This study (Tang & Marshall, 2011) suggested that using high-resolution crystal structures, separating enthalpy and entropy contributions, and including descriptors that can better capture the entropic contributions are effective strategies towards rapid and accurate calculation of binding affinity.

ID-Score

ID-Score (Li, Yang, Wang, Li, & Yang, 2013) was developed employing around 2278 protein-ligand complexes (training set), and a modified support vector regression (SVR) algorithm was employed to fit the experimental binding affinities. An extensive set of 50 descriptors related to protein–ligand interactions covering nine categories (vdW interaction, H-bonding interaction, electrostatic interaction, π-system interaction, metal–ligand bonding interaction, desolvation effect, entropic loss effect, shape matching, and surface property matching) was used. The quality of model was found to be better when compared with various other scoring functions in terms of Pearson's correlation coefficient (R_p =0.753), Spearman correlation coefficient (R_s= 0.779), and standard deviation (SD =1.63).

CONSENSUS SCORING FUNCTION

None of the scoring functions that have been developed is ideal for general applicability. Some researchers believe that the currently available scoring functions could be made more efficient when used in combination. A consensus scoring involves multiple scoring of a docked pose with different scoring functions or a combination of them (Liu & Wang, 2015; Yuriev, Agostino, & Ramsland, 2011). Some of the consensus-scoring functions reported in the literature are briefly discussed below:

XCScore

In this study, three empirical scoring functions were first developed to estimate the binding affinity of protein-ligand complexes. These scoring functions include terms accounting for vdW interaction, H-bonding, deformation penalty, and hydrophobic effect. Three different algorithms were implemented to calculate the hydrophobic effect term that resulted in three parallel scoring functions. All three scoring functions were then calibrated through multivariate regression analysis of a set of 200 protein-ligand complexes. At last, these three scoring functions were combined into a consensus scoring function (and termed as X-CSCORE) (Wang, Lai, & Wang, 2002), where X-CSCORE is the arithmetic average of all three functions. The X-CSCORE was tested on an independent set of 30 protein-ligand complexes and it was found that it predicts the binding free energies with R^2 = 0.59 and SD of 2.2 kcal/mol. The consensus model (X-CSCORE) was found to be better in terms of computed statistical parameters when compared to all three individual scoring functions.

Goldscore-CS and ChemScore-GS

In a study (Verdonk, Cole, Hartshorn, Murray, & Taylor, 2003), the ChemScore (CS) scoring function performance was compared to the original Goldscore (GS) function and two consensus docking protocols, namely, Goldscore-CS and ChemScore-GS, in terms of docking accuracy, prediction of binding affinities, and speed. Comparisons were studied on a set of 224 protein–ligand complexes. For larger ligands, Goldscore showed superior results, while docking with the ChemScore function was found to be up to three times faster than docking with the Goldscore function. However, combined docking protocols show significant improvements in terms of docking accuracy compared to the use of the Goldscore or ChemScore function alone. Goldscore- CS gives success rates of up to 81%, but is time consuming, while for high throughout virtual screening applications, ChemScore-GS appears optimal (Verdonk et al., 2003).

Recently, Cheng et al.(Cheng, Li, Li, Liu, & Wang, 2009) have combined ASP, piecewise linear potential (PLP1), DrugScore, GlideScore, LigScore, and ChemScore in a study carried out to assess various scoring functions. A set of 195 diverse high resolution crystal structures with reliable binding constants from PDBbind database was employed for evaluation. They reported that consensus scoring schemes have improved the success rate of pose prediction from 70% to 80% or even higher.

FUNDAMENTAL CHALLENGES IN SCORING FUNCTIONS

Although much development has been made in scoring functions in the past two decades, but there are still some components in scoring functions that are required to improve. For instance, sufficient success has been achieved for accurate prediction of pose, but the accuracy in prediction of absolute binding affinity is still lacking. However, it is also argued that absolute binding affinity is not as such required during screening of ligands, and relative binding affinity (also known as ranking score) is sufficient to select the best ligands for further optimization. Other important challenges includes entropy, solvation, scoring of metal coordination, ligands forming covalent complexes and macromolecules such as nucleic acids, that are also required to be considered for possible improvement. Many times, the problem is not that the solution for these challenges is not available, but the task is to find a solution which is not time consuming. For instance, to estimate the ligand-protein binding enthalpy based on quantum mechanical (QM) methods, or to calculate conformational entropies using normal-mode analysis (NMA) or quasi harmonic analysis (QHA) *etc.* are all fairly efficient method, but are time consuming. Thus, such time consuming methods are not useful for docking methodology, which is frequently utilized for high-throughput screening. Further, one way to speed up the process is to carefully introduce approximations while developing scoring functions. However, these approximations may inevitably introduce noise to the scoring functions, which results in reduced accuracy (Moitessier, Englebienne, Lee, Lawandi, Corbeil et al., 2008; Yuriev, Agostino, & Ramsland, 2011). Now, we are going to discuss here some challenges that are directly related to scoring functions.

Entropy

Entropic contributions form an important component of binding energy and are notoriously difficult to account for in docking applications. Computational estimation of the entropy is often based on time consuming approaches like NMA. Using such approaches would significantly reduce the high throughput applicability of docking methods.

Many scoring function (AutoDock, ChemScore, SCORE) as discussed before consider the number of rotatable bonds as an estimate of the entropic contribution to the binding free energy. However, all the bonds are not same and penalties should be assigned accordingly. For the same reason, ChemScore considers rotatable bonds in the context of their hydrophobicity; apolar portions are more flexible than the polar fragments in a binding site and hence are less penalized. But, entropic contribution may not be accurately predicted considering only rotatable bonds, which seems to be a crude method. Further, it is also reported by Kongsted and Ryde (Kongsted & Ryde, 2009) that the numbers of frozen rotatable bonds have no correlation to the conformational entropies calculated by NMA for biotin/avidin binding system.

Kongsted and Ryde (Kongsted & Ryde, 2009) have also attempted to improve the entropic calculations using MM/PBSA framework by introducing a buffer region of approximately 4Å outside the truncated protein (including solvent molecules), which is kept fixed during the minimization. This helps in reducing extensive changes in the molecular geometry and ensuring that the entropy term does not limit the precision of the MM/PBSA predictions. This method was tested on 17 complexes and found that it shows more stable results along with improved predictions of binding affinities.

Recently, Wang and Hou (Wang & Hou, 2012) have proposed a faster approach over NMA and QHA to calculate conformational entropy (based upon solvent accessible surface area calculations) that are needed to calculate the absolute binding free energies. In this approach, the conformational entropy of a molecule is estimated by summing up the contributions of all atoms, regardless of whether they are buried or exposed. Here, each atom has two types of surface areas, solvent accessible surface area (SAS) and buried SAS (BSAS). The two types of surface areas are weighted to estimate the contribution of an atom to conformational entropy and hence the approach is termed as weighted solvent accessible surface area (WSAS). In the reported study (Wang & Hou, 2012), WSAS has shown a comparable performance to NMA.

Solvation

Incorporation of solvent or water effects is important and there are some ways to treat solvent during calculations. One way is to treat the solvent in the same manner as the macromolecule, *i.e.*, considering individual explicit solvent molecules during simulations. However, explicit solvent can lead to high computational expense. A second way is an implicit solvent model that represents water as a continuum dielectric medium having the average properties of the real solvent, and it significantly reduces the computational expense of calculating the solute-solvent and solvent-solvent interactions. The most widely used implicit solvent models are PB/SA and the GB/SA.

Other ways applied to reduce computational time is to consider ligand lipophilicity as a measure of its desolvation, without taking into account the protein desolvation. However, larger ligand causes larger desolvation of the binding site and different poses for a single ligand can involve the desolvation of different residues. Further, solvation is also believed to be partly proportional to the change in solvent-exposed surface area. Based on this concept, DrugScore and ICM estimate the solvation contribution using a term proportional to this area (Moitessier et al., 2008).

Nucleic Acids

Generally, studying proteins have been the major goals of docking methods. However, nucleic acids such as DNA and RNA are also important targets in drug designing or pharmaceutical field. The major challenge is the differences in the binding site environment of nucleic acids compared to that of proteins. The binding sites of nucleic acids are highly charged and exposed to more solvent than the binding sites of proteins that are mostly deeply buried. Docking study of DNA intercalators (for *e.g.*,anti-cancer agents) is even more difficult, which involves a major opening of the bases that must occur to bind the ligand.

Additionally, flexibility of the nucleic acid strands and presence of key water molecules are also needed to consider during docking studies. For instance, Moitessier et al. (Moitessier, Westhof, & Hanessian, 2006) have developed a unique approach for docking aminoglycosides to RNA considering the flexibility of these macromolecules using conformational ensembles and accounting for the role of the first hydration shell. When this concept was implemented in AutoDock, and checked on 11 aminoglycosides, the results based on average RMSD values were found to be significantly better (average RMSD=1.41 Å) than docking in original AutoDock (average RMSD=3.25 Å). The reported docking studies have also been carried out using existing software on different nucleic acids sites such as the minor groove of DNA, a tetraloop structure of RNA, and the major groove of an RNA duplex, intercalation sites. For instance, Holt et al. (Holt, Chaires, & Trent, 2008) have reported molecular docking studies of intercalators and groove-binders to nucleic acids using AutoDock and Surflex docking. But still some dedicated software is required to perform docking studies on nucleic acids. Recently, some docking program have been developed that are dedicated to study ligand-nucleic acids or protein-nucleic acids interactions such as rDock (Ruiz-Carmona, Alvarez-Garcia, Foloppe, Garmendia-Doval, Juhos et al., 2014), NPDock (Tuszynska, Magnus, Jonak, Dawson, & Bujnicki, 2015), and HADDOCK (De Vries, van Dijk, & Bonvin, 2010), etc. showing the progress in this area.

Metal Interactions

Metalloenzymes are involved in various vital biological and chemical processes and they comprise of metal ions that are located in the active site, playing a crucial role in the catalytic process. The presence of metal ions in the active site of proteins has resulted in the use of coordination chemistry in docking study. Metal-ligand interactions became a challenge for docking methods concerning the parameterization and specific modeling of the metal-interaction geometry (like inter-atomic distances and angle deviations). Additionally, several issues, such as coordination geometry, atomic charge variability, and a potential proton transfer from ligands to a neighboring basic residue, are largely ignored.

In recent years, there have been considerable improvements in this area. For instance, Seebeck *et al.* (Seebeck, Reulecke, Kamper, & Rarey, 2008) developed a metal interaction model based on statistical analysis of metal interaction geometries that are relevant to protein–ligand complexes. A total of 43,061 metal sites of the PDB, containing metal ions such as magnesium, calcium, zinc, iron, manganese, copper, cadmium, cobalt, and nickel, were evaluated according to their metal coordination geometry. The resultant model was able to calculate and define the metal interaction geometries, which include the identification of the metal coordinating ligands, the calculation of the coordination geometry and the superposition of ideal polyhedra to identify the optimal positions for free coordination sites. This interaction model was integrated in the docking software FlexX and evaluated on a data set of 103 metalloprotein-ligand complexes. In 74% of the cases, the prediction of the coordination geometry was determined correct and in the majority of test cases, the interaction model improved the docking accuracy.

Recently, Pottel *et al.* (Pottel, Therrien, Gleason, & Moitessier, 2014) have developed specific functions and parameters to account for zinc–drug coordination in an attempt to discover novel HDAC inhibitors. These atom-type dependent, but atomic charge-independent, functions are implemented into FITTED 3.1. The developed specific functions and parameters significantly improved the pose prediction accuracy over the traditional electrostatic/Lennard-Jones treatment or the hydrogen bond-*like* treatment previously employed.

Covalent Bonds

Most of the docking techniques mainly focus on the non-covalent interactions, such as vdW interaction, the electrostatics interaction and H-bonding, because these interactions depict the mechanism of action for most of the mainstream drugs. But, there are another category of drugs that act via covalent bonding with the target protein. The existing computational tools do not account for covalent bonding event very well. Also, as many docking programs are designed for non-covalent docking, the preparation steps for covalent docking requires much effort, and hence it becomes even more difficult to perform high throughput screening due to the lack of automation in setting up the docking methodology (Moitessier et al., 2008; Ouyang, Zhou, Su, Ge, Li et al., 2013).

The scoring function of AutoDock 4(Morris, Huey, Lindstrom, Sanner, Belew et al., 2009) supports covalent docking feature, but have some limits in their functionality. AutoDock 4 employs a naive 'Gaussian well' on the score to reward the covalent bonding atoms. But the Gaussian well should be defined to a specified coordinate rather than on any target link atoms. The over-simplification of the scoring function and the lack of established guideline in parameter setting also compromise its performance (Ouyang et al., 2013).

In a recent study, Ouyang et al. (Ouyang et al., 2013) have developed a method named 'CovalentDock', to account for the energy contribution from covalent binding as a term compatible to the scoring function of non-covalent docking. This method is motivated from both AutoDock and GOLD that utilizes a 'dummy atom' (similar in purpose to the ''link atom'' in GOLD), while integrating the respective energy term into the scoring function of AutoDock. This method is also automated to perform high throughput screening via covalent docking.

In another study, Schröder et al. (Schroder, Klinger, Oellien, Marhofer, Duszenko et al., 2013) successfully performed docking-based virtual screening of covalently binding ligands using GOLD software utilizing modified GoldScore and ChemScore scoring functions.

CONCLUSION

The scoring functions have been evolving persistently during the past two decades. In this chapter, the authors have discussed various scoring functions that are employed in the docking experiments. Here, the authors have discussed the scoring functions in four main categories, *i.e.*, force field or physics-based methods, empirical scoring functions, knowledge-based potentials, and descriptor-based scoring functions. In each category, the authors have discussed a few scoring functions reported in the literature to get an idea about the composition of each function and to understand the difference between each category of scoring function. As of now, no scoring function is ideal and has more or less some limitations such as narrow applicability domain, or time consuming or less accuracy in binding affinity prediction or pose prediction *etc*. Consensus scoring functions as discussed, sometime provide better results, since they may show combined benefits of merged scoring functions. Also, tuned scoring functions are helpful to improve the results by modifying a particular scoring function according to the needs, for instance, to develop targeted scoring function. At last, the authors have also discussed the challenges faced by the scoring functions and their recent progress. This chapter will surely help readers to get familiar with the different types of scoring functions, their theoretical background, and essential components of each scoring function.

REFERENCES

Ballester, P. J., & Mitchell, J. B. O. (2010). A machine learning approach to predicting protein-ligand binding affinity with applications to molecular docking. *Bioinformatics (Oxford, England)*, *26*(9), 1169–1175. doi:10.1093/bioinformatics/btq112

Bohm, H. (1994). The development of a simple empirical scoring function to estimate the binding constant for a protein-ligand complex of known three-dimensional structure. *Journal of Computer-Aided Molecular Design*, *8*, 243–256. doi:10.1007/BF00126743

Bohm. (1998). Prediction of binding constants of protein ligands: a fast method for the prioritization of hits obtained from de novo design or 3D database search programs. *Journal of Computer-Aided Molecular Design*, *12*, 309-309.

Cheng, T., Li, X., Li, Y., Liu, Z., & Wang, R. (2009). Comparative assessment of scoring functions on a diverse test set. *Journal of Chemical Information and Modeling*, *49*(4), 1079–1093. doi:10.1021/ci9000053

De Vries, S. J., van Dijk, M., & Bonvin, A. M. J. J. (2010). The HADDOCK web server for data-driven biomolecular docking. *Nature Protocols*, *5*(5), 883–897. doi:10.1038/nprot.2010.32

DeWitte, R. S., & Shakhnovich, E. I. (1996). SMoG: De novo design method based on simple, fast, and accurate free energy estimates. 1. Methodology and supporting evidence. *Journal of the American Chemical Society*, *118*(47), 11733–11744. doi:10.1021/ja960751u

DiscoveryStudio. (2007). *Version 2.5, Software Package AI*. Retrieved from http://www.accelrys.com

Dobes, P., Fanfrlik, J., Rezac, J., Otyepka, M., & Hobza, P. (2011). Transferable scoring function based on semiempirical quantum mechanical PM6-DH2 method: CDK2 with 15 structurally diverse inhibitors. *Journal of Computer-Aided Molecular Design*, *25*(3), 223–235. doi:10.1007/s10822-011-9413-5

Eldridge, M. D., Murray, C. W., Auton, T. R., Paolini, G. V., & Mee, R. P. (1997). Empirical scoring functions: I. The development of a fast empirical scoring function to estimate the binding affinity of ligands in receptor complexes. *Journal of Computer-Aided Molecular Design*, *11*(5), 425–445. doi:10.1023/A:1007996124545

Friesner, R. A., Banks, J. L., Murphy, R. B., Halgren, T. A., Klicic, J. J., Mainz, D. T., & Perry, J. K. et al. (2004). Glide: A new approach for rapid, accurate docking and scoring. 1. Method and assessment of docking accuracy. *Journal of Medicinal Chemistry*, *47*(7), 1739–1749. doi:10.1021/jm0306430

Friesner, R. A., Murphy, R. B., Repasky, M. P., Frye, L. L., Greenwood, J. R., Halgren, T. A., & Mainz, D. T. et al. (2006). Extra precision glide: Docking and scoring incorporating a model of hydrophobic enclosure for protein-ligand complexes. *Journal of Medicinal Chemistry*, *49*(21), 6177–6196. doi:10.1021/jm051256o

Gohlke, H., Hendlich, M., & Klebe, G. (2000). Knowledge-based scoring function to predict protein-ligand interactions. *Journal of Molecular Biology*, *295*(2), 337–356. doi:10.1006/jmbi.1999.3371

Halgren, T. A. (1996). Merck molecular force field. I. Basis, form, scope, parameterization, and performance of MMFF94. *Journal of Computational Chemistry*, *17*(5-6), 490–519. doi:10.1002/(SICI)1096-987X(199604)17:5/6<490::AID-JCC1>3.0.CO;2-P

Hensen, C., Hermann, J. C., Nam, K., Ma, S., Gao, J., & Holtje, H.-D. (2004). A combined QM/MM approach to protein-ligand interactions: Polarization effects of the HIV-1 protease on selected high affinity inhibitors. *Journal of Medicinal Chemistry*, *47*(27), 6673–6680. doi:10.1021/jm0497343

Holt, P. A., Chaires, J. B., & Trent, J. O. (2008). Molecular docking of intercalators and groove-binders to nucleic acids using Autodock and Surflex. *Journal of Chemical Information and Modeling*, *48*(8), 1602–1615. doi:10.1021/ci800063v

Huang, N., Kalyanaraman, C., Bernacki, K., & Jacobson, M. P. (2006). Molecular mechanics methods for predicting protein-ligand binding. *Physical Chemistry Chemical Physics*, *8*(44), 5166–5177. doi:10.1039/B608269F

Huang, S.-Y., Grinter, S. Z., & Zou, X. (2010). Scoring functions and their evaluation methods for protein-ligand docking: Recent advances and future directions. *Physical Chemistry Chemical Physics*, *12*(40), 12899–12908. doi:10.1039/c0cp00151a

Huang, S.-Y., & Zou, X. (2014). A knowledge-based scoring function for protein-RNA interactions derived from a statistical mechanics-based iterative method. *Nucleic Acids Research*, *42*(7), e55–e55. doi:10.1093/nar/gku077

Huey, R., Morris, G. M., Olson, A. J., & Goodsell, D. S. (2007). A semiempirical free energy force field with charge-based desolvation. *Journal of Computational Chemistry*, *28*(6), 1145–1152. doi:10.1002/jcc.20634

Ishchenko, A. V., & Shakhnovich, E. I. (2002). Small molecule growth 2001 (SMoG2001): An improved knowledge-based scoring function for protein-ligand interactions. *Journal of Medicinal Chemistry*, *45*(13), 2770–2780. doi:10.1021/jm0105833

Kongsted, J., & Ryde, U. (2009). An improved method to predict the entropy term with the MM/PBSA approach. *Journal of Computer-Aided Molecular Design*, *23*(2), 63–71. doi:10.1007/s10822-008-9238-z

Kramer, C., & Gedeck, P. (2011). Global free energy scoring functions based on distance-dependent atom-type pair descriptors. *Journal of Chemical Information and Modeling*, *51*(3), 707–720. doi:10.1021/ci100473d

Krammer, A., Kirchhoff, P. D., Jiang, X., Venkatachalam, C. M., & Waldman, M. (2005). LigScore: A novel scoring function for predicting binding affinities. *Journal of Molecular Graphics & Modelling*, *23*(5), 395–407. doi:10.1016/j.jmgm.2004.11.007

Kroemer, R. T. (2007). Structure-based drug design: Docking and scoring. *Current Protein & Peptide Science*, *8*, 312–328. doi:10.2174/138920307781369382

Li, G.-B., Yang, L.-L., Wang, W.-J., Li, L.-L., & Yang, S.-Y. (2013). ID-Score: A new empirical scoring function based on a comprehensive set of descriptors related to protein-ligand interactions. *Journal of Chemical Information and Modeling*, *53*(3), 592–600. doi:10.1021/ci300493w

Liu, J., & Wang, R. (2015). On Classification of Current Scoring Functions. *Journal of Chemical Information and Modeling*, *55*(3), 475–482. doi:10.1021/ci500731a

Moitessier, N., Englebienne, P., Lee, D., Lawandi, J., & Corbeil, C. R. (2008). Towards the development of universal, fast and highly accurate docking/scoring methods: A long way to go. *British Journal of Pharmacology*, *153*(S1), S7–S26. doi:10.1038/sj.bjp.0707515

Moitessier, N., Westhof, E., & Hanessian, S. (2006). Docking of aminoglycosides to hydrated and flexible RNA. *Journal of Medicinal Chemistry*, *49*(3), 1023–1033. doi:10.1021/jm0508437

Momany, F. A., & Rone, R. (1992). Validation of the general purpose QUANTA® 3.2/CHARMm® force field. *Journal of Computational Chemistry*, *13*(7), 888–900. doi:10.1002/jcc.540130714

Mooij, W., & Verdonk, M. L. (2005). General and targeted statistical potentials for protein-ligand interactions. *Proteins: Structure, Function, and Bioinformatics*, *61*(2), 272–287. doi:10.1002/prot.20588

Morris, G. M., Goodsell, D. S., Halliday, R. S., Huey, R., Hart, W. E., Belew, R. K., & Olson, A. J. (1998). Automated docking using a Lamarckian genetic algorithm and an empirical binding free energy function. *Journal of Computational Chemistry*, *19*(14), 1639–1662. doi:10.1002/(SICI)1096-987X(19981115)19:14<1639::AID-JCC10>3.0.CO;2-B

Morris, G. M., Huey, R., Lindstrom, W., Sanner, M. F., Belew, R. K., Goodsell, D. S., & Olson, A. J. (2009). AutoDock4 and AutoDockTools4: Automated docking with selective receptor flexibility. *Journal of Computational Chemistry*, *30*(16), 2785–2791. doi:10.1002/jcc.21256

Muegge, I. (2006). PMF scoring revisited. *Journal of Medicinal Chemistry*, *49*(20), 5895–5902. doi:10.1021/jm050038s

Muegge, I., & Martin, Y. C. (1999). A general and fast scoring function for protein-ligand interactions: A simplified potential approach. *Journal of Medicinal Chemistry*, *42*(5), 791–804. doi:10.1021/jm980536j

Neudert, G., & Klebe, G. (2011). DSX: A knowledge-based scoring function for the assessment of protein-ligand complexes. *Journal of Chemical Information and Modeling*, *51*(10), 2731–2745. doi:10.1021/ci200274q

Ouyang, X., Zhou, S., Su, C. T. T., Ge, Z., Li, R., & Kwoh, C. K. (2013). CovalentDock: Automated covalent docking with parameterized covalent linkage energy estimation and molecular geometry constraints. *Journal of Computational Chemistry*, *34*(4), 326–336. doi:10.1002/jcc.23136

Pottel, J., Therrien, E., Gleason, J. L., & Moitessier, N. (2014). Docking ligands into flexible and solvated macromolecules. 6. Development and application to the docking of HDACs and other zinc metalloenzymes inhibitors. *Journal of Chemical Information and Modeling*, *54*(1), 254–265. doi:10.1021/ci400550m

Rappe, A. K., Casewit, C. J., Colwell, K. S., Goddard III, W. A., & Skiff, W. M. (1992). UFF, a full periodic table force field for molecular mechanics and molecular dynamics simulations. *Journal of the American Chemical Society*, *114*(25), 10024–10035. doi:10.1021/ja00051a040

Rarey, M., Kramer, B., Lengauer, T., & Klebe, G. (1996). A fast flexible docking method using an incremental construction algorithm. *Journal of Molecular Biology*, *261*(3), 470–489. doi:10.1006/jmbi.1996.0477

Rognan, D., Lauemoller, S. L., Holm, A., Buus, S., & Tschinke, V. (1999). Predicting binding affinities of protein ligands from three-dimensional models: Application to peptide binding to class I major histocompatibility proteins. *Journal of Medicinal Chemistry*, *42*(22), 4650–4658. doi:10.1021/jm9910775

Ruiz-Carmona, S., Alvarez-Garcia, D., Foloppe, N., Garmendia-Doval, A. B., Juhos, S., Schmidtke, P., & Morley, S. D. et al. (2014). rDock: A fast, versatile and open source program for docking ligands to proteins and nucleic acids. *PLoS Computational Biology*, *10*, e1003571.

Schroder, J., Klinger, A., Oellien, F., Marhofer, R. J., Duszenko, M., & Selzer, P. M. (2013). Docking-based virtual screening of covalently binding ligands: An orthogonal lead discovery approach. *Journal of Medicinal Chemistry*, *56*(4), 1478–1490. doi:10.1021/jm3013932

Seebeck, B., Reulecke, I., Kamper, A., & Rarey, M. (2008). Modeling of metal interaction geometries for protein-ligand docking. *Proteins: Structure, Function, and Bioinformatics*, *71*(3), 1237–1254. doi:10.1002/prot.21818

Sippl, M. J., Ortner, M., Jaritz, M., Lackner, P., & Flockner, H. (1996). Helmholtz free energies of atom pair interactions in proteins. *Folding & Design*, *1*(4), 289–298. doi:10.1016/S1359-0278(96)00042-9

Sotriffer, C. A., Sanschagrin, P., Matter, H., & Klebe, G. (2008). SFCscore: Scoring functions for affinity prediction of protein-ligand complexes. *Proteins: Structure, Function, and Bioinformatics*, *73*(2), 395–419. doi:10.1002/prot.22058

Tang, Y. T., & Marshall, G. R. (2011). PHOENIX: A scoring function for affinity prediction derived using high-resolution crystal structures and calorimetry measurements. *Journal of Chemical Information and Modeling*, *51*(2), 214–228. doi:10.1021/ci100257s

Trott, O., & Olson, A. J. (2010). AutoDock Vina: Improving the speed and accuracy of docking with a new scoring function, efficient optimization, and multithreading. *Journal of Computational Chemistry*, *31*, 455–461.

Tuszynska, I., Magnus, M., Jonak, K., Dawson, W., & Bujnicki, J. M. (2015). NPDock: A web server for protein-nucleic acid docking. *Nucleic Acids Research*, *43*(W1), W425–W430. doi:10.1093/nar/gkv493

Velec, H. F. G., Gohlke, H., & Klebe, G. (2005). DrugScoreCSD knowledge-based scoring function derived from small molecule crystal data with superior recognition rate of near-native ligand poses and better affinity prediction. *Journal of Medicinal Chemistry*, *48*(20), 6296–6303. doi:10.1021/jm050436v

Verdonk, M. L., Cole, J. C., Hartshorn, M. J., Murray, C. W., & Taylor, R. D. (2003). Improved protein-ligand docking using GOLD. *Proteins: Structure, Function, and Bioinformatics*, *52*(4), 609–623. doi:10.1002/prot.10465

Wang, J., & Hou, T. (2012). Develop and test a solvent accessible surface area-based model in conformational entropy calculations. *Journal of Chemical Information and Modeling*, *52*(5), 1199–1212. doi:10.1021/ci300064d

Wang, J., Wolf, R. M., Caldwell, J. W., Kollman, P. A., & Case, D. A. (2004). Development and testing of a general amber force field. *Journal of Computational Chemistry*, *25*(9), 1157–1174. doi:10.1002/jcc.20035

Wang, R., Lai, L., & Wang, S. (2002). Further development and validation of empirical scoring functions for structure-based binding affinity prediction. *Journal of Computer-Aided Molecular Design*, *16*(1), 11–26. doi:10.1023/A:1016357811882

Wang, R., Liu, L., Lai, L., & Tang, Y. (1998). SCORE: A new empirical method for estimating the binding affinity of a protein-ligand complex. *Molecular Modeling Annual*, *4*(12), 379–394. doi:10.1007/s008940050096

Yuriev, E., Agostino, M., & Ramsland, P. A. (2011). Challenges and advances in computational docking: 2009 in review. *Journal of Molecular Recognition*, *24*(2), 149–164. doi:10.1002/jmr.1077

Zilian, D., & Sotriffer, C. A. (2013). SFCscore RF: A random forest-based scoring function for improved affinity prediction of protein-ligand complexes. *Journal of Chemical Information and Modeling*, *53*(8), 1923–1933. doi:10.1021/ci400120b

KEY TERMS AND DEFINITIONS

Binding Mode: The orientation of the molecule (let say, ligand) relative to another molecule (let say, receptor) as well as the conformation of the ligand and receptor when bound to each other.

Binding Pose: A binding mode of a query/candidate molecule.

Descriptors: Descriptors are the set of numbers that encode the physical, chemical and/or biological properties of the molecules.

Docking: A computational method employed to predict the binding orientation or native pose of a molecule while interacting with another molecule to form a stable complex.

Enthalpy: The sum of the internal energy of the system plus the product of the pressure of the gas in the system and its volume.

Entropy: A measure of the amount of energy in a physical system which is not available to do work. It is also understood as a measure of disorder, since when a physical system becomes more disordered, then its energy becomes more evenly distributed and hence the energy becomes less available to do work.

Force Field: It refers to the functional form and parameter sets that are used to compute the potential energy of a system of atoms or coarse-grained particles in molecular mechanics and molecular dynamics simulations. The parameters of the energy functions are usually derived from quantum mechanical calculations and experimental work.

Scoring Function: In molecular docking studies, scoring function is a mathematical function that is used to calculate a representative score and further also to decide a binding pose, by evaluating the favorable and unfavorable inter-molecular interactions found in a docked complex.

Chapter 4
The Comparison of Docking Search Algorithms and Scoring Functions: An Overview and Case Studies

Marjana Novič
National Institute of Chemistry, Slovenia

Marko Anderluh
University of Ljubljana, Slovenia

Tjaša Tibaut
National Institute of Chemistry, Slovenia

Jure Borišek
National Institute of Chemistry, Slovenia

Tihomir Tomašič
University of Ljubljana, Slovenia

ABSTRACT

This chapter, composed of two parts, firstly provides molecular docking overview and secondly two molecular docking case studies are described. In overview, basic information about molecular docking are presented such as description of search algorithms and scoring functions applied in various docking programs. Brief description of methods utilized in some of the most popular docking programs also applied in our experimental work is provided. AutoDock, CDOCKER, GOLD, FlexX and FRED were used for docking studies of the DC-SIGN protein, while GOLD was further used for docking studies of cathepsin K protein. Methods and results of our studies with their contribution to science and medicine are presented. Content of the chapter is therefore appropriate for public of Medicinal and Organic Chemistry as an overview of docking studies, and also for Computational Chemists at the beginning of their work as the introduction to application of molecular docking programs.

INTRODUCTION

Those who devote their time to molecular science are well aware of the amount of molecular data available that has rapidly increased in the past few decades due to breakthrough in structural biology. Since 1975 development of high-throughput protein purification, crystallography, and nuclear magnetic resonance

DOI: 10.4018/978-1-5225-0115-2.ch004

spectroscopy techniques predominantly contributed to structural details of macromolecules and complexes with ligands. In 2014, after less than four decades, the number of structures exceeded the 100.000 with 10.000 structures per year being deposited in Protein Data Bank (PDB). PDB is the single worldwide repository of information about the 3D structures of large biological molecules, including proteins and nucleic acids (Berman et al., 2000; Berman, Kleywegt, Nakamura, & Markley, 2014).

These advances are responsible for computational methods to become indispensable in all aspects of drug discovery including the structure-based drug design and computer-based analysis of molecular interactions. Interactions between biomolecules are fundamental to biological processes; using these interactions, living organisms maintain complex regulatory and metabolic mechanisms responsible for the life course. Experimental research work is fundamental for a better understanding of biological processes; however, computer simulations and analysis are of mayor importance to extract information from large amount of multivariate experimental data. Application of these scientific methods provides exploration of molecules that can be utilized as bioactive substances able to modify and control molecular interactions. For bioactive substances, molecular recognition of the constituents is imperative. Binding mode, binding affinity, and kinetics play the crucial role in molecular recognition as they characterize intermolecular interaction.

Due to the limitations of experimental methods for determining the structure of molecular complexes, as they are expensive, time consuming and not always feasible, computational methods are preferable for predicting putative binding modes and binding affinities. Nowadays, the main advantages of rational drug discovery approaches, such as virtual screening, are low costs and effectiveness when compared to high-throughput screening. Molecular docking is one of the virtual screening methods used since 1980's. It is based on modeling of interactions between two molecules, a small molecule (ligand) and its receptor/protein (target). The predicted interactions allow us to characterize the binding mode and binding affinity of the ligand-target complex.

Currently there are more than 50 molecular docking programs available, which all have similar implementations. This chapter provides explanation and comparison of computational approaches, specifically different algorithms enclosed in software for molecular docking, considering several of the most commonly used program tools (AutoDock, FRED, CDOCKER, FlexX and GOLD). Authors also present the use of these algorithms in the comparative study for docking of small ligands to specific protein, lectin Dendritic Cell-Specific Intercellular adhesion molecule-3-Grabbing Non-integrin (DC-SIGN) and covalent docking of cathepsin K.

BACKGROUND

Molecular docking, in the case of small molecules also known as ligand docking, can be defined as computational method for prediction of the structure of ligand-receptor complexes. Ligand docking process involves two basic steps: prediction of ligand conformation and orientation (the pose) by search algorithms and assessment of binding affinity by scoring functions.

By search algorithms as mentioned above different ligand conformations and orientations are searched and therefore different modes (poses) are generated. Over time different search algorithms, all being defined by set of rules and parameters applied to predict conformations, were developed and became available for users. Algorithms differ regarding to constituent molecules flexibility consideration, therefore several approaches can be utilized that are classified as: methods for "rigid-body" docking and methods for "flexible docking". Depending on how much conformational space is covered, speed and effectiveness are two critical elements in search algorithms. Rigid-body docking methods use either a single conformation or multiple conformations implemented by user to consider ligand flexibility. Search algorithms here consider only geometrical complementarities between molecules and that is limiting factor concerning specificity and accuracy of prediction. Methods for flexible docking consider various conformations of ligand or receptor (or both molecules at the same time) generated by program itself, what makes prediction more effective at higher computational time cost. So far, molecular docking software generally uses one or more methods for binding mode prediction: matching algorithms, incremental construction, Monte Carlo simulations, genetic algorithms, etc.

Assessment of the different orientations and conformations of a ligand is the second step in molecular docking, where scoring functions delineate the correct poses from incorrect poses or binders from inactive compounds. It is necessary to mention that scoring function involve estimation adopting various assumptions and simplifications, rather than calculating accurate binding affinity between constituent molecules. Scoring functions can be applied for ligand binding mode identification, binding affinity prediction, and consequently virtual database screening. There are numerous available scoring functions that can be classified into three main categories: force-field based, empirical and knowledge-based. Force-field based scoring functions calculate the sum of intermolecular interaction and atomic forces for assessment of binding energy, whereas empirical scoring functions estimate binding energy of a complex by summing up the weighted energy terms. Knowledge-based scoring functions employ statistical analysis of molecular complexes crystal structures. Moreover there has been additional strategy introduced; consensus scoring combines several different scores from multiple scoring functions to assess a docking conformation/orientation.

There are lots of reviews about molecular docking and since the beginning of the new millennium practically every year new publications are available that report recent advances and challenges in the field, compare programs available and outline growing number of applications of molecular docking in drug discovery (Table 1).

SEARCH ALGORITHMS

Search algorithms for molecular docking originate from computer analysis studies of intermolecular interactions. Calculations of protein-protein interactions led to the development of first algorithms for computing molecular interactions starting from the structure of two components. These algorithms utilize approximate shapes of two molecules represented as rigid bodies. As these algorithms essentially consider the geometrical complementarities between molecules, they are termed shape matching algorithms. Although this approach does not take into account the conformational space of molecules it is still convenient for the applications in drug design and discovery, such as for initial stages of virtual screening studies and as a part of strategies of other more sophisticated docking algorithms.

Development of sophisticated docking algorithms practically coincided with the increasing power of computers, allowing the integration of more parameters into search algorithms. Considering more degrees of freedom a new methodology emerged. Flexible docking accounts for conformational space of molecules that is responsible for molecular dynamics essential for all biological systems. Since the molecular complex is composed of two molecules, it would be optimal to consider both of them as

Table 1. Molecular docking review publications outlined according to the year of publication

Year of Publication	Reference
1995	Jones & Willett, 1995
1995	Lybrand, 1995
1996	Gschwend, Good, & Kuntz, 1996
1996	Lengauer & Rarey, 1996
2002	Halperin, Ma, Wolfson, & Nussinov, 2002
2002	Taylor, Jewsbury, & Essex, 2002
2003	Brooijmans & Kuntz, 2003
2003	Kroemer, 2003
2004	Kitchen, Decornez, Furr, & Bajorath, 2004
2006	Sousa, Fernandes, & Ramos, 2006
2007	Kroemer, 2007
2008	Moitessier, Englebienne, Lee, Lawandi, & Corbeil, 2008
2010	Huang, Grinter, & Zou, 2010
2010	Youriev, Agostino, & Ramsland, 2010
2011	Meng, Zhang, Mezei, & Cui, 2011
2012	Mihasan, 2012
2013	Bello, Martinez-Archundia, & Correa-Basurto, 2013
2013	Sousa et al., 2013
2013	Youriev & Ramsland, 2013
2014	Grinter & Zou, 2014
2015	Chen, 2014
2015	Ferreira, dos Santos, Oliva, & Andricopulo, 2015
2015	Wang et al., 2015
2015	Yuriev, Holien, & Ramsland, 2015

flexible. Even though, due to the computational limitations, predominate algorithms considering only flexibility of the ligand, there are several examples of specialized algorithms and computational strategies that account for partial flexibility of the receptor.

Computational tools for molecular docking use different methods to deal with ligand or receptor flexibility. There are three categories of algorithms that consider ligand flexibility: systematic, random or stochastic and deterministic (Figure 1) and some of these are appropriate and applied also for exploring receptor flexibility. Difference between docking software depends on the nature of algorithm implementation. This subchapter describes basic search algorithm approaches that progressed from the time of first intermolecular interaction studies.

Search Algorithms that Consider Ligand and Receptor as Rigid Bodies

All primer algorithms calculating intermolecular interactions of molecular complexes considered molecules as rigid bodies. Rigid body, according to physics, is an idealization of a solid body whose deformation is neglected and its position determined by the position of its center of mass and by its al-

Figure 1. Search algorithms that consider receptor and ligand as rigid/flexible bodies

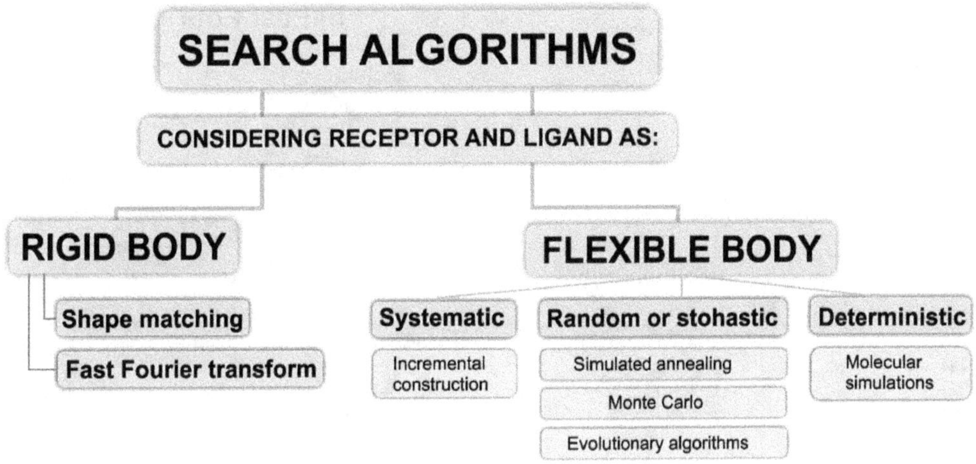

titude, taking into account 6 basic parameters referred as degrees of freedom. There are 3 rotational and 3 translational degrees of freedom that represent possible movements of rigid body in three-dimensional space (3D) (Figure 2). These algorithms consider geometry match between molecules and are based on Shape Matching (SM) and Fast Fourier Transform (FFT).

Shape Matching

First calculations and structural analysis of protein-protein interactions (Levinthal, Wodak, Kahn, & Dadivanian, 1975; Selemme, 1976; Wodak & Janin, 1978) led to development of shape matching algorithms applied to search for shape complementarities between two molecules. Kuntz et al. (1982) first

Figure 2. Demonstration of rigid body movement in 3D space considering 6 degrees of freedom (three translational and three rotational); molecule – ID 3 represents rigid body in 3D space
Borišek et al., 2015.

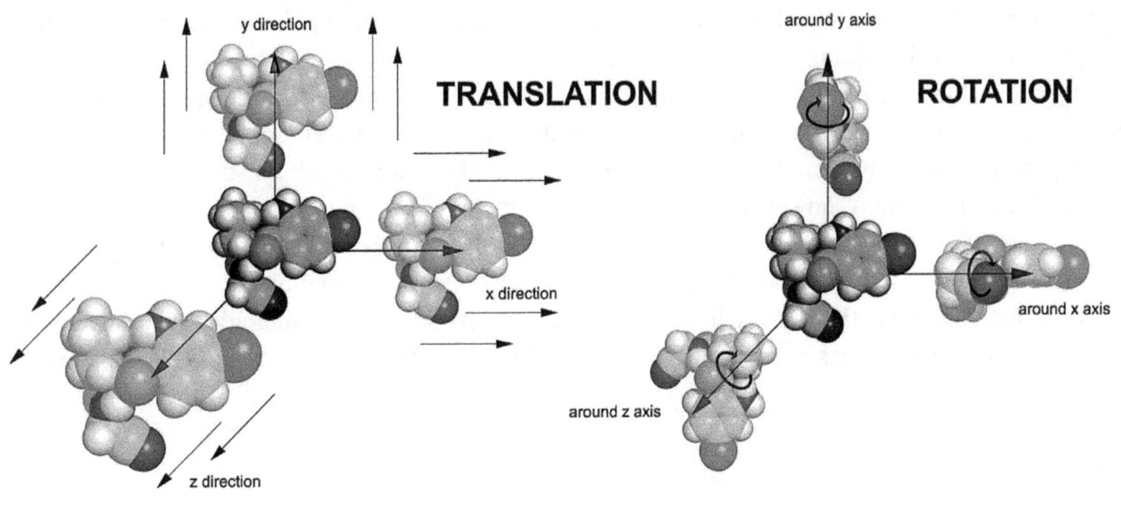

Figure 3. Schematic representation of shape matching algorithm and fast Fouirer transform

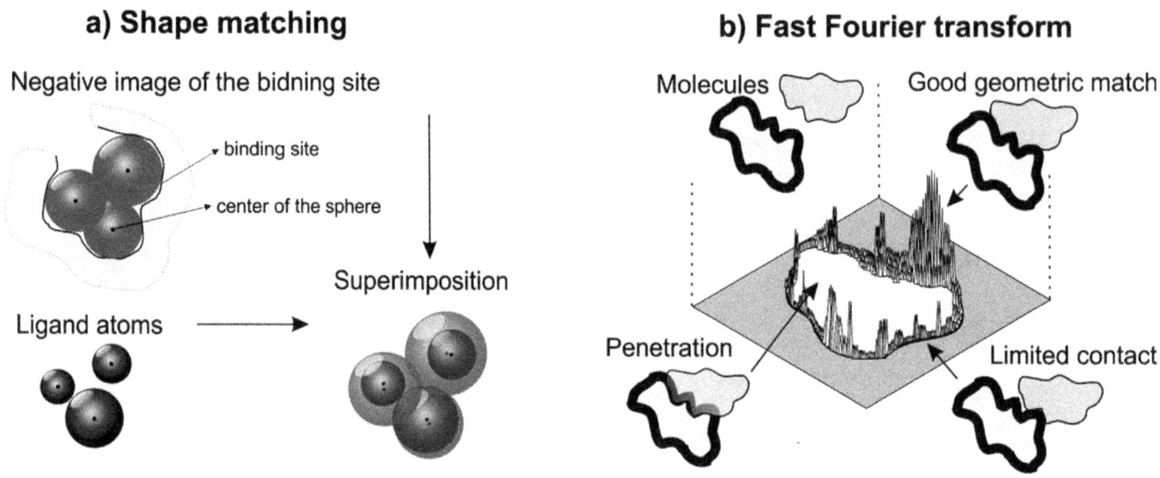

presented a method to explore geometrically feasible alignments of ligands and receptors of known structure, where negative images of the receptor were generated. Negative images are represented by spheres that fill the binding sites and algorithm attempts to superimpose the center of these spheres onto ligand atoms (Figure 3a). Several programs such as DOCK (Ewing, Makino, Skillman, & Kuntz, 2001; Allen et al., 2015) and MSDOCK (Sauton, Lagorce, Villoutreix, & Miteva 2008), utilize this approach as part of their search strategies of flexible docking algorithms.

Fast Fourier Transform

Ten years later, new geometric recognition approach was developed by Katchalski-Katzir et al. (1992). Algorithm involves atomic coordinates for the digital representation of molecule by 3D discrete functions that distinguish between surface and the interior of the molecule. Algorithm also calculates correlation function that assesses the degree of molecular surface overlap and penetration considering 6 degrees of freedom using Fourier transformation (Figure 3b). This approach, in comparison with shape matching algorithms, more rapidly evaluates possible associations between molecules. FTDOCK (Gabb, Jackson, & Sternberg, 1997) and ZDOCK (Chen & Weng, 2002, 2003) are examples of docking programs that apply Fourier transformation.

Algorithms that consider ligand and receptor as rigid bodies were initially developed for intermolecular interaction analysis. Latter these methods were optimized for searching binding sites and appropriate ligand pose prediction in ligand-receptor complex where 3D structures of constituent molecules were known. These algorithms, however, are not successful if there is conformational change upon binding of the ligand. Currently, practically all molecular docking software utilizes different approaches for considering conformational space of ligand (and in some cases receptor) molecule. Therefore, it can be suggested that nowadays there are no limitations concerning rigid body consideration in molecular docking since several approaches to cope with molecular flexibility are available (details discussed below).

Search Algorithms that Consider Ligand and Receptor as Flexible Bodies

Flexible bodies with consideration of rotational, translational and conformational degrees of freedom are somehow more realistic imitation of biological molecules. Therefore, molecular docking with flexible ligand and flexible receptor is rather accurate approximation of *in vivo* complexes formation, especially when conformational changes upon formation are present. At first flexible-ligand docking algorithms emerged, which considered partial or full flexibility of just ligand and afterwards, due to importance of 'induced fit' paradigm, some specialized algorithms were developed considering also partial flexibility of the receptor/protein. Depending on which molecule of the complex (ligand or receptor) is considered as flexible, the molecular docking is categorized in flexible-ligand docking and flexible-receptor/protein docking methods. Former consider only ligand (partial or full) flexibility, meanwhile latter consider ligand and partial receptor/protein flexibility.

This section of chapter provides description of search algorithms available that consider ligand and/or receptor as flexible bodies and their advantages and limitations. There are three categories of flexible-ligand docking methods (Brooijmans & Kuntz, 2003; Bras, Cerqueira, Sousa, Fernandes, & Ramos, 2014): systematic, random or stochastic, and deterministic.

Systematic Methods

Example of systematic search algorithm is incremental construction or "anchor and grow procedure" approach, (DesJarlais, Sheridan, Dixon, Kuntz, & Venkataraghavan, 1986; Leach & Kuntz, 1992) where ligand is split into several fragments that are separately docked in the receptor site and covalently linked to reassemble the ligand. This method is one of the most applied and employed by several docking programs such as DOCK (Ewing et al. 2001; Allen et al., 2015), FLOG (Miller, Kearsley, Underwood, & Sheridan, 1994), LUDI (Böhm, 1992), etc. FlexX (Rarey, Kramer, Lengauer, & Klebe, 1996; Rarey, Wefing, & Lengauer, 1996) uses this strategy, where chemical interactions are basis for placing the fragments. Therefore the generation of a large set of geometrically feasible solutions is avoided. For placing the base fragment into active site of the protein, FlexX uses pose clustering, where algorithm merges initial placements into clusters. Another systematic search algorithm is database search approach, where libraries of multiple conformations of each ligand structure are constructed. FLOG is the program that constructs Flexibases (Kearsley, Underwood, Sheridan, & Miller, 1994) for ligand flexibility consideration. OMEGA (Hawkins & Nicholls, 2012) is program that generates multi-conformer structure databases used by docking program FRED (McGann, 2011).

Random or Stochastic Methods

There are several types of docking programs that apply random approach: Monte Carlo, simulated annealing, evolutionary algorithms, particle swarm optimization, etc.

Monte Carlo (Caflisch, Niederer, & Anliker, 1992; Hart & Read, 1992) is method that searches for global energy minimum of the conformers. It generates random changes in the ligand conformation, thermodynamically explores different states and selects energetically favorable states to create more reliable protein-ligand complexes. Monte Carlo is standard minimization method and tends to be trapped in local minimum, therefore simulated annealing has been applied to avoid this issue. Simulated annealing (Kirkpatrick, Gelatt, & Vecchi, 1983; Goodsell & Olson, 1990; Yue, 1990) is a technique, where changes in temperature are involved during the run for every conformation and therefore more detailed conformational state (flexibility) at different thermodynamic states is considered. MCDOCK (Liu & Wang, 1999) is a program that utilizes Monte Carlo method for global minimum searching, while AutoDock (Morris et al., 2009) uses simulated annealing approach for more accurate conformations exploration.

Evolutionary algorithms (Morris, Olson, & Goodshell, 2000; Thomsen, 2003), same as Monte Carlo methods, generate and optimize low-energy conformations and orientations of the ligand and thus search for global energy minimum. They have the ability to simultaneously explore the search space with a population of candidate solutions represented by artificial "genes" and form new solutions by combining existing ones. Evolutionary algorithms encompasses a range of algorithms that are based on the principles of biological evolution and the most popular variant of these algorithms are genetic algorithms (GA) (Jones, Willet, & Gleen, 1995; Jones, Willet, Glen, Leach, & Taylor, 1997; Oshiro, Kuntz, & Dixon, 1995) implemented in docking program GOLD (Verdonk, Cole, Hartshorn, Murray, & Taylor, 2003).

Genetic algorithms mimic the process of evolution by manipulating a collection of data-structures called chromosomes. Chromosomes represent randomly generated population that are different ligand poses, where each ligand pose is characterized by a set of state variables that describe translation, orientation and conformation of the ligand in relation to the protein. Each state variable corresponds to a gene, ligand's state corresponds to the genotype and atomic coordinates of ligand correspond to the phenotype. Each of the chromosomes is evaluated with appropriate fitness score, where fitness represents the total interaction energy of the ligand and the protein. Genetic operators like mutations and crossovers, are applied to the population for offspring generation or, in the case of molecular docking, pose space sampling. Selection of the offspring of the current generation occurs based on the fitness of the individual, where the best chromosomes are saved, chromosomes with good scores are selected for propagation into new generation, whereas chromosomes with poor scores perish. The survival of the best chromosomes enables evolution of optimal solutions that represents the correct binding mode. AutoDock uses a hybrid of a genetic algorithm search method together with local search method, named Lamarckian genetic algorithm (LGA) (Morris et al., 1998), which has enhanced performance relative to genetic algorithm alone. This is due to the concept of local search during optimization, which can help to escape local minimum. Although this method is superior to standard genetic algorithm there are still several million evaluation steps needed for the prediction of a reliable protein-ligand pose.

Another stochastic method is particle swarm optimization (Eberhart & Kennedy, 1995), a bio-inspired evolutionary approach to model swarm intelligence. It simulates the social behavior among individuals (particles), with a population (swarm) moving through a multidimensional search space. Each particle of the swarm represents a possible solution described by its location in the search space. As particle between each other share their knowledge about their own best position, the swarm moves towards the best overall solution. In the case of molecular docking swarm represents population of the ligands and particles represent an individual pose, which are determined by translation, orientation and conformation. Each ligand pose moves through the search space and if one of them recognizes a region with low energy the rest of the particles will follow quickly and the best solution is memorized. This method is implemented in AutoDock program (PSO@AutoDock) (Namasivayam & Günther, 2007) as it outperforms default LGA and standard GA in terms of speed, robustness and accuracy of the predicted protein-ligand complex.

Deterministic Methods

Third category that considers ligand as flexible body includes deterministic methods, such as molecular simulations. Simulations provide fine detail that concerns the motions of individual particles as a function of time. Molecular simulations for example numerically solve Newton's equation of motion and therefor allow observation of structural fluctuations with respect to time. Therefore simulations can quantify the properties of a system at a precision and on a time scale that is otherwise inaccessible. Molecular dynamics, energy minimization and also Monte Carlo (Caflisch et al., 1992; Hart & Read, 1992) are methodologies that can be used to investigate and study the dynamics of the system with known atom coordinates (Adock & McCammon, 2006). AMBER (Case et al., 2005; Salomon-Ferrer, Case, & Walker, 2013) and CHARM (Brooks et al., 1983) are example of programs using molecular simulations for different poses generation. Molecular simulations computations are very demanding and require long computation time, which is the reason this method is not popular for virtual screening of large compound libraries. With more powerful computers, there are more and more docking programs that incorporate this method, for example CDOCKER (Wu, Robertson, Brooks, & Vieth, 2003).

SCORING FUNCTIONS

After the first step in molecular docking, where different ligands and ligand poses are determined or generated by various search algorithms, the assessment with scoring function follows, where correct poses from incorrect poses are delineated. The main goal of scoring function is the calculation of the energy that estimates the binding affinity between the protein and the ligand, which can be expressed by the binding free energy (ΔG_{bind}):

$$\Delta G_{bind} = G_{complex} - G_{protein} - G_{ligand} = \Delta H - T\Delta S \quad (1)$$

where ΔG_{ligand} is the free energy of the ligand, $\Delta G_{protein}$ is the free energy of the protein and $\Delta G_{complex}$ is the free energy of the complex formed by the constituent molecules. Binding free energy, also known as Gibbs free energy can be measured by determining the equilibrium binding constant K_{eq}. ΔG_{bind} is related to the K_{eq} by the following equation:

$$\Delta G_{bind} = -RT\ln\left(K_{eq}\right) \quad (2)$$

where R is the perfect gas constant and T is the temperature.

There are numerous methods that are capable to computationally predict the binding free energy of a molecular complex. They significantly differ in accuracy and speed. Scoring function has to follow several criteria for being appropriate for molecular docking applications. Despite being sufficiently fast, it must be capable to estimate interaction between molecules that is proportional to the binding free energy, correctly rank the poses, discriminate binders from nonbinders or active compound from inactive compounds, all in comparison with experimentally determined 3D structures of complexes. Free energy perturbation (Zwanzig, 1954) and thermodynamic integration (Straatsma & McCammon, 1991)

Figure 4. Scoring functions applied for assessment in molecular docking

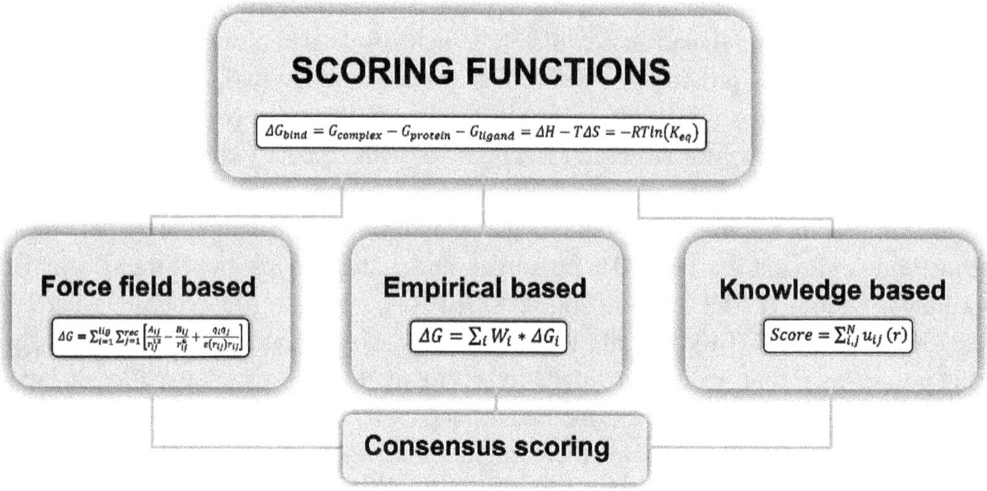

are methods, which enable very accurate calculation of binding free energies with very high calculation time consumption and that is the reason these are not applied for molecular docking. Other less accurate and therefore less time consuming methods rely on several assumptions and simplifications for binding free energy estimation. These scoring functions are generally classified into four main categories: force-field based (first principle based approach), empirical, knowledge-based and consensus scoring (Gohlke & Klebe, 2001; Brooijmans & Kuntz, 2003; Huang et al., 2010, Meng et al., 2011; Bras et al, 2014) (Figure 4). Below, brief description of every category is provided.

Force Field Based Scoring Functions

Force field based scoring functions basically take into account the non-bonded terms of the molecular mechanics force field, which means they estimate the interaction energy between the protein and the ligand. Non-bonded terms in this case represent electrostatic and van der Waals interactions, where electrostatic term is estimated by a Columbic formulation and van der Waals term by a Lennard-Jones potential function (Mehler & Solmajer, 1991; Meng, Shoichet, & Kuntz, 1992):

$$\Delta G = \sum_{i=1}^{lig}\sum_{j=1}^{rec}\left[\frac{A_{ij}}{r_{ij}^{12}} - \frac{B_{ij}}{r_{ij}^{6}} + \frac{q_i q_j}{\mu\left(r_{ij}\right) r_{ij}}\right] \tag{3}$$

where r_{ij} is the distance between protein atom i and ligand atom j, A_{ij} and B_{ij} are Lenard-Jones parameters, q_i and q_j are atomic charges and ε is distance dependent dielectric constant in the Coulombic term. DOCK (Ewing et al., 2001; Allen et al., 2015) and AutoDock (Huey, Morris, Olson, & Goodsell, 2007) are based on force field scoring functions.

Besides non-bonded forces there are forces present in protein-ligand binding, such as ionic bonds and hydrogen bonds and also conformational changes after ligand binding, which have to be considered for free energy estimation. Empirical force field scoring function used in AutoDock 4.0 for example (Huey et al., 2007) applies thermodynamic model of the binding process which includes intramolecular terms for the free energy estimation and also a full desolvation model, that includes terms for all atom types (favorable energetics of desolvating carbon atoms and unfavorable energetics of desolvating polar and charged atoms). This semiempirical approach estimates the energetics of the complex formations using pair-wise terms to evaluate interaction between molecules and an empirical method to estimate the contribution of surrounding water. In this aspect semiempirical approach differs from traditional molecular mechanics force field, which uses explicit water molecules for evaluation of solvation contributions and does not include entropic terms that are important for the interaction energy assessment. In some cases implicit solvation models are used for the estimation of desolvation energies of the ligand and the receptor. Poison-Boltzman-surface (Grant, Pickup, & Nicholls, 2001) and generalized Born-surface area (Still, Tempczyk, Hawley, & Hendrickson, 1990; Hawkins, Cramer, & Truhlar, 1995) are examples of models that treat water as continuum dielectric medium.

Empirical Based Scoring Functions

Approximation of binding free energy with empirical based scoring functions is composed of a sum of various energetic terms (Huang et al., 2010; Bras et al., 2014):

$$\Delta G = \sum_i W_i * \Delta G_i \tag{4}$$

where ΔG_i represents different energy terms such as van der Waals energy, electrostatics, desolvation, hydrogen bond, entropy, etc. The coefficients W_i that correspond to specific energy term are derived from regression analysis of known (protein-ligand) complex structures binding affinities.

Böhm (1994) developed first empirical scoring function, implemented in docking program LUDI, which takes into account hydrogen bonds, ionic interactions, lipophilic protein-ligand contact surface and the number of rotatable bonds in the ligand. Afterwards various docking programs implemented similar scoring functions. ChemScore (Eldridge, Murray, Auton, Paolini, & Mee, 1997) for example is parametrized form 82 complexes and is implemented in program GOLD (Cole, Nissink, & Taylor, 2005; Liebeschuetz, Cole, & Korb, 2012) and is also one of the functions available in FRED (McGann, 2011). In comparison with Böhm's scoring function ionic interactions term is absent in ChemScore, whereas metal-ligand binding contributions are taken into account. FlexX (Rarey et al., 1996) docking program also uses similar scoring function with additional aromatic group interactions term consideration. Various docking programs, despite applying similar energy terms typical for empirical scoring functions, implement different computations of specific term.

Empirical scoring functions are generally simpler than force filed based functions, therefor binding score calculations of former are faster. As these functions are dependent on experimental data set used for parameterization it is not necessary that the binding affinity will be predicted correctly for structurally different ligands from those exploited in the training set. With the increasing number of protein-ligand complexes with known 3D structures, functions can be trained with higher number of known binding affinities, which leads into development of scoring functions appropriate for various ligands.

Knowledge Based Scoring Functions

Knowledge based scoring functions are statistical methods, where the parameters of the potential functions are extracted from the structural information of experimentally determined atomic structures. This approach implicitly incorporates physical interactions (electrostatic, van der Walls, cation-π interactions), where high frequency of occurrence/non-occurrence of specific atom-atom interactions is assumed to represent favorable/unfavorable contact (Brooijmans & Kuntz, 2003; Li & Liang, 2006; Huang et al., 2010; Bras et al., 2014). Potential function is the weighted linear sum of pairwise contacts, presented in following formula:

$$Score = \sum_{i,j}^{N} u_{ij}\left(r\right) \tag{5}$$

where i and j represent protein and ligand atom type, r is the distance between atom pair, N is the number of all possible atom pairs in the system and u_{ij} corresponds to the pairwise potentials between atom types, obtained from the occurrence frequency of atom pairs in a database using inverse Boltzman relation:

$$u\left(r\right) = -k_B T\, ln\left(\frac{\acute{A}_{ij}\left(r\right)}{\acute{A}_{ref}\left(r\right)}\right) \tag{6}$$

where k_B is the Boltzmann constant, T is the absolute temperature of the system, $\rho_{ij}(r)$ is the normalized probability of protein-ligand atom pair to be in contact at distance r, and $\rho_{ref}(r)$ is the reference probability - the state where atomic interactions between protein and ligand are zero.

First ideas of statistical potentials originate from the field of protein folding and structure prediction (Huang & Zou, 2006). Tanaka and Scheraga (1976) were first that discussed about the statistical mechanical basis for obtaining the parameters for interactions between amino acids in proteins for prediction of the 3D structure. Since then, knowledge based approaches have been extensively used to develop the scoring functions for protein-ligand interactions and nowadays such scoring functions are implemented in various docking programs, e.g. DrugScore^{CSD} (Velec, Gohlke, & Klebe, 2005), AutoDock Vina (Trott & Olson, 2009) and also GOLD, where knowledge based Astex Statistical Potential (Mooij & Verdonk, 2005) is used as one of the scoring functions for docking. Another interesting method for scoring protein-ligand interaction is knowledge based quantitative structure-activity relationship (QSAR) (Ballester & Mitchell, 2010), where distance-dependent atom pair occurrences are utilized as descriptors for QSAR. Here is machine learning used instead of reverse Boltzman relationship and advantages of this methods are application of a very large training set of binding affinities besides structural data (traditional knowledge based methods use only structural data).

Knowledge based scoring functions, in comparison with force field based and empirical based, offer good balance between accuracy and speed, since the computation is relatively simple and the potentials are extracted from the known structures of protein-ligand complexes. These functions are also able to implicitly capture the binding effects that are difficult to model explicitly (e.g. sulfur-aromatic and cation-π interactions). The disadvantage of these functions is the accuracy of the reference state, where different methods were developed with attention to solve this problem, for example the adjustments of the pair potentials with iterative method (Huang & Zou, 2006).

Consensus Scoring

Consensus scoring combines results from various scoring functions to improve the performance of scoring in molecular docking. Charifson et al. (1999) first presented the results, where consensus scoring provided reduction in the number of false positives identified by individual scoring functions, leading to a significant enhancement in hit rates. Since then, several studies were published where consensus scoring that indicate improved quality of the results, was applied for evaluating ligand-protein interactions for molecular docking and ligand based virtual screening (Stahl & Rarey, 2001; Clark, Strizhev, Leonard, Blake, & Matthew, 2002; Baber, Shirley, Gao, & Feher, 2005). Improvements depend on the applications, the consensus method and the conditions of the computational tests. However, it should be mentioned that blind combination of arbitrarily chosen scoring functions will not necessarily lead to better results (Feher, 2006). For this purpose Oda et al. (2006) compared the performance of consensus scoring strategies, where multiple scoring functions were used simultaneously and demonstrated that choosing the most appropriate type of consensus score is essential for model selection in computational docking.

CASE STUDY: COMPARATIVE EVALUATION OF SEVERAL DOCKING TOOLS FOR DOCKING SMALL MOLECULE LIGANDS TO DC-SIGN

DC-SIGN: How to Dock Non-Drug-Like Saccharide Ligands into a Non-Druggable Target?

DC-SIGN or Dendritic Cell-Specific Intercellular adhesion molecule-3-Grabbing Non-integrin is a C-type lectin located specifically on dendritic cells (DCs). Several pathogens like HIV-1 enter DCs using DC-SIGN as an entry receptor, but escape the normal lytic pathway thus avoiding the host immune system (Anderluh, Jug, Švajger, & Obermajer, 2012). Potential anti-infectives that would inhibit DC-SIGN-pathogen interaction as the very first step in pathogen infection would act via completely new mechanism of action as compared to the current mode of anti-infectives' action. Therefore, the main scope of the research was the design of new DC-SIGN carbohydrate recognition domain (CRD) antagonists that could act as novel anti-infectives (Anderluh et al., 2012; Obermajer et al., 2011; Tomašič et al., 2014). Authors have envisaged docking studies as a great help in the design of new DC-SIGN antagonists to virtually preselect potential antagonists to be synthesized and characterized thus avoiding a vast effort in synthetic and biochemical laboratory (Jug, Anderluh, & Tomašič, 2015; Obermajer et al., 2011; Tomašič et al., 2014). Unfortunately, DC-SIGN proved to be a hard target for docking studies for several reasons and therefore authors decided to seek a docking tool that would address the below listed issues to a highest possible extent (Jug et al., 2015).

- As a lectin, DC-SIGN specifically binds glycosylated proteins, oligosaccharides and monosaccharides. The main determinant of its binding site is the hexacoordinated Ca^{2+} ion that forms two complex bonds with water molecules in the unbound state (Figure 5). Similar interactions are formed with the hydroxyl groups of the bound monosaccharide at positions 3 and 4 alone (i.e. D-mannose or L-fucose) or as a part of glycoconjugate with aglycone part attached to increase binding affinity. Accordingly, a key question was whether the docking algorithm can predict the coordination bonds between the core mannose and DC-SIGN and whether the scoring functions are able to evaluate the coordination bonds accurately.

- Desolvation is expected to be of paramount importance in docking studies of highly solvated ligands like (oligo)saccharides (Kroemer, 2007). DC-SIGN binds oligosaccharides and non-drug-like glycoconjugates that undergo massive desolvation of saccharide moieties upon binding, so different docking algorithms were tested to correctly evaluate their binding, even if desolvation effects are not considered explicitly or ignored.
- Docking of small molecules to a flat/shallow binding site is likely to result in a number of different (either correct or wrong) docked poses (Gowthaman, Deeds, & Karanicolas, 2013). Accordingly, the values of assigned binding energy could be of similarly diverse values because scoring functions are not able to accurately estimate all of the possible interactions and desolvation of the target protein and ligand poses. DC-SIGN CRD binding site is rather flat and shallow, and hence non-druggable, so there is a high probability of docking protocol failing to correctly predict the binding pose of a ligand.

Selected Docking Programs and Crystal Structures

Five docking tools that differ in the search algorithms and scoring functions have been validated in the study: FlexX (Rarey et al., 1996; Rarey et al., 1996) as available in LeadIT 2.1.2 (BioSolveIT GmbH), CDOCKER (Wu et al., 2003) as available in Accelrys Discovery Studio 3.0 (Accelrys Inc.), FRED 2.2.5 (McGann, 2011) (OpenEye Scientific Software, Inc.), AutoDock 4.2 (Morris et al., 2009) and GOLD 5.2 (Verdonk et al., 2003) (The Cambridge Crystallographic Data Centre). GOLD and AutoDock use a genetic algorithm (Kitchen et al., 2004), FRED uses a docking algorithm based on Gaussian function (McGann, 2011), CDOCKER uses a molecular dynamics simulated annealing-based docking algorithm (CHARMm) (Wu et al., 2003) and FlexX uses an incremental construction algorithm (Rarey et al., 1996). Re-scoring was performed with HYDE (Schröder et al., 2013) (BioSolveIT).

Figure 5. Binding site of DC-SIGN with bound ligand(s) from several crystal structures of DC-SIGN CRD; Ca^{2+} ion and flexible side chain of Phe313 are explicitly named. The protein is presented as a surface colored according to the electrostatic potential and the Ca^{2+} ion is presented as a yellow sphere. Ligands shown on this figure were taken from the above stated crystal structures and are presented as sticks in following color: ligand from 2IT6 (dimannoside Manα1-2Man) is presented in green, ligand from 2XR5 (pseudo-dimannoside mimic) is presented in yellow, and ligand from 2XR6 (pseudo-trimannoside mimic) is presented in violet (Jug et al., 2015)

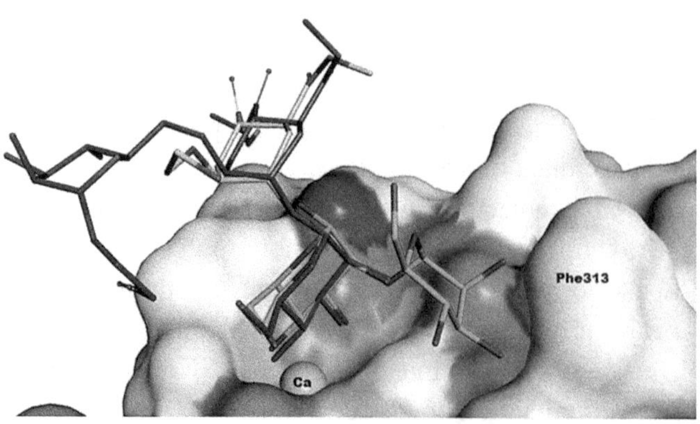

Re-docking of ligands from selected crystal structures has been performed to allow high ligand diversity. The 3D structures of all ligands used for docking validation were extracted from selected crystal structures of D-mannose containing ligands in complex with DC-SIGN CRD (Figure 5) (PDB codes: 2IT6, 2XR5 and 2XR6). Description of specific ligand preparation and docking protocols is beyond the scope of this chapter and may be found in the original publication (Jug et al., 2015).

RMSD (root-mean-square deviation) was selected as the key parameter for evaluation of docking success (Jain, 2008). Ligand poses from the above mentioned crystal structures were used as a reference for calculation of RMSD values. The upper RMSD limit which still fulfils criteria of acceptance was set to 2.00 Å. A visual comparison of crystal structure ligand pose and the docked poses was done after every successfully completed docking run. Afterwards, RMSD values for the top ten ranked poses were calculated and compared.

Results

The aim was to set up the docking protocols within all numbered programs in such a way that the top pose of docking will also have an RMSD less than 2.00 Å or, in the best possible situation, that the top pose will also have the lowest RMSD value. The calculated RMSD values for the top ranked ligand pose for all three crystal structures and all stated docking programs are listed in Table 2.

FRED gave the lowest RMSD values for the top ranked poses of all three ligands that were re-docked within the RMSD cut-off value of 2.00 Å (Figure 6). Furthermore, FRED performed the best among chosen docking tools when considering overall RMSD values - taking into account RMSD values of the top ten ranked poses for all three ligands. Promising results were also achieved using CDOCKER, which did not perform well only for the most complex ligand (2XR6). FlexX, which was used in previous study (Obermajer et al., 2011) showed exceptional performance with the first ligand (2IT6), but intriguingly failed later with the other two ligands. However, when FlexX docking poses were re-scored by HYDE, significant improvements of RMSD values were observed in these cases. GOLD and AutoDock performed exceptionally well in docking studies of targets with constrained binding pockets, but failed to perform satisfactorily even at re-docking of the least complex disaccharide ligand (2IT6). Authors presume that the major reason for different performance of the selected tools is the difference between their search algorithms. Majority of them were designed for non-carbohydrate ligands and druggable binding sites, where a small difference in the ligand conformation might result in a substantial difference in the score due to the steric clash and/or repulsive electrostatic interactions, but when faced with the large, flat and

Table 2. RMSD values obtained for re-docked ligands of different DC-SIGN CRD-ligand complexes (Jug et al., 2015)

	FRED	AutoDock	FlexX	HYDE	GOLD	CDOCKER
RMSD (Å)	DC-SIGN 2IT6					
	0.81 (Figure 6a)	3.50	0.61	0.79	1.68	0.97
	DC-SIGN 2XR5					
	1.32 (Figure 6b)	4.21	7.26	2.01	4.31	1.22
	DC-SIGN 2XR6					
	1.68 (Figure 6c)	5.48	7.44	6.49	6.82	2.34

Figure 6. Re-docked a) dimannoside (PDB entry: 2IT6, RMSD = 0.81 Å), b) pseudo-1,2-mannobioside (PDB entry: 2XR5, RMSD = 1.32 Å) and c) pseudo-trimannoside mimic (PDB entry: 2XR6, RMSD = 1.68 Å) using FRED docking protocol to DC-SIGN CRD (left). Zoomed view of the docking and X-ray determined pose of the core mannose residue. The position of the ligand/core mannose from X-ray determination is coloured green and the docked structure is coloured grey (right)

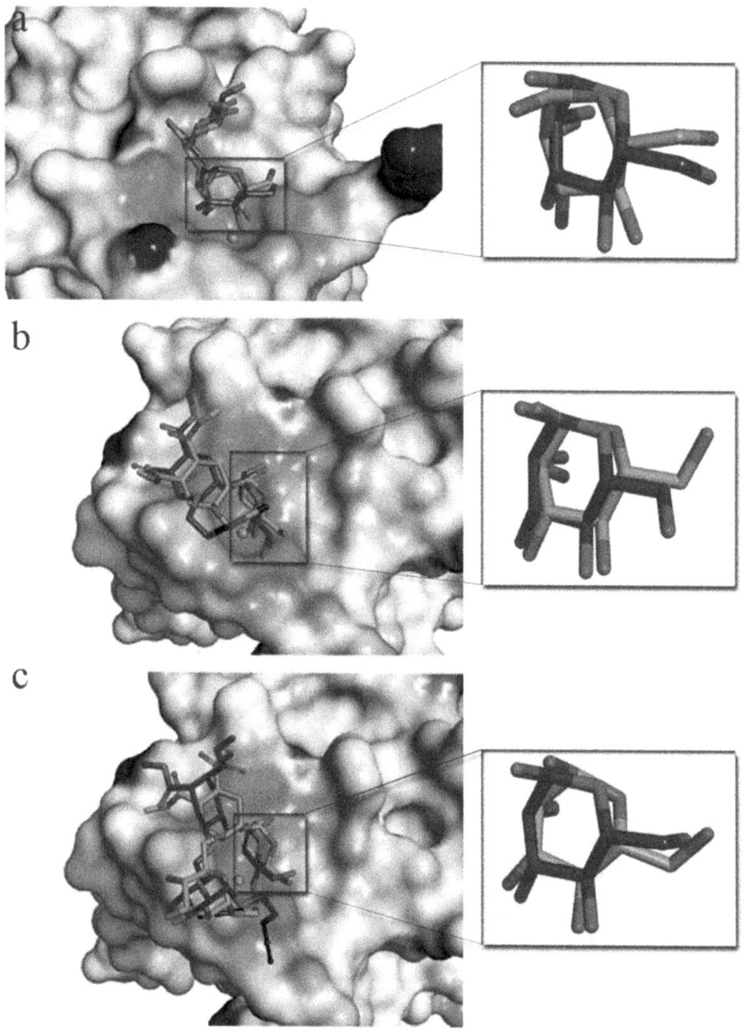

shallow binding sites, these algorithms presumably tend to overestimate the influence of small changes in the ligand conformation. For example, AutoDock and GOLD use a genetic algorithm (stochastic) that requires a large number of ligand conformations to explore all the conformational space available for the ligand. In this manner, a vast number of conformations combined with poorly defined (large, flat and shallow) binding site results in a combinatorial explosion that inevitably leads to a higher number of wrongly predicted docking poses. The results also suggest that the incremental construction algorithm used by FlexX should not be considered as the first choice for docking studies involving DC-SIGN and possibly to other non-druggable binding sites. On the other hand, FRED uses a systematic docking algorithm based on Gaussian function and a multi-conformer structure database pre-generated by OMEGA

(Hawkins & Nicholls, 2012) that is limited to a much smaller number of ligand conformations. As a consequence, FRED's scoring function is faced with an easier task to distinguish between rather different conformations, so fewer errors are made. Furthermore, observed differences in docking tool performance might be a consequence of inadequate estimation of desolvation penalties and/or complex bonds. One rather simple and expected general rule could be extracted from the results; re-docking of all the ligands with all five docking programs showed that the RMSD increases with greater complexity of the ligand. However, relative differences between all 3 ligands do not correlate among docking tools as these are smaller when using FRED and CDOCKER compared to other docking tools, so again it is presumed that their search algorithms perform better when faced with non-druggable targets and/or large ligands.

The second aim of the study was to check how well the selected docking programs are able to reproduce the pose of the core mannose in comparison with its pose in the crystal structure. The core mannose pose is presumed to be even more important than the overall position of the ligand, as is strongly related to establishment of the coordination bond with Ca^{2+} and the hydrogen bonds with the amino acid residues in the active site. The RMSD values for core mannose for all of the three crystal structures are presented in the Table 3.

Most programs successfully predicted core mannose pose (when considering RMSD for the mannose only), but not when RMSD values for the whole ligands were considered. Once again FRED performed well (Figure 6), with an average RMSD for the core mannose of 0.79 Å. CDOCKER outperformed FRED with an average RMSD for the core mannose of 0.62 Å, which might imply that CDOCKER calculates complex interactions better than FRED. Taking into account the positioning of core mannose only, the fair performance of FlexX should be noted. Authors presume that FlexX possibility of defining Ca^{2+} ion as a "pharmacophore" allows better estimation of complex bonding and thus correct prediction of the core mannose pose. GOLD also showed promising results for the pose prediction, while AutoDock was not satisfactory. HYDE re-scoring improved the RMSD in the case of the least complex ligand (2IT6), while in other two cases RMSD surprisingly slightly increased. The graphic presentations of the docking results can be obtained in the original publication (Jug et al., 2015).

To conclude, results in this study show that even widely used docking programs have certain limitations when faced with a rather shallow and flat (i.e. non-druggable) binding site, as is the case of DC-SIGN. FRED and CDOCKER were found to be appropriate for docking studies on DC-SIGN, as evaluated by the calculated RMSD values for each of the re-docked ligands and for the core mannose only, while other programs showed more or less weaknesses. Among these two, FRED was found as the best tool for docking ligands to DC-SIGN and therefore authors suggest it as a suitable docking program for docking studies on DC-SIGN and similar targets with non-druggable binding sites.

Table 3. RMSD values obtained for re-docked core mannose

	FRED	AutoDock	FlexX	HYDE	GOLD	CDOCKER
RMSD (Å)	**DC-SIGN 2IT6**					
	0.65 (Figure 6a)	3.40	0.79	0.55	1.04	0.73
	DC-SIGN 2XR5					
	0.97 (Figure 6b)	1.36	0.79	0.92	0.80	0.34
	DC-SIGN 2XR6					
	0.76 (Figure 6c)	2.35	2.96	3.11	8.71	0.78

(Jug et al., 2015).

CASE STUDY: COVALENT DOCKING OF CATHEPSIN K

Covalent Docking

There are several drugs on the market that act by covalent inhibition (e.g. acetylsalicylic acid, amoxicillin, omeprazole, rivastigmine). Their mechanism of action is connected to their electrophilic "warheads", which target nucleophilic protein residues such as serine and cysteine, and they can form reversible or irreversible covalent linkage (London et al., 2014; Weerapana et al., 2010).

The main aim of designing covalent inhibitors is to increase selectivity and potency, particularity by targeting non-conservative protein residues and shallow binding pockets. However, covalent inhibitors could also possess promiscuous behavior causing off target effects and can be toxic due to prolonged duration off biological effects in case of irreversible binding mode. Therefore, specific noncovalent interactions are crucial for warhead orientation relative to protein nucleophile, stabilizing ligand-target complex and increasing selectivity in order to achieve desired selectivity and potency. Thus the main goal of covalent docking is to identify scaffolds that optimally orientate electrophilic warhead, and as in the case of noncovalent docking, to minimize the number of compounds that have to be prepared and biochemically/biologically evaluated (Kumalo, Bhakat, & Soliman, 2015; London et al., 2014).

In covalent docking studies, the reaction between ligand and receptor has to be taken into account. That could be achieved by "link atom" approach, which utilizes predefined link atom for both, the ligand and the target. It mimics the covalent binding event by forcing ligand link atom to occupy the same steric volume as the target link atoms (GOLD docking software) (Verdonk et al., 2003). On the other hand, "grid based method" and "modification of the flexible side chain" methods are utilized in covalent docking approach of Autodock software (Morris et al., 2009). Another covalent docking approach is based on treating covalent bond as a noncovalent interaction with the sum of additional energy contribution from the covalent bond formation - program CovalentDock (Ouyang et al., 2013).

It has to be mentioned that covalent docking has also its disadvantages; similar to noncovalent docking, poor scoring functions, limited protein flexibility, speed and accuracy are the main issues. Besides that, lack off automation and ligand preparation steps are time consuming (Kumalo et al., 2015).

There are several software packages and webservers suitable for covalent docking:

- **Software:** GOLD (Verdonk et al., 2003), CovalentDock (Ouyang et al., 2013), Autodock (Cosconati et al., 2010), Glide (Friesner et al., 2006), CovDock (Zhu et al., 2014), CovDock-VS (Warshaviak, Golan, Borrelli, Zhu, & Kalid, 2014), etc.
- **Webservers:** CovalentDock Cloud (Ouyang, Zhou, Ge, Li, & Kwoh, 2013), Dockovalent (Irwin et al., 2009), etc.

Covalent Docking of Potential Cathepsin K Inhibitors by GOLD Docking Software

The covalent molecular docking studies were performed on 254-membered benzamide-containing aminonitriles dataset by GOLD Suite v5.1 docking package (Verdonk et al., 2003). Crystal structure of cathepsin K (Cat K) was derived from PDB with entry code 4DMX (Dossetter et al., 2012). Binding site was defined around experimental coordinates of the 0LB301 (Dossetter et al., 2012) bound ligand in radius of 6Å. In order to perform covalent docking, dummy sulfur linking atom was attached to each studied ligand. Subsequently, ligands were renumbered in a way that previously added sulfur's atom number was set to 1. Cat K sulfur atom on Cys25 was then matched with dummy sulfur atom of ligand. Preparing ligands in this way was necessary for GOLD to recognize and set the correct covalent bond between predefined ligands and proteins linking atoms (Schröder et al., 2013).

Re-docking procedure was performed by docking reference ligand 0LB301 into the binding site with early termination allowed if the top three solutions were within 1.5 Å of the root-mean-squared-deviation value in order to define optimal parameter and scoring function (Kirchmair, Markt, Distinto, Wolber, & Langer, 2008). Finally, each ligand was docked into the binding site by the selected scoring function and applying the following parameters of the GOLD genetic algorithm: population size = 100, selection pressure = 1.1, no. of operations = 100,000, no. of islands = 5, Niche Size = 2, migrate = 10, mutate = 95, crossover = 95 (Jones et al., 1997).

Results

Co-crystallized Cat K inhibitor 0LB301 was used to access visual fitness and orientation of docked compounds. Re-docking ligand validation gave satisfactory results as GOLD managed to reproduce accurate binding poses of experimentally bounded ligand 0LB301 with good accuracy. Scoring function ChemScore (CSF) yielded the lowest RMSD value (0.19 Å) for re-docking procedure and was chosen for the assessment of the binding affinities.

Post docking analyses were divided into two parts. In the first part, docking dataset was visually examined as visual fitness and orientation were compared to experimental pose of the 0LB301 ligand. If visual fitness and orientation of a studied compound were correct, the second level post docking analyses were performed by measuring interatomic distances between docked compound features and key amino acid residues of 4DMX.Visual inspection yielded 113/254 compounds from docking dataset with orientation and fitness in concordance with experimentally bound ligand.

In the second part of post-docking analyses the key distances between ligands features and binding site of protein residues 4DMX were measured. According to previously published data (Palmer et al., 2005), three H-bond interactions between ligand and protein were used: (R- N**H**-CH_3 - CN)-(Asn161-C**O**) [d_{4DMX-1} = 3.28 Å], (R-C**O**-NH-CH_3-CN)-(Gly66-N**H**) [d_{4DMX-2} = 3.22 Å], (R-Ph-CO-N**H**-R\`)-(Gly66-C**O**) [d_{4DMX-3} = 2.9 Å]. Because of dynamic nature of biological systems, the tolerance of ±1Å for each interaction distance is used: (R- N**H**-CH_3 - CN)-(Asn161-C**O**) [d_{4DMX-1} = 2.28 - 4.28 Å], (R-C**O**-NH-CH_3-CN)-(Gly66-N**H**) [d_{4DMX-2} = 2.22 – 4.22 Å], (R-Ph-CO-N**H**-R\`)-(Gly66-C**O**) [d_{4DMX-3} = 1.9 - 3.9 Å]. Interatomic distance measurement yielded 12/254 correct compounds (Figure 7).

Additionally, QSAR model for prediction of biological activities were applied (Borišek, Drgan, Minovski, & Novič, 2014) which resulted in 4/254 finally valid compounds (Table 4).

Resulted compounds from the docking study were thereafter synthesized and proved to be reversible and selective inhibitors of Cat K in a nanomolar range (Borišek et al., 2015).

Authors decided to employ GOLD docking software for covalent docking since the familiarness with this program and positive experience relating to noncovalent docking in GOLD. There is also easy to follow step by step description of covalent docking procedure in GOLD tutorial. It has to be stressed that ligand preparation for covalent docking was not automatized and was more time consuming compared to noncovalent docking. Additional studies are planned to compare several docking programs and webservers for covalent docking in order to find most suitable one for particular ligand-target system.

Figure 7. Close view of the 12 GOLD-calculated covalently bound combinatorial Cat K ligands; the key attachment point for covalent bond formation is Cys25 and the amino acid residues Gly66 and Asn161 (depicted in cyan) are important for the stabilization of the ligand through H-bonding.

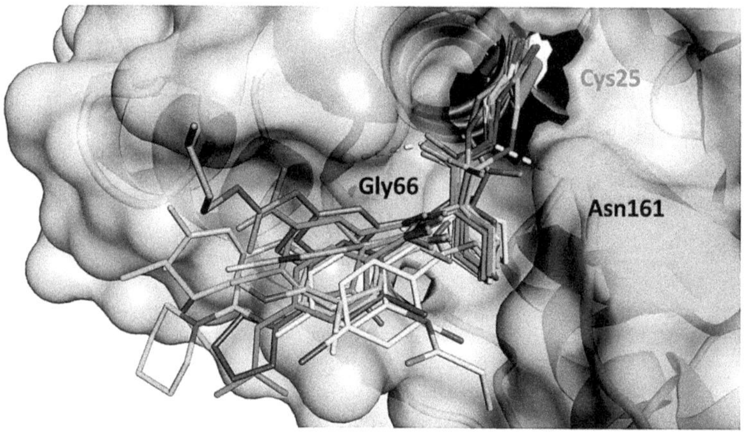

Table 4. Resulted 4/254 compounds after combining molecular docking and QSAR modeling results

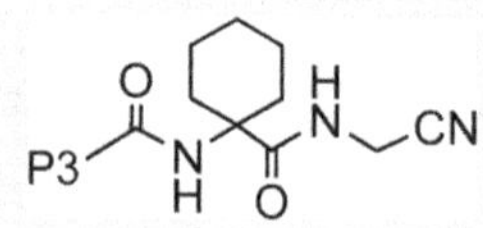

ID	P3	CSF	d_{4DMX-1}	d_{4DMX-2}	d_{4DMX-3}	K_i (nM)
1	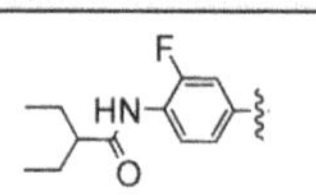	35.0	3.25	2.91	2.80	70
2	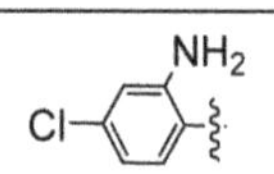	32.4	3.23	2.75	2.98	16
3	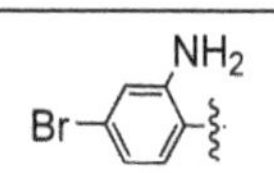	33.6	3.29	2.83	2.92	10
4	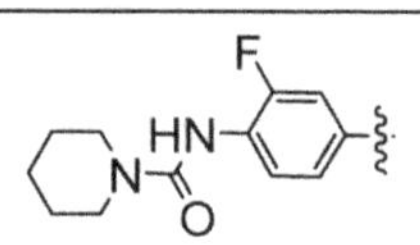	33.3	3.25	2.76	2.95	55

Borišek et al., 2015.

CONCLUSION

This chapter provides overview of computational approaches for molecular docking, considering several of the most commonly used program tools (AutoDock, FRED, CDOCKER, FlexX and GOLD). The application of these algorithms authors present in the comparative study for docking of small ligands to specific protein, lectin Dendritic Cell-Specific Intercellular adhesion molecule-3-Grabbing Non-integrin (DC-SIGN) and covalent docking of cathepsin K with GOLD. As explained in the chapter, five docking tools have been validated in the case study based on the in-house availability, but above all because of the intrinsic difference among them. They differ in the search algorithms and scoring functions, and those with most different algorithms and scoring function were selected. Of course, the selection of various other docking software would be more appropriate (like ADAM, DARWIN, DOCK, eHiTS, FLIPDock, Glide, YUCCA, etc.), but it is practically impossible to evaluate all of them. It was presumed that the mentioned intrinsic differences would most probably result in significant differences in performance of various software, which indeed turned out to be true.

REFERENCES

Adock, S. A., & McCammon, J. A. (2006). Molecular dynamics: Survey of methods for simulating the activity of proteins. *Chemical Reviews*, *106*(5), 1589–21615. doi:10.1021/cr040426m

Allen, W. J., Balius, T. E., Mukherjee, S., Brozell, S. R., Moustakas, D. T., Lang, P. T., & Rizzo, R. C. et al. (2015). DOCK 6: Impact of new features and current docking performance. *Journal of Computational Chemistry*, *36*(15), 1132–1156. doi:10.1002/jcc.23905

Anderluh, M., Jug, G., Švajger, U., & Obermajer, N. (2012). DC-SIGN antagonists, a potential new class of anti-infectives. *Current Medicinal Chemistry*, *19*(7), 992–1007. doi:10.2174/092986712799320664

Baber, J. C., Shirley, W. A., Gao, Y., & Feher, M. (2005). The use of consensus scoring in ligand-based virtual screening. *Journal of Chemical Information and Modeling*, *46*(1), 277–288. doi:10.1021/ci050296y

Ballester, P. J., & Mitchell, J. B. (2010). A machine learning approach to predicting protein–ligand binding affinity with applications to molecular docking. *Bioinformatics (Oxford, England)*, *26*(9), 1169–1175. doi:10.1093/bioinformatics/btq112

Bello, M., Martinez-Archundia, M., & Correa-Basurto, J. (2013). Automated docking for novel drug discovery. *Expert Opinion on Drug Discovery*, *8*(7), 821–834. doi:10.1517/17460441.2013.794780

Berman, H. M., Kleywegt, G. J., Nakamura, H., & Markley, J. L. (2014). The protein data bank archive as an open data resource. *Journal of Computer-Aided Molecular Design*, *28*(10), 1009–1014. doi:10.1007/s10822-014-9770-y

Berman, H. M., Westbrook, J., Feng, Z., Gilliland, G., Bhat, T. N., Weissig, H., & Bourne, P. E. et al. (2000). The protein data bank. *Nucleic Acids Research*, *28*(1), 235–242. doi:10.1093/nar/28.1.235

Böhm, H. J. (1992). The computer program LUDI: A new method for the de novo design of enzyme inhibitors. *Journal of Computer-Aided Molecular Design, 6*, 61-78.

Böhm, H. J. (1994). The development of a simple empirical scoring function to estimate the binding constant for a protein-ligand complex of known three-dimensional structure. *Journal of Computer-Aided Molecular Design, 8*, 243-256.

Borišek, J., Drgan, V., Minovski, N., & Novič, M. (2014). Mechanistic interpretation of artificial neural network-based QSAR model for prediction of cathepsin K inhibition potency. *Journal of Chemometrics*, *28*(4), 272–281. doi:10.1002/cem.2617

Borišek, J., Vizovišek, M., Sosnowski, P., Turk, B., Turk, D., Mohar, B., & Novič, M. (2015). Development of N-(Functionalized Benzoyl)-homocycloleucyl-glycinonitriles as Potent Cathepsin K Inhibitors. *Journal of Medicinal Chemistry*, *58*(17), 6928–6937. doi:10.1021/acs.jmedchem.5b00746

Bras, N. F., Cerqueira, N. M. F. S. A., Sousa, S. F., Fernandes, P. A., & Ramos, M. J. (2014). Protein ligand docking in drug discovery. In N. S. Gabor (Ed.), *Protein modeling* (pp. 249–286). Switzerland: Springer International Publishing.

Brooijmans, N., & Kuntz, I. D. (2003). Molecular recognition and docking algorithms. *Annual Review of Biophysics and Biomolecular Structure*, *32*(1), 335–373. doi:10.1146/annurev.biophys.32.110601.142532

Brooks, B. R., Bruccoleri, R. E., Olafson, B. D., States, D. J., Swaminathan, S., & Karplus, M. (1983). CHARMM: A Program for Macromolecular Energy, Minimization, and Dynamics Calculations. *Journal of Computational Chemistry*, *4*(2), 187–217. doi:10.1002/jcc.540040211

Caflisch, A., Niederer, P., & Anliker, M. (1992). Monte Carlo docking of oligopeptides to proteins. *Proteins*, *13*(3), 223–230. doi:10.1002/prot.340130305

Case, D. A., Cheatham, T. E. III, Darden, T., Gohlke, H., Luo, R., Merz, K. M. Jr, & Woods, R. et al. (2005). The Amber biomolecular simulation programs. *Journal of Computational Chemistry*, *26*(16), 1668–1688. doi:10.1002/jcc.20290

Charifson, P. S., Corkery, J. J., Murcko, M. A., & Walters, W. P. (1999). Consensus scoring: A method for obtaining improved hit rates from docking databases of three-dimensional structures into proteins. *Journal of Medicinal Chemistry*, *42*(25), 5100–5109. doi:10.1021/jm990352k

Chen, R., & Weng, Z. (2002). Docking unbound proteins using shape complementarity, desolvation, and electrostatics. *Proteins*, *47*(3), 281–294. doi:10.1002/prot.10092

Chen, R., & Weng, Z. (2003). ZDOCK: An initial stage protein docking algorithm. *Proteins*, *52*(1), 80–87. doi:10.1002/prot.10389

Chen, Y.-C. (2015). Beware of docking! *Trends in Pharmacological Sciences*, *36*(2), 78–95. doi:10.1016/j.tips.2014.12.001

Clark, R. D., Strizhev, A., Leonard, J. M., Blake, J. F., & Matthew, J. B. (2002). Consensus scoring for ligand/protein interactions. *Journal of Molecular Graphics & Modelling*, *20*(4), 281–295. doi:10.1016/S1093-3263(01)00125-5

Cole, J. C., Nissink, J. W. M., & Taylor, R. (2005). Protein–ligand docking and virtual screening with GOLD. In J. Alvarez & B. Shoichet (Eds.), *Virtual screening in drug discovery* (pp. 379–416). Boca Raton, FL: Taylor & Francis Group. doi:10.1201/9781420028775.ch15

Cosconati, S., Forli, S., Perryman, A. L., Harris, R., Goodsell, D. S., & Olson, A. J. (2010). Virtual screening with AutoDock: Theory and practice. *Expert Opinion on Drug Discovery*, *5*(6), 597–607. doi:10.1517/17460441.2010.484460

DesJarlais, R. L., Sheridan, R. P., Dixon, J. S., Kuntz, I. D., & Venkataraghavan, R. (1986). Docking flexible ligands to macromolecular receptors by molecular shape. *Journal of Medicinal Chemistry*, *29*(11), 2149–2153. doi:10.1021/jm00161a004

Dossetter, A. G., Beeley, H., Bowyer, J., Cook, C. R., Crawford, J. J., Finlayson, J. E., & Vickers, M. et al. (2012). (1R,2R)-N-(1-cyanocyclopropyl)-2-(6-methoxy-1,3,4,5-tetrahydropyrido[4,3-b]indole -2-carbonyl)cyclohexanecarboxamide (AZD4996): A potent and highly selective cathepsin K inhibitor for the treatment of osteoarthritis. *Journal of Medicinal Chemistry*, *55*(14), 6363–6374. doi:10.1021/jm3007257

Eberhart, R., & Kennedy, J. (1995). A new optimizer using particle swarm theory. In *Proceedings of the Sixth International Symposium on Micro Machine and Human Science (MHS '95)*. Nagoya, Japan. IEEE. doi:10.1109/MHS.1995.494215

Eldridge, M. D., Murray, C. W., Auton, T. R., Paolini, G. V., & Mee, R. P. (1997). Empirical scoring functions: I. The development of a fast empirical scoring function to estimate the binding affinity of ligands in receptor complexes. *Journal of Computer-Aided Molecular Design*, *11*(5), 425–445. doi:10.1023/A:1007996124545

Ewing, T. J. A., Makino, S., Skillman, A. G., & Kuntz, I. D. (2001). DOCK 4.0: Search strategies for automated molecular docking of flexible molecule databases. *Journal of Computer-Aided Molecular Design*, *15*(5), 411–428. doi:10.1023/A:1011115820450

Feher, M. (2006). Consensus scoring for protein–ligand interactions. *Drug Discovery Today*, *11*(9), 421–428. doi:10.1016/j.drudis.2006.03.009

Ferreira, L. G., dos Santos, R. N., Oliva, G., & Andricopulo, A. D. (2015). Molecular Docking and Structure-Based Drug Design Strategies. *Molecules (Basel, Switzerland)*, *20*(7), 13384–13421. doi:10.3390/molecules200713384

Friesner, R. A., Murphy, R. B., Repasky, M. P., Frye, L. L., Greenwood, J. R., Halgren, T. A., & Mainz, D. T. et al. (2006). Extra precision glide: Docking and scoring incorporating a model of hydrophobic enclosure for protein-ligand complexes. *Journal of Medicinal Chemistry*, *49*(21), 6177–6196. doi:10.1021/jm051256o

Gabb, A. H. A., Jackson, R. M., & Sternberg, M. J. E. (1997). Modeling protein docking using shape complementarity, electrostatics and biochemical information. *Journal of Molecular Biology*, *272*(1), 106–120. doi:10.1006/jmbi.1997.1203

Gohlke, H., & Klebe, G. (2001). Statistical potentials and scoring functions applied to protein–ligand binding. *Current Opinion in Structural Biology*, *11*(2), 231–235. doi:10.1016/S0959-440X(00)00195-0

Goodsell, S., & Olson, A. J. (1990). Automated docking of substrates to proteins by simulated annealing. *Proteins*, *8*(3), 195–202. doi:10.1002/prot.340080302

Gowthaman, R., Deeds, E. J., & Karanicolas, J. (2013). Structural properties of non-traditional drug targets present new challenges for virtual screening. *Journal of Chemical Information and Modeling*, *53*(8), 2073–2081. doi:10.1021/ci4002316

Grant, J. A., Pickup, B. T., & Nicholls, A. (2001). A smooth permittivity function for Poisson–Boltzmann solvation methods. *Journal of Computational Chemistry*, *22*(6), 608–640. doi:10.1002/jcc.1032

Grinter, S. Z., & Zou, X. (2014). Challenges, Applications, and Recent Advances of Protein-Ligand Docking in Structure-Based Drug Design. *Molecules (Basel, Switzerland)*, *19*(7), 10150–10176. doi:10.3390/molecules190710150

Gschwend, D. A., Good, A. C., & Kuntz, I. D. (1996). Molecular docking towards drug discovery. *Journal of Molecular Recognition*, *9*(2), 175–186. doi:10.1002/(SICI)1099-1352(199603)9:2<175::AID-JMR260>3.0.CO;2-D

Halperin, I., Ma, B., Wolfson, H., & Nussinov, R. (2002). Principles of docking: An overview of search algorithms and a guide to scoring functions. *Proteins*, *47*(4), 409–443. doi:10.1002/prot.10115

Hart, T. N., & Read, R. J. (1992). A multiple-start Monte Carlo docking method. *Proteins*, *13*(3), 206–222. doi:10.1002/prot.340130304

Hartshorn, M. J., Verdonk, M. L., Chessari, G., Brewerton, S. C., Mooij, W. T. M., Mortenson, P. N., & Murray, C. W. (2007). Diverse, high-quality test set for the validation of protein-ligand docking performance. *Journal of Medicinal Chemistry*, *50*(4), 726–741. doi:10.1021/jm061277y

Hawkins, G. D., Cramer, C. J., & Truhlar, D. G. (1995). Pairwise solute descreening of solute charges from a dielectric medium. *Chemical Physics Letters*, *246*(1-2), 122–129. doi:10.1016/0009-2614(95)01082-K

Hawkins, P. C. D., & Nicholls, A. (2012). Conformer generation with OMEGA: Learning from the data set and the analysis of failures. *Journal of Chemical Information and Modeling*, *52*(11), 2919–2936. doi:10.1021/ci300314k

Huang, S.-Y., Grinter, S. Z., & Zou, X. (2010). Scoring functions and their evaluation methods for protein–ligand docking: Recent advances and future directions. *Physical Chemistry Chemical Physics*, *12*(40), 12899–12908. doi:10.1039/c0cp00151a

Huang, S.-Y., & Zou, X. (2006). An iterative knowledge-based scoring function to predict protein–ligand interactions: I. Derivation of interaction potentials. *Journal of Computational Chemistry*, *27*(15), 1866–1875. doi:10.1002/jcc.20504

Huey, R., Morris, G. M., Olson, A. J., & Goodsell, D. S. (2007). Software news and update: A semiempirical free energy force field with charge-based desolvation. *Journal of Computational Chemistry*, *28*(6), 1145–1152. doi:10.1002/jcc.20634

Irwin, J. J., Shoichet, B. K., Mysinger, M. M., Huang, N., Colizzi, F., Wassam, P., & Cao, Y. (2009). Automated docking screens: A feasibility study. *Journal of Medicinal Chemistry*, *52*(18), 5712–5720. doi:10.1021/jm9006966

Jain, A. (2008). Bias, reporting, and sharing: Computational evaluations of docking methods. *Journal of Computer-Aided Molecular Design*, *22*(3-4), 201–212. doi:10.1007/s10822-007-9151-x

Jones, G., & Willett, P. (1995). Docking small-molecule ligands into active sites. *Current Opinion in Biotechnology*, *6*(6), 652–656. doi:10.1016/0958-1669(95)80107-3

Jones, G., Willett, P., & Glen, R. C. (1995). Molecular recognition of receptor sites using genetic algorithm with a description of desolvation. *Journal of Molecular Biology*, *245*(1), 43–53. doi:10.1016/S0022-2836(95)80037-9

Jones, G., Willett, P., Glen, R. C., Leach, A. R., & Taylor, R. (1997). Development and validation of a genetic algorithm for flexible docking. *Journal of Molecular Biology*, *267*(3), 727–748. doi:10.1006/jmbi.1996.0897

Jug, G., Anderluh, M., & Tomašič, T. (2015). Comparative evaluation of several docking tools for docking small molecule ligands to DC-SIGN. *Journal of Molecular Modeling*, *21*(6), 164–176. doi:10.1007/s00894-015-2713-2

Katchalski-Katzir, E., Shariv, I., Eisenstein, M., Friesem, A. A., Aflalo, C., & Vakser, I. A. (1992). Molecular surface recognition: Determination of geometric fit between proteins and their ligands by correlation techniques. *Proceedings of the National Academy of Sciences of the United States of America*, *89*(6), 2195–2199. doi:10.1073/pnas.89.6.2195

Kearsley, S. K., Underwood, D. J., Sheridan, R. P., & Miller, M. D. (1994). Flexibases: A way to enhance the use of molecular docking methods. *Journal of Computer-Aided Molecular Design*, *8*(5), 565–582. doi:10.1007/BF00123666

Kirchmair, J., Markt, P., Distinto, S., Wolber, G., & Langer, T. (2008). Evaluation of the performance of 3D virtual screening protocols: RMSD comparisons, enrichment assessments, and decoy selection - What can we learn from earlier mistakes? *Journal of Computer-Aided Molecular Design*, *22*(3-4), 213–228. doi:10.1007/s10822-007-9163-6

Kirkpatrick, S., Gelatt, C. D., & Vecchi, M. P. (1983). Optimization by simulation annealing. *Science*, *220*(4598), 671–680. doi:10.1126/science.220.4598.671

Kitchen, D. B., Decornez, H., Furr, J. R., & Bajorath, J. (2004). Docking and scoring in virtual screening for drug discovery: Methods and applications. *National Review*, *3*, 935–949.

Kroemer, R. T. (2003). Molecular modelling probes: Docking and scoring. *Biochemical Society Transactions*, *31*(5), 980–984. doi:10.1042/bst0310980

Kroemer, R. T. (2007). Structure-based drug design: Docking and scoring. *Current Protein & Peptide Science*, *8*, 312–328. doi:10.2174/138920307781369382

Kumalo, H. M., Bhakat, S., & Soliman, M. E. (2015). Theory and applications of covalent docking in drug discovery: Merits and pitfalls. *Molecules (Basel, Switzerland)*, *20*(2), 1984–2000. doi:10.3390/molecules20021984

Kuntz, I. D., Blaney, J. M., Oatley, S. J., Langridge, O. R., & Ferrin, T. E. (1982). A geometric approach to macromolecule-ligand interactions. *Journal of Molecular Biology*, *161*(2), 269–288. doi:10.1016/0022-2836(82)90153-X

Leach, A. R., & Kuntz, I. D. (1992). Conformational analysis of flexible ligands in macromolecular receptor sites. *Journal of Computational Chemistry*, *13*(6), 730–748. doi:10.1002/jcc.540130608

Lengauer, T., & Rarey, M. (1996). Computational methods for biomolecular docking. *Current Opinion in Structural Biology*, *6*(3), 402–406. doi:10.1016/S0959-440X(96)80061-3

Levinthal, C., Wodak, S. J., Kahn, P., & Dadivanian, A. K. (1975). Hemoglobin interactions in sickle cell fibers I: Theoretical approach to the molecular contacts. *Proceedings of the National Academy of Sciences of the United States of America*, *72*(4), 1330–1334. doi:10.1073/pnas.72.4.1330

Li, X., & Liang, J. (2006). Knowledge-based energy functions for computational studies of proteins. In Y. Xu., D. Xu & J. Liang (Eds.), Computational methods for protein structure prediction and modeling (pp. 71-124). New York, NY: Springer.

Liebeschuetz, J. W., Cole, J. C., & Korb, P. (2012). Pose prediction and virtual screening performance of GOLD scoring function in a standardized test. *Journal of Computer-Aided Molecular Design*, *26*(6), 737–748. doi:10.1007/s10822-012-9551-4

Liu, M., & Wang, S. (1999). MCDOCK: A Monte Carlo simulation approach to the molecular docking problem. *Journal of Computer-Aided Molecular Design*, *13*(5), 435–451. doi:10.1023/A:1008005918983

London, N., Miller, R. M., Krishnan, S., Uchida, K., Irwin, J. J., Eidam, O., & Taunton, J. et al. (2014). Covalent docking of large libraries for the discovery of chemical probes. *Nature Chemical Biology*, *10*(12), 1066–1072. doi:10.1038/nchembio.1666

Lybrand, T. P. (1995). Ligand-protein docking and rational drug design. *Current Opinion in Structural Biology*, *5*(2), 224–228. doi:10.1016/0959-440X(95)80080-8

McGann, M. (2011). FRED pose prediction and virtual screening accuracy. *Journal of Chemical Information and Modeling*, *51*(3), 578–596. doi:10.1021/ci100436p

Mehler, E. L., & Solmajer, T. (1991). Electrostatic effects in proteins: Comparison of dielectric and charge models. *Protein Engineering*, *4*(8), 903–910. doi:10.1093/protein/4.8.903

Meng, E. C., Shoichet, B. K., & Kuntz, I. D. (1992). Automated docking with grid-based energy evaluation. *Journal of Computational Chemistry*, *13*(4), 505–524. doi:10.1002/jcc.540130412

Meng, X.-Y., Zhang, H.-X., Mezei, M., & Cui, M. (2011). Molecular docking: A powerful approach for structure-based drug discovery. *Current Computer-aided Drug Design*, *7*(2), 146–157. doi:10.2174/157340911795677602

Mihasan, M. (2012). What in silico molecular docking can do for the 'bench-working biologists'. *Journal of Biosciences*, *37*(6), 1089–1095. doi:10.1007/s12038-012-9273-8

Miller, M. D., Kearsley, S. K., Underwood, D. J., & Sheridan, R. P. (1994). FLOG: A system to select 'quasi-flexible' ligands complementary to a receptor of known three-dimensional structure. *Journal of Computer-Aided Molecular Design*, *8*(2), 153–174. doi:10.1007/BF00119865

Moitessier, N., Englebienne, P., Lee, D., Lawandi, D., & Corbeil, C. R. (2008). Towards the development of universal, fast and highly accurate docking/scoring methods: A long way to go. *British Journal of Pharmacology*, *153*(S1), 7–26. doi:10.1038/sj.bjp.0707515

Mooij, W. T. M., & Verdonk, M. L. (2005). General and targeted statistical potentials for protein–ligand interactions. *PROTEINS: Structure, Function, and Bioinformatics*, *61*(2), 272–287. doi:10.1002/prot.20588

Morris, G. M., Goodsell, D. S., Halliday, R. S., Huey, R., Hart, W. E., Belew, R. K., & Olson, A. J. (1998). Automated Docking Using a Lamarckian genetic algorithm and an empirical binding free energy function. *Journal of Computational Chemistry, 19*(14), 1639–1662. doi:10.1002/(SICI)1096-987X(19981115)19:14<1639::AID-JCC10>3.0.CO;2-B

Morris, G. M., Huey, R., Lindstrom, W., Sanner, M. F., Belew, R. K., Goodsell, D. S., & Olson, A. J. (2009). AutoDock4 and AutoDockTools4: Automated docking with selective receptor flexibility. *Journal of Computational Chemistry, 30*(16), 2785–2791. doi:10.1002/jcc.21256

Morris, G. M., Olson, A. J., & Goodsell, D. S. (2000). Protein-Ligand Docking. In D. E. Clark (Ed.), *Evolutionary Algorithms in Molecular Design* (pp. 31–48). New York, NY: Wiley. doi:10.1002/9783527613168.ch3

Namasivayam, V., & Günther, R. (2007). PSO @ AUTODOCK: A fast flexible molecular docking program based on swarm intelligence. *Chemical Biology & Drug Design, 70*(6), 475–484. doi:10.1111/j.1747-0285.2007.00588.x

Obermajer, N., Sattin, S., Colombo, C., Bruno, M., Švajger, U., Anderluh, M., & Bernardi, A. (2011). Design, synthesis and activity evaluation of mannose-based DC-SIGN antagonists. *Molecular Diversity, 15*(2), 347–360. doi:10.1007/s11030-010-9285-y

Oda, A., Tsuchida, K., Takakura, T., Yamaotsu, N., & Hirono, S. (2006). Comparison of consensus scoring strategies for evaluating computational models of protein-ligand complexes. *Journal of Chemical Information and Modeling, 46*(1), 380–391. doi:10.1021/ci050283k

Oshiro, C. M., Kuntz, I. D., & Dixon, J. S. (1995). Flexible ligand docking using a genetic algorithm. *Journal of Computer-Aided Molecular Design, 9*(2), 113–130. doi:10.1007/BF00124402

Ouyang, X., Zhou, S., Ge, Z., Li, R., & Kwoh, C. K. (2013). CovalentDock Cloud: A web server for automated covalent docking. *Nucleic Acids Research, 329-332*. doi:10.1093/nar/gkt406

Ouyang, X., Zhou, S., Su, C. T. T., Ge, Z., Li, R., & Kwoh, C. K. (2013). CovalentDock: Automated covalent docking with parameterized covalent linkage energy estimation and molecular geometry constraints. *Journal of Computational Chemistry, 34*(4), 326–336. doi:10.1002/jcc.23136

Palmer, J. T., Bryant, C., Wang, D.-X., Davis, D. E., Setti, E. L., Rydzewski, R. M., & Rodan, G. et al. (2005). Design and synthesis of tri-ring P3 benzamide-containing aminonitriles as potent, selective, orally effective inhibitors of cathepsin K. *Journal of Medicinal Chemistry, 48*(24), 7520–7534. doi:10.1021/jm058198r

Rarey, M., Kramer, B., Lengauer, T., & Klebe, G. (1996). A fast flexible docking method using an incremental construction algorithm. *Journal of Molecular Biology, 261*(3), 470–489. doi:10.1006/jmbi.1996.0477

Rarey, M., Wefing, S., & Lengauer, T. (1996). Placement of medium-sized molecule fragments into active site of proteins. *Journal of Computer-Aided Molecular Design, 10*(1), 41–54. doi:10.1007/BF00124464

Salomon-Ferrer, R., Case, D. A., & Walker, R. C. (2012). An overview of the Amber biomolecular simulation package. *WIREs Computational Molecular Science, 3*(2), 198–210. doi:10.1002/wcms.1121

Sauton, N., Lagorce, D., Villoutreix, B. O., & Miteva, M. A. (2008). MS-DOCK: Accurate multiple conformation generator and rigid docking protocol for multi-step virtual ligand screening. *BMC Bioinformatics*, *9*(1), 184. doi:10.1186/1471-2105-9-184

Schneider, N., Hindle, S., Lange, G., Klein, R., Albrecht, J., Briem, H., & Rarey, M. et al. (2012). Substantial improvements in large-scale redocking and screening using the novel HYDE scoring function. *Journal of Computer-Aided Design*, *26*(6), 701–723. doi:10.1007/s10822-011-9531-0

Schröder, J., Klinger, A., Oellien, F., Marhöfer, R. J., Duszenko, M., & Selzer, P. M. (2013). Docking-based virtual screening of covalently binding ligands: An orthogonal lead discovery approach. *Journal of Medicinal Chemistry*, *56*(4), 1478–1490. doi:10.1021/jm3013932

Selemme, F. R. (1976). An hypothetical structure for an intermolecular electron transfer complex of Cytochromes c and b_5. *Journal of Molecular Biology*, *102*(3), 563–568. doi:10.1016/0022-2836(76)90334-X

Sousa, S. F., Fernandes, P. A., & Ramos, M. J. (2006). Protein–ligand docking: Current status and future challenges. *PROTEINS: Structure, Function, and Bioinformatics*, *65*(1), 15–26. doi:10.1002/prot.21082

Sousa, S. F., Ribeiro, A. J. M., Coimbra, J. T. S., Neves, R. P. P., Martins, S. A., Moorthy, N. S. H. N., & Ramos, M. J. et al. (2013). Protein-ligand docking in the new millennium – a retrospective of 10 years in the field. *Current Medicinal Chemistry*, *20*(18), 2296–2314. doi:10.2174/0929867311320180002

Stahl, M., & Rarey, M. (2001). Detailed analysis of scoring functions for virtual screening. *Journal of Medicinal Chemistry*, *44*(7), 1035–1042. doi:10.1021/jm0003992

Still, W. C., Tempczyk, A., Hawley, R. C., & Hendrickson, T. (1990). Semianalytical treatment of solvation for molecular mechanics and dynamics. *Journal of the American Chemical Society*, *112*(16), 6127–6129. doi:10.1021/ja00172a038

Straatsma, T. P., & McCammon, J. A. (1991). Multiconfiguration thermodynamic integration. *The Journal of Chemical Physics*, *95*(2), 1175–1118. doi:10.1063/1.461148

Tanaka, S., & Scheraga, H. A. (1976). Medium- and long-range interaction parameters between amino acids for predicting three-dimensional structures of proteins. *Macromolecules*, *9*(6), 945–950. doi:10.1021/ma60054a013

Taylor, R. D., Jewsbury, P. J., & Essex, J. W. (2002). A review of protein-small molecule docking methods. *Journal of Computer-Aided Molecular Design*, *16*(3), 151–166. doi:10.1023/A:1020155510718

Thomsen, R. (2003). Flexible ligand docking using evolutionary algorithms: Investigating the effects of variation operators and local search hybrids. *Bio Systems*, *72*(1-2), 57–73. doi:10.1016/S0303-2647(03)00135-7

Tomašič, T., Hajšek, D., Švajger, U., Luzar, J., Obermajer, N., Petit-Haertlein, I., & Anderluh, M. et al. (2014). Monovalent mannose-based DC-SIGN antagonists: Targeting the hydrophobic groove of the receptor. *European Journal of Medicinal Chemistry*, *75*, 308–326. doi:10.1016/j.ejmech.2014.01.047

Trott, O., & Olson, A. J. (2009). AutoDock Vina: Improving the speed and accuracy of docking with a new scoring function, efficient optimization, and multithreading. *Journal of Computational Chemistry*, *31*, 455–461.

Velec, H. F., Gohlke, H., & Klebe, G. (2005). DrugScore^{CSD} - knowledge-based scoring function derived from small molecule crystal data with superior recognition rate of near-native ligand poses and better affinity prediction. *Journal of Medicinal Chemistry*, *48*(20), 6296–6303. doi:10.1021/jm050436v

Verdonk, M. L., Cole, J. C., Hartshorn, M. J., Murray, C. W., & Taylor, R. D. (2003). Improved Protein–Ligand Docking Using GOLD. *Proteins*, *52*(4), 609–623. doi:10.1002/prot.10465

Wang, T., Wu, M.-B., Chen, Z.-J., Chen, H., Lin, J.-P., & Yang, L.-R. (2015). Fragment-Based Drug Discovery and Molecular Docking in Drug Design. *Current Pharmaceutical Biotechnology*, *16*(1), 11–25. doi:10.2174/1389201015666141122204532

Warshaviak, D., Golan, G., Borrelli, K. W., Zhu, K., & Kalid, O. (2014). Structure-based virtual screening approach for discovery of covalently bound ligands. *Journal of Chemical Information and Modeling*, *54*(7), 1941–1950. doi:10.1021/ci500175r

Weerapana, E., Wang, C., Simon, G. M., Richter, F., Khare, S., Dilon, M. B. D., & Cravatt, B. F. et al. (2010). Quantitative reactivity profiling predicts functional cysteines in proteomes. *Nature*, *10*(12), 790–795. doi:10.1038/nature09472

Wodak, S. J., & Janin, J. (1978). Computer analysis of protein-protein interactions. *Journal of Molecular Biology*, *124*(2), 323–342. doi:10.1016/0022-2836(78)90302-9

Wu, G., Robertson, D. H., Brooks, C. L. III, & Vieth, M. (2003). Detailed analysis of grid-based molecular docking: A case study of CDOCKER—A CHARMm-based MD docking algorithm. *Journal of Computational Chemistry*, *24*(3), 1549–1562. doi:10.1002/jcc.10306

Yue, S. Y. (1990). Distance-constrained molecular docking by simulated annealing. *Protein Engineering*, *4*(2), 177–184. doi:10.1093/protein/4.2.177

Yuriev, E., Agostino, M., & Ramsland, P. A. (2010). Challenges and advances in computational docking: 2009 in review. *Journal of Molecular Recognition*, *24*(2), 149–164. doi:10.1002/jmr.1077

Yuriev, E., Holien, J., & Ramsland, P. A. (2015). Improvements, trends, and new ideas in molecular docking: 2012–2013 in review. *Journal of Molecular Recognition*, *28*(10), 581–604. doi:10.1002/jmr.2471

Yuriev, E., & Ramsland, P. A. (2013). Latest developments in molecular docking: 2010–2011 in review. *Journal of Molecular Recognition*, *26*(5), 215–239. doi:10.1002/jmr.2266

Zhu, K., Borrelli, K. W., Greenwood, J. R., Day, T., Abel, R., Farid, R. S., & Harder, E. (2014). Docking covalent inhibitors: A parameter free approach to pose prediction and scoring. *Journal of Chemical Information and Modeling*, *54*(7), 1932–1940. doi:10.1021/ci500118s

Zwanzig, R. W. (1954). High-temperature equation of state by a perturbation method. I. Nonpolar gases. *The Journal of Chemical Physics*, *22*(8), 1420–1426. doi:10.1063/1.1740409

Chapter 5
Protein Ligand Interaction Fingerprints

Ali HajiEbrahimi
Shiraz University of Medical Sciences, Iran

Mohsen Ranjbar
Shiraz University of Medical Sciences, Iran

Hamidreza Ghafouri
Shiraz University of Medical Sciences, Iran

Amirhossein Sakhteman
Shiraz University of Medical Sciences, Iran

ABSTRACT

A most challenging part in docking-based virtual screening is the scoring functions implemented in various docking programs in order to evaluate different poses of the ligands inside the binding cavity of the receptor. Precise and trustable measurement of ligand–protein affinity for Structure-Based Virtual Screening (SB-VS) is therefore, an outstanding problem in docking studies. Empirical post-docking filters can be helpful as a way to provide various types of structure-activity information. Different types of interaction have been presented between the ligands and the receptor so far. Based on the diversity and importance of PLIF methods, this chapter will focus on the comparison of different protocols. The advantages and disadvantages of all methods will be discussed explicitly in this chapter as well as future sights for further progress in this field. Different classifications approaches for the protein-ligand interaction fingerprints were also discussed in this chapter.

INTRODUCTION

The number of protein-ligand complex structures from both in silico and experimental methods have highly increased during the past decade. This was due to advances in structure-based drug design in pharmaceutical industry, high-throughput NMR technology, crystallography and structural genomics. In the public domain, availability of over 5000 small molecule complexes within databases is the result of many effective studies performed in this field (Deng, Chuaqui, & Singh, 2004).

Utilizing this experimental and in silico information depends highly on the ability to organize, analyze and mine the structural data, based on virtual chemical library screening (Deng et al., 2004).

DOI: 10.4018/978-1-5225-0115-2.ch005

The most important merit of virtual screening as a complementary approach to the traditional high-throughput screening is to give insights into molecular recognition in biological systems as well as to facilitate the design of novel therapeutics (Kelly & Mancera, 2004). However, some pitfalls are observed in virtual screening studies, which can be summarized as bellow (Scior et al., 2012):

1. **Detecting High Affinity Molecules by Virtual Screening:** Virtual screening is not considered as a successful tool for identifying high affinity substances. Since reaching highly attracted compounds to particular target is possible during the optimization process.
2. **Certainty of Inquiries:** The setting chosen to select potent materials from a large number of substances are very arbitrary and may depend on the performance of screening.
3. **An Expectation the False Binding Site:** Possibility of binding to exactly correct poses is low in fact. Actually, a few studies revealed exactly true prediction of binding modes.
4. **Predicting Water Mediated Interactions:** Possibly there is many water mediated hydrogen bonds between ligand and receptor. Prediction of count, situation and orientation of these interactions is very ambitious.
5. **Allosteric or Multiple Binding Sites vs. Single Ones:** Many receptors possess allosteric or multiple binding sites, but VS approaches are unable to predict the affinity to an allosteric binding site.

Retrieving useful and crucial information for rational drug discovery is not facile due to an overwhelming number of 3D X-ray and NMR structures. In order to perform virtual screening, these prospective ligands need to be filtered in order to reduce their number for more precise and intensive studies. (Kelly & Mancera, 2004).

Up to now, a variety of approaches have been developed to capture structural knowledge ranging from 2D ligand filters and pharmacophores to 3D interaction constraints and pharmacophoric features (Nandigam, Kim, Singh, & Chuaqui, 2009). All these methods can be used to limit poses generated from ligand receptor docking. Although analyzing a few structures by in silico methods is now considered as a usual practice, mining and making comparisons between a huge number of protein-ligand complexes needs special care such as simplification of the 3D information. Among the most useful simplification processes for analyzing protein-ligand interactions is the conversion of atomic coordinates into simpler 1D or 2D fingerprints (Desaphy, Raimbaud, Ducrot, & Rognan, 2013). Moreover, because binding information is encoded as 1D fingerprint strings, pattern identification of the fingerprints can be done using machine learning and clustering methods. This approach is beneficial in enhancing the ability to make useful implications that are not apparent by looking at individual structures (Nandigam et al., 2009).

BACKGROUND

Fingerprints are easy to generate and can manipulate, compare and, therefore, enable a systematic analysis of large data sets. They are largely used to describe and compare molecular objects (small molecular weight ligands, pharmacophores, proteins and protein-ligand binding sites) and can be used as descriptors. This feature of fingerprints makes it possible to use them by drug design programs as well as in silico screening tools. These fingerprints are frequently manipulated in either ligand based, structure based and combined approaches of drug design. They can solve problems related to finding conserved interaction patterns, explain interaction features qualitatively and perform efficient post docking processing. (Desaphy et al., 2013)

Implementations of fingerprinting methods assist scoring functions to rank the predicted ligand-protein interactions in order to select those candidates that have a greater chance of success in vitro. It also permits us to perform further analyses and similarity comparisons between these structures.(Kelly & Mancera, 2004)

Different protein ligand interaction fingerprints have been developed so far. Based on this approach, the residues involved in the interaction must be determined. In the first step, it is of great importance to consider different interactions of the ligand poses within the target for generation of a suitable algorithm. (Deng et al., 2004; Singh, Deng, Narale, & Chuaqui, 2006) Different types of interaction have been presented between the ligands and the receptor. (Chourasia, Sastry, & Sastry, 2011; Crowley & Golovin, 2005; Dalkas, Teheux, Kwasigroch, & Rooman, 2014) A common interaction observed between the ligand and protein is called cation-π interactions.(Crowley & Golovin, 2005) This interaction is made between a positively charge group such as quaternary nitrogen inside ligands with aromatic side chains of the protein such as PHE or TYR. (Crowley & Golovin, 2005). In addition, the positive charge amino acids such as LYS or ARG can form this type of interaction with aromatic fragments of the ligands. Cation-pi interaction refers to a cation fragment on the axis perpendicular to aromatic ring. (Crowley & Golovin, 2005; Dalkas et al., 2014) The hydrogen bonding interaction can be formed between a hydrogen donor and an acceptor. Two distance and angle terms normally are being considered in calculation of this interaction. π-π interaction is another interaction between the aromatic fragment of the ligand and aromatic side chain. (Dalkas et al., 2014) This interaction is among the most discussed interactions in the literature. Two different types of face to face and edge to face π- π can be investigated in docking studies. (Dalkas et al., 2014) The distance between two aromatic planes centers is normally defined as 4 Å in both types. Ionic interaction is an interaction between two groups of different charges. This interaction as well as hydrophobic and metal complex interactions have only a distance criterion. (Marcou & Rognan, 2007)

The next step for generating the bit strings is highly dependent to interactions criteria. As depicted in Figure 1, each residue is represented by a seven long bit string based on the following criteria. (Singh et al., 2006)

1. Is it in contact with the ligand or not?
2. Are any main chain atoms in contact with the ligand?
3. Are any side chain atoms involved in the binding?
4. Are there any polar interactions?
5. Are there any non-polar interactions?
6. Is the residue providing hydrogen bond acceptor?
7. Is the residue providing hydrogen bond donor?

And for nine-bit strings described as 1-any contact 2-backbone 3-side chain 4-polar 5-hydrophobic 6-H-donor 7- H-acceptor 8-aromatic 9-charged interactions (Figure 2).(Mordalski, Kosciolek, Kristiansen, Sylte, & Bojarski, 2011)

Figure 1. A seven bit string generated on interaction criteria

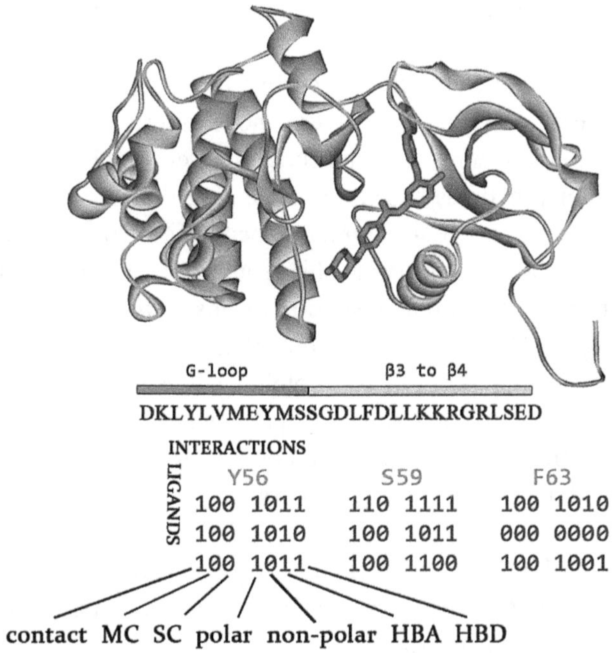

Figure 2. An example of a nine bit string

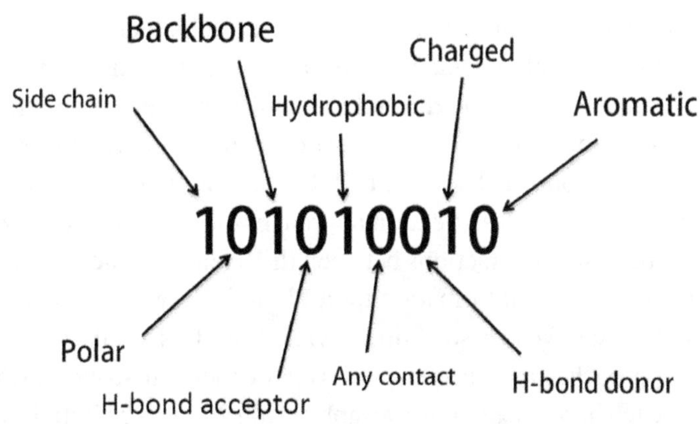

PROTEIN-LIGAND INTERACTION DESCRIPTORS

The primary methods for presenting molecular fingerprints were based upon chemical composition and structures of the molecules, First approach in ligand-protein interaction fingerprints was explained by Bajorath et al (Bajorath, 2008). The idea behind this project was that all atoms in a ligand do not take a similar role for the activity (Tan, Batista, & Bajorath, 2010). As a powerful method, interaction annotated fingerprint was more successful than the previous 2D or 3D descriptors since descriptors related to protein cavity, and those related to ligands were merged as a single fingerprint profile (Crisman, Sisay, & Bajorath, 2008). Different machine learning methods were able discriminate between true and false complex of ligand and protein (Wang et al., 2011). As an instance, this method was used in identification of novel ligands for oxytocin receptor and showed to be more useful than ligand based 2D and 3D descriptors (Weill et al., 2011).

Further studies by Rognan and Deng introduced two novel topological scoring methods based on molecular interaction fingerprints (IFPs). This was a novel approach to represent the properties of a ligand and a better descriptor to quantify docking success. (Deng et al., 2004; Desaphy et al., 2013)

In contrast with traditional methods based on calculating the properties of the ligands from their chemical and structural components, the structural interaction fingerprints (SIFTs) represent a ligand by its interaction in the binding site of the protein. This requires knowledge for the binding mode of each ligand inside its target protein. Basic implementation of this method has revealed effectiveness in the post processing of docking studies and as a filter during a virtual chemical library screening process. (Mordalski et al., 2011)Standard descriptors for protein cavities and their cognate ligands can also be concatenated into a single fingerprint and then used as input to train machine learning algorithms. The results can be used to discriminate true from false complexes. The structural interaction fingerprints show several promising features (Desaphy et al., 2013):

1. Enhancing the quality of pose prediction in docking experiments.
2. Clustering protein-ligand interactions for a panel of related structures according to the diversity of their interactions with a target subfamily.
   3 Assisting target-based library design.

In fact, IFPs are simple bit strings that convert 3D information of protein-ligand interactions into 1D bit vector representations that can be quickly compared with each other.

In this regard, the first step includes identification of all the residues involved in protein-ligand inter-molecular interactions. Next step is dedicated to classify the calculated interactions. Each bit can denote either the presence (1) or absence (0) of a particular discrete property such as hydrogen bond, hydrophobic or van der Waals contact. The obtained data might be utilized for ligand analysis as well describing protein binding sites and classification approaches (Nandigam et al., 2009).

Since these bit strings represent interactions between the ligand and the receptor, they can be divided as data pieces containing the contact information for all individual amino acids. In some works, seven-bit strings were used to describe those associations, while nine-bits are used in others (Desaphy et al., 2013). The fingerprint approaches are different in terms of their bit string definition and the type of interaction considered in each implementation (Desaphy et al., 2013). The initial fingerprinting approach at the residue level considered only hydrophobic and hydrogen bond contacts. This concept which was based on residues has later evolved to a new one based on the role of atoms for hydrogen-bond sites. A new concept of weighting for showing the importance of detected interactions was introduced by Kelly and Mancera (Kelly & Mancera, 2004).

The idea of this approach was to use a ligand from a co-crystal structure as a reference to score docking poses of other ligands by fingerprint similarity. The SIFT approach was shown to outperform traditional scoring functions in virtual screening and was applied successfully to cluster docking poses of individual ligands. Several extensions of the algorithms were implemented up to now, extending the accuracy of the definition of the interactions used for the fingerprint, evaluating the approach for virtual screening, scaffold docking, or the use of profile SIFTs for virtual screening (Deng et al., 2004). The largest obstacle for this approach is that a reference ligand is mandatory (Renner, Derksen, Radestock,

& Mörchen, 2008).Docking based screening were capable of generating impressive results and many novel hits were introduced by the time being. The performance of these scoring functions can nonetheless be improved through the inclusion of tools such as clustering methods based on structural similarity.

DIFFERENT IMPLEMENTATIONS, DIFFERENT SOLUTIONS

A virtual screening can cover the ligand protein interaction for more than thousands of small molecules in a data set. The key point in performing a successful virtual screening is to comprehend the correct binding sites of the receptor as well as the most favorable binding mode for the ligand (Lyne, 2002). As described the most common method to visualize the 2D ligand-protein interaction was using softwares such as LIGPLOT (Wallace, Laskowski, & Thornton, 1995). As the number of docking poses are increasing in virtual screening studies, using other automated methods for analysis of ligand protein interactions becomes crucial (Stierand, Maaß, & Rarey, 2006). The methods must be smart enough to compare the interaction patterns of each ligand and distinguish those poses with more interactions in the active site.

Structural Interaction Fingerprints (SIFT)

This is a fingerprinting approach for mining, analyzing, organizing as well as visualizing huge number of structural binding data. The basic algorithm for generating this fingerprint is to convert structural data of ligand protein interactions in to a vector which represents every bonding as well as its nature (Deng et al., 2004). Since the binding information for each ligand is encoded as 1D string, advanced filters such as machine learning clustering methods can be implemented for further analysis of the bit strings. The statistical analysis used in any fingerprinting approach can be readily used in other methods for more comparison (Learning, 2010). In SIFT method, at first key residues that are involved in all ligand protein complexes of that target are identified. The next step includes representing all key residues by means of bit strings based on their interactions with the ligands. The first bit is the master bit which determines whether any interactions exist. The second and the third bits are related to the interactions with the main chain and the side chain, respectively. The next four bits are identifying the type of the interactions. The fourth and the fifth bits are "on" or "off" according to polar and non-polar type of the residues involved in the binding. "On" values at positions 6 and 7 of the bits are determined based on the presence of donor or acceptor functional groups in the residues (Chuaqui, Deng, & Singh, 2005). The combination of all bit strings generates a unique pattern for each complex of the ligand-protein.

In order to validate this method, docking simulations were performed by four popular docking programs (FlexX, Glide, Gold, and Surflex) for generation of poses in case of three data sets. A set of 40 compounds with low-molecular-weight from the Protein Data Bank, a collection of 40 scaffolds from protein ligands, and a database of 19 scaffolds extracted from true cdk2 inhibitors.(Deng et al., 2004)

Drugscore

Drugscore-FP is also a ligand-protein interaction fingerprint that is generated by atom score partitioning of docked inhibitors as 1D vector. Similar to the previous methods, simple distance metrics allow measuring the similarities between the fingerprints of docked structures from co-crystal structures. Furthermore, drugscore makes it possible to produce family based fingerprint profiles similar to SIFT method. The profiles of interaction by drugscore identify similarities and dissimilarities related to each target. The most advantage of Drugscore-FP is that near-native-docking-poses are better identified compared to SIFT and Drugscore-CSD methods. This finding was discovered using a customized dataset by Wang et al. Finally, it was seen that drugscore can be considered as a powerful filter to recognize similar binding profiles and its efficiency was verified using cross-validation studies (Pfeffer, Neudert, & Klebe, 2008).

Based on this fingerprint many x-ray structures were studied to find out their interaction patterns, which are effective for designing inhibitors with potency and selectivity. As described, the primary fingerprinting methods presented a simple overview for the protein ligand interactions, wherein all interactions are presumed to have an equal impact. But in fact, the roles of different residues are not similar in the interactions. For example, in kinase targets the interactions at hinge regions are more important for the activity compared to those in other parts of the enzyme. This was therefore a great pitfall for fingerprinting approaches and a motivating factor for the researchers to invent new algorithms for interaction fingerprinting. To overcome this limitation, Nandigan et al presented weighted structure interaction fingerprint (w-SIFT) (Nandigam et al., 2009).

Weighted Structure Interaction Fingerprint (w-SIFT)

Based on this method the ligand-protein interactions are generated as a weighted profile in such a way that the weight of each interaction determines its importance. For example, hydrogen bonding might be different weighted from polar or non-polar interactions. Therefore, by considering the interaction data distinct from each other and determining the importance of each interaction a more realistic fingerprint can be obtained. It was also possible to consider the weight of each interaction so that a reasonable correlation between experimental activity and docking results would be obtained. These equations can be used for further analysis of drug design. To obtain the profile, the weight values must be calculated in a training set based on the correlation between experimental activity and interaction fingerprints. The weighted profile includes non-negative values within the range 0 to 1. The value, 0 belong to positions which do not have any 1 values in any of its SIFTs while 1 goes to position with at least one 1 value in that position. This type of weighting is so that those interactions which are frequently observed in high potent compounds are taking more weights. On the other hand, the interactions which are frequently observed in less active compounds are taking fewer weights in the w-SIFT profile. w-SIFTS with more weights are therefore denoting the more potent inhibitors of protein. In this method tanimoto distances are finally used to compare between different w-SIFTs (Nandigam et al., 2009).

As seen in Table 1, another version of SIFT (R-SIFT) is a fragment based method for screening combinatorial libraries.

Structural Protein-Ligand Interaction Fingerprints (SPLIF)

Another algorithm is based on calculating structural protein-ligand interaction fingerprints (SPLIF) for each docking pose and comparing it with the reference ligand-protein complex (Da & Kireev, 2014).

SPLIF presents quantitative measurement of whether a docking pose interacts with the protein target in a similar way as its known ligand and can diminish the false negative active compounds with poor initial docking scores.(Da & Kireev, 2014) A trial benchmark study of 10 diverse data sets extracted from the DUD-E database has been carried out to evaluate the ability and efficiency of this protocol.

Table 1. Different fingerprinting methods; used algorithms and advantages

	Method	Algorithm	Features
Binding site dependent IFP	w-SIFt (weighted IFP)	• Considering bit frequency in SIFT • Weighted profiles	• Finds interactions that are critical for compound potency • Interpretable position dependent scoring function for ligand protein interactions
	r-SIFt (R-group IFP)	• Incorporating binding information based on different fragments of a combinatorial library • Bits "on" or "off" changes based on their interaction of a ligand with either a specific R group or the core fragment	• Very sensitive and effective in separating different binding modes • Made combinatorial library analysis possible
Binding site independent IFP	APIF	• Map distance ranges for quadruplet interaction based on interaction patterns • Prepare an illustration of binding site as a 294 bits string • Calculates the fingerprints based on the relative positions of the interacting protein -ligand atom pairs	• Independent on the binding site • Comparable to current scoring functions • Showed better screening efficiency performance than the residue-based IF
	Pharm-IF	• Pharm-IF is calculated from the distances of pairs of ligand • Pharmacophore features that interact with protein atoms	• Independent on the binding site • Comparable to current scoring functions • Detected patterns of ligand pharmacophores would • Facilitate an understanding of the SAR of the target protein

SPLIF can be considered as an interaction fingerprint with the ability to explicitly encode structural information as well as capturing all types of ligand–protein interactions (e.g., stacking, polarization, cation-π, CH-π). (Da & Kireev, 2014)

In this regard, if the distances between two atoms are less than the predefined threshold, they are considered as being in contact with each other. Afterwards, for each protein-ligand pair, the protein and the ligand atoms are expanded to circular fragments. The third step is retrieving the 3D coordinates for the ligand-protein complex and assigning "on" and "off" bit for the contacts. A privilege for this type of fingerprinting compared to that discussed before is that in the previous algorithm the interactions types need to be encoded by empirical rules at each step. For example, in case of two parallel aromatic fragments a pi-pi interaction is anticipated.(Crowley & Golovin, 2005; Dalkas et al., 2014)

These types of interactions are merely considered if the specific criteria for them are accepted by the application. On the other hand, SPLIF is calculating any contacts between the receptor and the ligand. The ability to generate 3D fingerprints is nowadays implemented in many softwares (Vilar, Cozza, & Moro, 2008). These 3D fingerprints are utilized to express the binding mode of the ligand in the binding cavity of the receptor. These 3D fingerprints are useful in studying small sets of ligands. In case of virtual screening or studying the binding mode for a large data set, 1D bit strings must be taken.

In addition, due to analysis efficiency of IFs, they are considered as an important toolbox in post processing of docking results. Many softwares such as Schrodinger and MOE have implemented these criteria for analysis of docking results (Vilar et al., 2008).

Protein Ligand Interaction Fingerprint (PLIF)

This is a residue based interaction fingerprint developed by chemical computing group implemented in MOE software (Clark & Labute, 2007). The interaction fingerprint is generated by a presence or absence of 6 types of intermolecular interactions including hydrogen bond with donor side chain, hydrogen bond with side chain acceptor, hydrogen bond with backbone donor, and hydrogen bond with backbone acceptor, ionic attraction and surface contact. Each type of interaction can be defined as either weak or strong. Accordingly, for each residue a 12 bit string is encoded based on interaction type and strength. As described one of the most important applications of PLIF studies is their ability to be used in virtual screening or binding mode detection. Different interaction fingerprints have been developed so far. The efficiency and applicability of a PLIF algorithm are dependent upon the bit string definition implemented in the software (Clark & Labute, 2007).

A pitfall of many fingerprint methods is that a bit is related to each active site residue atom. This limits the analysis to highly homologue active sites. To overcome this problem cavity independent fingerprinting methods have emerged (Pérez-Nueno, Rabal, Borrell, & Teixidó, 2009).

Atom Pairs Based Interaction Fingerprint (APIF)

A novel fingerprinting approach called APIF (Atom Pairs based Interaction Fingerprint) which is based on atom pairs contacts for encoding protein-ligand interaction fingerprints is an example of cavity independent fingerprinting approach (Clark & Labute, 2007).

APIF has been validated by means of enrichment factor wherein poses with a similar pattern of contact towards crystallographic data were ranked as top poses. Its application as a post processing strategy using four targets HIV protease, Carboxypeptidase, trypsin and estrogen receptor α has been revealed. Comparison of GOLD score and APIF by means of enrichment factor has demonstrated its applicability in docking studies (Clark & Labute, 2007).

APIF algorithm was developed in SVL language as implemented in MOE package.(Pérez-Nueno et al., 2009) To generate the 1D fingerprint, first the center site is defined based on a 10 Å radius. In the next step the interactions between the protein and the ligands are calculated. Hydrogen bond acceptor, hydrogen bond donor and hydrophobic contact features of the ligand or receptor atoms can produce 6 combination types (Figure 3). Different distance zones are also expressed in the next step. Accordingly, based on different zone clusters and contact maps a 7*7*6=294 bit string is generated for each pose of the ligand (Figure 3) (Clark & Labute, 2007).

Fingerprints for Ligand and Protein (FLAP)

FLAP is a fast and efficient algorithm for describing small molecules in the active site. It includes using a reference frame work of 4 point pharamcophores for the ligand conformations which represent the shape of the molecule in the active site. In the first step, GRID force field is used for calculation of interaction fields for identification of target location for small molecules with favorable energy values. The second step includes generation of a 4 point pharmacophore based on molecular interaction fields (MIFs). This technique can be used for comparison of protein families without aligning them or further knowledge. It can be also used for descriptor generation in chemometrics analysis as well as docking procedure. On average the optimum number of the interactions between the receptor and ligand is considered to be at

Figure 3. An illustration of how APIF produces a 294 bit string for each pose of the ligand

Ranks of distances (d1_d2 in Å) : 1) [0-2.5] 2) [2.5-4] 3) [4-6] 4) [6-9] 5) [9-13] 6) [13-18] 7) [>18]

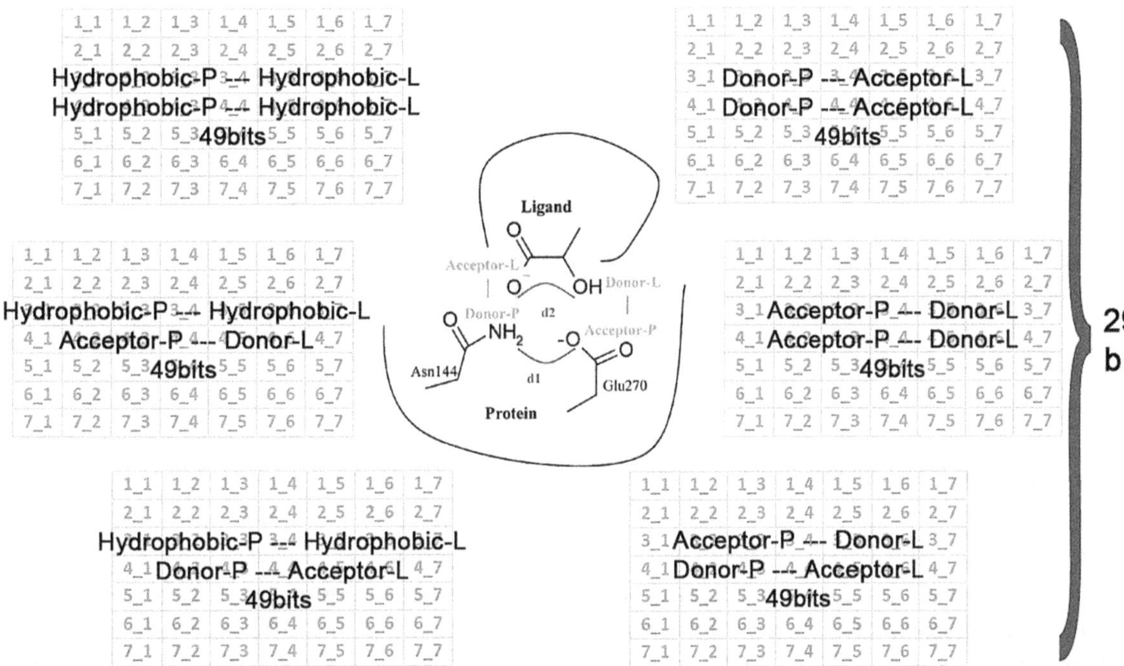

least 4. The first example of using FLAP in structure based virtual screening (SBVS) was in designing ligands for xa factor, where its efficiency and speed were shown to be more than the scoring functions implemented in GOLD and GLIDE (Baroni, Cruciani, Sciabola, Perruccio, & Mason, 2007).

Pharmacophore Interaction Fingerprint (Pharm-IF)

Pharm-IF is another method which calculates ligand protein interaction in three steps. In the first step, the interactions are identified in the complex of ligand and receptor. Then the interaction pairs are extracted based on pharmacophore features for each ligand. The most important interactions types in this method are hydrogen bond with ligand acceptor, hydrogen bond with ligand donor, ligand anion with ionic interaction, ligand cation, with ionic interaction and hydrophobic interactions. Similar to MOE in PLIF method it is also possible to consider the weak interactions.

Pharm-if is different from the algorithm implemented in MOE in terms of the two following criteria:

1. Hydrophobic interactions are recognized when two atoms have an overlap greater than 1% of their total surface.
2. Since it is possible that the hydroxyl groups inside either ligand or receptor take role in formation of hydrogen bonds, the role of each atom is ignored. It is believed that the superior results of this method are caused by these reasons.

Pharm-if can be considered as an extended method based on 3D geometry of ligand pharmacophores for docking poses. As a perspective, the area of research in the field of protein ligand interaction fingerprints is focused on the bellow subjects:

- The orientation of interactions (for example, hydrogen bonding from ligand or protein)
- Interaction strength
- Considering atom sites for generation of bit strings rather than residues (Sato, Honma, & Yokoyama, 2009).

As summarized in Table 1, different binding site dependent and independent methods have been developed for ligand-protein interaction fingerprints. While binding site dependent methods are very sensitive and effective, cavity independent methods can be efficient in developing structure activity relationship and virtual screening.

ANALYSIS OF THE FINGERPRINTS

After construction of the bit strings using any fingerprinting approaches for every docking poses, some methods should be recruited to measure their similarities for further analysis. Traditionally, four kinds of similarity coefficient have been used to evaluate equivalence of bit stings: 1) Euclidean distance 2) Manhattan distance 3) Tanimoto coefficient 4) Simple matching coefficient. A and B are numbers of bit sets in the two interaction finger prints and C is the number of common bits (Pérez-Nueno et al., 2009). Equation of Euclidean distance is:

$$\sqrt{A+B-2C} \tag{1}$$

Equation of Manhattan distance is:

$$A+B-2C \tag{2}$$

Simple matching coefficient, as called, simply shows the number of common bits in two bit strings so simple matching coefficient is C.(Pérez-Nueno et al., 2009)

Tanimoto coefficient (Tc) is the most regularly executed method to obtain similarity of bit stings among aforesaid approaches. Tc for evaluation of similarity between two sets of bits is:

$$Tc(A,B) = \; A \cap B \; / \; \; AUB \tag{3}$$

which $|A \cap B|$ is number of common ON bits in both A and B (previously defined as C), while $|AUB|$ is the number of bits in either A or B.(Deng et al., 2004) We can rearrange (3) as below: (Kelly & Mancera, 2004)

$$\frac{C}{A+B-C} \tag{4}$$

Since this method only considers ON bits, the results can be affected by unwanted size dependent effects. Therefor to bypass this disadvantage, Tc was developed as Modified Tanimoto coefficient (MT) (Kelly & Mancera, 2004)

To assess OFF bits, Tanimoto complement coefficient (T_0) can be defined as:

$$\frac{D}{A+B-2C+D} \tag{5}$$

where A, B and C are for the Tc and D shows the number of OFF bits in both A and B

ON bits are considered by Tc and OFF bits are considered by T_0, so by combining these two equations, modified Tanimoto (MT) is achieved as below:

$$\alpha Tc + (1-\alpha)T \tag{6}$$

where α is:

$$\frac{2-P}{3} \tag{7}$$

where P is:

$$\frac{A+B}{2n} \tag{8}$$

where n shows the total number of bits in A or B.(Kelly & Mancera, 2004)For example, in case of two bit strings 1100110 and 1010110. The calculation of similarity co-efficient for these bit strings are as below:

Euclidean distance (1) $\rightarrow \sqrt{7+7-(2\times 3)} = 2.82$

Manhattan distance (2) $\rightarrow 7+7-6 = 8$

Simple matching coefficient = C → = 3

Tanimoto coefficient (3) $\rightarrow \dfrac{3}{7+7-3} = 0.27$

Modified Tanimoto (6) $\rightarrow \mathrm{T0} = \dfrac{2}{7+7-4} = 0.2 \rightarrow \mathrm{p} = \dfrac{7+7}{14} \rightarrow \alpha = \dfrac{2-1}{3} = 0.33 \rightarrow \mathrm{Mt} = 0.224$

Some types of scoring functions as designed for GOLD software can be obtained from Tanimoto coefficient and the original GOLD score presented by the software. According to the bellow equation:

$$SCORE\ 2 - Tanimoto = SCORE1 - Tanimoto \times GOLD\ score\ function \qquad (9)$$

Based on this scoring system, after docking procedure, compounds that don't have any relation to expected binding mode will be allocated (Mpamhanga, Chen, McLay, & Willett, 2006).

Clustering Methods

After calculation of similarity distances for the fingerprints in case of every docking poses, they should be subjected to further analysis.(Renner et al., 2008) One way to visualize the data is to represent them as lower-dimensional models. Projection methods are hired to represent the distribution of the poses in high dimensional descriptor space into lower dimensional ones; e.g. two dimensional.(Renner et al., 2008) There are some methods called *clustering methods* for this purpose. In other words, these methods arrange data based on their similarity.

The classification methods can be categorized as two types unsupervised (clustering methods) and supervised classification methods (Höppner, 1999).

Based on the separation patterns clustering methods can be categorized as below:

- **Well-Separated:** In this type, each cluster contains objects that are similar to each other and are completely different from the objects from another cluster. Sometimes a threshold should be considered to define rate of similarity among objects. However, this type of clusters can be utilized when data has the potential of generating natural clusters with reasonable distances. In other words, similarity distance between objects of two different clusters should be more than the distances between objects of one cluster.
- **Prototype Based:** In this type, each cluster has a prototype that is representative for other objects of its cluster. In other words members of a cluster congregate around these main objects. For continuous data, prototype is generated via a *centroid* that is calculated through averaging members of a cluster. On the contrary, for categorical data where a centroid cannot be defined, a *medoid* should be hired which is the most representative member of its cluster. Prototype based cluster is also called a center-based cluster.
- **Graph Based:** The data are presented in a graph, where the nodes of the graph are objects of each cluster and relation between these nodes are illustrated by links. In this regard, a cluster can be defined as a connected component. In fact, a cluster contains members which are well connected with each other but completely irrelevant with members of other clusters. This type of cluster is useful when clusters are irregular and intertwined. But the pitfall of this method is that two clusters can be displayed as a merged one when noise is present in the data set. In other words, noise can play the role of a bridge between two distinct clusters.
- **Density Based:** In this type, a cluster can be defined as a dense region of data that is bordered by a region of low density. This clustering approach can be applied even when outliers and noises are present since they will be vanished into a low density region so any relation cannot be recognized between two distinct clusters.

Two of the most important tools used for clustering of docking fingerprints are presented as below: SOM –HCA – PCA – K-means – DBSCAN.(Bouvier, Evrard-Todeschi, Girault, & Bertho, 2010; Singh et al., 2006; Yan, Nie, Wang, & Wang, 2013)

Self-Organizing Map (SOM)

It is a sort of artificial neural network (ANN) presented by Kohonen, trained to organized complex, high dimensional, nonlinear data into a regular, understandable, low-dimensional (usually two dimensional) display. In other hand, this tool represents some kind of abstraction which reveals data more intelligible (Kohonen, 1998).

SOM often includes a two dimensional grid of nodes. This method orders models into a graph where similar models are closer than dissimilar ones (Kohonen, 1998).

For example YAN et al. used SOM for classification of Aurora kinase inhibitors. They collected 512 compounds of Aurora-A and -B inhibitors and classified them into three classes using Self-Organizing Map (SOM). These classes were dual Aurora-A and Aurora-B inhibitors, selective inhibitors of Aurora-A and selective inhibitors of Aurora-B. The prediction accuracy of the model (based on the training/test set splitting using SOM method) for the test set was 92.2% (Yan et al., 2013).

Hierarchical Clustering Analysis (HCA)

It's a method for clustering data as a dendrogram by state of similarity (Maimon & Rokach, 2005). This method follows two general strategy to organize the data:

- **Agglomerative:** It's a bottom-up approach that begins from each cluster and then they merge as one moves up *hierarchy*.
- **Divisive:** It's a top-down one which is started in one cluster and then divide into detailed clusters. In other words, there is an evolution from less detailed data into more ones.

A measure of dissimilarity between sets is needed for ordering clusters and deciding how clusters should be merged or divided, usually a metric measure of distance is hired (Maimon & Rokach, 2005).

For instance, Singh et al. utilized this approach in computing the similarity metric (Tanimoto coefficient) between all of the fingerprints and then used a hierarchical clustering algorithm to group similar interaction fingerprints together (Singh et al., 2006).

There are three methods for merging of clusters in HCA:

- **Single Link (Neighbor Joining):** This method as the most common strategy calculates distance between two closest points of two neighbor clusters. This metric is then considered as distance between these clusters.
- **Average Link:** This method calculates distances between the centroids of the clusters.
- **Complete Link:** It considers distance between farthest points of two neighbor cluster.

A limitation of HCA is that it is sensitive to outlier points and some pretreatment of the data set is therefore needed before analysis.

K Means

K-means is widely utilized as a suitable clustering method. In this method at first, K cluster centroids should be defined. After that, data will be clustered by means of vicinity of two clusters centroids. This procedure is iteratively repeated until best reasonable clusters are obtained based on the selected K. Selection of K values must be done with care based on validation methods such as cross-validation. The aforementioned method can be merely utilized for numerical data and the samples should be compared using metrics such as Euclidean distance or city-block.

As seem for HCA, K-means is also sensitive to outlier points. This sensitivity can lead to disorganized cluster centroids.

The results obtained by K-means method can be inserted into a diagram called Voronoi. The cells in this diagram are called voronoi cells (Mpahnga-CHAP 8- celera).

PCA (Principal Component Analysis)

It is considered as an analysis approach for detection of patterns in the sets of data and represents data in a way which makes it possible to reveal similarities and differences between the samples.

Since finding pattern in high dimensional space is difficult, PCA can be useful in analyzing data at a low dimensional space. In other words, PCA is a tool for decreasing number of numerical data without any distract cluster structures.

In this method, major component can be determined. After that, data will be decomposed and transferred from a big high dimensional space into a lower one without loss of data.

DBSCAN (Density Based Spatial Clustering of Applications with Noise)

In this method, space should be divided into low and high density region. Number of cluster will be determined automatically by means of some special algorithms. Low density spaces will be considered as noise and should be removed. Since some parts of data will be missed, this method is not considered as a complete clustering approach. (Höppner, 1999)

APPLICATIONS OF FINGERPRINTS AND FUTURE PERSPECTIVES

Accordingly, IFs have great potentials in scaffold hopping based on the interaction pattern of the reference ligand. This feature can be used in virtual screening to find ligands with different chemotype from the reference ligands but similar interaction patterns (Venhorst, Núñez, Terpstra, & Kruse, 2008).

Two complementary tools namely Ishape and Grim have been also developed to align protein-ligand complexes based on the intermolecular interactions obtained by the described software.

While Ishape is working by the overlap of the interactions using Gaussian function, Grim is utilizing a standard detection for providing pattern graphs. The fingerprint has focused on three main areas (Desaphy et al., 2013):

- Alignment of protein-ligand complexes based on their interaction profiles.
- Making algorithms for post-processing of docking poses according to known interaction.
- Virtual screening of bioisosteric derivatives with similar interaction behaviors.

As the applicability of IF in many studies was revealed, they became popular to score and rank docking results for more accruable hit identification. Many docking programs such as DOCK are based on using energy values of electrostatic (columbic) and steric (vanderwaals) terms. Since energy values could not always explain the best pose of the ligand foot print similarity can be a useful measure for comparison between different poses. In a study performed by Balius et al. FPS (footprint similarity) was developed for DOCK software (Balius, Mukherjee, & Rizzo, 2011). Validation of FPS was done by means of pose identification, cross-docking, enrichment and virtual screening. FPS can be explained as residue based decomposition of electrostatic, steric and hydrogen bond interactions between the ligand and the protein. Generation of bit strings was designed in such a way to make a 2D pharmacophore based on contribution of each residue in the binding cavity. Calculation of FPS was according to calculation of xi (interaction energy between the ligand and the ith residue). Where N denotes the number of residues in the receptor ($\vec{x} = \left\{x1, \ldots\ldots xN\right\}$). Different methods such as pearson and Euclidean distance were used for comparison of the generated bit-strings (Balius et al., 2011). A feature of this method is that it is also based on energy terms compared to those methods using interactions. For example in SIFT the tanimoto comparison are being used for the bit-strings which are merely based on presence or lack of the seven type interactions for each residue. SIFT was designed to be used in library design, dataset enrichment and identification of the natural ligands.

Most scoring functions, implemented in different docking programs, are based on energy terms rather than considering the right orientation, location and geometry of the poses. It's noticeable that these scoring functions alone are not sufficient to obtain accurate energy values for ligand-protein interaction. Identification of true binding modes for the ligands is also another challenging task in docking studies. Actually, most of them are based on approximations in their calculation. Many current limitations of docking are the results of implemented solvation, entropic and electrostatic terms for different protein systems. One strategy to overcome these problems is therefore developing different algorithms for post-processing of docking simulation results.(Kelly & Mancera, 2004)

An interaction fingerprint based scoring function is AuposSOM (Mantsyzov, Bouvier, Evrard-Todeschi, & Bertho). The efficiency of AuposSOM for Surflex-Dock 2.0 from the Sybyl 1.2 package was reported. In this implementation, the contacts between the ligands and the protein include hydrophobic, hydrogen bonding and coulombic interactions. The resulted vectors of contacts will be subsequently analyzed using SOM (self-organizing map) as implemented in AuposSOM software (Bouvier et al., 2010; Mantsyzov et al.). Since a bit is defined for an active site/atom/residue, the previously described methods are therefore limited to analyze interactions with highly homologous active sites sharing a fixed number of cavity-lining atoms/residues. To overcome this limitation, cavity independent fingerprints do not consider absolute but only the relative positions of pairs of protein-ligand interacting atoms. Based on this method for the case of each interaction (hydrophobic, aromatic, hydrogen bond, ionic bond, metal complexation) a psuedoatom is centered either on the interacting ligand atom, the interacting protein atom, or the geometric center of both interacting atoms. (Desaphy et al., 2013) Finally, a 210 digits fingerprint is obtained by counting all possible triplets of interaction for pseudoatoms within the six distance ranges and gathering the data of the most frequent triplets. This bit string can be used to compare any pair of protein-ligand complexes. This fingerprinting approach has been tested in case of 10 000 drug like protein-ligand complexes and made it possible to compare the interaction pattern similarity and ligand or binding site pairwise similarity. Based on this approach it was verified that the interaction pattern similarity depends on binding site similarity.

CONCLUSION

Rather than the two and three-dimensional visualization of the ligand-protein interactions, generation of bit-strings are implemented in some applications. Different protocols of protein-ligand interaction fingerprints (PLIF) have been proposed so far. While some algorithms like SIFT are dependent to the binding site of the receptor, other methods such as Pharm-IF and APIF are independent from the binding cavity. Binding site dependent methods are very sensitive and can be used for detection of the correct binding mode of the ligands. On the other hand binding site independent methods can be utilized for developing pharmacophore models for the ligands. PLIF patterns can be applied in developing more 3D descriptors for QSAR analysis as well as designing post-processing filters for virtual ligand screening projects. For a more convenient analysis, classification methods should be taken to provide an overview of the similarities between different docking results. Validation results of these methods with respect to experimental data seem to be crucial in making decisions about the usefulness of any of these presented algorithms in drug design studies.

REFERENCES

Bajorath, J. (2008). Computational analysis of ligand relationships within target families. *Current Opinion in Chemical Biology*, *12*(3), 352–358. doi:10.1016/j.cbpa.2008.01.044 PMID:18312862

Balius, T. E., Mukherjee, S., & Rizzo, R. C. (2011). Implementation and evaluation of a docking-rescoring method using molecular footprint comparisons. *Journal of Computational Chemistry*, *32*(10), 2273–2289. doi:10.1002/jcc.21814 PMID:21541962

Baroni, M., Cruciani, G., Sciabola, S., Perruccio, F., & Mason, J. S. (2007). A common reference framework for analyzing/comparing proteins and ligands. Fingerprints for Ligands and Proteins (FLAP): Theory and application. *Journal of Chemical Information and Modeling*, *47*(2), 279–294. doi:10.1021/ci600253e PMID:17381166

Bouvier, G., Evrard-Todeschi, N., Girault, J.-P., & Bertho, G. (2010). Automatic clustering of docking poses in virtual screening process using self-organizing map. *Bioinformatics (Oxford, England)*, *26*(1), 53–60. doi:10.1093/bioinformatics/btp623 PMID:19910307

Chourasia, M., Sastry, G. M., & Sastry, G. N. (2011). Aromatic–Aromatic Interactions Database, A 2 ID: An analysis of aromatic π-networks in proteins. *International Journal of Biological Macromolecules*, *48*(4), 540–552. doi:10.1016/j.ijbiomac.2011.01.008 PMID:21255607

Chuaqui, C., Deng, Z., & Singh, J. (2005). Interaction profiles of protein kinase-inhibitor complexes and their application to virtual screening. *Journal of Medicinal Chemistry*, *48*(1), 121–133. doi:10.1021/jm049312t PMID:15634006

Clark, A. M., & Labute, P. (2007). 2D depiction of protein-ligand complexes. *Journal of Chemical Information and Modeling*, *47*(5), 1933–1944. doi:10.1021/ci7001473 PMID:17715911

Crisman, T. J., Sisay, M. T., & Bajorath, J. (2008). Ligand-target interaction-based weighting of substructures for virtual screening. *Journal of Chemical Information and Modeling*, *48*(10), 1955–1964. doi:10.1021/ci800229q PMID:18821751

Crowley, P. B., & Golovin, A. (2005). Cation–π interactions in protein–protein interfaces. *Proteins: Structure, Function, and Bioinformatics*, *59*(2), 231–239. doi:10.1002/prot.20417 PMID:15726638

Da, C., & Kireev, D. (2014). Structural Protein–Ligand Interaction Fingerprints (SPLIF) for Structure-Based Virtual Screening: Method and Benchmark Study. *Journal of Chemical Information and Modeling*, *54*(9), 2555–2561. doi:10.1021/ci500319f PMID:25116840

Dalkas, G. A., Teheux, F., Kwasigroch, J. M., & Rooman, M. (2014). Cation–π, amino–π, π–π, and H-bond interactions stabilize antigen–antibody interfaces. *Proteins: Structure, Function, and Bioinformatics*, *82*(9), 1734–1746. doi:10.1002/prot.24527 PMID:24488795

Deng, Z., Chuaqui, C., & Singh, J. (2004). Structural interaction fingerprint (SIFt): A novel method for analyzing three-dimensional protein-ligand binding interactions. *Journal of Medicinal Chemistry*, *47*(2), 337–344. doi:10.1021/jm030331x PMID:14711306

Desaphy, J., Raimbaud, E., Ducrot, P., & Rognan, D. (2013). Encoding protein–ligand interaction patterns in fingerprints and graphs. *Journal of Chemical Information and Modeling*, *53*(3), 623–637. doi:10.1021/ci300566n PMID:23432543

Höppner, F. (1999). *Fuzzy cluster analysis: methods for classification, data analysis and image recognition*. John Wiley & Sons.

Kelly, M. D., & Mancera, R. L. (2004). Expanded interaction fingerprint method for analyzing ligand binding modes in docking and structure-based drug design. *Journal of Chemical Information and Computer Sciences*, *44*(6), 1942–1951. doi:10.1021/ci049870g PMID:15554663

Kohonen, T. (1998). The self-organizing map. *Neurocomputing*, *21*(1), 1–6. doi:10.1016/S0925-2312(98)00030-7

Learning, C. M. (2010). Pharmacophore-Based Interaction Fingerprint for in Silico Screening Sato, Tomohiro; Honma, Teruki; Yokoyama. *Journal of Chemical Information and Modeling*, *50*(1), 170–185. doi:10.1021/ci900382e PMID:20038188

Lyne, P. D. (2002). Structure-based virtual screening: An overview. *Drug Discovery Today*, *7*(20), 1047–1055. doi:10.1016/S1359-6446(02)02483-2 PMID:12546894

Maimon, O., & Rokach, L. (2005). *Data mining and knowledge discovery handbook* (Vol. 2). Springer.

Marcou, G., & Rognan, D. (2007). Optimizing fragment and scaffold docking by use of molecular interaction fingerprints. *Journal of Chemical Information and Modeling*, *47*(1), 195–207. doi:10.1021/ci600342e PMID:17238265

Mordalski, S., Kosciolek, T., Kristiansen, K., Sylte, I., & Bojarski, A. J. (2011). Protein binding site analysis by means of structural interaction fingerprint patterns. *Bioorganic & Medicinal Chemistry Letters*, *21*(22), 6816–6819. doi:10.1016/j.bmcl.2011.09.027 PMID:21974955

Mpamhanga, C. P., Chen, B., McLay, I. M., & Willett, P. (2006). Knowledge-based interaction fingerprint scoring: A simple method for improving the effectiveness of fast scoring functions. *Journal of Chemical Information and Modeling*, *46*(2), 686–698. doi:10.1021/ci050420d PMID:16562999

Nandigam, R. K., Kim, S., Singh, J., & Chuaqui, C. (2009). Position Specific Interaction Dependent Scoring Technique for Virtual Screening Based on Weighted Protein– Ligand Interaction Fingerprint Profiles. *Journal of Chemical Information and Modeling*, *49*(5), 1185–1192. doi:10.1021/ci800466n PMID:19415918

Pérez-Nueno, V. I., Rabal, O., Borrell, J. I., & Teixidó, J. (2009). APIF: A new interaction fingerprint based on atom pairs and its application to virtual screening. *Journal of Chemical Information and Modeling*, *49*(5), 1245–1260. doi:10.1021/ci900043r PMID:19364101

Pfeffer, P., Neudert, G., & Klebe, G. (2008). DrugScoreFP: Profiling protein-ligand interactions using fingerprint simplicity paired with knowledge-based potential fields. *Chemistry Central Journal*, *2*(Suppl 1), S16. doi:10.1186/1752-153X-2-S1-S16

Renner, S., Derksen, S., Radestock, S., & Mörchen, F. (2008). Maximum common binding modes (MCBM): Consensus docking scoring using multiple ligand information and interaction fingerprints. *Journal of Chemical Information and Modeling*, *48*(2), 319–332. doi:10.1021/ci7003626 PMID:18211051

Sato, T., Honma, T., & Yokoyama, S. (2009). Combining machine learning and pharmacophore-based interaction fingerprint for in silico screening. *Journal of Chemical Information and Modeling*, *50*(1), 170–185. doi:10.1021/ci900382e PMID:20038188

Scior, T., Bender, A., Tresadern, G., Medina-Franco, J. L., Martínez-Mayorga, K., Langer, T., & Agrafiotis, D. K. et al. (2012). Recognizing pitfalls in virtual screening: A critical review. *Journal of Chemical Information and Modeling*, *52*(4), 867–881. doi:10.1021/ci200528d PMID:22435959

Singh, J., Deng, Z., Narale, G., & Chuaqui, C. (2006). Structural interaction fingerprints: A new approach to organizing, mining, analyzing, and designing protein–small molecule complexes. *Chemical Biology & Drug Design*, *67*(1), 5–12. doi:10.1111/j.1747-0285.2005.00323.x PMID:16492144

Stierand, K., Maaß, P. C., & Rarey, M. (2006). Molecular complexes at a glance: Automated generation of two-dimensional complex diagrams. *Bioinformatics (Oxford, England)*, *22*(14), 1710–1716. doi:10.1093/bioinformatics/btl150 PMID:16632493

Tan, L., Batista, J., & Bajorath, J. (2010). Computational methodologies for compound database searching that utilize experimental protein–ligand interaction information. *Chemical Biology & Drug Design*, *76*(3), 191–200. PMID:20636330

Venhorst, J., Núñez, S., Terpstra, J. W., & Kruse, C. G. (2008). Assessment of scaffold hopping efficiency by use of molecular interaction fingerprints. *Journal of Medicinal Chemistry*, *51*(11), 3222–3229. doi:10.1021/jm8001058 PMID:18447336

Vilar, S., Cozza, G., & Moro, S. (2008). Medicinal chemistry and the molecular operating environment (MOE): Application of QSAR and molecular docking to drug discovery. *Current Topics in Medicinal Chemistry*, *8*(18), 1555–1572. doi:10.2174/156802608786786624 PMID:19075767

Wallace, A. C., Laskowski, R. A., & Thornton, J. M. (1995). LIGPLOT: A program to generate schematic diagrams of protein-ligand interactions. *Protein Engineering*, *8*(2), 127–134. doi:10.1093/protein/8.2.127 PMID:7630882

Wang, F., Liu, D., Wang, H., Luo, C., Zheng, M., Liu, H., & Jiang, H. et al. (2011). Computational screening for active compounds targeting protein sequences: Methodology and experimental validation. *Journal of Chemical Information and Modeling*, *51*(11), 2821–2828. doi:10.1021/ci200264h PMID:21955088

Weill, N., Valencia, C., Gioria, S., Villa, P., Hibert, M., & Rognan, D. (2011). Identification of Nonpeptide Oxytocin Receptor Ligands by Receptor-Ligand Fingerprint Similarity Search. *Molecular Informatics*, *30*(6-7), 521–526. doi:10.1002/minf.201100026

Yan, A., Nie, X., Wang, K., & Wang, M. (2013). Classification of Aurora kinase inhibitors by self-organizing map (SOM) and support vector machine (SVM). *European Journal of Medicinal Chemistry*, *61*, 73–83. doi:10.1016/j.ejmech.2012.06.037 PMID:22796044

Chapter 6
Different Types of Molecular Docking Based on Variations of Interacting Molecules:
Variations of Molecular Docking

Amit Das
University of Kalyani, India

Simanti Bhattacharya
University of Kalyani, India

ABSTRACT

Molecular docking plays an important role in drug discovery research by facilitating target identification, target validation, virtual screening for lead identification and lead optimization. Depending upon the nature of the disease of interest, targets can be either protein or DNA while drugs are mostly organic small molecules. Different types of molecular docking techniques like protein-protein or protein-DNA or protein-small molecule or DNA-small molecule are employed for achieving the above mentioned objectives. This chapter provides a clear idea of the position of molecular docking in drug discovery with detailed discussion on different types of molecular docking based on the varieties of interacting partners. Subsequently the authors provide a detailed list of tools that can be used for docking in drug discovery and discus some examples of molecular docking in drug discovery before concluding with a remark on future areas of improvement in molecular docking related to drug discovery.

1. INTRODUCTION

Present day drug discovery faces several major issues like rising cost, high development time and high failure rate at different stages of development. On average it takes 10-12 years and around one billion US dollar to market a drug but less than 5% of the compounds which are selected to undergo the drug-discovery pipeline, only end up being marketable. This high failure rate is associated with every stages of drug development starting from initial screening to final patient trial. Since the marketed drugs are

DOI: 10.4018/978-1-5225-0115-2.ch006

only available for revenue generation which has to cover the development cost of all the trials, including the ones which failed at different developmental stages. Thus it is easily understandable that optimizing strategies that save both time and money are of significant interest to the drug-discovery world. It should also be kept in mind that subsequent stages of drug development are also associated with exponential increase of cost. To counteract these issues, computer aided drug development (CADD) has played significant role to reduce cost and time for the last two decades and has established itself as an essential part of modern day drug discovery.

Combined with the above mentioned issues, the ever rising compound library has imposed another challenge in terms of selection of proper compounds which can be used for initial screening. While a few thousand compounds can be screened using standard high throughput screening (HTS) methods, the challenge remains to find these few thousands from the immensely large compound library which houses several tens of thousands of compounds, sometimes in the ranges of hundreds of thousands. CADD plays an important role not only in this aspect but also in two other ways – a) by computationally generating secondary modifications of compounds which saves both time and money in comparison to synthesizing them in real world, b) by analyzing the quantitative structure-activity relationship (QSAR) of the compounds of interest. All the three above mentioned ways, in which CADD plays an important role in drug discovery, require the association of the compound with the target of interest which can be either DNA or protein and the technique which facilitates this association in a computational way is known as 'Molecular Docking'. However the benefit of molecular docking extends beyond this small molecule–DNA/protein interaction as it facilitates the docking of DNA with proteins as well as of proteins with proteins.

The benefit of molecular docking is exhibited at several stages of drug-discovery. Even before the beginning of drug-discovery pipeline, molecular docking facilitates the proper understanding of several potential target molecules by facilitating DNA-protein and protein-protein interactions. Later this technique facilitates virtual screening which enables screening of compound libraries to find the potentially promising ones as well as helps in understanding of the best suitable secondary modifications. Thus it can be clearly envisioned that different types of molecular docking depends on the types of interacting molecules. Keeping in mind the scope of this chapter, further discussion on molecular docking with respect to drug discovery will be limited to different types of molecular docking.

As we have previously discussed, small molecules or ligands, DNA, proteins and short peptides are four different individual components with which molecular docking can be performed. Based on the combinations of participating components, drug discovery can be of several types. In case of protein-ligand docking, molecular docking is performed with a small molecule ligand and a protein receptor. Interaction of 'dorzolamide', used for treatment of glaucoma, with carbonic anhydrase is an example of protein-ligand docking. It should be highlighted at this point that most of the known or developed drugs fall in to this category. Platinum containing cancer drugs interact with DNA and thereby results in cell death. Studies of interactions of such drugs with DNA are facilitated through DNA-ligand type molecular docking. Azacitidine and decitabine containing drugs which are used for treatment of myelodysplastic syndrome and acute myeloid leukemia act through several mechanisms that ultimately inhibit DNA-methyltransferases. Understanding the mechanism of actions of such drugs require proper knowledge of interaction of DNA, it's bases and base analogues with DNA-methyltransferases.DNA-protein docking analyses such interactions. The tyrosine protein kinase BCR-Abl when mutated can become constitutively active and thereby causes uncontrolled target phosphorylation which ultimately causes abnormal cell growth. Imatinib effectively blocks this constitutive BCR-Abl phosphorylation activity by binding to the enzyme which also blocks the binding of target proteins. Effective understanding of imatinib actions require proper understanding of BCR-Abl and its target proteins interactions, both in the form of wild type and mutant enzyme. Protein-protein docking facilitates such studies while the interactions of imatinib with BCR-Abl kinase as well as with the enzyme-target protein complexes can be analyzed through protein-ligand and protein complex-ligand types of docking respectively. This chapter provides a detailed discussion on these several types of molecular docking based on the variations of interacting molecules.

The chapter begins with very brief discussion on basic molecular docking with an introduction to ligand, receptor and inhibitor concept. Then detailed discussions are followed on different types of molecular docking that are mentioned above. Subsequently different tools that can be used to perform molecular docking are listed to aid readers in order to make them free from the tedious work of finding out the right tool to perform their job. Thereafter this chapter discusses on case studies i.e. different known drugs which have been reported to be significantly benefitted from molecular docking. Towards the end an emerging aspect of highly potential alternative therapeutic approach, termed as 'peptidomimetics' is introduced. Finally the chapter concludes with highlighting the present day challenges which are required to overcome in order to make molecular docking and CADD more effective in drug discovery.

2. BACKGROUND

The aim of rational drug discovery is to find a small molecule which when made available within the body, reacts with its target molecule of interest to achieve some desired modification of the target molecule and thereby exert a positive effect toward disease cure. Historically drug discovery was a trial and error type of work. Although several lifesaving drugs were discovered in this way, modern day drug discovery is a much more systematic effort (Figure 1). As shown in Figure 1, present day drug discovery is broadly grouped in to two parts – 1. in-house drug research and development (R&D) and 2. clinical trials. Both of these two parts constitute of several sub parts (Figure 1). Success at the previous sub part only ensures progress towards the next sub part. It is true that several drugs have been develop by this systematic developmental pipeline but hundreds of disease are yet to find a cure even when underlying molecular reasons are known (Bhattacharya, Das, Dasgupta, & Bagchi, 2014). Thus there is an ever growing need for development of new drugs for un-treatable disease as well as development of more efficient drugs for those diseases with some level of existing treatment. To address these urgent requirements, potential drug molecules are required to be identified which can be sent for clinical trials. In house R&D is responsible for generation of these potential drug molecules. As detailed in Figure 1, the process of generation of these molecules starts with background data analysis for target identification and validation, followed by screening of compound library to find lead molecules and subsequent lead optimization which runs in conjugation with pre-clinical animal model studies.

As discussed in previous sections, drug design is a very time and resource intensive process. Thus there was always a need to apply as much automation as possible. Initially HTS proved to be a significant step forward by allowing automatization of the screening process. However preparing the resources required for HTS along with assay development is another challenge. Hence computers were brought in to this drug design and optimization arena to make the whole scenario faster, cheaper and more streamlined. Figure 1 highlights the role of CADD in this aspect. Molecular docking is an integral part of CADD and has found its application at almost all levels of in house drug R&D (Kapetanovic, 2008; Meng, Zhang, Mezei, & Cui, 2011; Shoichet, McGovern, Wei, & Irwin, 2002). Molecular docking has made it possible

Figure 1. Involvement of CADD and molecular docking in the drug discovery pipeline; drug discovery primarily constitutes of two parts – in-house R&D and clinical trials. Virtual screening and QSAR works for lead identification and optimization part of in-house R&D takes the advantage of molecular docking for better management of time and cost of drug discovery.

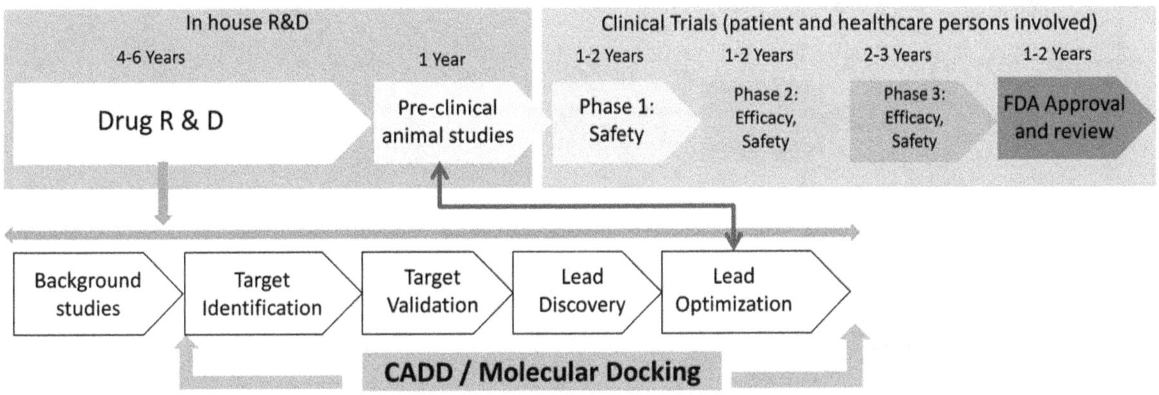

to find out suitable target, screen initial compound set to find a handful of lead molecules along with enabling QSAR analysis of optimized lead molecules. However different types of docking are required at different level of drug R&D. The following section provides a detailed discussion on different types of molecular docking, their underlying principles, advantages and fundamental limitations.

3. DIFFERENT TYPES OF DOCKING

Molecular docking requires two interacting partners. In biological, especially drug discovery context, these two partners can either be proteins or DNA or small organic molecules. Based on the type of two interacting molecules, molecular docking can be of 4 different types which are discussed in details in the following sections.

3.1 Protein-Protein Docking

As the name implies, protein-protein docking is the type of molecular docking in which both the interacting partners are proteins in nature (Figure 2). Proteins in this context are a three dimensional stable conformation of an amino acid chain. Essentially protein-protein docking algorithms do not specifically discriminate between peptides, truncated proteins, protein domain structures and full proteins (Bhattacharya, Das, Ghosh, Dasgupta, & Bagchi, 2014; Das, Bhattacharya, Bagchi, & Dasgupta, 2013) which allows uniformity of tools and algorithms. The main challenge associated with protein-protein docking is the inherent flexibility of proteins and its constituting bonds. However protein-protein docking being the major point of interest across different types of docking, several tools have been designed to take care of these issues. However the speed and output of these tools are essentially a tradeoff between the computing resources and time against the degree of flexibility allowed. Thus it is fairly easy to understand that rigid body protein-protein docking is significantly faster and less resource intensive than flexible docking while the later ensures better quality of output with a time penalty.

Figure 2. Schematic illustration of protein-protein docking; two protein molecules (one in blue, another in green) were docked with each other (A). The binding surface of blue color protein is shown in figure 2B while 2D shows the binding surface of green color protein. Intermolecular hydrogen bonds are shown in red dotted lines (A and C). Elucidation of contact surface and its properties along with residues of significance often come handy in drug discovery research.

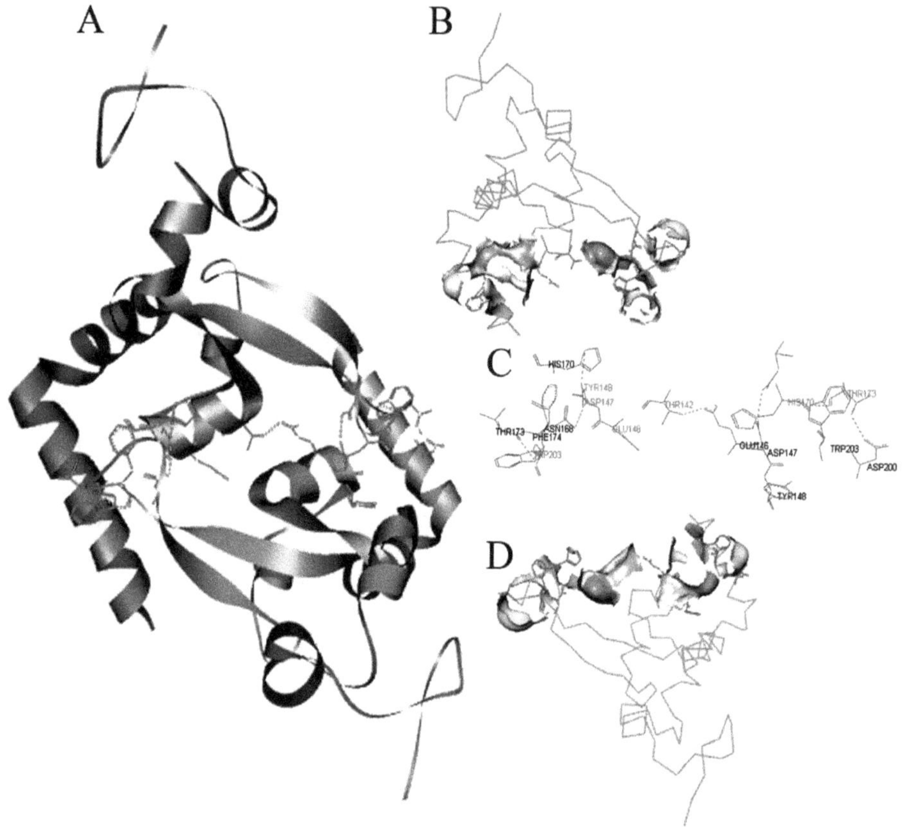

In the drug discovery context protein-protein interaction plays major role in target identification and validation i.e. at the very beginning of drug discovery pipeline (Figure 1). Before the screening of compounds for a potential lead molecule, it is essential to identify a target molecule which in most cases a protein like enzyme, receptor or membrane associated pump or similar other proteins of significant functional interest. Protein-protein docking helps in understanding the various interactions of protein molecules with their interacting partners (Figure 2B-2D). With the increase in protein structure data, protein-protein docking is becoming more and more important in unraveling significant details of bimolecular interactions which are very essential for proper target identification. Not only target identification, protein-protein docking also helps in identification of potential drug binding pocket and elucidation of several characteristics of binding surfaces and interacting residues (S. Grosdidier & Fernández-Recio, 2012; Sable & Jois, 2015).

Several tools that can be used for protein-protein docking with different degree of flexibility are discussed in a later section.

3.2 Protein-DNA Docking

DNA carries the genetic information of a cell whereas proteins are the molecules of execution. DNA-protein interactions hence are an important hub to regulate the mechanisms of replication, transcription (or gene expression) and recombination. Proteins with helix-turn-helix motif, zinc binding sites, leucine zipper form a large cluster of proteins that can bind DNA. Several transcription factors bind to DNA at promoter regions to regulate the expression of certain genes. Mutation in the promoter region or in the DNA binding region of the transcription factor can result in the improper interaction between these two partners, resulting in improper regulation of the concerned genes. A number of diseases have been associated with mutation in either of the two or in both cases (Heidenreich, Rachakonda, Hemminki, & Kumar, 2014; Songdej & Rao, 2015). However in many cases the underlying molecular mechanism either unknown or poorly understood. Protein – DNA docking can play a significant role in unraveling these unknown issues. With the progress in researches that showed the importance of protein DNA interaction in living cell environment, the research on DNA-protein docking has gained new attention (Gao & Skolnick, 2009; Takeda, Corona, & Guo, 2013; van Dijk & Bonvin, 2008).

As the name implies, in case of protein-DNA docking, there is no receptor-ligand complex since both the interacting partners are bio-macromolecules (Figure 3). However depending upon the nature of algorithm or the requirements of the docking program, sometimes it may be necessary to declare either of the molecules as a receptor and the other as a ligand but theoretically this should not make much difference in to the output. The main challenge associated with protein-DNA docking is the size of the interacting partners and their associated molecular and structural flexibility. DNA has its inherent flexibility and also there remains a lack of proper information for the particular interacting residues on DNA structures. These have made the situation unfavorable to generate effective docking methods for proper docking of DNA with protein. Computationally it is a huge task to take into account all the rotational and vibrational flexibility of each and every atom and bond during the search for the best docking pose. While some docking algorithms take the advantage of associated information to counteract this issue (Takeda et al., 2013), few other programs only allow partial flexibility (Banitt & Wolfson, 2011). However some docking tools take the advantage of advanced computing architecture to allow greater flexibility and more parameters like explicit addition of water (van Dijk, Visscher, Kastritis, & Bonvin, 2013; Wu, Hong, Takeda, & Guo, 2012).

The structural basis of protein DNA recognition process has been deciphered with X-Ray crystallography, NMR and many other biochemical and biophysical experiments. With the increasing availability of structural data on protein-DNA interaction, it can be expected that molecular docking will play greater role in discovering the underlying mechanisms of disease causing mutations of transcription factors, DNA promoters and other DNA binding proteins. Such knowledge will help to develop new drugs for several diseases.

3.3 DNA-Small Molecule Docking

Molecular interaction of ligand to its targeted macromolecules plays pivotal role in designing sensible therapeutic agents. Being the genetic material, DNA has always been the subject of interest and have been targeted with small molecules or drugs to inhibit replication, transcription, in other words, the central dogma to kill cancerous cells or tumor cells, auto-immuned cells as well as fungal, viral and pathogenic microbial growths (Rohs, Bloch, Sklenar, & Shakked, 2005). DNA is negative in charge dis-

Figure 3. Protein-DNA docking: (A) schematic illustration of docking between a protein and a DNA molecule. The region of significance of the DNA molecule is highlighted in yellow. Figure 3B highlights the binding surface of the protein while Figure 3c also shows the energy map of those residues. Identification of DNA interacting residues and their surface properties of a protein are of vital interest from drug discovery point of view while the binding region of the DNA (yellow in Figure 3A) provides the consensus sequence required for proper protein binding.

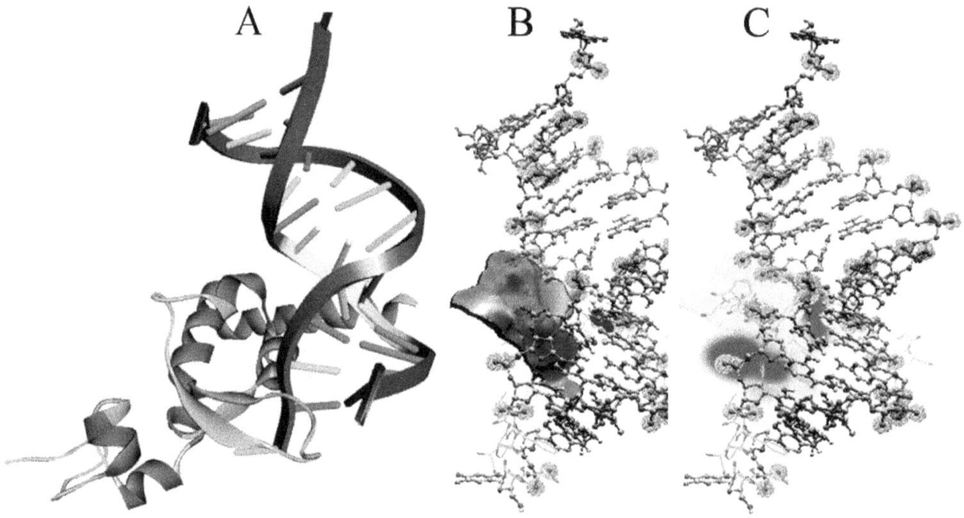

tribution and hence it attracts molecules with positive charges (Figure 4). Based on structural scaffolds and orientation flexibility, small molecules bind to either major or minor groove of DNA molecule and therefore interfere with replication and transcriptional machineries which in turn mostly activates the cellular apoptotic pathway (Pandya, Islam, Kumar, Jayaram, & Kumar, 2010). The biological activity of these molecules depend on slow rate of dissociation from DNA complexes which affects intra and intermolecular hydrogen bonds, planar interaction between DNA bases and small molecules and several other van der Waal interactions.

Molecular docking between DNA and small molecule takes in to account the above mentioned fundamental aspects of DNA-small molecule interaction during drug development and optimization. Selection of cation carrier molecules, their orientations (cis-trans), poses, DNA binding groove, binding affinity etc can be analyzed from DNA-small molecule docking (Figure 4B-4C). Similar to Protein-DNA docking, availability of proper tools is also a challenge with DNA-small molecule docking.

3.4 Protein-Small Molecular Docking

Proteome world in a cell depends on the central dogma: sequence, structure and function. The last trail of this pattern requires the knowledge whether a structurally fit protein can interact successfully with its partner proteins. If the former question gets its answer then determining the active sites and residues responsible for interaction play central role to build up the stage to study protein interaction world. To cure diseases, medical chemistry needs drugs or small molecules to be delivered successfully to the ailed system. Their bio-activity, stability and reduced toxicity form the key parameters for proper development of a lead compound. But at the same time, the mode of interaction of these small molecules with their

Figure 4. DNA-small molecule interaction: (A) schematic diagram of two platinum containing small molecules after being docked to a DNA molecule. The small molecules are shown in blue spheres while the interacting guanine residues of the DNA are highlighted in yellow. (B-C) Two different representations of the Ligand map of one the small molecules with DNA. Understanding the interaction map of small molecules with DNA is of prime importance for the development of drugs that target DNA.

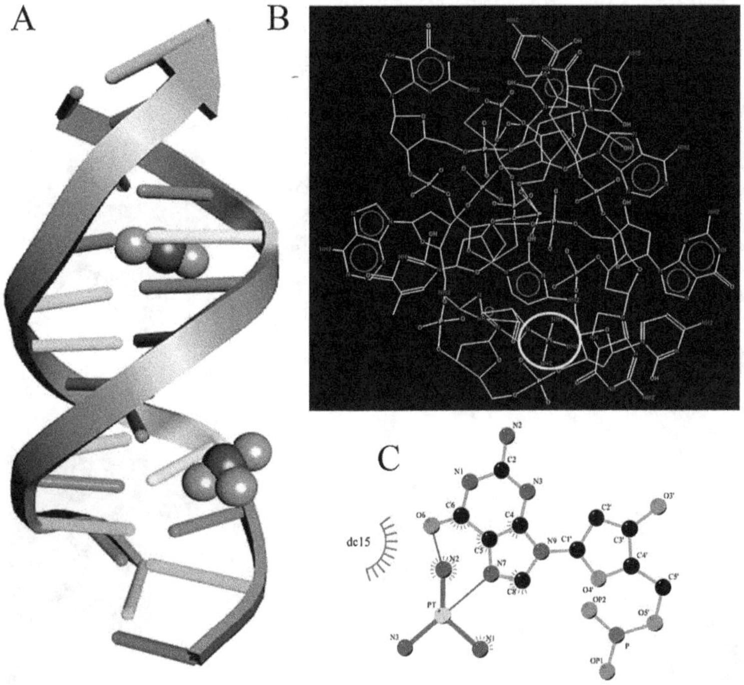

respective cellular receptor or target protein form the golden stone to pave the path in curing diseases. Not only working as therapeutics, protein small molecule interactions are key factors in enzymatic reactions and signaling cascades.

Protein-small molecule docking takes the center stage of modern day drug discovery as most of drugs available at market are organic small molecule in nature and bind to protein targets like enzymes, cell receptors, membrane pumps etc (Figure 5). This form of docking is associated with virtual screening, lead discovery, QSAR studies and lead optimization. Thus several tools are available to dock proteins with small molecules as well as to analyze different parameters of their interactions (such tools are mentioned in a later section).

In general, protein-small molecule docking is classified into two categories: bound docking where a complex is reproduced based on the template known complex and unbound/ de-novo docking where scoring functions and probability estimation generate the idea to dock the residue to target protein (Taylor, Jewsbury, & Essex, 2002). All small molecules docking program have three principal components: understanding the molecular architecture, generating search algorithms and assigning scoring functions for energy evaluation. While variety of search algorithms have been developed to study protein small molecule docking, less attention has been given to the relative side chain flexibility of the proteins involved in these interactions (Meiler & Baker, 2006). Side chain conformational changes frequently occur upon small molecule (ligand/drug) binding. But recent studies have come up with plausible con-

Figure 5. Example of protein-small molecule docking: (A) an organic small molecule is docked to a multi pass trans-membrane protein (A) side view, (B) top view. The small molecule is shown in yellow ball and stick while the protein is colored according to its secondary structure. C-D. Energy plot of this interaction (C-side view, D-top view). E. Interaction map between this small molecule and the interacting residues of the protein. F-H. Ligand interaction surface of the protein (F) top view, (G-H) two different side views. Deciphering the ligand interacting residues of a protein and characteristics of these important residues have pivotal role in optimization of drug molecules. Also the characteristics of this binding pocket determine the nature of the interacting ligand.

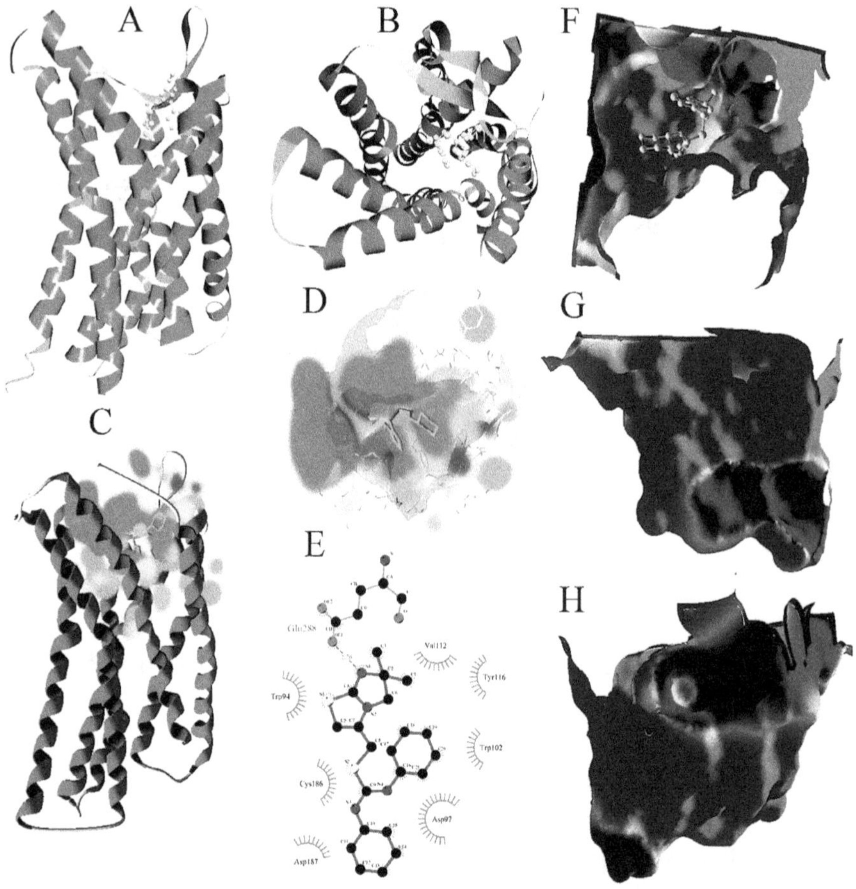

sideration to simulate the real world interactions which has brought the protein-small molecule research at its ease. Moreover, ligand flexibility has also been taken into account with the application of Monte Carlo sampling of torsion angles in small molecules. Search algorithms have dealt with the scoring functions majorly, but de-solvation and entropy functions remain the two major problems to solve. Finally an effective benchmark system for the evaluation of the robustness in scoring functions also needs be taken care of (Huang & Zou, 2010).

Beyond the computational and algorithmic view of protein-small molecule docking, there lies another physiological view of understanding protein-small molecule docking. Based on the nature and outcome of an interaction between a protein and a small molecule, small molecule ligands can be broadly classified in to two groups – agonists and antagonists.

Figure 6. Relationship between physiological effects of agonists, antagonists and inverse agonists in a concentration dependent manner; while agonists help in activation of its receptor, inverse agonists exert opposite effects. Antagonists attenuates the effect of any agonist.

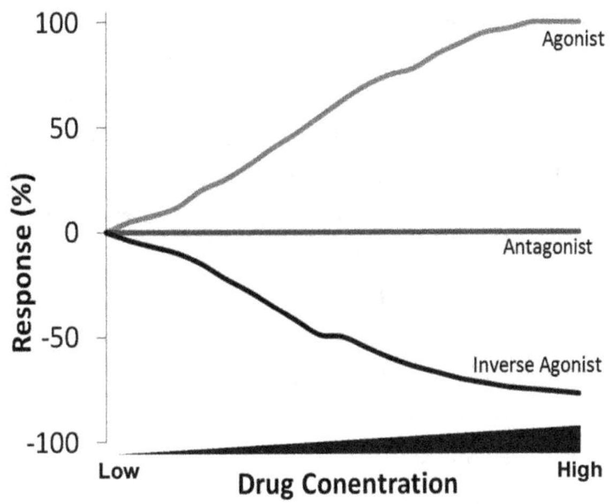

Agonists are nature identical small molecules that upon binding to its receptor, activates it while inverse agonists exert opposite effect (Figure 6). For example a drug meant for vasodilation would be acting as an agonist if upon administration it dilates blood vessels while an inverse agonist will cause vasoconstriction. Both agonists and inverse agonists bind to the same receptor binding site and its effects can be either full or partial.

Antagonists are small molecules that attenuate the effect of an agonist (Figure 6). They can be either competitive or non-competitive in nature and in each case can be reversible or irreversible in nature. Competitive antagonists bind to the same agonist binding site of a receptor while non-competitive ones bind to some other allosteric site of the receptor. Reversible antagonists bind non-covalently and can be out-competed in a concentration dependent manner while non-reversible antagonists form covalent bonds and often exert permanent effects on the receptor. For example Naloxone acts as an opiate antagonist by binding to the same receptor binding site of opiate agonists.

Information available on drug agonists and antagonists have been widely used by researchers for different purposes (Kim, Gao, Jeong, & Jacobson, 2006; Kortagere et al., 2010; Lu et al., 2010; Ng et al., 2014). Not only basic research usage, drug agonist and antagonist information are of equally high importance in the drug discovery scenario. During screening of molecules as well as during lead development and optimization, agonist and antagonist information (pose, rotamers, binding affinity, nature of interaction etc.) are taken in to consideration to select and optimize a potential drug molecule base on its intended purpose.

4. TOOLS FOR DIFFERENT TYPES OF MOLECULAR DOCKING

To facilitate molecular docking, several tools have been developed by researchers and companies over the years. Table 1 provides a detailed list of such tools which have proven to be reliable and have been popular across wide range of researchers which include basic science scientists as well as researchers or experts specifically involved in drug discovery projects. It should be noted that not all tools facilitate all the previously mentioned types of molecular docking. Table 1 also highlights the type of docking that can be performed by each tool. Most of these tools are online where the users can either provide their custom receptor and ligands (mostly in the .pdb format) or can select the receptors or ligands from standard known databases like the protein data bank (PDB, available at www.rcsb.org). Online tools or servers enable the users to make the most of the facility without any added cost (both hardware and software wise) along with enabling multiple users to be able to submit their jobs which can be run simultaneously. However several online servers are only accessible to academic users and thereby require an academic affiliation (in most cases an academic email id is sufficient). Standalone programs either free or paid are required to be installed or compiled on local computers, often a very high end desktop computer or a standard workstation. These programs provide the users the freedom to modify it as per their requirements as well as ensure data security.

As we have already described, molecular docking, in drug discovery context, is significantly involved in virtual screening of compounds. Several docking programs have been developed with this added screening facility. Among the softwares or Standalone programs mentioned in Table 1, some of them feature this added facility. Such programs are mentioned as 'virtual screening' enabled in the remarks section of Table 1. There are some other programs like MSDOCK (Sauton, Lagorce, Villoutreix, & Miteva, 2008), EUDOC (Pang, Perola, Xu, & Prendergast, 2001) etc. which have been specifically designed to aid drug discovery by facilitating virtual screening and drug interaction site identification.

A list of useful resources related to docking and virtual screening is available at the Click2Drug repository which can be accessed via http://www.click2drug.org/bibliography/docking/webservices.html and http://www.click2drug.org/bibliography/screening/webservices.html.

5. INDUCED FIT DOCKING: OPTIMAL BALANCE BETWWEN FLEXIBILITY AND RESOURCE COST

In principle, to obtain the best possible ligand receptor pose (in the form of complex or docked structure), an explicit solvent MD simulations with the presence of ligand and receptor structures would prove to be best solution. However it is quite evident from the previous discussions that significant time and resource cost would implemented to obtain such level of precision. Additionally different force field parameters, optimization of simulation environment add further level of complexity. To reduce such complex setup without much sacrifice in the quality of docking outcome, induced fit docking has been established as an optimal solution (Xu & Lill, 2013).

The biological principle behind induced fit docking is that in real life, protein, unlike computer structures (for example pdb structures), are flexible molecules and their active site residues often incorporate certain degrees of rotational freedom to accommodate the ligand in most effective conformation. Even there are certain receptors which only attain ultimate stable confirmation when bound to ligand. So impact of presence of ligand on receptor structure and conformation (especially of active site) is beyond doubt.

Induced fit docking takes in to account the above mentioned ligand impacts to figure out best ligand and receptor active site poses to facilitate the following benefits:

1. Generation of active ligand conformation that cannot be simply docked in to a rigid body receptor.
2. Rescue of poorly scoring true ligands (false negatives) during virtual screening experiments.

Table 1. List of docking servers and softwares

Sl. No.	Server / Software Name	Type of Docking Performed	Online Server / Standalone Program (O/S)	Free/Paid (F/P)	Remark	Reference
1	AutoDock / AutoDockVina + AutoDock Tools	Protein-Ligand	S	F	Flexible docking	(Morris et al., 2009; Trott & Olson, 2010)
2	FlipDock	Protein-ligand	O	F	Flexible docking	(Zhao & Sanner, 2007)
3	HADDOCK	Protein-protein, protein-DNA	O & S	F	Data / information driven flexible docking	(de Vries et al., 2007; Dominguez, Boelens, & Bonvin, 2003; van Dijk, van Dijk, Hsu, Boelens, & Bonvin, 2006)
4	ClusPro	Protein-protein	O	F	Best outputs are filtered from several poses generated by the program.	(Comeau, Gatchell, Vajda, & Camacho, 2004a, 2004b)
5	PyDock	Protein-protein	O	F	Rigid body docking, require academic email	(Cheng, Blundell, & Fernandez-Recio, 2007; Gabb, Jackson, & Sternberg, 1997; Jiménez-García, Pons, & Fernández-Recio, 2013)
6	Patchdock	Protein-protein, protein-ligand	O	F	Rigid body docking, outputs can be further refined by Firedock	(Andrusier, Nussinov, & Wolfson, 2007; Duhovny, Nussinov, & Wolfson, 2002; Mashiach, Schneidman-Duhovny, Andrusier, Nussinov, & Wolfson, 2008; Paquet & Viktor, 2013)
7	Hex	Protein-protein, protein-ligand, DNA-ligand, DNA-protein	O & S	F	Rigid ligand	(Macindoe, Mavridis, Venkatraman, Devignes, &Ritchie, 2010)
8	GRAMM-X	Protein-protein	O	F		(Tovchigrechko & Vakser, 2006)
9	SwarmDock	Protein-protein	O	F	Flexible docking	(Torchala & Bates, 2014; Torchala, Moal, Chaleil, Fernandez-Recio, & Bates, 2013)
10	rDock	DNA/RNA-ligand, protein-ligand	S	F	Can be used for virtual screening	(Ruiz-Carmona et al., 2014)
11	GOLD	Protein-ligand	S	P	Can be used for virtual screening, lead optimization	
12	ArgusLab	Protein-ligand	S	F	Can also be used for molecular modeling and graphics purposes	http://www.arguslab.com/arguslab.com/ArgusLab.html_ (last accessed June 2015)
13	DockingServer	Protein-ligand	O	P	Restricted usage with free registrations	http://www.dockingserver.com/web_(last accessed June 2015), (Bikadi & Hazai, 2009)
14	1-Click docking/mcule.com	Protein-ligand	O	P	Choice of ligand can be drawn or created, limited free usage	
15	EADock DSS/ SwissDock	Protein-ligand	O	F		(A. Grosdidier, Zoete, & Michielin, 2011a, 2011b)

continued on following page

Table 1. Continued

Sl. No.	Server / Software Name	Type of Docking Performed	Online Server / Standalone Program (O/S)	Free/Paid (F/P)	Remark	Reference
16	Molsoft ICM Docking (Pro package)	Protein-ligand, protein-protein	S	P	Additional features like structure analysis, 3D editor, virtual screening, structure prediction etc.	
17	Glide	Protein-ligand	S	P	Flexible docking, virtual screening possible.	(Friesner et al., 2004; Repasky, Shelley, & Friesner, 2007)
18	Molegro Virtual Docker	Protein-ligand	S	P	Flexible docking, virtual screening possible.	
19	FlexX-HYDE	Protein-ligand	S	P	Flexible docking, virtual screening possible, can integrate metal ions and flexible water molecules	
20	FRED	Protein-ligand	S	P	Rigid docking, virtual screening, command line interface	
21	Surflex-Dock (Tripos SYBYL-X package)	Protein-ligand	S	P	Virtual screening possible.	(Jain, 2003)
22	MOE (from CCG)	Protein-ligand	S	P	Combined with several other packages for drug discovery.	
23	Pardock	Protein-ligand	O	F	Requires some initial molecule editing like hydrogen addition	(Gupta, Gandhimathi, Sharma, & Jayaram, 2007)
24	Discovery Studio – Biovia (formerly Accelrys)	Protein-protein	S	P	Combined with several other packages for drug discovery, uses Z-DOCK algorithm for molecular docking	
25	RosettaDock/ ROSIE	Protein-ligand, protein-protein	O	F		(Lyskov & Gray, 2008; Lyskov et al., 2013)
26	Paradock	Protein-DNA	O	F	Rigid protein, flexible DNA	Not yet reported
27	iGEMDOCK v2.1	Protein-ligand	O	S	Purely flexible or Hybrid mode	(Yang & Chen, 2004)
28	Lead Finder	Protein-ligand	P	S		

continued on following page

Table 1. Continued

Sl. No.	Server / Software Name	Type of Docking Performed	Online Server / Standalone Program (O/S)	Free/Paid (F/P)	Remark	Reference
29	LigandFit	Protein-ligand	O	S	Flexible ligand	(Terwilliger, Adams, Moriarty, & Cohn, 2006; Terwilliger, Klei, Adams, Moriarty, & Cohn, 2006; Venkatachalam, Jiang, Oldfield, & Waldman, 2003)
30	LigDockCSA	Protein-ligand				(Shin et al., 2011)
31	MADAMM	Protein-ligand			Flexible docking	(Cerqueira, Bras, Fernandes, & Ramos, 2009)
32	PLANTS (Protein-Ligand ANT System)	Protein-ligand	O (for academic users)	S	Side chain flexibility, additional facilities like explicit water, virtual screening etc.	(Korb, Möller, & Exner, 2010; Korb, Stützle, & Exner, 2007, 2009)
33	SODOCK	Protein-ligand	O	S	Flexible docking	(Chen, Liu, Huang, Hwang, & Ho, 2007)
34	3D-Garden	Protein-protein	O	F	Flexible docking	(Lesk & Sternberg, 2008)
35	GPU.proton. DOCK	Protein-protein	O	F	Flexible docking	(Kantardjiev, 2011)

An induced fit docking algorithm primarily undertakes following steps:

1. Preparation of receptor molecule,
2. Identification of flexible residues, including potential ones,
3. Rotamer sampling of those flexible residues,
4. Interaction analysis and sampling,
5. Docking and complex/pose refinement,
6. Scoring of complexes,
7. Selection of best outputs

There have been numerous studies taking advantage of induced fit docking (Nabuurs, Wagener, & de Vlieg, 2007; Zhong, Tran, & Stang, 2009) and now a day it has become a pretty common feature of many softwares (example: Schrodinger Glide) or servers (example: Fleksy) that facilitate virtual screening to incorporate these protocols for generation of better outputs.

6. DOCKING AND DRUG DISCOVERY

Since molecular docking is an integral part of present day drug discovery procedure, almost all the drugs that come to market or clinical trial are designed by CADD to a significant extent. Docking being an essential part of CADD, contributes to a great extent in the development of such drugs. Notably development of several anti-HIV drugs has been associated with CADD and molecular docking (Esquieu et al.,

2003; Liao & Nicklaus, 2010). A number of drugs like Dorzolamide, Captopril, Oseltamivir, Aliskiren, Saquinavir, Zanamivir, Rupintrivir, Boceprevir, Nolatrexed etc have been designed through CADD and progressed to the level of clinical trials or marketed after regulatory approval (Talele, Khedkar, & Rigby, 2010; Zhang et al., 2013). These molecules are very well representatives of protein-small molecule docking.

With the advent of multi-drug resistant (MDR) varieties of *Mycobacterium tuberculosis* and other MDR type causative agents for visceral leishmaniasis and several other pathogenic diseases; development of new drugs are becoming essential for the survival of a significant portion of human population. Not only these few diseases, several cancers often develop MDR properties after a few doses of standard treatment. The ever changing influenza viruses are also causing greater number of causalities with each passing year. The two main benefits of molecular docking and CADD are its abilities to save time and money and these two apparently seems to be the best weapon in hand to deliver new drugs to address the above mentioned challenges within a foreseeable future.

7. FUTURE RESEARCH DIRECTIONS

Drug discovery, especially in today's context, is becoming even more dependent on computational techniques that ensure best utilization of time and money. But it should be acknowledged that neither computers can alone deliver a magic bullet nor computers, however advanced it may be, can take in to account all parameters at a time. All the biological interactions of a cell cannot be simulated by any computer in a reasonable time frame. Thus the entire process of a drugs entry to a human body till it is metabolized and subsequently excreted can only be mathematically modeled, not simulated (with the consideration of all degrees of freedom). Thus the future areas of development which bear the potential of efficient use of molecular docking in drug development can be broadly classified in to two parts – 1. technological advancement and 2. methodological changes.

Technologically there are two areas of development. First efficient use of present day computer architecture, especially the general purpose graphics processing unit (GPGPU) promises a significant level of advancement in the speed of molecular dynamics (MD) simulations. It is beyond any point of argument that flexible docking with MD simulations ensures best output. However additional time requirements and hardware associated costs have often been limiting factors in flexible docking and MD simulations. To counter these issues, rigid body docking was the choice of many. However with the advent of advanced computing facilities like GPGPU, MD simulation jobs have become significantly faster. These present day multi-core (often in the range of several thousands) GPGPUs can significantly accelerate MD simulations by offloading a huge portion of load from the computer/workstation CPU (with essentially run in serial mode and only contain a few cores, often at a maximum number of 20 in high end standalone workstations). Present day MD simulation programs like GROMACS or NAMD (Phillips et al., 2005) are intrinsically equipped to harness the power of a computer architecture created by pairing of GPGPU with CPUs. These GPGPU-CPU combinations (both as single node and cluster) not only significantly lowers the simulation time but also enables users to take in to account greater number of molecular parameters without much time penalty and added cost. GPU based algorithms of molecular docking should be developed to generate better quality of outputs compared to rigid body docking. The added benefit of all atom flexible docking with MD simulations which can easily be performed at a GPGPU-CPU combined system, is that it can be employed to docking studies with all kinds interacting molecules i.e. DNA, protein and small molecules or ligands. If technological advancements (mainly in the form of efficient algorithms) can be directed to harness the power of upcoming computing architectures, molecular docking would be able to provide greater in-sights in to the interaction patterns of biological macromolecules and/or their interacting small molecules in much greater details and that too within a reduced time frame.

Methodologically docking in drug discovery has been traditionally associated with virtual screening, lead discovery and optimization steps. While the contribution of molecular docking in saving valuable time and billions of dollars cannot be ignored in any way, it should also be acknowledged that the much hyped benefit of CADD, QSAR or docking has failed to a significant extent to keep up to their expectations. However it is very important to understand the reason behind such gap of expectations and achievements of molecular docking in drug discovery. One fundamental point in this aspect is what to dock with which i.e. what molecule to dock to which molecule. Much like the exponential increase in protein or DNA structural data, compound libraries have also been increasing exponentially over the years. Despite such remarkable achievements, not many drugs with significant potentials have made it to the market in recent years. Thus it can be clearly understood that there lies some issues with the best utilization of this increased structural and chemical data. Several factors like administrative, financial, intellectual property etc. can be linked to the causes of such issues. It is very essential to overcome such factors in order to make best use of present day and upcoming computational facilities and docking algorithms to achieve increased success with molecular docking in drug discovery. Greater collaboration between industry and academia as well as sharing of molecule libraries are potentially the avenues of achieving higher success with molecular docking based lead discovery and optimization aspects. Further distributed computing architecture can be employed globally to increase the rate of docking with higher degree of flexibility.

Finally it should be noted that the greatest benefit of molecular docking has been in the area of virtual screening which has significantly reduced the time, effort and expenditure of assay development, result interpretation and infrastructural requirement of standard high throughput screening. However it should be noted that to make more reliable outputs with maximum specificity, both virtual screening and HTS should be employed to a reasonable level while virtual screening should be allowed to use as many parameters as possible (Ferreira et al., 2010; Gschwend, Good, & Kuntz; Okimoto et al., 2009)

Beyond such apparently traditional solutions of making molecular docking more efficient in drug discovery, another approach which promises much greater potential is the use of peptidomimetics. It makes great use of protein-protein docking during its design, optimization and efficacy analysis part.

7.1 Peptidomimetics and Drug Discovery (Role of QSAR and Molecular Docking)

A part of present day drug discovery aims at developing modified peptides with increased stability and better pharmacokinetic trends. In short this technique is known as peptidomimetics and can be defined as synthetic molecules that are developed by modifying the natural peptides or proteins. Peptidomimetics are applied in targeted chemotherapy, in mitochondria apoptosis, as antimicrobial and antiviral agents as well as in cosmetics production (Walensky et al., 2004). Peptidomimetics can be synonymous with Molecular mimicry to a certain extent (Vagner, Qu, & Hruby, 2008). Recently, bioactive peptides are used as powerful tools for initial investigation in disease biology. However, these natural peptides are associated with certain problems: 1. physiological stability, 2. lack of oral bio-availability, 3. poor DMPK properties and 4. prone to peptidase digestion and susceptibility to proteolysis (Vagner et al.,

2008). Peptidomimetics is employed to broadly circumvent these problems associated with natural peptides. The pharmacophores (hotspots which are necessary for the molecular recognition of a ligand by a biological macromolecule) of peptidomimetics retain the ability to interact with the target proteins and help to mediate the natural biological activities.

Various peptide bond modifications are designed and synthesized to obtain natural peptide analogues with improved pharmacological properties. These added advantages in these compounds come from significant differences in their structures. Such modifications contribute to protect the peptidomimetics from protease cleavage and significantly improve their metabolic stability. Structure activity relationship (SAR) is the founding step in designing a peptidomimetic compounds. It defines minimal active region and key residues required for identical biological effects. The binding relationships of target molecule and the peptide analogues are evaluated and this evaluation is done majorly through amino acid scan, truncations and N- and C- terminal modifications. This is followed by stable bioactive conformation generation. The stepwise structural modifications require computational approaches, whereas the confirmation for bioactive set generation largely depends on experimental evaluation. In all these steps of peptidomimetic design and optimization, molecular docking plays crucial role as it technically eliminates the experimental part of synthesizing each and every modifications and performing their binding assays. By enabling virtual screening and in-silico analysis, molecular docking allows fast and efficient screening of peptidomimetic compounds (Hruby, al-Obeidi, & Kazmierski, 1990; Kharb, Meenakshi, Sharma, & Yar, 2011; Singh, Dolphin, Razkin, & Dumy, 2006).

8. CONCLUSION

CADD has been well established as an approach to accelerate in house R&D related to drug discovery. Among several aspects of CADD, target identification and validation as well as lead identification and optimization are the most important parts for couple of reasons. First maximum numbers of failures occur at these steps of drug discovery. Secondly these steps being the early steps of drug discovery pipeline, failure at these steps incur less time and economical penalty. Computerization and automatization of these steps to a significant extent using molecular docking techniques have further reduced the above mentioned penalties in case of failures. Depending upon the type of disease of interest, protein-protein or protein-DNA or protein-ligand or DNA-ligand types of docking programs can be employed for proper identification of the potential drug target. Thereafter in-silico mapping techniques can be employed to figure out potential drug binding pockets which in turn save lot of experimental time and other resources. Subsequently computer based virtual screening, which essentially employs protein-ligand / DNA-ligand docking algorithms, can be used to screen tens of thousands of molecules to pick up only a few lead molecules. Such lead molecules can be used, in conjunction with limited experimental studies, for efficacy analysis using QSAR. Based on the outputs of QSAR and other efficacy analyses, in-silico secondary modifications of lead molecules can be generated which can further be subjected to molecular docking with its receptor of interest. Jointly these molecular modeling and docking techniques enable faster generation of all possible secondary modifications as well as analyses of potential impact on efficacy due to the introduced changes. All these steps, if were performed in traditional ways, would have posted several challenges like – synthesis and purification of all the modified lead molecules and their subsequent storage and structural validation using NMR or similar techniques, experimental analyses of their efficacy and manual comparison of efficiency of all the incorporated modifications. Thus it is fairly easy to understand the benefit of different types of molecular docking techniques on different segments of the R&D part of drug discovery. Especially the ability to virtually interact all possible types of bio-macromolecules with each other and other organic ligands has made molecular docking and indispensable part of present day drug discovery. However time has come to explore more efficient docking algorithms for better accuracy of output prediction as well as identification of newer avenues of implementations of various docking techniques to further accelerate drug discovery.

REFERENCES

Andrusier, N., Nussinov, R., & Wolfson, H. J. (2007). FireDock: Fast interaction refinement in molecular docking. *Proteins*, *69*(1), 139–159. doi:10.1002/prot.21495 PMID:17598144

Banitt, I., & Wolfson, H. J. (2011). ParaDock: A flexible non-specific DNA--rigid protein docking algorithm. *Nucleic Acids Research*, *39*(20), e135. doi:10.1093/nar/gkr620 PMID:21835777

Bhattacharya, S., Das, A., Dasgupta, R., & Bagchi, A. (2014). Analyses of the presence of mutations in Dystrophin protein to predict their relative influences in the onset of Duchenne Muscular Dystrophy. *Cellular Signalling*, *26*(12), 2857–2864. doi:10.1016/j.cellsig.2014.09.006 PMID:25220406

Bhattacharya, S., Das, A., Ghosh, S., Dasgupta, R., & Bagchi, A. (2014). Hypoglycosylation of dystroglycan due to T192M mutation: A molecular insight behind the fact. *Gene*, *537*(1), 108–114. doi:10.1016/j.gene.2013.11.071 PMID:24361964

Bikadi, Z., & Hazai, E. (2009). Application of the PM6 semi-empirical method to modeling proteins enhances docking accuracy of AutoDock. *Journal of Cheminformatics*, *1*(1), 15. doi:10.1186/1758-2946-1-15 PMID:20150996

Cerqueira, N. M. F. S. A., Bras, N. F., Fernandes, P. A., & Ramos, M. J. (2009). MADAMM: A multistaged docking with an automated molecular modeling protocol. *Proteins*, *74*(1), 192–206. doi:10.1002/prot.22146 PMID:18618708

Chen, H.-M., Liu, B.-F., Huang, H.-L., Hwang, S.-F., & Ho, S.-Y. (2007). SODOCK: Swarm optimization for highly flexible protein–ligand docking. *Journal of Computational Chemistry*, *28*(2), 612–623. doi:10.1002/jcc.20542 PMID:17186483

Cheng, T. M.-K., Blundell, T. L., & Fernandez-Recio, J. (2007). pyDock: Electrostatics and desolvation for effective scoring of rigid-body protein-protein docking. *Proteins*, *68*(2), 503–515. doi:10.1002/prot.21419 PMID:17444519

Comeau, S. R., Gatchell, D. W., Vajda, S., & Camacho, C. J. (2004a). ClusPro: a fully automated algorithm for protein-protein docking. *Nucleic Acids Research*, *32*, W96–9. http://doi.org/10.1093/nar/gkh354

Comeau, S. R., Gatchell, D. W., Vajda, S., & Camacho, C. J. (2004b). ClusPro: An automated docking and discrimination method for the prediction of protein complexes. *Bioinformatics (Oxford, England)*, *20*(1), 45–50. doi:10.1093/bioinformatics/btg371 PMID:14693807

Das, A., Bhattacharya, S., Bagchi, A., & Dasgupta, R. (2013). Structural insight into the formin binding property of human formin binding protein 4. *Journal of Proeins and Proteomics*, *4*(December), 93–100.

De Vries, S. J., van Dijk, A. D. J., Krzeminski, M., van Dijk, M., Thureau, A., Hsu, V., & Bonvin, A. M. J. J. et al. (2007). HADDOCK versus HADDOCK: New features and performance of HADDOCK2.0 on the CAPRI targets. *Proteins*, *69*(4), 726–733. doi:10.1002/prot.21723 PMID:17803234

Dominguez, C., Boelens, R., & Bonvin, A. M. J. J. (2003). HADDOCK: A protein-protein docking approach based on biochemical or biophysical information. *Journal of the American Chemical Society*, *125*(7), 1731–1737. doi:10.1021/ja026939x PMID:12580598

Duhovny, D., Nussinov, R., & Wolfson, H. (2002). Efficient Unbound Docking of Rigid Molecules. In *Algorithms in Bioinformatics* (pp. 185–200). Springer Berlin Heidelberg; doi:10.1007/3-540-45784-4_14

Esquieu, D., Péloponèse, J.-M., Opi, S., Gregoire, C., de Mareuil, J., Watkins, J., & Loret, E. P. et al. (2003). Discovery of a Tat HIV-1 inhibitor through computer-aided drug design. *Spectroscopy*, *17*(4), 639–645. doi:10.1155/2003/283731

Ferreira, R. S., Simeonov, A., Jadhav, A., Eidam, O., Mott, B. T., Keiser, M. J., & Shoichet, B. K. et al. (2010). Complementarity between a docking and a high-throughput screen in discovering new cruzain inhibitors. *Journal of Medicinal Chemistry*, *53*(13), 4891–4905. doi:10.1021/jm100488w PMID:20540517

Friesner, R. A., Banks, J. L., Murphy, R. B., Halgren, T. A., Klicic, J. J., Mainz, D. T., & Shenkin, P. S. et al. (2004). Glide: A new approach for rapid, accurate docking and scoring. 1. Method and assessment of docking accuracy. *Journal of Medicinal Chemistry*, *47*(7), 1739–1749. doi:10.1021/jm0306430 PMID:15027865

Gabb, H. A., Jackson, R. M., & Sternberg, M. J. (1997). Modelling protein docking using shape complementarity, electrostatics and biochemical information. *Journal of Molecular Biology*, *272*(1), 106–120. doi:10.1006/jmbi.1997.1203 PMID:9299341

Gao, M., & Skolnick, J. (2009). From nonspecific DNA-protein encounter complexes to the prediction of DNA-protein interactions. *PLoS Computational Biology*, *5*(3), e1000341. doi:10.1371/journal.pcbi.1000341 PMID:19343221

Grosdidier, A., Zoete, V., & Michielin, O. (2011a). Fast docking using the CHARMM force field with EADock DSS. *Journal of Computational Chemistry*, *32*(10), 2149–2159. doi:10.1002/jcc.21797 PMID:21541955

Grosdidier, A., Zoete, V., & Michielin, O. (2011b). SwissDock, a protein-small molecule docking web service based on EADock DSS. *Nucleic Acids Research, 39*, W270–7. http://doi.org/10.1093/nar/gkr366

Grosdidier, S., & Fernández-Recio, J. (2012). Protein-protein docking and hot-spot prediction for drug discovery. *Current Pharmaceutical Design*, *18*(30), 4607–4618. doi:10.2174/138161212802651599 PMID:22650255

Gschwend, D. A., Good, A. C., & Kuntz, I. D. Molecular docking towards drug discovery. *Journal of Molecular Recognition : JMR, 9*(2), 175–86. http://doi.org/10.1002/(SICI)1099-1352(199603)9:2<175::AID-JMR260>3.0.CO;2-D

Gupta, A., Gandhimathi, A., Sharma, P., & Jayaram, B. (2007). ParDOCK: An all atom energy based Monte Carlo docking protocol for protein-ligand complexes. *Protein and Peptide Letters*, *14*(7), 632–646. doi:10.2174/092986607781483831 PMID:17897088

Heidenreich, B., Rachakonda, P. S., Hemminki, K., & Kumar, R. (2014). TERT promoter mutations in cancer development. *Current Opinion in Genetics & Development*, *24*, 30–37. doi:10.1016/j.gde.2013.11.005 PMID:24657534

Hruby, V. J., al-Obeidi, F., & Kazmierski, W. (1990). Emerging approaches in the molecular design of receptor-selective peptide ligands: Conformational, topographical and dynamic considerations. *The Biochemical Journal*, *268*(2), 249–262. doi:10.1042/bj2680249 PMID:2163604

Huang, S.-Y., & Zou, X. (2010). Advances and challenges in protein-ligand docking. *International Journal of Molecular Sciences*, *11*(8), 3016–3034. doi:10.3390/ijms11083016 PMID:21152288

Jain, A. N. (2003). Surflex: Fully automatic flexible molecular docking using a molecular similarity-based search engine. *Journal of Medicinal Chemistry*, *46*(4), 499–511. doi:10.1021/jm020406h PMID:12570372

Jiménez-García, B., Pons, C., & Fernández-Recio, J. (2013). pyDockWEB: A web server for rigid-body protein-protein docking using electrostatics and desolvation scoring. *Bioinformatics (Oxford, England)*, *29*(13), 1698–1699. doi:10.1093/bioinformatics/btt262 PMID:23661696

Kantardjiev, A. A. (2011). GPU.proton.DOCK: Genuine Protein Ultrafast proton equilibria consistent DOCKing. *Nucleic Acids Research*, *39*, W223–8. http://doi.org/10.1093/nar/gkr412

Kapetanovic, I. M. (2008). Computer-aided drug discovery and development (CADDD): In silico-chemico-biological approach. *Chemico-Biological Interactions*, *171*(2), 165–176. doi:10.1016/j.cbi.2006.12.006 PMID:17229415

Kharb, R., Meenakshi, R., Sharma, P. C., & Yar, M. S. (2011). Therapeutic importance of peptidomimetics in medici nal chemistry. *Journal of Chemical and Pharmaceutical Research*, *3*(6), 173–186.

Kim, S.-K., Gao, Z.-G., Jeong, L. S., & Jacobson, K. A. (2006). Docking studies of agonists and antagonists suggest an activation pathway of the A3 adenosine receptor. *Journal of Molecular Graphics & Modelling*, *25*(4), 562–577. doi:10.1016/j.jmgm.2006.05.004 PMID:16793299

Korb, O., Möller, H. M., & Exner, T. E. (2010). NMR-guided molecular docking of a protein-peptide complex based on ant colony optimization. *ChemMedChem*, *5*(7), 1001–1006. doi:10.1002/cmdc.201000090 PMID:20486157

Korb, O., Stützle, T., & Exner, T. E. (2007). An ant colony optimization approach to flexible protein–ligand docking. *Swarm Intelligence*, *1*(2), 115–134. doi:10.1007/s11721-007-0006-9

Korb, O., Stützle, T., & Exner, T. E. (2009). Empirical scoring functions for advanced protein-ligand docking with PLANTS. *Journal of Chemical Information and Modeling*, *49*(1), 84–96. doi:10.1021/ci800298z PMID:19125657

Kortagere, S., Krasowski, M. D., Reschly, E. J., Venkatesh, M., Mani, S., & Ekins, S. (2010). Evaluation of Computational Docking to Identify Pregnane X Receptor Agonists in the ToxCast Database. *Environmental Health Perspectives*, *118*(10), 1412–1417. doi:10.1289/ehp.1001930 PMID:20558333

Lesk, V. I., & Sternberg, M. J. E. (2008). 3D-Garden: A system for modelling protein-protein complexes based on conformational refinement of ensembles generated with the marching cubes algorithm. *Bioinformatics (Oxford, England)*, *24*(9), 1137–1144. doi:10.1093/bioinformatics/btn093 PMID:18326508

Liao, C., & Nicklaus, M. C. (2010). Computer tools in the discovery of HIV-1 integrase inhibitors. *Future Medicinal Chemistry*, *2*(7), 1123–1140. doi:10.4155/fmc.10.193 PMID:21426160

Lu, S.-Y., Jiang, Y.-J., Lv, J., Wu, T.-X., Yu, Q.-S., & Zhu, W.-L. (2010). Molecular docking and molecular dynamics simulation studies of GPR40 receptor–agonist interactions. *Journal of Molecular Graphics & Modelling*, *28*(8), 766–774. doi:10.1016/j.jmgm.2010.02.001 PMID:20227312

Lyskov, S., Chou, F.-C., Conchúir, S. Ó., Der, B. S., Drew, K., Kuroda, D., & Das, R. et al. (2013). Serverification of molecular modeling applications: The Rosetta Online Server that Includes Everyone (ROSIE). *PLoS ONE*, *8*(5), e63906. doi:10.1371/journal.pone.0063906 PMID:23717507

Lyskov, S., & Gray, J. J. (2008). The RosettaDock server for local protein-protein docking. *Nucleic Acids Research, 36*, W233–8. http://doi.org/10.1093/nar/gkn216

Macindoe, G., Mavridis, L., Venkatraman, V., Devignes, M.-D., & Ritchie, D. W. (2010). HexServer: an FFT-based protein docking server powered by graphics processors. *Nucleic Acids Research, 38*, W445–9. http://doi.org/10.1093/nar/gkq311

Mashiach, E., Schneidman-Duhovny, D., Andrusier, N., Nussinov, R., & Wolfson, H. J. (2008). FireDock: a web server for fast interaction refinement in molecular docking. *Nucleic Acids Research, 36*, W229–32. http://doi.org/10.1093/nar/gkn186

Meiler, J., & Baker, D. (2006). ROSETTALIGAND: Protein-small molecule docking with full side-chain flexibility. *Proteins*, *65*(3), 538–548. doi:10.1002/prot.21086 PMID:16972285

Meng, X.-Y., Zhang, H.-X., Mezei, M., & Cui, M. (2011). Molecular docking: A powerful approach for structure-based drug discovery. *Current Computer-aided Drug Design*, *7*(2), 146–157. doi:10.2174/157340911795677602 PMID:21534921

Morris, G. M., Huey, R., Lindstrom, W., Sanner, M. F., Belew, R. K., Goodsell, D. S., & Olson, A. J. (2009). AutoDock4 and AutoDockTools4: Automated docking with selective receptor flexibility. *Journal of Computational Chemistry*, *30*(16), 2785–2791. doi:10.1002/jcc.21256 PMID:19399780

Nabuurs, S. B., Wagener, M., & de Vlieg, J. (2007). A flexible approach to induced fit docking. *Journal of Medicinal Chemistry*, *50*(26), 6507–6518. doi:10.1021/jm070593p PMID:18031000

Ng, H. W., Zhang, W., Shu, M., Luo, H., Ge, W., Perkins, R., & Hong, H. et al. (2014). Competitive molecular docking approach for predicting estrogen receptor subtype α agonists and antagonists. *BMC Bioinformatics*, *15*(Suppl 1), S4. doi:10.1186/1471-2105-15-S11-S4 PMID:25349983

Okimoto, N., Futatsugi, N., Fuji, H., Suenaga, A., Morimoto, G., Yanai, R., & Taiji, M. et al. (2009). High-performance drug discovery: Computational screening by combining docking and molecular dynamics simulations. *PLoS Computational Biology*, *5*(10), e1000528. doi:10.1371/journal.pcbi.1000528 PMID:19816553

Pandya, P., Islam, M. M., Kumar, G. S., Jayaram, B., & Kumar, S. (2010). DNA minor groove binding of small molecules: Experimental and computational evidence. *Journal of Chemical Sciences*, *122*(2), 247–257. doi:10.1007/s12039-010-0029-4

Pang, Y.-P., Perola, E., Xu, K., & Prendergast, F. G. (2001). EUDOC: A computer program for identification of drug interaction sites in macromolecules and drug leads from chemical databases. *Journal of Computational Chemistry*, *22*(15), 1750–1771. doi:10.1002/jcc.1129 PMID:12116409

Paquet, E., & Viktor, H. L. (2013). Macromolecular structure comparison and docking: An algorithmic review. *Current Pharmaceutical Design*, *19*(12), 2183–2193. doi:10.2174/1381612811319120006 PMID:23016846

Phillips, J. C., Braun, R., Wang, W., Gumbart, J., Tajkhorshid, E., Villa, E., & Schulten, K. et al. (2005). Scalable molecular dynamics with NAMD. *Journal of Computational Chemistry*, *26*(16), 1781–1802. doi:10.1002/jcc.20289 PMID:16222654

Repasky, M. P., Shelley, M., & Friesner, R. A. (2007). Flexible ligand docking with Glide. Current Protocols in Bioinformatics. http://doi.org/ doi:10.1002/0471250953.bi0812s18

Rohs, R., Bloch, I., Sklenar, H., & Shakked, Z. (2005). Molecular flexibility in ab initio drug docking to DNA: Binding-site and binding-mode transitions in all-atom Monte Carlo simulations. *Nucleic Acids Research*, *33*(22), 7048–7057. doi:10.1093/nar/gki1008 PMID:16352865

Ruiz-Carmona, S., Alvarez-Garcia, D., Foloppe, N., Garmendia-Doval, A. B., Juhos, S., Schmidtke, P., & Morley, S. D. et al. (2014). rDock: A fast, versatile and open source program for docking ligands to proteins and nucleic acids. *PLoS Computational Biology*, *10*(4), e1003571. doi:10.1371/journal.pcbi.1003571 PMID:24722481

Sable, R., & Jois, S. (2015). Surfing the Protein-Protein Interaction Surface Using Docking Methods: Application to the Design of PPI Inhibitors. *Molecules (Basel, Switzerland)*, *20*(6), 11569–11603. doi:10.3390/molecules200611569 PMID:26111183

Sauton, N., Lagorce, D., Villoutreix, B. O., & Miteva, M. A. (2008). MS-DOCK: Accurate multiple conformation generator and rigid docking protocol for multi-step virtual ligand screening. *BMC Bioinformatics*, *9*(1), 184. doi:10.1186/1471-2105-9-184 PMID:18402678

Shin, W.-H., Heo, L., Lee, J., Ko, J., Seok, C., & Lee, J. (2011). LigDockCSA: Protein-ligand docking using conformational space annealing. *Journal of Computational Chemistry*, *32*(15), 3226–3232. doi:10.1002/jcc.21905 PMID:21837636

Shoichet, B. K., McGovern, S. L., Wei, B., & Irwin, J. J. (2002). Lead discovery using molecular docking. *Current Opinion in Chemical Biology*, *6*(4), 439–446. doi:10.1016/S1367-5931(02)00339-3 PMID:12133718

Singh, Y., Dolphin, G. T., Razkin, J., & Dumy, P. (2006). Synthetic Peptide templates for molecular recognition: Recent advances and applications. *ChemBioChem*, *7*(9), 1298–1314. doi:10.1002/cbic.200600078 PMID:16892470

Songdej, N., & Rao, A. K. (2015). Hematopoietic transcription factor mutations and inherited platelet dysfunction. *F1000prime Reports*, *7*, 66. doi:10.12703/P7-66 PMID:26097739

Takeda, T., Corona, R. I., & Guo, J.-T. (2013). A knowledge-based orientation potential for transcription factor-DNA docking. *Bioinformatics (Oxford, England)*, *29*(3), 322–330. doi:10.1093/bioinformatics/bts699 PMID:23220572

Talele, T. T., Khedkar, S. A., & Rigby, A. C. (2010). Successful applications of computer aided drug discovery: Moving drugs from concept to the clinic. *Current Topics in Medicinal Chemistry*, *10*(1), 127–141. doi:10.2174/156802610790232251 PMID:19929824

Taylor, R. D., Jewsbury, P. J., & Essex, J. W. (2002). A review of protein-small molecule docking methods. *Journal of Computer-Aided Molecular Design*, *16*(3), 151–166. doi:10.1023/A:1020155510718 PMID:12363215

Terwilliger, T. C., Adams, P. D., Moriarty, N. W., & Cohn, J. D. (2006). Ligand identification using electron-density map correlations. *Acta Crystallographica. Section D, Biological Crystallography*, *63*(1), 101–107. doi:10.1107/S0907444906046233 PMID:17164532

Terwilliger, T. C., Klei, H., Adams, P. D., Moriarty, N. W., & Cohn, J. D. (2006). Automated ligand fitting by core-fragment fitting and extension into density. *Acta Crystallographica. Section D, Biological Crystallography*, *62*(Pt 8), 915–922. doi:10.1107/S0907444906017161 PMID:16855309

Torchala, M., & Bates, P. A. (2014). Predicting the structure of protein-protein complexes using the SwarmDock Web Server. *Methods in Molecular Biology (Clifton, N.J.)*, *1137*, 181–197. doi:10.1007/978-1-4939-0366-5_13 PMID:24573482

Torchala, M., Moal, I. H., Chaleil, R. A. G., Fernandez-Recio, J., & Bates, P. A. (2013). SwarmDock: A server for flexible protein-protein docking. *Bioinformatics (Oxford, England)*, *29*(6), 807–809. doi:10.1093/bioinformatics/btt038 PMID:23343604

Tovchigrechko, A., & Vakser, I. A. (2006). GRAMM-X public web server for protein-protein docking. *Nucleic Acids Research, 34*, W310–4. http://doi.org/10.1093/nar/gkl206

Trott, O., & Olson, A. J. (2010). AutoDock Vina: Improving the speed and accuracy of docking with a new scoring function, efficient optimization, and multithreading. *Journal of Computational Chemistry*, *31*(2), 455–461. doi:10.1002/jcc.21334 PMID:19499576

Vagner, J., Qu, H., & Hruby, V. J. (2008). Peptidomimetics, a synthetic tool of drug discovery. *Current Opinion in Chemical Biology*, *12*(3), 292–296. doi:10.1016/j.cbpa.2008.03.009 PMID:18423417

Van Dijk, M., & Bonvin, A. M. J. J. (2008). A protein-DNA docking benchmark. *Nucleic Acids Research*, *36*(14), e88. doi:10.1093/nar/gkn386 PMID:18583363

Van Dijk, M., van Dijk, A. D. J., Hsu, V., Boelens, R., & Bonvin, A. M. J. J. (2006). Information-driven protein-DNA docking using HADDOCK: It is a matter of flexibility. *Nucleic Acids Research*, *34*(11), 3317–3325. doi:10.1093/nar/gkl412 PMID:16820531

Van Dijk, M., Visscher, K. M., Kastritis, P. L., & Bonvin, A. M. J. J. (2013). Solvated protein-DNA docking using HADDOCK. *Journal of Biomolecular NMR*, *56*(1), 51–63. doi:10.1007/s10858-013-9734-x PMID:23625455

Venkatachalam, C. M., Jiang, X., Oldfield, T., & Waldman, M. (2003). LigandFit: A novel method for the shape-directed rapid docking of ligands to protein active sites. *Journal of Molecular Graphics & Modelling*, *21*(4), 289–307. http://www.ncbi.nlm.nih.gov/pubmed/12479928 doi:10.1016/S1093-3263(02)00164-X PMID:12479928

Walensky, L. D., Kung, A. L., Escher, I., Malia, T. J., Barbuto, S., Wright, R. D., & Korsmeyer, S. J. et al. (2004). Activation of apoptosis in vivo by a hydrocarbon-stapled BH3 helix. *Science*, *305*(5689), 1466–1470. doi:10.1126/science.1099191 PMID:15353804

Wu, J., Hong, B., Takeda, T., & Guo, J. (2012). High performance transcription factor-DNA docking with GPU computing. *Proteome Science*, *10*(Suppl 1), S17. doi:10.1186/1477-5956-10-S1-S17 PMID:22759575

Xu, M., & Lill, M. A. (2013). Induced fit docking, and the use of QM/MM methods in docking. *Drug Discovery Today. Technologies*, *10*(3), e411–e418. doi:10.1016/j.ddtec.2013.02.003 PMID:24050138

Yang, J.-M., & Chen, C.-C. (2004). GEMDOCK: A generic evolutionary method for molecular docking. *Proteins*, *55*(2), 288–304. doi:10.1002/prot.20035 PMID:15048822

Zhang, X., Song, Z., Qin, B., Zhang, X., Chen, L., Hu, Y., & Yuan, Z. (2013). Rupintrivir is a promising candidate for treating severe cases of enterovirus-71 infection: Evaluation of antiviral efficacy in a murine infection model. *Antiviral Research*, *97*(3), 264–269. doi:10.1016/j.antiviral.2012.12.029 PMID:23295352

Zhao, Y., & Sanner, M. F. (2007). FLIPDock: Docking flexible ligands into flexible receptors. *Proteins*, *68*(3), 726–737. doi:10.1002/prot.21423 PMID:17523154

Zhong, H., Tran, L. M., & Stang, J. L. (2009). Induced-fit docking studies of the active and inactive states of protein tyrosine kinases. *Journal of Molecular Graphics & Modelling*, *28*(4), 336–346. doi:10.1016/j.jmgm.2009.08.012 PMID:19767223

KEY TERMS AND DEFINITIONS

Computer Aided Drug Design: Computer aided drug design or CADD means the use of computer programs or softwares or online resources for the purpose of identification and optimization of potential drug molecules. The functionality of CADD also extends to identification of potential drug targets.

Docking: The process of finding a thermodynamically and physiologically relevant pose of interaction between two biological macromolecules or between a small molecule and a biological macromolecule. Surface complementarities and energetic parameters are often taken in to account to find out such poses.

HTS: HTS stands for High Throughput Screening, a process which can be automated to a very large extent to facilitate the screening of molecules of interest without much human intervention. In present day, often robotic liquid handling systems facilitate HTS.

Ligand: In molecular docking, ligand is referred to a organic small molecule that can bind to another large biological macromolecule like a DNA or a protein and upon binding, often causes structural and functional alterations of the concerned macromolecule.

Online Server: An online facility that hosts certain website or program and makes it available to global users. In the context of this chapter, several online docking servers enable the users to dock their molecules of interest by making the docking algorithm available in the form of a website.

QSAR: QSAR stands for Quantitative Structure Activity Relationships. QSAR is the process of quantitative identification of relationship between a chemical structure and its functional activity, primarily pharmacological. QSAR is often used in drug discovery for the efficacy analysis of the molecule of interest.

Receptor: A biological macromolecule (mostly a protein) to which a ligand or small molecule binds.

Small Molecule: In drug discovery or docking context, small molecules are synonymous with ligand.

Standalone Program: A computer program that is required to be installed or compiled in a local computer or workstation to make it usable. In docking terms, several docking programs are available as source codes or binaries which the users can download from the web and thereafter can either compile or install on their local systems. Depending upon the program type it can be either GUI (graphic user interface) based or command line based.

Chapter 7
Protein-Protein Docking:
Are We There Yet?

Horia Jalily Hasani
University of Alberta, Canada

Khaled Barakat
University of Alberta, Canada

ABSTRACT

Protein-protein docking algorithms are powerful computational tools, capable of analyzing the protein-protein interactions at the atomic-level. In this chapter, we will review the theoretical concepts behind different protein-protein docking algorithms, highlighting their strengths as well as their limitations and pointing to important case studies for each method. The methods we intend to cover in this chapter include various search strategies and scoring techniques. This includes exhaustive global search, fast Fourier transform search, spherical Fourier transform-based search, direct search in Cartesian space, local shape feature matching, geometric hashing, genetic algorithm, randomized search, and Monte Carlo search. We will also discuss the different ways that have been used to incorporate protein flexibility within the docking procedure and some other future directions in this field, suggesting possible ways to improve the different methods.

INTRODUCTION

Protein-protein interactions play key roles in several biological processes. These processes involve many essential mechanisms ranging from signal transduction and cellular transport to gene expression and immune responses. All these processes are mediated by selective and potent protein-protein interactions (Waksman & Sansom, 2005). Furthermore, many diseases have been associated with either an over-activated or an under-regulated protein-protein interaction and the cure for these diseases has been focused on either inhibiting or stimulating these interactions, respectively. For example, the p53-MDM2 interaction is associated with a severe down regulation of the p53 pathway. An inhibitor for this interaction (e.g. nutlin3) can reactivate the p53 pathway, forcing cancer cells to undergo apoptosis (Barakat, Gajewski, & Tuszynski, 2012; Barakat, Issack, Stepanova, & Tuszynski, 2011; Barakat, Mane, Friesen,

DOI: 10.4018/978-1-5225-0115-2.ch007

& Tuszynski, 2010; Chène, 2003; Kojima et al., 2006). The more we know about such crucial interactions, the more we can build vital protein networks and apply this knowledge to identify treatments for many diseases. Moreover, characterizing these interactions at the atomic level can help in rationally designing new therapeutic agents that can either enhance or inhibit these interactions. Constructing a three dimensional structure of such protein complexes is an essential step toward identifying their binding interface and recognizing any hot spots that can be targeted for their regulation (Elcock et al. 2001; Kann, 2007; Kortemme & Baker, 2004).

For the last few decades X-ray crystallography, nuclear magnetic resonance (NMR) spectroscopy and electron microscopy have been the main source to predict such structures. Despite their accuracy, efficiency and the huge amount of details they can provide, they are expensive and very labour and skill demanding. A simple comparison between protein structure and gene sequence databases would simply reveal the great discrepancy between the two. That is, although hundreds of thousands of gene expressions have been characterized, only less than thirty thousand protein structures have been determined so far and most of these structures are either redundant or describe only apo (unbound) proteins (Villoutreix et al., 2014). Moreover, protein complexes are more difficult to crystallize than the individual proteins, consequently, they are less represented in Protein Data Bank (PDB) (Berman et al., 2000) and constitute only a small fraction of the experimentally determined structures. This huge discrepancy and lack of structural details motivated many computational groups to fill this gap and suggest a new, rapid and cheap way to predict these interactions (Barakat & Tuszynski, 2011; Barakat, Houghton, Tyrrell, & Tuszynski, 2014; Barakat, Mane, & Tuszynski, 2011; Gógl et al., 2015; Nillegoda et al., 2015; Pedotti, Simonelli, Livoti, & Varani, 2011; Taylor et al., 2015). One solution they provided which is also the focus of this chapter is protein-protein docking.

Protein-protein docking simulations date back to the early 90s (Katchalski-Katzir et al., 1992), soon after the development of many protein-ligand docking techniques. Although the concept of docking in both cases is similar, there are still key differences that prevented the direct technology transfer between the two problems. For example, since proteins are continually dynamic and usually undergo huge structural changes, their modeling is not as easy as that of a small molecule and requires more precise predictions. Furthermore, proteins are large and have complex shapes making them harder to simulate than ligand-protein interactions (Ravikant & Elber, 2011; Sheinerman et al. 2000). On the other hand, the concept of a binding site is usually inapplicable to protein-protein interactions and in many cases proteins interact through 'hot spot' areas on their surfaces. Identifying these binding locations is particularly difficult and requires highly innovative methods. In this context, the search methods in protein-protein docking simulations are somehow different from those being used in small molecule docking (Keskin, Ma, & Nussinov, 2005). However, the reader should keep in mind that although protein-protein docking is in many ways different from protein-ligand docking, some of its reflections might be mentioned in this chapter, if applicable.

In this chapter, we will first outline the concept of docking in general, with an emphasis on the different approaches. We then briefly review the theoretical concepts, principles and specific features of the search strategies and scoring functions implemented in different programs, highlighting their strengths as well as their limitations. We will also discuss the different methods of incorporating protein flexibility within the docking procedure. After these descriptive review sections, we will guide the reader through a number of case studies and recent successes that illustrate some of the significant applications of protein-protein docking algorithms. We will also discuss the existing challenges and limitations in current algorithms and highlight some potential future directions and possible improvements in this field. We hope that this brief, yet comprehensive review chapter provides researchers interested in this area with current state-of-the-art technologies in this field.

THE "DOCKING PROBLEM"

The so-called protein-protein "docking problem" refers to the task of docking two individual protein structures and predicting their biologically relevant complex. It can be classified into two different approaches; "bound docking" where two proteins from a co-crystallized complex are separated and then reassembled using a computational docking algorithm or "unbound docking" (Wiehe, Peterson, Pierce, Mintseris, & Weng, 2008) where individually solved structures of two proteins are docked together to predict their most probable mode of binding. The bound docking approach has no investigational significance, but it is frequently employed to test and validate the different docking algorithms while developing a new method. Unbound docking, on the other hand, is the approach employed by the current docking programs and is a valuable tool that can predict complexes from individual proteins for producing new knowledge at the atomic level. Unbound docking is far more complicated than bound docking. This is because the two proteins, which come together to form a complex, may undergo some conformational changes upon binding, which needs to be predicted correctly (Vakser, 2014; Ravikant & Elber, 2011). In this section, we will focus mainly on the unbound docking approach involving two different methods, namely rigid-body docking and flexible docking. Initially, we will have a closer look at these two approaches and then will focus more on rigid-body docking and its different aspects.

Rigid-body docking treats the two proteins as rigid entities. In this case, the side chains and backbone of the two interacting proteins are kept fixed with no bond angles or distances allowed to change upon docking. Although this method offers a rapid way to predict the formed complex, it completely ignores any possible conformational rearrangements that may occur upon complex formation. This sampling approach adopts a simplistic, yet structurally meaningful, technique. It projects the protein coordinates onto a low-resolution three-dimensional grid and then calculates the overlap between the two grids of the interacting proteins (Ehrlich & Wade, 2001). Flexible docking, on the other hand, takes into account such conformational changes by allowing minimal movements of the side chains or the backbone of the proteins at high resolution. Rigid-body docking includes six degrees of freedom; three translations and three rotational in the Cartesian coordinates, whereas flexible docking is usually preceded by transforming the Cartesian coordinates into internal coordinates, which involves more variables compared to rigid docking (Guo, Li, Ma, & Wang, 2013; Vakser, 2014).

It is noteworthy to mention that in many cases, no significant conformational changes take place upon the binding of the two proteins and a "lock-and-key" approach would be sufficient to address their interaction (Betts & Sternberg, 1999). This usually occurs when the two protein structures are not fluctuating too much, particularly at their binding interface. In this case, employing a rigid docking protocol to study these interactions would be adequate and can predict, with an acceptable degree of accuracy, their mode of binding. However, to predict higher resolution structural details at the interface some degree of flexibility is required, and the more protein dynamics incorporated during the docking simulation, the more structural details can be revealed. In this case, flexible docking comes into play and more details on that are described in the next sections.

It is important to emphasize the significance of incorporating experimental data within the docking procedure. This type of data not only helps in validating the docking outcomes, but also can help in identifying the potential binding interface of the two proteins, reducing the search space to a manageable region (Grinter & Zou, 2014; Wodak & Méndez, 2004). In this context, a number of docking algorithms can restrict the search space to a specific region and therefore feeding these algorithms with a minimal knowledge of the binding site residues called, 'hot spots' can significantly improve the search process (Grimm et al., 2005; Grosdidier & Fernández-Recio, 2012).

DOCKING STEPS: SAMPLING AND SCORING

Within the scope of the two classes (rigid-body docking & flexible docking) discussed above, the process of docking is implemented through two main interconnected steps: *sampling* which will search for all possible binding modes between the two proteins and *scoring* for the differentiation between correct, near-correct or incorrect poses. In the following two sections, we will describe these two processes in details, giving examples of different software packages employing the different methods. We will then describe how protein flexibility is incorporated during docking and post-docking approaches. Finally, a brief evaluation of the different docking techniques will follow.

Sampling Techniques

We will begin by briefly introducing the sampling stage. The reader is referred to the references included her and the references within for a more details on this area of research. The sampling stage of docking involves generating a large number of alternative poses for the proteins by sampling all the possible binding modes. The correct prediction of the complex is more of a near-native approximation than an accurate native (co-crystalized) complex of the two proteins, which is computationally impossible to generate (Smith & Sternberg, 2002). Standard search methods are computationally expensive and lead to a huge number of solutions; most of them are physiologically irrelevant. It is for this reason more sophisticated sampling approaches and strategies are usually incorporated to correctly predict realistic solutions at this stage. This includes exhaustive global search methods as adopted by fast Fourier transform (FFT) algorithms, local shape feature matching used in geometrical hashing and genetic algorithms, randomized search employed in Monte Carlo (MC) search or Boolean operations and effective energy gradients to determine new atomic positions (Ehrlich & Wade, 2001; Huang, 2014). Table 1 lists the most commonly used software packages used for protein-protein docking with their respective sampling techniques. Although the focus of this chapter is particularly on protein-protein docking, there are a number of small-molecule docking programs optimized to become applicable to protein-protein docking and therefore, they will be discussed in various sections to illustrate the different aspects of docking.

Some recently developed and/or optimized protein-protein docking programs are DockTrina (docking triangular protein trimmers) (Popov, Ritchie, & Grudinin, 2014), ATTRACT (Vries & Zacharias, 2013), MEGA-DOCK (Matsuzaki et al., 2013), F(2)Dock 2.0 and GB-Rerank (Chowdhury et al., 2013), Swarm-Dock (Moal, Torchala, Bates, & Fernández-Recio, 2013). In addition to the programs listed in Table 1, there are numerous docking programs, which facilitate the use of docking algorithms by availing online servers. Some of them include: ClusPro (Comeau, Gatchell, Vajda, & Camacho, 2004), GRAMM-X (Tovchigrechko & Vakser, 2006), ZDOCK SERVER (B. G. Pierce et al., 2014), 3D-GARDEN (Lesk & Sternberg, 2008), PatchDock & SymmDock (Schneidman-Duhovny, Inbar, Nussinov, & Wolfson, 2005), HADDOCK (Domin-

Table 1. The most commonly used protein-protein docking programs with their method of sampling

Name of Software	Method of Sampling	References
ZDOCK, FTDOCK, GRAMM, F2DOCK, DOT, Cell-Dock	FFT-correlation	(Chen et al., 2003; Gabb et al.,1997; Vakser, 1995; Bajaj et al., 2011; Mandell et al., 2001; Pons et al., 2012)
HEX, FRODOCK	FFT- Spherical Harmonics	(Ritchie & Kemp, 2000; Garzon et al., 2009)
PatchDock, LZerD	Geometric Hashing	(Duhovny et al., 2002; Venkatraman et al., 2009)
BIGGER, SKE-DOCK, SoftDock	Direct Search in Cartesian Space	Palma et al., 2000; Terashi et al., 2007; Jiang & Kim 1991)
DOCK	Distance Geometry Algorithm	(Kuntz et al., 1982)
DARWIN	Genetic Algorithm	(Taylor & Burnett, 2000)
RosettaDock, ATTRACT, ICM	Monte Carlo Search	(Gray et al., 2003; Zacharias, 2003; Fernandez-Recio et al., 2002)
HADDOCK	Based on both Biochemical/ Biophysical Information and MC Search	(Dominguez et al., 2003)

guez, Boelens, & Bonvin, 2003), RosettaDock (Gray et al., 2003), FIBERDOCK (Mashiach, Nussinov, & Wolfson, 2010), FIREDOCK (Mashiach, Schneidman-Duhovny, Andrusier, Nussinov, & Wolfson, 2008), pyDockWEB (Jiménez-García, Pons, & Fernández-Recio, 2013), pyDockSAXS (Jimenez-Garcia, Pons, Svergun, Bernado, & Fernandez-Recio, 2015) and ATTRACT web interface (Vries, Schindler, Chauvot de Beauchêne, & Zacharias, 2015).

Fast Fourier Transform (FFT) correlation approach is regarded as one of the most popular and widely used techniques of rigid-body search, developed in 1992 by Katchalski-Katzir and coworkers (1992). This technique allows an accelerated exhaustive global search of the binding orientations over six degrees of freedom in 3D translational plus 3D rotational space. The popularity and wide usage of this technique is attributed to its ability to rapidly perform an exhausted sampling of all the possible binding modes of the interacting proteins. In this method, the two proteins (receptor protein and ligand protein) are discretized onto two different 3-dimensional grids each with the size N * N * N (where N is the number of grid points), and each point on the grid with coordinates *l,m,n* is assigned a value based on its geometric characteristics ($a_{l,m,n}$). This includes a value of one for grid points on the surface of the protein, a value of zero for points outside of the protein surface and a value of ρ for points located within the protein core structure (where ρ usually has a negative value). The complementarity of shape function *C* (p,q,r) between the two proteins is then evaluated from the Equation 1, shown below:

$$C(p,q,r) = \sum_{l=1}^{N}\sum_{m=1}^{N}\sum_{n=1}^{N} R(l,m,n) \times L(l+p, m+q, n+r)$$

The translational space is then sampled by calculating a correlation between the two discrete grids. FFT correlation has been employed in a number of docking programs: FTDock (Gabb, Jackson, & Sternberg, 1997), GRAMM (Vakser, 1995), DOT (Mandell et al., 2001), ZDOCK (R. Chen & Weng, 2002), F2DOCK (Bajaj, Chowdhury, & Siddahanavalli, 2011) and Cell-Dock (Pons et al., 2012) . The basic approach is the same for all of them and the difference lies in the way the proteins are projected onto the grids.

A successful modification of the FFT algorithm is the spherical Fourier transform-based search, initially introduced in the program HEX (Ritchie & Kemp, 2000), which employs spherical harmonics to rapidly search the complete conformational space of the proteins. Another exhaustive sampling technique employs a direct search in Cartesian space to identify the correct binding mode between two proteins using 3D Cartesian grid space. This technique has been adopted in many docking packages including SOFTDOCK (Jiang & Kim, 1991), BIGGER (Palma et al., 2000), SKE-DOCK (Terashi et al.,2007). The program BIGGER (Palma et al., 2000) for example, employs a surface implicit method based on Boolean operations, in which the molecular shape of the two proteins is mapped onto a 3D grid in Cartesian space. Each point on the grid is assigned a value of '1' if it is on the protein core and a '0' if it is outside. Geometric fitting between the two proteins is then performed by shape matching directly in the Cartesian space. The advantage of this technique over FFT-based methods is the ability to have more control over the search process because of the direct operation in Cartesian space, which allows inclusion of protein flexibility and biological information (Huang, 2014).

Geometric hashing techniques are another type of global protein-protein docking search using a shape-explicit algorithm. The technique is based on finding a set of shape descriptors on the molecular surface of the proteins (one of them called the receptor and the other the ligand). Following that, a set of local coordinate frames is defined, for which the positions of nearby critical points are used as indices hashed into a table. The hash table stores the current coordinate frame of each protein. After superimposing the local coordinate frames of the ligand and the receptor, the positions of the critical points of the ligand in proximity of the receptor are then used to scan the hash table for the corresponding ligand reference frames (Smith & Sternberg, 2002). Several programs employ different shape descriptors. For example patch descriptors are used in PatchDock (Duhovny, Nussinov, & Wolfson, 2002) and 3D Zernike descriptors are adopted by LZerD (Venkatraman, Yang, Sael, & Kihara, 2009). High shape similarity is revealed by the frequency of correspondences to a specific frame. The method has been improved by introducing surface softness as a tolerance value of about 1.5 A in the coordinates.

Genetic algorithm is the basis for a number of other successful docking algorithms. This generally involves a real-space search method, which is based on a local shape feature matching. The DARWIN (J. S. Taylor & Burnett, 2000), for example, includes a solvent accessible area (SAA) for the proteins with normals, surface curvature and associated hydrogen bonding character. The complementarity between the curvatures and normals are scanned for the two proteins to find the most probable binding mode. The 'chromosome' of the genetic algorithm (GA) consists of the relative positions and orientations of the molecules (6D space).

Randomized search is another broad category of sampling techniques. Monte Carlo (MC) search is one of the approaches in which sampling of all possible conformations with a technique like FFT correlation is replaced with a random approach. As such, random translational and rotational orientations for the two proteins are performed and the final positions are selected based on a Metropolis criterion. This is then followed by a multistage optimization of the side-chains and rigid-body orientations based on MC simulations (Gray et al., 2003). The advantages of MC methods over FFT methods includes the allowance of more physical decoy distribution, arbitrary energy functions and structural flexibility to some additional extent (Lorenzen & Zhang, 2007). RosettaDock (Gray et al., 2003) is an efficient protein-protein docking software that is based on the MC method. Some other software based on randomized search are HADDOCK (Dominguez et al., 2003), ICM-DISCO (Fernandez-Recio, Totrov, & Abagyan, 2002) and ATTRACT (Zacharias, 2003).

More recent practical successes in sampling techniques, include the recent work of Zhang et al. (2015), who identified a variation of MC search for the RosettaDock package to improve the sampling stage towards a more efficient identification of the near native docking geometries. This variation involves combining the well-tempered ensemble method with a 2-dimensional temperature and Hamiltonian replica exchange scheme (WTE-REMC). A comparative study of the results from different docking protocols of Rosetta, using the same set of targets has shown the enhanced performance of WTE-REMC in exploring the phase space.

HADDOCK (High Ambiguity Driven Protein-Protein Docking) software (Dominguez et al., 2003) is implemented with a different scheme for the randomized search process. It employs a data-driven approach and supports various types of biochemical and/or biophysical data sets such as chemical shift perturbation data from NMR titration experiments or mutagenesis data. HADDOCK is able to handle almost full flexibility (side-chain and backbone) and incorporate explicit water during the modeling process. In addition, the software provides a user-friendly web server (De Vries, Van Dijk, & Bonvin, 2010), which is accessible through four interfaces: Easy, Expert, Guru and parameter file upload interfaces, for different levels of expertise and type of application allowing more control and modification on the docking process. Several CAPRI experiments (Janin, 2013), have confirmed the profound ability of HADDOCK in precise predictions of protein-protein complexes based on experimental information. In addition it is stated that HADDOCK has generated about 70 protein-protein complexes in PDB (Grosdidier & Fernández-Recio, 2012).

Scoring Functions

The large numbers of possible conformations (poses) generated from the sampling stage are ranked using a scoring function that involves the assessment of the overall favorability of a protein-protein complex and/or a score between the two proteins in the bound form. The final docking results comprise of the top ranked complexes, which are in order of their binding score (Huang, 2014). These different scoring methods can be categorized into different groups based on the basis they are derived from; for example the *forcefield-based* scoring functions employ non-bonded interaction terms as the scoring method. *Empirical* scoring functions, on the other hand make use of regression methods to differentiate between the different poses. *Knowledge-based* scoring functions use statistical atom pair potentials from the structural databases as the scoring method. Yet, another category is the *heuristic* scheme of scoring, which makes use of chemical scores or geometrical scores such as contact/surface scores, or shape complementarity scores (Ehrlich & Wade, 2001; Muegge & Rarey, 2001)

Docking programs use different scoring functions that largely vary from one algorithm to the other. Usually, a docking program makes use of a composition of different scoring functions to increase the efficacy and accuracy of the ranking process. For instance, although shape complementarity between the interacting proteins is regarded as a simple, inexpensive and powerful scoring function, it does not necessarily distinguish near-native from non-native complexes and is thus combined with other functions such as electrostatic field, hydrophobic complementarity, desolvation free energy, etc. (Szilagyi et al., 2005). An example of implementing such composite scoring functions is the ZDOCK program (R. Chen & Weng, 2002). FTDOCK (Gabb et al., 1997) and DOT (Mandell et al., 2001), which combine electrostatic and geometric scores but in different ways. FTDOCK uses geometric matching in the initial stage and then filters the docked poses based on their electrostatics scores in later stages. In the DOT algorithm, two types of scoring methods are used in conjunction. The GRAMM package performs the scoring through a geometric surface matching approach by involving the hydrophobic contacts between the two proteins. PyDock (Cheng, Blundell, & Fernandez-Recio, 2007) uses an energy function based on ICM potentials, composed of van der Waals, Coloumbic electrostatics and Atomic Solvation Parameter (ASP)-based desolvation energy. The scoring function in ICM-DISCO (Fernandez-Recio et al., 2002) takes into consideration the intermolecular grid-based energy terms and the Accessible Surface Area (ASA)-based desolvation energy. RosettaDock program also employs a similar scoring scheme.

Incorporating Protein Flexibility

Currently, all modern docking techniques use different approximations to include protein flexibility within the docking process. This may include rotating the protein side chains and possibly involve simple dynamics of the backbone. Flexibility in general is treated in two different ways: "implicit flexibility" and "explicit flexibility". Rigid-body approaches incorporate protein flexibility in an implicit way, by either using soft docking (surface softening) or through more advanced techniques such as ensemble-based docking. Surface softening techniques include low resolution docking methods, simplified models for the side chains and surface layer thickening (Szilagyi et al., 2005). Rigid-body docking methods show successful results in many cases (Viricel, Ahmed, & Barakat, 2015; Xu et al., 2013). Nevertheless, an accurate and biologically relevant prediction of these protein-protein interactions which involve considerable conformational changes of backbone and/or side chains upon complex formation is merely not possible through rigid-body docking (Grosdidier & Fernández-Recio, 2012; Wang & Schueler-furman, 2005).

Explicit treatment of the protein flexibility is only possible in docking algorithms that involve the atomic coordinates of the different atoms, rather than those that implicitly represent the protein structure on a grid. Each approach treats flexibility to a specific extent in a different way, such as accounting for the dynamic changes of the protein domains, amino acid side-chains and less frequently the backbone of the protein. The different approaches to include protein flexibility are based on energy minimization, atomic-level simulations or optimization of the shape-based characteristics and/or complementarity. Others include considerations related to electrostatic interactions and hydrogen bonding which have been known for a long time. All the mentioned approaches are time-consuming and take several hours to days for an effective treatment in a docking process and this represents the main reason for which flexibility remains a major challenge in the field of protein-protein docking (Ehrlich & Wade, 2001). For example, FiberDock (Mashiach et al., 2010) includes protein flexibility through several energetic calculations and ranking through normal mode analysis (NMA) along with side-chain optimizations (using a rotamer library) and restrained energy minimization with limited degrees of freedom. Another randomized search algorithm, RosettaDock (Gray et al., 2003) that is based on Monte Carlo principles, employs side-chain repacking, followed by optimizing the docked poses using a MC search with rigid-body displacements. Hinge-bending treatment is used in FlexDock (Schneidman-Duhovny, Ruth, & Wolfson, 2007) and FlexDoBi (Guo et al., 2013) programs in which the hinge regions (regions of the protein causing large-scale motions) are specified on the proteins prior to the docking simulation and the docking procedure is then performed, taking these specifications into account. HADDOCK (Dominguez et al., 2003) is regarded as an experimental data-driven program which performs a semi-flexible simulated annealing. Biochemical & biophysical interaction data from NMR experiments, mutagenesis investigations or bioinformatics predictions are used to introduce ambiguous interaction restraints (AIRs). These are then used to control the docking process with allowance for both backbone and side-chain flexibility on the interface. Average interaction energies are then used as a basis for the clustering and ranking of top predictions.

Post Docking Refinement

Docking procedures can be followed by a second ranking/scoring stage to refine the docking results and remove any steric clashes that may emerge form the docking algorithm. This is generally required due to the lack of precision in docking scoring functions. There are many different methods that have been used in the post-docking refinement stage. This includes a biased probability side-chain optimization approach as implemented in the ICM program (Abagyan, Totrov, & Kuznetsov, 1994) or side-chain minimization as employed by Multidock (Multiple copy side-chain refinement Dock) (Jackson, Gabb, & Sternberg, 1998) algorithm. The RosettaDock program (Gray et al., 2003) uses another effective method, which involves the correction of main-chain displacements. In addition, some post-docking algorithms are specialized in refining the results to a higher extent. These include ZRANK (Pierce & Weng, 2007), EMPIRE (Liang, Liu, Zhang, & Zhou, 2007), DARS (Chuang, Kozakov, Brenke, Comeau, & Vajda, 2008), DECK (Liu & Vakser, 2011), RDOCK (Li, Chen, & Weng, 2003), pyDock (Cheng et al., 2007), Eigen-Hex (Venkatraman & Ritchie, 2012), RPScore (Moont, Gabb, & Sternberg, 1999), Multidock (Jackson et al., 1998) among others. DOCKSCORE (Huang et al., 2015) and FiltRest3D (Gajda, Tuszynska, Kaczor, Bakulina, & Bujnicki, 2010) are two of the online webservers which allow ranking of poses from protein-protein docking.

Docking Evaluation

When a new docking technique is developed and is claimed to be good at predicting the structure of a protein-protein complex, it needs to be evaluated to provide reliable data and information to scientists and research groups to obtain the highest quality results from their investigations. The Critical Assessment of Predicted Interactions (CAPRI) experiment (Janin et al., 2003) is created to compare and evaluate the performance of various docking algorithms developed by different groups. CAPRI is modeled after Critical Assessment of Structural Prediction (CASP), which started in 1994 for comparing the performance of protein-folding algorithms (Moult et al., 2014). The way CAPRI aims at evaluating the docking algorithms is through using a set of standard benchmark proteins (unbound pairs), with the correct match known (co-crystallized complex) which is kept confidential. These are assigned to the computational research groups and the results obtained are further assessed and compared. CAPRI has attracted significant attention of the scientific community and the results have led to great improvements to the docking techniques (Vakser, 2014). Furthermore, systematic studies are being constantly conducted to provide up to date insights into different aspects of docking programs (Chen, 2015; Moal et al., 2013).

The variability in the conformational sampling methods and scoring employed by the different docking algorithms, makes selecting a particular docking software a hard task. Each docking tool discussed so far has its own strengths as well as shortcomings. A fair way to compare these different tools is to test them against the same set of data using exactly the same set of parameters with identical amount of elapsed time (Cole, et al., 2005). This has been the case employed by CAPRI evaluations, which provide reliable and periodical comparison among the different groups developing PPI docking software packages and servers.

Table 2 summarizes a comparative overview of protein-protein docking software and/ servers with an emphasis on the most recent CAPRI experimental outcomes. For the complete references for each docking tool, please refer to Table 1. The latest CAPRI experiment (5th Evaluation) which was carried out in 2013 (Bonvin, 2013), reported the ClusPro server as the best in the automatic protein docking server category. In particular, the server's performance was equivalent to the best human predictor group.

Table 2. Comparative overview of different docking software and tools with respect to 2013 CAPRI experiment

Docking Software and /Server	Strength	Weakness
HADDOCK	• Semi-flexible simulated annealing permitting almost full protein flexibility • Allows for ambiguous interaction restraints • Efficient scoring performance •CAPRI 2013: software ranked 1st, server ranked 2nd	•Reduced accuracy without structural information
ClusPro	• Efficient prediction •CAPRI 2013: Server ranked 1st	•Rigid-body docking; no/less inclusion of flexibility
RosettaDock	• Flexible protein docking • Post-docking correction of main-chain displacements.	•Σχορινγ φυνχτιονσ require improvement
ZDOCK	• Efficient scoring through a combination of functions. • Allows for specifying residual restraints.	• Rigid-body docking; no/less inclusion of flexibility • Inaccurate for docking of homology models
SwarmDock	• Εφφιχιεντ πρεδιχτιον using normal modes and flexibility during docking •CAPRI 2013: Server ranked 3rd	•Inconsistency ιν σχορινγ φυνχτιονσ.
PIE-DOCK (software and server	•Εφφιχιεντ πρεδιχτιον •CAPRI 2013: server ranked 4th.	•Poor scoring as indicated by CAPRI 2013
PatchDock	•Scalable ΦΦΤ βασεδ αλγοριτημ •Efficient computer run time	•Poor scoring as indicated by CAPRI 2013 •Takes minimal flexibility into account •Poor refinement

HADDOCK (De Vries et al., 2010), SwarmDock (Moal et al., 2013) and PIE-DOCK (Ravikant & Elber, 2010) servers were ranked next. In the human predictor category, HADDOCK (Dominguez et al., 2003) was ranked first followed by SwarmDock (Venkatraman & Ritchie, 2012), Vakser group, Vajda/Kozakov group (Camacho & Vajda, 2002; Chuang et al., 2008) and ICM (Fernandez-Recio et al., 2002) in the 2nd to 5th position (Kozakov et al., 2013) (Lensink & Wodak, 2013).

However, it is worth noting that a single docking algorithm cannot provide the ultimate answer for a particular problem and it would be better to combine at least two algorithms for the same research question. This will ensure a high quality of the final results, which should also make use of additional computational tools along with the docking methods to fulfill their gaps and shortcomings. Likewise, other numerical measures such as root-mean-square deviation (RMSD), careful analysis of interaction-based measures and visual examinations may provide additional credibility to the results (Hernández-santoyo et al., 2013).

Figure 1. A depiction of two proteins shown individually and when docked together in complex

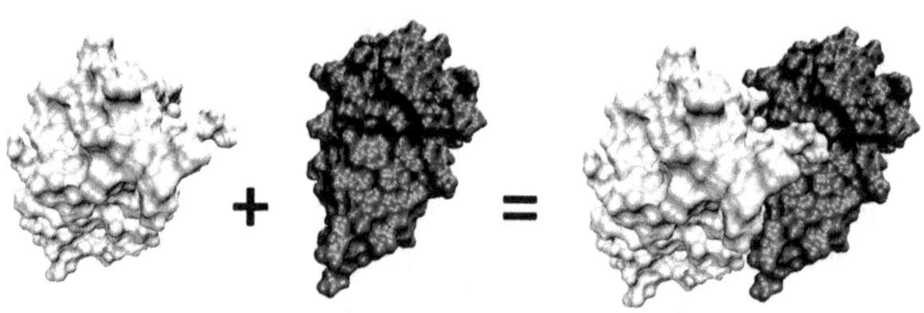

Figure 2. General scheme of docking and the different stages involved. The docking process usually starts with rigid-body docking, covering only the six dimensional space of the two proteins in 3D translational and 3D rotational. This is followed by ranking the generated complexes using a simple scoring function (e.g. shape complementarity). A more complex energy-based scoring method is next employed. Flexibility is then introduced to allow side chains or backbone dynamics. Incorporating protein flexibility usually leads to a more realistic and detailed picture of interaction. Rescoring stage is sometimes implemented for a more sophisticated ranking of poses.

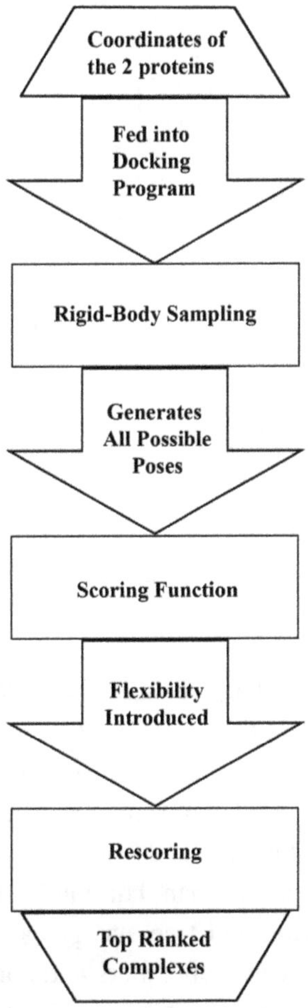

Furthermore, we carried out a quick citations' survey for the top ranked docking algorithms (using Google Scholar; http://scholar.google.com), which resulted in the pie chart shown in Figure 3. This chart includes the citations of only the software packages. All docking servers were excluded to avoid discrepancies in the presented data. Accordingly, HADDOCK software was considered to be one of the most popular protein-protein docking tools, which confirms its proven accuracy and credibility amongst scientific groups.

Figure 3. Pie chart presenting the relative number of citations to each docking tool; for the sake of consistency, only the docking software are mentioned and servers are excluded.

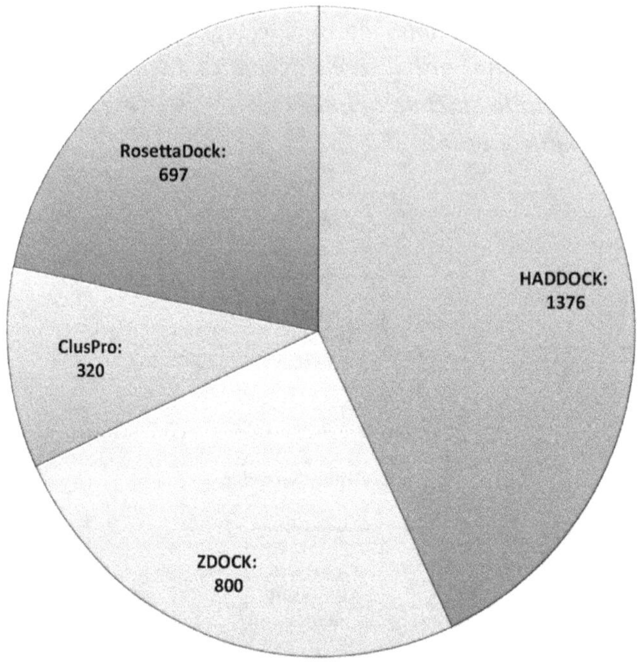

CASE STUDIES

In our recent work (Ahmed & Barakat, 2015; Viricel et al., 2015), we have applied ensemble-based docking technique to understand the interaction between Programmed Death- 1 (PD-1) and PD-Ligand-1 proteins using the ZDOCK software (R. Chen, Li, & Weng, 2003) and the FFT search. These proteins belong to the immune system and play important functions in the T-cell inhibition pathway, leading to immune tolerance and inactivity against cancer and infectious cells. The novel results produced in this study predicted how these proteins interact in human. This interesting modeling approach, which combined molecular dynamics simulations, docking and free energy calculations, provided a good example for how protein-protein docking techniques can help understand complex protein interactions at the atomic level. In another interesting study (Zhang et al., 2005), the ZDOCK server (Pierce et al., 2014) was used for finding the possible protein-protein complex formed between the severe acute respiratory syndrome coronavirus (SARS-CoV) S1 protein and its receptor, human angiotensin-converting enzyme 2 (ACE2). This study illustrates successful identification of the interfacial key residues involved in the process of complex formation and has provided valuable information for further study of the mentioned system.

Protein-protein docking tools are frequently employed for investigating the structural details of various types of ion channels. These multimeric biological complexes are difficult to crystallize and the available literature indicates wide use of computational techniques for their study. A good example of such successful attempts is the work by Smith et al. (2007), who employed Rosetta-based docking for modeling the transmembrane domain of KCNQ1/KCNE1 cardiac ion channel. The docking approach in this study has produced one of the most accurate starting models and has provided significant information on the structural basis for this ion channel.

The HADDOCK software (Dominguez et al., 2003) has been quite successful in resolving a large number of accurate models for protein-protein complexes. One good example is the study of the complex formed between plectasin, a member of the innate immune system and the bacterial cell-wall precursor Lipid II (Schneider et al., 2010). The bacterial cell wall biosynthesis is targeted by plectasin, through complex formation with Lipid II. The study has clearly identified the residues involved at the binding site between the two proteins by using the HADDOCK protein-protein docking tool, providing valuable information for the design of novel antibiotics.

CHALLENGES AND FUTURE DIRECTIONS

Despite the outstanding success of many protein-protein docking algorithms and their application throughout the last decade, there are still a number of challenges that need to be addressed within the next few years to advance the field for more realistic predictions. The initial sampling stage of most of docking algorithms is generally rapid and efficient, however there are certain shortcomings of protein-protein docking approaches, which fall into different categories mainly, related to scoring and flexibility issues.

The scoring functions are regarded as one of the major challenges of docking algorithms and frequently fail to accurately rank the resultant complexes. Even with the considerable improvements so far, detection of correct conformations is still considered to be beyond the scope of current docking algorithms. Using post-docking filtering techniques usually conceals this drawback, although it complicates the overall docking procedure. Accordingly, it is crucial to reach a point where the docking algorithms themselves gain the ability of completely refining the complexes correctly. The scoring functions also need to focus on solvation, entropy and the effects of protonation states of the charged residues. Development of new implicit and explicit models for further investigation of the role of water in the proteins interfaces during protein-protein docking, requires more effort. (Moal et al., 2013)

The treatment of proteins' structural flexibility remains an active area of research and is still lagging behind. The problem is even more puzzling when dealing with the proteins' backbone flexibility and large structural changes as seen in antigen-antibody complexes (Chen & Tou, 2013).

Furthermore, advancements in the structural genomics studies are leading to more number of therapeutically significant 3D structures being determined and thus, opening doors for docking studies to explore such interactions at the atomic level. However, it is predicted that there would be more number of 3D structures obtained by homology modeling rather than experimental techniques. This would produce more controversies and errors in the resulting complexes, due to uncertainties and limitations of homology modeling in correct prediction of 3D structures (Wass, Fuentes, Pons, Pazos, & Valencia, 2011).

The success of the current docking algorithms is largely dependent on the availability of experimental data (such as binding sites, interaction partners and conformational changes), without which there are high chances of obtaining incorrect conformations. This reliance on experimental data would always remain a challenge for many docking studies (Smith & Sternberg, 2002).

Improvements to the docking techniques in general, may include the optimization of existing algorithms or the development of new ones to overcome these limitations. With the aid of community efforts such as CAPRI, docking algorithms are being successfully evaluated to identify the potential limitations and the benchmark testing has led to overcoming a large number of them. The remaining challenges such as flexibility incorporation are still under investigation and with the accelerated pace of research in this area, the problem of flexibility will be soon tackled. Other tricky challenges like the prediction of multi-protein complex structures could be addressed and investigated more in the future (Grimm et al., 2005; Wodak & Méndez, 2004).

CONCLUSION

Understanding protein-protein interactions is an essential step to comprehend many important biological processes and to identify new treatments for many diseases. The huge discrepancy between the amount of genomic data deposited in the literature and the experimentally determined protein structures place a huge pressure on the computational groups to predict the structure of many proteins and understand how they interact. Amongst the current computational tools in use is protein-protein docking. This technique involves several algorithms that have been developed to analyze the protein-protein interactions at the atomic-level. They predict the mode of binding between two interacting proteins and involve two interconnected stages, namely sampling of protein conformations and scoring the docked poses. More advanced algorithms include the proper incorporation of protein flexibility and sophisticated scoring methods, referred to as post-docking refinement.

This chapter briefly introduced the concepts behind protein-protein docking, with an emphasis on the different approaches. It also reviewed the theoretical concepts, principles and specific features of the search strategies and scoring functions implemented in different programs, highlighting their strengths as well as their limitations. The main sampling techniques reviewed here include FFT correlation, direct search in Cartesian space, randomized search and MC algorithm. Scoring of the protein-protein docking solutions span different methods including force field-based methods, empirical scoring, knowledge-based and heuristic scoring functions.

Despite the great efforts of many research groups in this exciting area of research, there are still many challenges that need to be addressed to improve the method. This includes mainly developing more accurate scoring methods and the proper incorporation of protein flexibility. Although many algorithms are on the edge of overcoming these problems, more work need to be done to fully solve these issues. With the accelerated pace of improvements in the field of computational approaches and their successful applications in understanding many biological systems, we believe that the field of protein-protein docking will evolve rapidly towards a more accurate method to predict protein-protein interactions. We hope this chapter provides a brief, yet comprehensive, overview to researchers working in this area.

REFERENCES

Abagyan, R., Totrov, M., & Kuznetsov, D. (1994). ICM - A new method for protein modeling and design: Applications to docking and structure prediction from the distorted native conformation. *Journal of Computational Chemistry*, *15*(5), 488–506. doi:10.1002/jcc.540150503

Ahmed, M., & Barakat, K. (2015). Baby steps toward modelling the full human programmed Death-1 (PD-1) pathway. *Receptors & Clinical Investigation*, *1*, 1–5.

Bajaj, C. L., Chowdhury, R., & Siddahanavalli, V. (2011). F2dock: Fast fourier protein-protein docking. *IEEE/ACM Transactions on Computational Biology and Bioinformatics*, *8*(1), 45–58. doi:10.1109/TCBB.2009.57 PMID:21071796

Barakat, K., Gajewski, M., & Tuszynski, J. (2012). DNA Repair Inhibitors: Our Last Disposal to Improve Cancer Therapy. *Current Topics in Medicinal Chemistry*. doi:10.2174/156802612801319070 PMID:22690684

Barakat, K., Issack, B. B., Stepanova, M., & Tuszynski, J. (2011). Effects of temperature on the p53-DNA binding interactions and their dynamical behavior: Comparing the wild type to the R248Q mutant. *PLoS ONE*, *6*(11), e27651. doi:10.1371/journal.pone.0027651 PMID:22110706

Barakat, K., Mane, J., Friesen, D., & Tuszynski, J. (2010). Ensemble-based virtual screening reveals dual-inhibitors for the p53-MDM2/MDMX interactions. *Journal of Molecular Graphics & Modelling*, *28*(6), 555–568. doi:10.1016/j.jmgm.2009.12.003 PMID:20056466

Barakat, K., & Tuszynski, J. (2011). Virtual Screening for DNA Repair Inhibitors. *DNA Repair-On the Pathways to Fixing DNA Damage and Errors, 1*, 287–312.

Barakat, K. H., Houghton, M., Tyrrell, D. L., & Tuszynski, J. (2014). Rational Drug Design : One Target, Many Paths to It. *International Journal of Computational Models and Algorithms in Medicine*, *4*(1), 59–85. doi:10.4018/ijcmam.2014010104

Barakat, K. H., Mane, J. Y., & Tuszynski, J. A. (2011). Virtual Screening : An Overview on Methods and Applications. In L. Liu, D. Wei, Y. Li, & H. Lei (Eds.), *Handbook of Research on Computational and Systems Biology: Interdisciplinary Applications* (pp. 28–60). Hershey, PA: Medical Information Science Reference. doi:10.4018/978-1-60960-491-2.ch002

Berman, H. M., Westbrook, J., Feng, Z., Gilliland, G., Bhat, T. N., Weissig, H., & Bourne, P. E. et al. (2000). The Protein Data Bank. *Nucleic Acids Research*, *28*(1), 235–242. doi:10.1093/nar/28.1.235 PMID:10592235

Betts, M. J., & Sternberg, M. J. (1999). An analysis of conformational changes on protein-protein association: Implications for predictive docking. *Protein Engineering*, *12*(4), 271–283. doi:10.1093/protein/12.4.271 PMID:10325397

Bonvin, A. (2013). Coming to peace with protein complexes? 5th CAPRI evaluation meeting, April 17–19th 2013 – Utrecht. *Proteins: Structure, Function, and Bioinformatics, 81*(12), 2073–2074.

Boyd, D. B., & Lipkowitz, K. B. (2001). *Reviews in Computational Chemistry* (Vol. 17). New York: Wiley-VCH.

Camacho, C. J., & Vajda, S. (2002). Protein-protein association kinetics and protein docking. *Current Opinion in Structural Biology*, *12*(1), 36–40. doi:10.1016/S0959-440X(02)00286-5 PMID:11839487

Chen, C. Y.-C., & Tou, W. (2013). How to design a drug for the disordered proteins? *Drug Discovery Today*, *18*(19-20), 910–915. doi:10.1016/j.drudis.2013.04.008 PMID:23643489

Chen, R., Li, L., & Weng, Z. (2003). ZDOCK: An Initial-Stage Protein-Docking Algorithm. *Proteins, 52*, 80–87.

Chen, R., & Weng, Z. (2002). Docking unbound proteins using shape complementarity, desolvation, and electrostatics. *Proteins: Structure, Function and Genetics, 47*, 281–294.

Chen, Y. C. (2015). Beware of docking! *Trends in Pharmacological Sciences, 36*(2), 78–95. doi:10.1016/j.tips.2014.12.001 PMID:25543280

Chène, P. (2003). Inhibiting the p53–MDM2 interaction: An important target for cancer therapy. *Nature Reviews. Cancer, 3*(2), 102–109. doi:10.1038/nrc991 PMID:12563309

Cheng, T. M.-K., Blundell, T. L., & Fernandez-Recio, J. (2007). pyDock: Electrostatics and Desolvation for Effective Scoring of Rigid-Body Protein–Protein Docking. *Proteins, 68*(2), 503–515. doi:10.1002/prot.21419 PMID:17444519

Chowdhury, R., Rasheed, M., Keidel, D., Moussalem, M., Olson, A., Sanner, M., & Bajaj, C. (2013). Protein-Protein Docking with F2Dock 2.0 and GB-Rerank. *PLoS ONE, 8*(3), e51307. doi:10.1371/journal.pone.0051307 PMID:23483883

Chuang, G.-Y., Kozakov, D., Brenke, R., Comeau, S. R., & Vajda, S. (2008). DARS (Decoys As the Reference State) potentials for protein-protein docking. *Biophysical Journal, 95*(9), 4217–4227. doi:10.1529/biophysj.108.135814 PMID:18676649

Cole, J. C., Murray, C. W., Nissink, J. W. M., Taylor, R. D., & Taylor, R. (2005). Comparing protein-ligand docking programs is difficult. *Proteins: Structure, Function, and Bioinformatics, 60*(3), 325–332. doi:10.1002/prot.20497 PMID:15937897

Comeau, S. R., Gatchell, D. W., Vajda, S., & Camacho, C. J. (2004). ClusPro: An automated docking and discrimination method for the prediction of protein complexes. *Bioinformatics (Oxford, England), 20*(1), 45–50. doi:10.1093/bioinformatics/btg371 PMID:14693807

De Vries, S., & Zacharias, M. (2013). Flexible docking and refinement with a coarse-grained protein model using ATTRACT. *Proteins: Structure. Function and Bioinformatics, 81*(12), 2167–2174. doi:10.1002/prot.24400

De Vries, S. J., Schindler, C. E. M., Chauvot de Beauchêne, I., & Zacharias, M. (2015). A Web Interface for Easy Flexible Protein-Protein Docking with ATTRACT. *Biophysical Journal, 108*(3), 462–465. doi:10.1016/j.bpj.2014.12.015 PMID:25650913

De Vries, S. J., Van Dijk, M., & Bonvin, A. M. J. J. (2010). The HADDOCK web server for data-driven biomolecular docking. *Nature Protocols, 5*(5), 883–897. doi:10.1038/nprot.2010.32 PMID:20431534

Dominguez, C., Boelens, R., & Bonvin, A. M. J. J. (2003). HADDOCK: A protein-protein docking approach based on biochemical or biophysical information. *Journal of the American Chemical Society, 125*(7), 1731–1737. doi:10.1021/ja026939x PMID:12580598

Duhovny, D., Nussinov, R., & Wolfson, H. (2002). Efficient Unbound Docking of Rigid Molecules. *Algorithms in Bioinformatics*, 185–200.

Ehrlich, L. P., & Wade, R. C. (2001). Protein–Protein Docking. In *Reviews in Computational Chemistry* (17th ed.; pp. 61–92). New York: Wiley-VCH. doi:10.1002/0471224413.ch2

Elcock, A. A. H., Sept, D., & McCammon, J. A. (2001). Computer simulation of protein-protein interactions. *The Journal of Physical Chemistry B*, *105*(8), 1504–1518. doi:10.1021/jp003602d

Fernandez-Recio, J., Totrov, M., & Abagyan, R. (2002). Soft protein – protein docking in internal coordinates. *Protein Science*, *11*(2), 280–291. doi:10.1110/ps.19202 PMID:11790838

Gabb, H., Jackson, R. M., & Sternberg, M. J. E. (1997). Modelling protein docking using shape complementarity, electrostatics and biochemical information. *Journal of Molecular Biology*, *272*(1), 106–120. doi:10.1006/jmbi.1997.1203 PMID:9299341

Gajda, M. J., Tuszynska, I., Kaczor, M., Bakulina, A. Y., & Bujnicki, J. M. (2010). FILTREST3D: Discrimination of structural models using restraints from experimental data. *Bioinformatics (Oxford, England)*, *26*(23), 2986–2987. doi:10.1093/bioinformatics/btq582 PMID:20956242

Gógl, G., Schneider, K. D., Yeh, B. J., Alam, N., Nguyen Ba, A. N., Moses, A. M., & Weiss, E. L. et al. (2015). The Structure of an NDR/LATS Kinase–Mob Complex Reveals a Novel Kinase–Coactivator System and Substrate Docking Mechanism. *PLoS Biology*, *13*(5), e1002146. doi:10.1371/journal.pbio.1002146 PMID:25966461

Gray, J. J., Moughon, S., Wang, C., Schueler-Furman, O., Kuhlman, B., Rohl, C., & Baker, D. (2003). Protein–Protein Docking with Simultaneous Optimization of Rigid-body Displacement and Side-chain Conformations. *Journal of Molecular Biology*, *331*(1), 281–299. doi:10.1016/S0022-2836(03)00670-3 PMID:12875852

Grinter, S. Z., & Zou, X. (2014). Challenges, applications, and recent advances of protein-ligand docking in structure-based drug design. *Molecules (Basel, Switzerland)*, *19*(7), 10150–10176. doi:10.3390/molecules190710150 PMID:25019558

Grosdidier, S., & Fernández-Recio, J. (2012). Protein-protein docking and hot-spot prediction for drug discovery. *Current Pharmaceutical Design*, *18*(30), 4607–4618. doi:10.2174/138161212802651599 PMID:22650255

Guo, F., Li, S. C., Ma, W., & Wang, L. (2013). Detecting protein conformational changes in interactions via scaling known structures. Lecture Notes in Computer Science, 7821(10), 58–74.

Hernández-santoyo, A., Tenorio-barajas, A. Y., Altuzar, V., Vivanco-cid, H., & Mendoza-barrera, C. (2013). Protein-Protein and Protein-Ligand Docking, Protein Engineering - Technology and Application (D. T. Ogawa, Ed.). Academic Press.

Huang, S. (2014). Search strategies and evaluation in protein – protein docking : Principles, advances and challenges. *Drug Discovery Today*, *19*(8), 1081–1096. doi:10.1016/j.drudis.2014.02.005 PMID:24594385

Huang, T.-T., Hwang, J.-K., Chen, C.-H., Chu, C.-S., Lee, C.-W., & Chen, C.-C. (2015). (PS)2: protein structure prediction server version 3.0. *Nucleic Acids Research, 454*(Ext 56921), 1–5.

Jackson, R. M., Gabb, H., & Sternberg, M. J. E. (1998). Rapid refinement of protein interfaces incorporating solvation: Application to the docking problem. *Journal of Molecular Biology*, *276*(1), 265–285. doi:10.1006/jmbi.1997.1519 PMID:9514726

Janin, J. (2013). The targets of CAPRI rounds 20-27. *Proteins: Structure, Function, and Bioinformatics*, *81*(12), 2075–2081. doi:10.1002/prot.24375

Janin, J., Henrick, K., Moult, J., Eyck, L., Ten, , Sternberg, M. J. E., & Vajda, S. (2003). CAPRI: A critical assessment of PRedicted interactions. *Proteins*, *52*(1), 2–9. doi:10.1002/prot.10381 PMID:12784359

Jiang, F., & Kim, S. H. (1991). "Soft docking": Matching of molecular surface cubes. *Journal of Molecular Biology*, *219*(1), 79–102. doi:10.1016/0022-2836(91)90859-5 PMID:2023263

Jiménez-García, B., Pons, C., & Fernández-Recio, J. (2013). pyDockWEB: A web server for rigid-body protein-protein docking using electrostatics and desolvation scoring. *Bioinformatics (Oxford, England)*, *29*(13), 1698–1699. doi:10.1093/bioinformatics/btt262 PMID:23661696

Jimenez-Garcia, B., Pons, C., Svergun, D. I., Bernado, P., & Fernandez-Recio, J. (2015). pyDockSAXS: Protein-protein complex structure by SAXS and computational docking. *Nucleic Acids Research*, 1–6. PMID:25897115

Kann, M. G. (2007). Protein interactions and disease: Computational approaches to uncover the etiology of diseases. *Briefings in Bioinformatics*, *8*(5), 333–346. doi:10.1093/bib/bbm031 PMID:17638813

Katchalski-Katzir, E., Shariv, I., Eisenstein, M., Friesem, A. A., Aflalo, C., & Vakser, I. A. (1992). Molecular surface recognition: determination of geometric fit between proteins and their ligands by correlation techniques. *Proceedings of the National Academy of Sciences of the United States of America*, *89*(6), 2195–2199. doi:10.1073/pnas.89.6.2195

Keskin, O., Ma, B., & Nussinov, R. (2005). Hot regions in protein-protein interactions: The organization and contribution of structurally conserved hot spot residues. *Journal of Molecular Biology*, *345*(5), 1281–1294. doi:10.1016/j.jmb.2004.10.077 PMID:15644221

Kojima, K., Konopleva, M., McQueen, T., O'Brien, S., Plunkett, W., & Andreeff, M. (2006). Mdm2 inhibitor Nutlin-3a induces p53-mediated apoptosis by transcription-dependent and transcription-independent mechanisms and may overcome Atm-mediated resistance to fludarabine in chronic lymphocytic leukemia. *Blood*, *108*(3), 993–1000. doi:10.1182/blood-2005-12-5148 PMID:16543464

Kortemme, T., & Baker, D. (2004). Computational design of protein-protein interactions. *Current Opinion in Chemical Biology*, *8*(1), 91–97. doi:10.1016/j.cbpa.2003.12.008 PMID:15036162

Kozakov, D., Beglov, D., Bohnuud, T., Mottarella, S. E., Xia, B., Hall, D. R., & Vajda, S. (2013). How good is automated protein docking? *Proteins: Structure. Function and Bioinformatics*, *81*(12), 2159–2166. doi:10.1002/prot.24403

Lensink, M. F., & Wodak, S. J. (2013). Docking, scoring, and affinity prediction in CAPRI. *Proteins: Structure, Function, and Bioinformatics*, *81*(12), 2082–2095. doi:10.1002/prot.24428 PMID:24115211

Lesk, V. I., & Sternberg, M. J. E. (2008). 3D-Garden: A system for modelling protein-protein complexes based on conformational refinement of ensembles generated with the marching cubes algorithm. *Bioinformatics (Oxford, England)*, *24*(9), 1137–1144. doi:10.1093/bioinformatics/btn093 PMID:18326508

Li, L., Chen, R., & Weng, Z. (2003). RDOCK:Refinement of Rigid-body Protein Docking Predictions. *Proteins: Struc. Funct. Gen*, *53*(061/14), 693–707.

Liang, S., Liu, S., Zhang, C., & Zhou, Y. (2007). A simple reference state makes a significant improvement in near-native selections from structurally refined docking decoys. *Proteins*, *69*(2), 244–253. doi:10.1002/prot.21498 PMID:17623864

Liu, S., & Vakser, I. (2011). DECK: Distance and environment-dependent, coarse-grained, knowledge-based potentials for protein-protein docking. *BMC Bioinformatics*, *12*(1), 280. doi:10.1186/1471-2105-12-280 PMID:21745398

Lorenzen, S., & Zhang, Y. (2007). Monte Carlo refinement of rigid-body protein docking structures with backbone displacement and side-chain optimization. *Protein Science : A Publication of the Protein Society, 16*(12), 2716–2725.

Mandell, J. G., Roberts, V., Pique, M. E., Kotlovyi, V., Mitchell, J. C., Nelson, E., & Ten Eyck, L. F. et al. (2001). Protein docking using continuum electrostatics and geometric fit. *Protein Engineering*, *14*(2), 105–113. doi:10.1093/protein/14.2.105 PMID:11297668

Mashiach, E., Nussinov, R., & Wolfson, H. J. (2010). FiberDock: Flexible induced-fit backbone refinement in molecular docking. *Proteins: Structure. Function and Bioinformatics*, *78*(6), 1503–1519.

Mashiach, E., Schneidman-Duhovny, D., Andrusier, N., Nussinov, R., & Wolfson, H. J. (2008). FireDock: a web server for fast interaction refinement in molecular docking. *Nucleic Acids Research, 36*, W229–32.

Matsuzaki, Y., Uchikoga, N., Ohue, M., Shimoda, T., Sato, T., Ishida, T., & Akiyama, Y. (2013). MEGADOCK 3.0: A high-performance protein-protein interaction prediction software using hybrid parallel computing for petascale supercomputing environments. *Source Code for Biology and Medicine*, *8*(1), 18. doi:10.1186/1751-0473-8-18 PMID:24004986

Moal, I. H., Torchala, M., Bates, P. A., & Fernández-Recio, J. (2013). The scoring of poses in protein-protein docking: Current capabilities and future directions. *BMC Bioinformatics*, *14*(1), 286. doi:10.1186/1471-2105-14-286 PMID:24079540

Moont, G., Gabb, H., & Sternberg, M. J. E. (1999). Use of pair potentials across protein interfaces in screening predicted docked complexes. *Proteins*, *35*(3), 364–373. doi:10.1002/(SICI)1097-0134(19990515)35:3<364::AID-PROT11>3.0.CO;2-4 PMID:10328272

Moult, J., Fidelis, K., Kryshtafovych, A., Schwede, T., & Tramontano, A. (2014). Critical assessment of methods of protein structure prediction (CASP) - round x. *Proteins: Structure. Function and Bioinformatics*, *82*(SUPPL.2), 1–6. doi:10.1002/prot.24452

Muegge, I., & Rarey, M. (2001). Small Molecule Docking and Scoring. In *Reviews in Computational Chemistry* (pp. 1–60). John Wiley & Sons, Inc. doi:10.1002/0471224413.ch1

Nillegoda, N. B., Kirstein, J., Szlachcic, A., Berynskyy, M., Stank, A., Stengel, F., & Bukau, B. et al. (2015). Crucial HSP70 co-chaperone complex unlocks metazoan protein disaggregation. *Nature*, *524*(7564), 247–251. PMID:26245380

Palma, P. N., Krippahl, L., Wampler, J. E., & Moura, J. J. (2000). BiGGER: A new (soft) docking algorithm for predicting protein interactions. *Proteins*, *39*(4), 372–384. doi:10.1002/(SICI)1097-0134(20000601)39:4<372::AID-PROT100>3.0.CO;2-Q PMID:10813819

Pedotti, M., Simonelli, L., Livoti, E., & Varani, L. (2011). Computational Docking of Antibody-Antigen Complexes, Opportunities and Pitfalls Illustrated by Influenza Hemagglutinin. *International Journal of Molecular Sciences*, *12*(1), 226–251. doi:10.3390/ijms12010226 PMID:21339984

Pierce, B., & Weng, Z. (2007). ZRANK: Reranking Protein Docking Predictions With an Optimized Energy Function. *Proteins*, *67*(4), 1078–1086. doi:10.1002/prot.21373 PMID:17373710

Pierce, B. G., Wiehe, K., Hwang, H., Kim, B.-H., Vreven, T., & Weng, Z. (2014). ZDOCK server: Interactive docking prediction of protein–protein complexes and symmetric multimers. *Bioinformatics (Oxford, England)*, *30*(12), btu097. doi:10.1093/bioinformatics/btu097 PMID:24532726

Pons, C., Jiménez-González, D., González-Álvarez, C., Servat, H., Cabrera-Benítez, D., Aguilar, X., & Fernández-Recio, J. (2012). Cell-Dock: High-performance protein-protein docking. *Bioinformatics (Oxford, England)*, *28*(18), 2394–2396. doi:10.1093/bioinformatics/bts454 PMID:22815362

Popov, P., Ritchie, D. W., & Grudinin, S. (2014). DockTrina: Docking triangular protein trimers. *Proteins: Structure. Function and Bioinformatics*, *82*(1), 34–44. doi:10.1002/prot.24344

Ravikant, D. V. S., & Elber, R. (2010). PIE-efficient filters and coarse grained potentials for unbound protein-protein docking. *Proteins*, *78*(2), 400–419. doi:10.1002/prot.22550 PMID:19768784

Ravikant, D. V. S., & Elber, R. (2011). Energy design for protein-protein interactions. *The Journal of Chemical Physics*, *135*(6), 065102. doi:10.1063/1.3615722 PMID:21842951

Ritchie, D. W., & Kemp, G. J. L. (2000). Protein docking using spherical polar Fourier correlations. *Proteins*, *39*(2), 178–194. doi:10.1002/(SICI)1097-0134(20000501)39:2<178::AID-PROT8>3.0.CO;2-6 PMID:10737939

Schneider, T., Kruse, T., Wimmer, R., Wiedemann, I., Sass, V., Pag, U., & Kristensen, H.-H. et al. (2010). Plectasin, a Fungal Defensin, Targets the Bacterial Cell Wall Precursor Lipid II. *Science*, *328*(5982), 1168–1172. doi:10.1126/science.1185723 PMID:20508130

Schneidman-Duhovny, D., Inbar, Y., Nussinov, R., & Wolfson, H. J. (2005). PatchDock and SymmDock: Servers for rigid and symmetric docking. *Nucleic Acids Research*, *33*(Web ServerSUPPL. 2), 363–367. doi:10.1093/nar/gki481 PMID:15980490

Schneidman-Duhovny, D., Ruth, N., & Wolfson, H. J. (2007). Automatic prediction of protein interactions with large scale motion. *Proteins*, *69*(4), 764–773. doi:10.1002/prot.21759 PMID:17886339

Sheinerman, F. B., Norel, R., & Honig, B. (2000). Electrostatic aspects of protein-protein interactions. *Current Opinion in Structural Biology*, *10*(2), 153–159. doi:10.1016/S0959-440X(00)00065-8 PMID:10753808

Smith, G. R., & Sternberg, M. J. E. (n.d.). *Prediction of protein – protein interactions by docking methods*. Academic Press.

Smith, J., Vanoye, C. G., George, A. L., Meiler, J., & Sanders, C. R. (2007). Structural Models for the KCNQ1 Voltage-Gated Potassium Channel. *Biochemistry*, *46*(49), 14141–14152. doi:10.1021/bi701597s PMID:17999538

Szilagyi, A., Grimm, V., Arakaki, K., & Skolnick, J. (2005). Prediction of physical protein – protein. *Most, 2*, 1–16. PMID:16204844

Taylor, D. W., Zhu, Y., Staals, R. H. J., Kornfeld, J. E., Shinkai, A., van der Oost, J., & Doudna, J. A. et al. (2015). Structures of the CRISPR-Cmr complex reveal mode of RNA target positioning. *Science, 348*(6234), 581–585. doi:10.1126/science.aaa4535 PMID:25837515

Taylor, J. S., & Burnett, R. M. (2000). DARWIN: A program for docking flexible molecules. *Proteins, 41*(2), 173–191. doi:10.1002/1097-0134(20001101)41:2<173::AID-PROT30>3.0.CO;2-3 PMID:10966571

Terashi, G., Takeda-Shitaka, M., Kanou, K., Iwadate, M., Takaya, D., & Umeyama, H. (2007). The SKE-DOCK server and human teams based on a combined method of shape complementarity and free energy estimation. *Proteins, 69*(4), 866–872. doi:10.1002/prot.21772 PMID:17853449

Tovchigrechko, A., & Vakser, I. A. (2006). GRAMM-X public web server for protein-protein docking. *Nucleic Acids Research, 34*, 310–314.

Vakser, I. (1995). Protein docking for low-resolution structures. *Protein Engineering, 8*(4), 371–377. doi:10.1093/protein/8.4.371 PMID:7567922

Vakser, I. A. (2014). Protein-Protein Docking: From Interaction to Interactome. *Biophysical Journal, 107*(8), 1785–1793. doi:10.1016/j.bpj.2014.08.033 PMID:25418159

Venkatraman, V., & Ritchie, D. W. (2012). Flexible protein docking refinement using pose-dependent normal mode analysis. *Proteins: Structure. Function and Bioinformatics, 80*(9), 2262–2274. doi:10.1002/prot.24115

Venkatraman, V., Yang, Y. D., Sael, L., & Kihara, D. (2009). Protein-protein docking using region-based 3D Zernike descriptors. *BMC Bioinformatics, 10*(1), 407. doi:10.1186/1471-2105-10-407 PMID:20003235

Villoutreix, B. O., Kuenemann, M., Poyet, J. L., Bruzzoni-Giovanelli, H., Labbé, C., Lagorce, D., & Miteva, M. A. et al. (2014). Drug-like protein-protein interaction modulators: Challenges and opportunities for drug discovery and chemical biology. *Molecular Informatics, 33*(6-7), 414–437. doi:10.1002/minf.201400040 PMID:25254076

Viricel, C., Ahmed, M., & Barakat, K. (2015). Human PD-1 Binds Differently To Its Human Ligands: A Comprehensive Modelling Study. *Journal of Molecular Graphics & Modelling, 57*, 131–142. doi:10.1016/j.jmgm.2015.01.015 PMID:25723350

Waksman, G., & Sansom, C. (2005). Introduction : Proteomics and Protein – Protein Interactions : Biology, Chemistry, Bioinformatics, and Drug Design. In G. Waksman (Ed.), *Proteomics and Protein – Protein Interactions : Biology, Chemistry, Bioinformatics, and Drug Design* (pp. 1–18). New York: Springer. doi:10.1007/0-387-24532-4_1

Wang, C. H. U., Schueler-furman, O. R., & Baker, D. (2005). Improved side-chain modeling for protein – protein docking. *Protein Science, 14*(5), 1328–1339. doi:10.1110/ps.041222905 PMID:15802647

Wass, M. N., Fuentes, G., Pons, C., Pazos, F., & Valencia, A. (2011). Towards the prediction of protein interaction partners using physical docking. *Molecular Systems Biology, 7*(469), 469. PMID:21326236

Wiehe, K., Peterson, M. W., Pierce, B., Mintseris, J., & Weng, Z. (2008). Protein – Protein Docking : Overview and Performance Analysis. In M. Zaki & C. Bystroff (Eds.), Protein Structure Prediction (2nd ed.; vol. 413, pp. 283–314). Humana Press.

Wodak, S. J., & Méndez, R. (2004). Prediction of protein-protein interactions: The CAPRI experiment, its evaluation and implications. *Current Opinion in Structural Biology*, *14*(2), 242–249. doi:10.1016/j.sbi.2004.02.003 PMID:15093840

Xu, Y., Wang, Y., Meng, X. Y., Zhang, M., Jiang, M., Cui, M., & Tseng, G. N. (2013). Building KCNQ1/KCNE1 channel models and probing their interactions by molecular-dynamics simulations. *Biophysical Journal*, *105*(11), 2461–2473. doi:10.1016/j.bpj.2013.09.058 PMID:24314077

Zacharias, M. (2003). Protein-protein docking with a reduced protein model accounting for side-chain flexibility. *Protein Science : A Publication of the Protein Society*, *12*(6), 1271–1282.

Zhang, Y., Zheng, N., Hao, P., Cao, Y., & Zhong, Y. (2005). A molecular docking model of SARS-CoV S1 protein in complex with its receptor, human ACE2. *Computational Biology and Chemistry*, *29*(3), 254–257. doi:10.1016/j.compbiolchem.2005.04.008 PMID:15979045

Zhang, Z., Schindler, C. E. M., Lange, O. F., & Zacharias, M. (2015). Application of Enhanced Sampling Monte Carlo Methods for High-Resolution Protein-Protein Docking in Rosetta. *PLoS ONE*, *10*(6), e0125941. doi:10.1371/journal.pone.0125941 PMID:26053419

ADDITIONAL READING

Barakat, K. H., Mane, J. Y., & Tuszynski, J. A. (2011). Virtual Screening : An Overview on Methods and Applications. In *Handbook of Research on Computational and Systems Biology: Interdisciplinary Applications* (pp. 28–60). Hershey, PA: Medical Information Science Reference. doi:10.4018/978-1-60960-491-2.ch002

Ehrenberger, T., Cantley, L., & Yaffe, M. (2015). Computational Prediction of Protein-Protein Interactions. In *Protein-Protein Interactions SE - 4* (Vol. 1278, pp. 57–75). New York: Springer.

Ehrlich, L. P., & Wade, R. C. (2001). Protein–Protein Docking. In *Reviews in Computational Chemistry* (17th ed.; pp. 61–92). New York: Wiley-VCH. doi:10.1002/0471224413.ch2

Huang, S. (2014). Search strategies and evaluation in protein – protein docking : Principles, advances and challenges. *Drug Discovery Today*, *19*(8), 1081–1096. doi:10.1016/j.drudis.2014.02.005 PMID:24594385

Waksman, G., & Sansom, C. (2005). Introduction : Proteomics and Protein – Protein Interactions : Biology, Chemistry, Bioinformatics, and Drug Design. In *Proteomics and Protein–Protein Interactions: Biology, Chemistry, Bioinformatics, and Drug Design* (pp. 1–18). New York: Springer. doi:10.1007/0-387-24532-4_1

KEY TERMS AND DEFINITIONS

Docking: The computational task of predicting a biologically relevant and stable complex formed by two individual molecules.

Hot Spot Regions: The critical residues at the interface regions of proteins, where the binding takes place. These are the main components responsible for protein interactions and contribute the most to the binding free energy in bound forms of proteins.

Molecular Dynamic Simulations: Computer simulation techniques for prediction of the dynamic behaviour of molecules based on integration of Newton's equations of motion.

Post-Docking Refinement: A number of methods employed for more accurate ranking of the conformational poses from the docking process. Due to lack of precision in docking scoring functions, this stage is implemented to ensure the correctness of predictions.

Protein Flexibility: Proteins are dynamic entities and continuously undergo deviations from their equilibrium state. This is caused either due to the motion of the protein side-chains or sometimes the backbone structure. The issue of flexibility is often considered a challenge in docking studies and requires special treatments.

Sampling: A set of techniques in docking process for generating a large number of alternative poses for the two molecules (receptor and ligand) by sampling all the possible binding modes.

Scoring: Mathematical methods to rank the poses from docking with different types of scoring functions based on the binding free energy (non-bonded interactions) between the two docked molecules.

Chapter 8
Protein-Ligand Docking Methodologies and Its Application in Drug Discovery

Sanchaita Rajkhowa
Tezpur University, India

Ramesh C. Deka
Tezpur University, India

ABSTRACT

Molecular docking is a key tool in structural biology and computer-assisted drug design. Molecular docking is a method which predicts the preferred orientation of a ligand when bound in an active site to form a stable complex. It is the most common method used as a structure-based drug design. Here, the authors intend to discuss the various types of docking methods and their development and applications in modern drug discovery. The important basic theories such as sampling algorithm and scoring functions have been discussed briefly. The performances of the different available docking software have also been discussed. This chapter also includes some application examples of docking studies in modern drug discovery such as targeted drug delivery using carbon nanotubes, docking of nucleic acids to find the binding modes and a comparative study between high-throughput screening and structure-based virtual screening.

1. INTRODUCTION

Traditionally, the discovery of new drugs has arisen from observations that various plant extracts possess interesting biological effects. However, early users of such plant extracts did not understand or realize which components in the material were responsible for achieving these therapeutic benefits. The main difference between modern and age-old drug discovery is that the identification of the composition of matter, or the active form, within the medicine itself. Drug discovery and development is mostly a complex, expensive, time consuming and market-driven process with very few novel drug candidates actually making it through the Food and Drug Administration (FDA) for approval. Modern drug dis-

DOI: 10.4018/978-1-5225-0115-2.ch008

covery process involves rational approach with knowledge about enzymes and receptors and the use of complex synthetic methods and technologies such as combinatorial chemistry, microwave assisted organic synthesis and high-throughput (HTS) biological screening methods (Patil, 2012).

In recent years, there has been an increase in the number of new therapeutic targets for modern drug discovery with enormous contribution to many structural details of proteins and protein-ligand complexes. This has been possible as a result of the completion of the human genome project, along with the development of high-throughput protein purification, crystallography and nuclear magnetic resonance spectroscopy. Experimental high-throughput screening (HTS) and combinatorial chemistry were developed in order to overcome the various problems faced by lead discovery in drug development. Some exceptional number of novel leads was anticipated using sophisticated large-scale techniques, which would result in a substantial increase in novel entities launched to the market per year. However, this process had some drawbacks which included lack of validation and further optimization of these compounds into actual leads and preclinical candidates.

Such situations fuel the consideration and development of alternative techniques. These techniques have allowed the computational strategies to establish a strong hold in all aspects of drug development process (Jorgensen, 2004; Kitchen, Decornez, Furr, & Bajorath, 2004; Bajorath, 2002; Langer, & Hoffmann, 2001; Walters, Stahl, & Murcko, 1998), such as, virtual screening techniques (Gohlke, & Klebe, 2002) for hit identification and methods for lead optimization. Virtual screening (VS) is a direct and rational drug discovery approach and has the advantage of low cost and effective screening (Moitessier, Englebienne, Lee, Lawandi, & Corbeil, 2008; Shoichet, McGovern, Wei, & Irwin, 2002; Bailey, & Brown, 2001). The expression 'virtual screening' was coined in the late 1990s; however, the techniques involved are much older (Lahana, 1999). In contrast to high-throughput screening (HTS), which is largely driven by phenomenon and technology, in VS, compounds are selected by predicting their binding to a macromolecular target using computer programs. The term 'target' or 'receptor' is used to describe the macromolecule to which a drug binds, which is usually a protein but can also be DNA or RNA. The small molecules or "ligands" may bind to the targets of biological interest. These ligands which are considered for study do not necessarily exist and their 'testing' does not include consumption of valuable resources. Thus, experimental hindrances like low solubility, etc. that can interfere with experimentally applied assay conditions are not needed to be considered for the initial computational screening (Klebe, 2006). VS require a prerequisite knowledge of the spatial arrangement and energetic conditions which are responsible for the binding of a particular ligand molecule to the receptor under investigation. Therefore, either the three-dimensional (3D) structure of the target or, at the very least, a rigid reference ligand with a known bioactive conformation mapping out the putative receptor binding site must be available (Langer, & Krovat, 2003). This defines VS as a knowledge-driven approach. VS can be classified into ligand-based and structure-based methods. When a set of active ligand molecules is known but very little or no information of the structure is available, then the ligand-based methods are employed. This includes pharmacophore modelling (Yu, et al., 2015; Sindhu, & Srinivasan, 2015; Vuorinen, et al., 2014) and quantitative structure activity relationship (QSAR) methods (Rajkhowa, & Deka, 2014; Rajkhowa, Hussain, Hazarika, Sarmah, & Deka, 2013). When the three-dimensional (3D) structure or model of the target is available, then the structure-based virtual screening (SBVS) methods (Istyastono, et al., 2015; Muralidharan, et al. 2015; Lionta, Spyrou, Vassilatis, & Cournia, 2014) and employed. This process is known as the molecular docking process and is used to place computer-generated representations of a small molecule into a target structure in a variety of positions, conformations and orientations (Kroemer, 2007).

Molecular docking is a method which predicts the preferred orientation of a ligand when bound in an active site to form a stable complex. It is the most common method used as a structure-based drug design. Molecular docking is considered as a very important tool in computer-aided drug discovery and has secured a solid position in structure-based drug design. Programs based on different algorithms have been developed to perform molecular docking studies, which have made docking an increasingly important tool in pharmaceutical research. Many reviews on docking have been published in the past (Meng, Zhang, Mezei, & Cui, 2011; Kontoyianni, Madhav, Suchanek, & Seibel, 2008; Coupez, & Lewis, 2006; Kitchen, et al., 2004; Brooijmans, & Kuntz, 2003; Halperin, Ma, Wolfson, & Nussinov, 2002) and many comparative studies have been conducted to evaluate the relative performances of the programs (Plewczynski, Lazniewski, Augustyniak, & Ginalski, 2010; Li, Li, Cheng, Liu, & Wang, 2010; ten Brink, & Exner, 2009; Cross, et al. 2009).

The molecular docking approach can be used to study the interaction of a small molecule with a protein at the atomic level, which helps in characterizing the behaviour of the small molecule in the binding site of the protein as well as to understand the fundamental biochemical processes (McConkey, Sobolev, & Edelman, 2002). The docking process basically involves two steps: (a) prediction of the conformation of a ligand, its position and orientation within the sites, and (b) assessment of the binding affinity. These two steps are related to sampling methods and scoring schemes, respectively

In many cases, the binding site is known before the docking of the ligands, which significantly increases the docking efficiency. Information about the binding site may also be gathered by comparison of the target protein with protein families having similar function or with proteins which are co-crystallized with other ligands. Many cavity detection programs and online servers are available which can be utilized to identify probable active sites within the proteins. Some of the programs that are available are GRID (Kastenholz, Pastor, Cruciani, Haaksma, & Fox, 2000; Goodford, 1985), POCKET (Levitt, & Banaszak, 1992), SURFNET (Glaser, Morris, Najmanovich, Laskowski, & Thornton, 2006; Laskowski, 1995), PASS (Brady, & Stouten, 2000) and MMC (Mezei, 2003). A docking performed without any assumption about the binding site is known as blind docking.

The lock- and -key theory is considered as the first theory which explains the ligand-receptor binding mechanism. It was proposed by Emil Fischer in 1890 (Fischer, 1894). This theory proposes that the ligand fits into the active site of a receptor just like a key fits into a lock (Figure 1). Biological 'locks' have unique stereochemical features that are necessary for their functioning. Some of the earliest reported docking methods (Kuntz, Blaney, Oatley, Langridge, & Ferrin, 1982) were based on this theory where both the ligand and the receptor were treated as rigid bodies.

Figure 1. Lock- and -key model of Emil Fischer

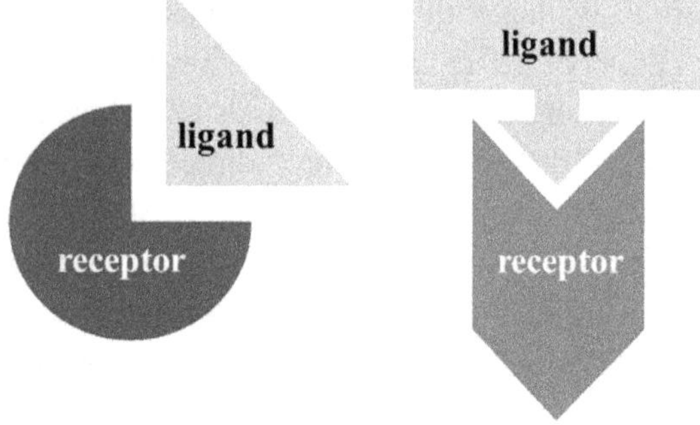

Figure 2. Induced-fit model of Daniel Koshland

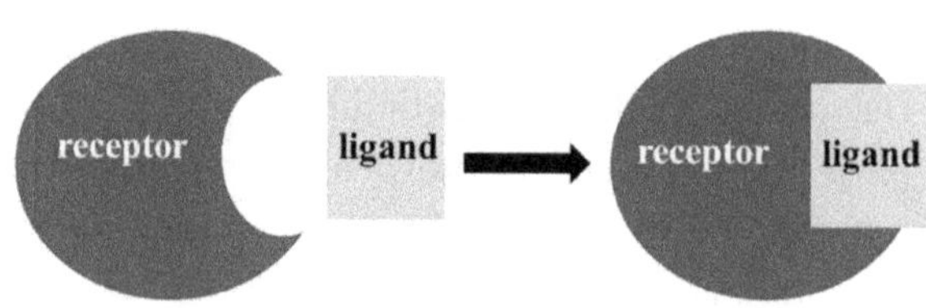

Figure 3. The conformation ensemble model showing the different conformational states of a protein

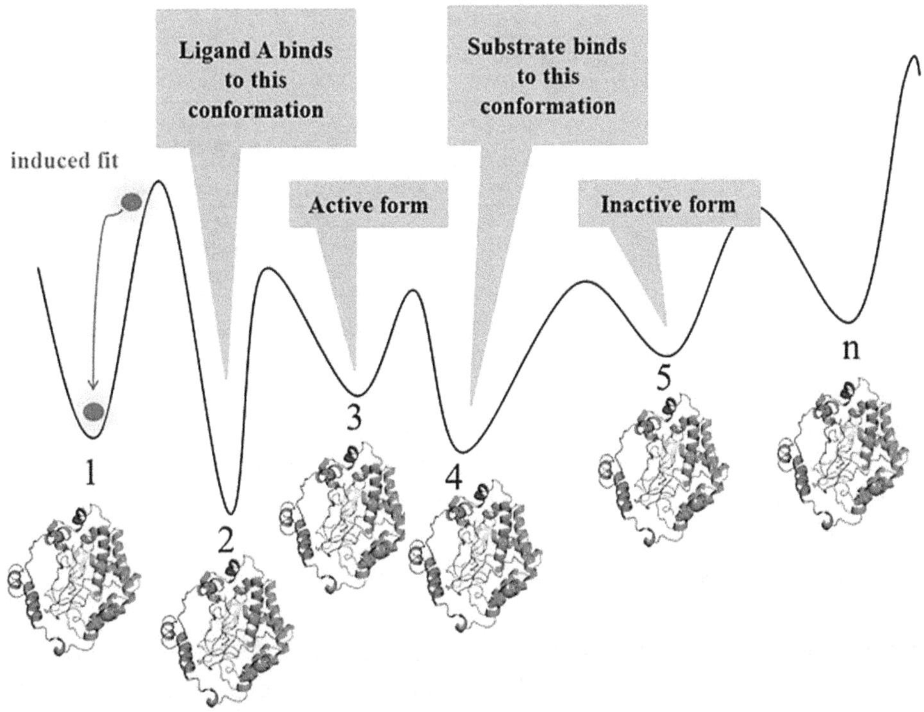

In 1958, Daniel Koshland introduced the 'induced-fit theory (Koshland, 1963). This theory proposed that the active site of the protein is continually reshaped by interactions with the ligands as the ligands interact with the protein. Here, both the ligand and target mutually adapt to each other through small conformational changes until an optimal fit is achieved as shown in Figure 2. It suggests that both the ligand and the target should be considered as flexible during docking.

Recently, it has been observed that proteins pre-exist as ensemble of conformational states (Boehr, Nussinov, & Wright, 2009). They can undergo much larger conformational changes in addition to the small induced-fit adaptations (Figure 3). The plasticity of the protein allows it to switch from one state to another.

The algorithms which are available to assess and justify protein ligand interactions is very large. They provide a wide range of techniques to tackle modern structure-based drug design. Despite the large size of this field, we have attempted to summarise and classify the types of docking methods. Types of docking methodologies can be divided into: (a) Rigid ligand and rigid receptor docking, (b) Flexible ligand and rigid receptor docking, and (c) Flexible ligand and flexible receptor docking. This chapter will, therefore, highlight the types of docking methods and some of its application in drug discovery.

2. THEORY OF MOLECULAR DOCKING

Molecular docking is a method which predicts the preferred orientation of one molecule (ligand) to a second (protein) when bound to each other to form a stable complex. Docking can be achieved through two interrelated steps: first by sampling conformations of the ligand in the active site of the protein; then

Table 1. Some sampling algorithms discussed in the chapter

Algorithms	Characteristic	Ref.
Matching algorithms	Geometry-based, suitable to VS and database enrichment for its high speed	(Norel, et al., 1994; Fischer, et al., 1993; Brint, & Willett, 1987)
Incremental construction	Fragment-based and docking incrementally	(Rarey, et al., 1996; Kuntz, & Leach, 1992; DesJarlais, et al., 1986)
LUDI	Fragment-based methods for the *de novo* design	(Bohm, 1992)
Monte Carlo and Genetic algorithms	Stochastic search	(Morris, et al., 1998; Jones, et al., 1997; Oshiro, et al., 1995; Goodsell, et al., 1993; Hart, & Read, 1992)
Molecular dynamics	For further refinement after docking	(Cornell, et al., 1995 ; Weine, et al., 1984)

ranking these conformations via a scoring function. The sampling algorithms reproduce the experimental binding mode and the scoring functions rank the highest among all generated conformations. From these two viewpoints, we give a brief overview of the basic docking theory.

a. Sampling Algorithms

A huge number of binding modes are possible between a ligand and a protein when their six degrees of translational and rotational freedom as well as conformational degrees of freedom are considered. As a result, it would be computationally too expensive to generate all the possible conformations. To overcome this problem, various sampling algorithms have been developed and widely used in molecular docking softwares (Table 1).

i. Matching Algorithms

Matching algorithms (MA) (Norel, Fischer, Wolfson, & Nussinov, 1994; Fischer, Norel, Wolfson, & Nussinov, 1993; Brint, & Willett, 1987) are based on the molecular shapes. They use the shape features and chemical information of a ligand to map it into an active site of a protein. Here, the protein and ligand are represented as pharmacophores and the distance of these pharmacophores within the protein and the ligand is calculated for a match wherein new ligand conformations are governed by the distance matrix between the pharmacophore and the corresponding ligand atoms. Some chemical properties such as hydrogen-bond donors, hydrogen-bond acceptors and aromatic rings can be taken into account during the match. Matching algorithms for ligand docking are available in programs such as DOCK (Knegtel, Kuntz, & Oshiro, 1997), FLOG (Miller, Kearsley, Underwood, & Sheridan, 1994), LibDock (Diller, & Merz, 2001) and SANDOCK (Burkhard, Taylor, & Walkinshaw, 1998).

ii. Incremental Construction

Incremental construction (IC) (Rarey, Kramer, Lengauer, & Klebe, 1996; Kuntz, & Leach, 1992; DesJarlais, Sheridan, Dixon, Kuntz, & Venkataraghavan, 1986) methods are available in programs such as DOCK 4.0 (Ewing, Makino, Skillman, & Kuntz, 2001), FlexX (Kramer, Rarey, & Lengauer, 1999), Hammerhead (Welch, Ruppert, & Jain, 1996), SLIDE (Schnecke, & Kuhn, 2000) and eHiTS (Zsoldos,

Reid, Simon, Sadjad, & Johnson, 2006). This method puts the ligand into an active site in a fragmental and incremental fashion. The rotatable bonds are broken so that ligand is divided into several fragments. One of these fragments is selected to dock into the active site first. This anchor is usually the largest fragment or the piece which may have significant functional role or interaction with the protein. The remaining fragments are added incrementally. The flexibility of the ligand is realized by generating different orientations to fit in the active site.

iii. LUDI

LUDI (Bohm, 1992) is a fragment-based method for the *de novo* design and modifications of ligands that may enhance their binding to the target protein. It focuses on the hydrogen bonds and hydrophobic interactions which are formed between the ligand and the protein. This program works a three stage protocol-by calculating the interaction sites of the target protein and ligand that are discrete positions in space, which can either, form hydrogen bonds or fill hydrophobic pockets. The next step involves the fitting of molecular fragments onto the interaction sites. The final stage of the protocol involves the joining of fragments using a database of bridge fragments and the previous fitting algorithm.

iv. Monte Carlo and Genetic Algorithm

Monte Carlo (MC) (Goodsell, Lauble, Stout, & Olson, 1993; Hart, & Read, 1992) and *Genetic Algorithm* (GA) (Morris, et al., 1998; Jones, Willett, Glen, Leach, & Taylor, 1997; Oshiro, Kuntz, & Dixon, 1995) belong to the class of stochastic methods. They search the conformational space by modifying a ligand conformation or a population of ligands randomly. The MC method has been applied to earlier versions of AutoDock (Goodsell, & Olson, 1990), ICM (Abagyan, Totrov, & Kuznetsov, 1994), QXP (McMartin, & Bohacek, 1997) and Affinity (Accelrys Inc., USA.), where it generates the poses of ligands through bond rotation, rigid-body translation or rotation. In order to test the conformations generated by the transformation, an energy-based selection criterion is used. These conformations, when passes the criterion, are then saved and modified to generate the next conformations. This procedure is repeated until the pre-defined quantity of conformations is collected. An advantage of MC technique is that a simple energy function can be used. This energy function does not require any derivative information. On the other hand, genetic algorithm (GA) encodes the degrees of freedom of the ligand as genes or binary strings. The collection of genes, or chromosome, represents the pose of the ligand. Mutation and crossover are two kinds of genetic operators in GA. Mutation makes random changes to the genes, whereas, crossover exchanges genes between two chromosomes. GA has been used in programs such as AutoDock (Goodsell, & Olson, 1990), GOLD (Verdonk, Cole, Hartshorn, Murray, & Taylor, 2003), DIVALI (Clark, & Ajay, 1995) and DARWIN (Taylor, & Burnett, 2000).

v. Molecular Dynamics

Molecular dynamics (MD) (Cornell, et al., 1995; Weine, et al., 1984) involves the calculation of solutions to Newton's equations of motions. A number of programs are available to perform molecular dynamics simulations such as AMBER (Kollman, 1996), GROMACS (van der Spoel, et al., 2005; Berendsen, van der Spoel, & van Drunen, 1995) and CHARMM (Brooks, et al., 1983). From docking point of view, MD simulation represents the flexibility of both the ligand and the protein considering the movement of each atom separately in the field of the rest atoms. However, the movement proceeds in very small steps and thus, have difficulties in stepping over high energy conformational barriers. The quality of the results from a standard MD simulation is extremely dependent on the starting conformation of the system.

b. Scoring Functions

Scoring functions distinguish between the current poses from incorrect ones or good binding ligands from inactive compounds in a reasonable computation time. The scoring function should be able to distinguish the experimental binding modes from all other modes explored through the searching algorithm. Since, a rigorous scoring function will be computationally very expensive, so often the function's complexity is reduced with some loss of accuracy. Scoring functions can be divided into force-field-based, empirical and knowledge-based scoring functions.

The docking programs that are available nowadays utilise the scoring functions in either of the two ways. In the first approach, it uses the full scoring function to rank a protein-ligand conformation. The system is then modified by the search algorithm, and the same scoring function is again applied to rank the new structure. In the second approach, it uses a two stage scoring function. Here, a reduced function is used to direct the search strategy and a more rigorous scoring function is then used to rank the resulting structures (Taylor, Jewsbury, & Essex, 2002). A serious problem faced by scoring functions is that of affinity prediction partly because of the limited treatment of solvation effect. One of the ways to solve this problem is physics-based scoring, which is involved in rescoring or lead optimization to improve the accuracy of binding affinity prediction.

3. TYPES OF DOCKING METHODOLOGIES

a. Rigid Ligand and Rigid Receptor Docking

When the ligand and receptor are both considered as rigid bodies, with the search space being very narrow and considering only the three translational and three rotational degrees of freedom, it is known as rigid docking. Here the protein is treated as rigid and the ligand flexibility is attained by using a pre-computed set of ligand conformations, or by allowing the overlap of the atoms between the protein and the ligand. Programs like HEX (Frishman, & Argos, 1995), and early versions of DOCK (Fischer, et al., 1993), FLOG (Miller, et al., 1994) and some protein-protein docking programs, such as FTDOCK (Gabb, Jackson, & Sternberg, 1997) has adopted this method which keep the ligand and receptor rigid during docking.

HEX is an interactive molecular graphics program for calculating and displaying feasible docking modes of pairs of protein and DNA molecules. Hex calculates the protein-ligand docking by assuming that the ligand and protein are rigid. Here, 3D expansions of real orthogonal spherical polar basis functions are used to model each molecule. This is done in order to encode both surface shape and electrostatic charge and potential distributions. This allows each property to be represented by a vector of coefficients. Hex represents the surface shapes of proteins using a two-term surface skin plus van der Waals steric density model, whereas the electrostatic model is derived from classical electrostatic theory. An overall docking score can be obtained as a function of the six degrees of freedom in a rigid body docking search by writing expressions for the overlap of pairs of parametric functions. This docking score can be interpreted as an interaction energy using suitable scaling factors.

DOCK is the first automated procedure for docking a molecule into a receptor site. DOCK characterizes the ligand and receptor as sets of spheres which could be overlaid my means of a clique detection procedure (Bron, & Kerbosch, 1973). DOCK uses geometrical and chemical matching algorithms as sampling algorithms and the scoring is done by considering steric fit, chemical complementation or pharmacophore similarity. Incremental construction method and exhaustive search are added to the improved versions in order to consider the ligand flexibility.

FLOG generates ligand conformations on the basis of distance geometry. The sets of distances are calculated using a clique-finding algorithm. In FLOG, conformations are scored with a function considering van der Waals, electrostatics, hydrogen bonding and hydrophobic interactions.

b. Flexible Ligand and Rigid Receptor Docking

When the ligand is kept flexible whereas the receptor is considered as rigid, the system follows the flexible ligand and rigid receptor docking. This system follows the induced-fit paradigm where the flexibilities of ligand and receptor are considered. However, keeping the receptor flexible would make the calculation very expensive. Thus, the common method, also a balance between accuracy and computational time, treats the ligand as flexible while receptor is kept rigid during docking. Almost all the docking programs available nowadays have adopted this method, such as AutoDock (Goodsell, & Olson, 1990), Molegro Virtual Docker (Thomsen, & Christensen, 2006) and FlexX (Kramer, et al., 1999).

Molegro Virtual Docker (MVD) is an integrated environment for studying and predicting how ligands interact with macromolecules. The identification of ligand binding modes is done by iteratively evaluating a number of ligand conformations and estimating the energy of their interactions with the target. The highest scoring solutions are returned for further analysis. This program considers the flexibility of the ligand and allows partial flexibility of the receptor (important amino acids present in the binding cavity), considering the cost of computation. MVD uses MolDock as a scoring function. The MolDock scoring function is based on a new hybrid search algorithm, called glided differential evolution. This glided differential evolution technique uses the tripos force-field and combines the differential evolution optimization technique with a cavity prediction algorithm (Storn, & Price, 1995). It is based on a piecewise linear potential (PLP) (Yang, & Chen, 2004).

AutoDock 3.0 (Morris, et al., 1998) uses a genetic algorithm as a global optimiser combined with energy minimisation as a local search method. AutoDock considers the ligand as flexible and the receptor as rigid and represents it as a grid. The scoring function used by AutoDock is based on AMBER forcefield which includes the van der Waals, hydrogen bonding, electrostatic interactions, conformational entropy and desolvation terms. These five terms are weighted using an empirical scaling factor determined using linear regression analysis from a set experimental data. AutoDock 4.0 (Morris, et al., 2009), however, is able to model receptor flexibility by allowing side-chains to move. As a result, interaction of protein-protein docking can be evaluated using this version of AutoDock. AutoDock Vina has been found to significantly improve the average accuracy of the binding mode predictions as compared to AutoDock 4.

FlexX uses an incremental construction algorithm and is one of the most popular programs to perform fragment docking. The base fragment is first docked into the active site by matching hydrogen bond pairs and metal and aromatic ring interactions between the ligand and the protein. The remaining components are then incrementally added according to the predefined rotatable torsion angles which results in the flexibility of the ligand. The scoring function of FlexX is based on Böhm's work (Bohm, 1994). The current version of FlexX includes placing explicit waters into the binding site during the docking procedure (Rarey, Kramer, & Lengauer, 1999) using pre-computed water positions. Some other terms such as electrostatic interactions, directional hydrogen bonds, rotational entropy, and aromatic and lipophilic interactions are also added in the latest extensions. The interactions between functional groups are also taken into account through assigning the type and geometry for groups. Furthermore, receptor flexibility has been included using discrete alternative protein conformations (Claussen, Buning, Rarey, & Lengauer, 2001).

c. Flexible Ligand and Flexible Receptor Docking

Receptor (protein) flexibility plays a very important role in receptor–ligand binding processes and should be considered in computational drug design. However, incorporating the receptor flexibility is a significant challenge in the field of docking. Several approaches have been reviewed (Teodoro, & Kavraki, 2003) in order to incorporate receptor flexibility in ligand docking studies where they have proposed a classification of methods that can be divided into five categories: (a) 'soft' receptors that limit penalties due to steric clashes, (b) selection of a few critical degrees of freedom in the binding site, c) use of multiple receptor structures, d) use of modified molecular simulation methods and e) use of collective degrees of freedom as a new basis of representation for protein flexibility (Totrov, & Abagyan, 2008). A number of methodologies have been proposed to treat the conformational changes in the structure-based docking procedure, which have a significant impact on the accuracy of docking and scoring and they have been extensively reviewed (Kokh, Wade, & Wenzel, 2011).

Molecular dynamics simulations are most commonly used to model all the degrees of freedom in the ligand-receptor complex. However, due to the problem of inadequate sampling and high computational expense, MD simulations cannot be used in the screening of large chemical database. The intrinsic mobility of proteins is related to the ligand binding behaviour (Teague, 2003). Several theoretical models such as conformer selection and conformational induction have been proposed to illustrate the flexible ligand-protein binding process. When a ligand selectively binds to a favourable conformation from a number of protein conformations, it is referred to as conformer selection (Teague, 2003), whereas conformational induction refers to a process in which the ligand converts the protein into a conformation that it would not spontaneously adopt in its unbound state. Various methods are currently available to implement the receptor flexibility (Table 2). They can be classified as:

i. Soft Receptors

The use of soft receptors is the simplest solution to represent some degree of receptor flexibility. The receptor is made softer by reducing the van der Waals contributions to the total energy which allows a larger ligand to fit in a binding site determined experimentally for a small molecule. The reason behind this approach is that the structure of the receptor has some inherent flexibility which allows it to adapt to slightly differently shaped ligands by resorting to small variations in the orientation of binding site chains and backbone positions. If the change in the receptor conformation is small enough, it is assumed that the receptor is capable of such a conformational change, given its large number of degrees of freedom, even though the conformational change itself is not modeled explicitly. The main advantage of using

Table 2. Some basic methods for including receptor flexibility along with their advantages and disadvantages

Method	Description	Advantage	Disadvantage	Program
Soft potential	Change vdW to allow for overlap between receptor and ligand atoms	Computational efficiency. Easy to implement and used in combination with other metheds.	Inadequate flexibility. Describe flexibility in an implicit, rude and non-quantitative way.	AutoDock (Goodsell, & Olson, 1990)
Rotamer library	Search side chain library to obtain possible conformations	Relative computational efficiency. Avoid minimization barriers.	Strong dependence on the database used. No backbone flexibility.	ICM (Abagyan, et al., 1994)
Receptor side chain flexibility	Sample both side chain and ligand conformations simultaneously using GA	Relative computational efficiency. Model the effect that ligand make on binding site residues.	Only selected side chains are involved. No backbone flexibility.	GOLD (Verdonk, et al., 2003) AutoDock 4.0 (Morris, et al., 2009)
Ensemble of protein conformations	Docking ligand to a series of receptor structures which represent different conformational states.	Include full and explicit flexibility.	Expensive computational cost. Limited by protein conformations used in sampling.	DOCK (Knegtel, et al., 1997) FlexE (Claussen, et al., 2001)

soft receptors is the ease of implementation (docking algorithms stay unchanged) and speed (the cost of evaluating the scoring function is the same as for the rigid case). Examples of soft docking are the LJ 8-4 potential in GOLD and smooth potential in AutoDock 3.0 belong to this class.

ii. Selection of Specific Degrees of Freedom

In order to reduce the complexity of modeling the very large dimensional space representing the full flexibility of the protein, it is possible to obtain an approximate solution by selecting only a few degrees of freedom to model explicitly. The degrees of freedom chosen usually correspond to rotations around single bonds. Rotations around bonds lead to deviations from ideal geometry that resulted in a small energy penalty when compared to deviations from ideality in bond lengths and bond angles. Such rotomer libraries (Leach, 1994; Desmet, Maeyer, Hazes, & Lasters, 1992) include a set of side-chain conformations which are usually determined from statistical analysis of structural experimental data. The main advantage of using rotomer libraries is the relative speed in sampling, and avoiding of minimization barriers. Some examples of programs using rotomer libraries are ICM (Internal Coordinates Mechanics) (Abagyan, et al., 1994) coupled with MC search of the ligand conformation and GOLD (Verdonk, et al., 2003) coupled with GA to dock a flexible ligand to a semi-flexible protein.

iii. Multiple Receptor Structures

One possible way to represent a flexible receptor for drug design applications is the use of multiple static receptor structures. This concept is supported by the currently accepted model that proteins in solution do not exist in a single minimum energy static conformation but are in fact constantly jumping between low energy conformational substrates (Kitao, Hayward, & Go, 1998; Frauenfelder, Sligar, & Wolynes, 1991; Noguti, & Go, 1989). In this way the best description for a protein structure is that of a conformational ensemble (Rich, Bursavich, & Estiarte, 2002) of slightly different protein structures coexisting in a low energy region of the potential energy surface. Here, the ligand is docked separately into a set of rigid protein conformations instead of a single one, and the results are merged depending on the method of choice (Cavasotto, & Abagyan, 2004). One of the main advantages of using multiple structures instead of using a selection of degrees of freedom to represent protein flexibility is that the flexible region is not limited to a specific small region of the protein. Multiple structures allow the consideration of the full flexibility of the protein without the exponential blow up in terms of computational cost that would derive from including all the degrees of freedom of the protein.

FlexE (Claussen, Buning, Rarey, & Lengauer, 2001) represents a flexible binding site by combining multiple receptor structures. The algorithm starts by superimposing the set of conformations available for a given receptor and merging similar parts of the structures. The best combination of substructures is selected with respect to the scoring function based on a comparison between the ligand and each alternative.

iv. Molecular Simulations

To simulate the binding process with as much detail as possible and avoid some of the limitations of previous flexibility models one can use force field based atomistic simulation methods such as MC or MD. MD applies the laws of classical mechanics to compute the motion of the particles in a molecular system, whereas, MC methods are so called because they are based on a random sampling of the conformational space. The main advantage of MC or MD flexibility representations in docking studies is that they are very accurate and can model explicitly all degrees of freedom of the system including the solvent if necessary (Teodoro, & Kavraki, 2003). MC based methods takes into account receptor flexibility by the use of a flexible enzyme binding site whose side chains are submitted to random perturbations.

QXP (Quick Explore) (McMartin, & Bohacek, 1997) is part of the Flo+ program and contains two conceptually different docking algorithms, namely, MCDock and ZipDock. The MCDock algorithm goes through repeated cycles of MC followed by energy minimization in order to generate and refine an ensemble of low-energy ligand poses. The new poses which have low energy and which are dissimilar to the poses already present are added to the ensemble. With the increase in the number of cycles, the size of the perturbations is reduced. This makes the MCDock procedure very efficient in finding low-energy solutions.

v. Collective Degrees of Freedom

An alternative representation for protein flexibility is the use of collective degrees of freedom which can be determined using different methods such as calculation of normal mode of the receptor (Levy, & Karplus, 1979). The collective degrees of freedom can also be calculated using dimensional reduction method such as principal component analysis (PCA). The receptor flexibility can also be achieved by using the concept of molecular hinges (Sandak, Nussinov, & Wolfson, 1995). These hinges are articulation points placed at specific locations in the protein which allow for relative movement of domains or substructural parts. These hinges are the articulations that are allowed to rotate in three dimensions, implicitly representing rotations about consecutive or nearby bonds. The advantage of collective degrees of freedom is that the protein flexibility is not limited to a specific small region of the protein as was the case when using only selected degrees of freedom. Also, since only a few independent degrees of freedom are used in the optimization procedure, the computational cost is similar to using only selected degrees of freedom and is much less than the cost of techniques that considers all degrees of freedom, such as traditional MD or MC.

Glide (Friesner, et al., 2004) is uses hybrid method to model receptor flexibility by designing a series of hierarchical filters to search the possible poses and orientations of the ligand within the binding site of the receptor. An exhaustive search of the ligand torsion angle space handles the ligand flexibility and the initial ligand conformations are selected based on torsion energies, which are then docked into receptor binding sites with soft potentials. To model the receptor flexibility, a rotamer exploration is used (Sherman, Day, Jacobson, Friesner, & Farid, 2006).

4. APPLICATION EXAMPLES OF MOLECULAR DOCKING IN MODERN DRUG DISCOVERY

Molecular docking has been one of the most widely used techniques in the field of drug design. Although the main application of docking lies in structure-based virtual screening for identification of new compounds towards a particular target protein (Kubinyi, 2006), it is not a stand-alone technique. It should be embedded in a workflow of different *in silico* as well as experimental techniques (Kroemer, 2007). Molecular docking can be combined with other computational techniques and experimental data to analyse the drug metabolism in drug discovery. A few successful applications of docking are discussed in the following.

In recent years, drug delivery by nanomaterials has gained considerable potential application due to its unique quasi one-dimensional structure and electronic properties (Reich,Thomsen, & Maultzsch, 2004). Single-wall carbon nanotubes have been selectively employed to deliver biologically active molecules across the cell membranes. In this context, Deka and co-workers (Saikia, Rajkhowa, & Deka, 2013) have carried out combined density functional theory (DFT) calculations and molecular docking studies on pyrazinamide (PZA) loaded functionalized single-wall carbon nanotube (fSWCNT) with wild type *Mycobacterium tuberculosis* pncA enzyme. They studied the interaction of pyrazinamide (PZA) with fSWCNT as function of nanotube chirality and length using two different approaches of covalent functionalization. They have investigated using DFT formalism the structural stability and electronic properties of prestine zigzag (*n*,0) SWCNT (n = 9, 10, 11, and 12) and functionalized nanotubes with PZA anti-tubercular drug molecules. Their studies revealed that the nanotubes with narrow diameter were thermodynamically more favourable during the course of drug delivery. The decrease in binding energies of PZA/(n,0) SWCNT with increasing diameter indicated that the curvature of nanotube sidewall played a major role in determining the reactivity. Moreover, they also found that highest occupied molecular orbitals (HOMO) contributed from the edge carbon atoms; except for (9,0) SWCNT, wherein it was delocalized along the nanotube sidewall. Their studies further showed that organic moieties like carboxylic and other functional groups could be attached at the reactive ends of the nanotube to enhance the solubility and dispersability of the nanotube in the human body. A detailed analysis on the contribution of HOMO, lowest unoccupied molecular orbitals (LUMO), HOMO – n and LUMO + n have been done which provided valuable insight and significant implications towards the orbital distribution in pristine and functionalized SWCNT systems. The docking studies performed by them suggested that SWCNTs facilitated the effective and target specific delivery of PZA drug within the active site of pncA and that narrow diameter nanotube are well docked rather than the larger diameter tubes (Figure 4). It was also found, that the length of PEG chain is reasonably adequate for the drug delivery of PZA drug within the specific active binding site of pncA protein. Furthermore, it was found that encapsulation facilitated in multiple loading and delivery of PZA molecules within the protein and for all the studied nanosystems conjugated with PZA molecule SWCNT did not bring any conformational change within the protein.

DNA is an important target for the treatment of multiple pathologies such as cancer. In particular, DNA intercalators have often been used as anticancer drugs. However, only a few computational studies have been performed on DNA-intercalator complexes. Very recently, Gilad and Senderowitz (Gilad, & Senderowitz, 2014) have analysed ligand binding site preferences in 63 high resolution DNA-intercalator

Figure 4. (I) The front and side view depiction of HOMO plots in pristine (9,0) SWCNT (a, b), pristine (10,0) SWCNT (c, d), pristine (11,0) SWCNT (e, f) each comprising of 5 unit cells, (II) The front and side view depiction of HOMO plots in pristine (9,0) SWCNT (a, b), pristine (10,0) SWCNT (c, d), pristine (11,0) SWCNT (e, f) each comprising of 5 unit cells, (III) The HOMO isosurface of (a) side view and (b) front view depiction of edge functionalized PZA/(9,0) SWCNT, (c) side view and (d) front view depiction of edge functionalized PZA/(10,0) SWCNT, (IV) (a) docked conformation of two PZA drug molecules encapsulated within (6,6) SWCNT comprising of 6 unit cells of SWCNT within the active binding site of pncA protein, (b) the hydrophobic surface of two PZA/(6,6) SWCNT docked within pncA protein., (V) (a) The docked conformation of sidewall functionalized PZA/(9,0) SWCNT comprising of 3 unit cells of SWCNT within the active binding site of pncA protein, (b) hydrophobic surface of PZA/(9,0) SWCNT docked within pncA protein, and (VI) (a) docked conformation of edge functionalized PZA/(9,0) SWCNT comprising of 5 unit cells of SWCNT within the active binding site of pncA protein, (b) the hydrophobic surface of PZA/(9,0) SWCNT docked within pncA protein

Source: Saikia, Rajkhowa, & Deka, 2013.

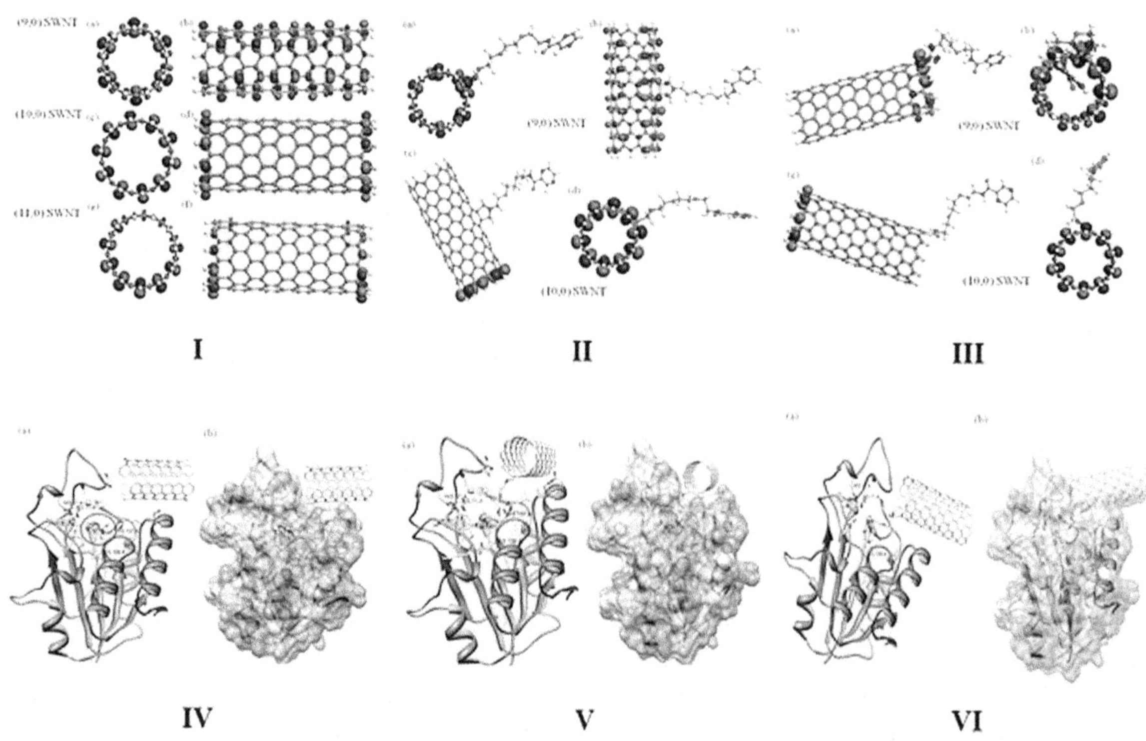

Figure 5. (a) Docked conformations of DNA - ligand complexes. (b) Decision trees for predicting DNA
Source: Gilad, & Senderowitz, 2014.

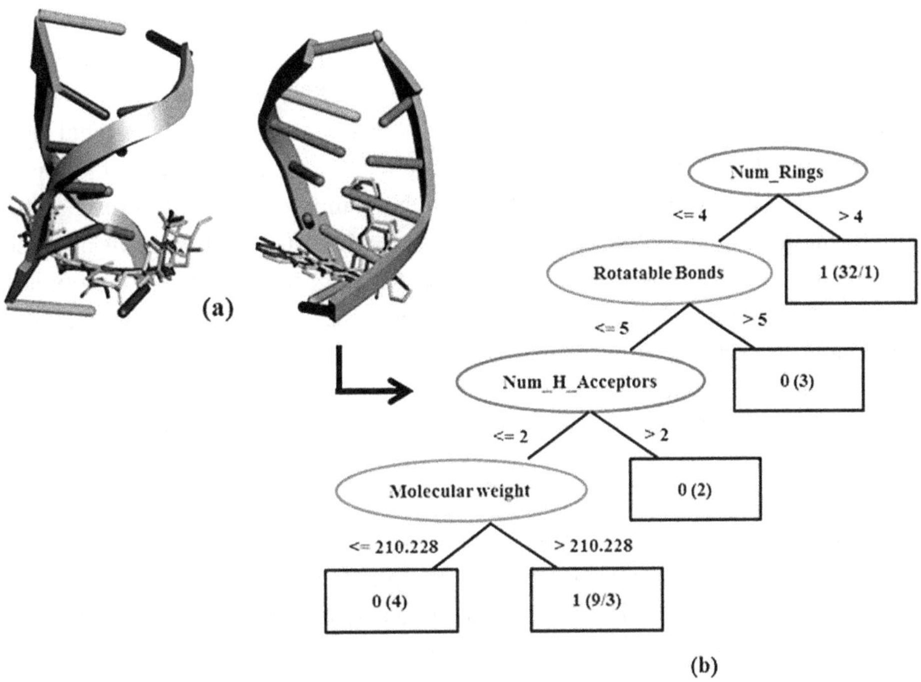

complexes available in the PDB and found that ligands binds preferentially between G and C and between the C and A base pairs (70% and 11%, respectively). They have also examined the ability of the docking program AUTODOCK to reproduce DNA-ligand complexes. They were able to distinguish between the two major binding modes, namely intercalation and groove binding modes. Using the docking results and a set of ligand-derived molecular descriptors, they were able to develop decision trees to predict the likelihood of successful docking (Figure 5).

High-throughput screening (HTS) and computer-based screening using molecular docking such as VS has been rarely used together on the same target. In this context, Doman *et. al.*(Doman, et al., 2002) have applied both HTS and VS to screen the potential inhibitors of the protein tyrosine phosphatase-1B. For HTS, they have considered a library of approximately 400,000 compounds from a corporate collection. After screening these compounds, they found that only 85 of them have IC_{50} value of less than 100 µM, corresponding to a hit rate of 0.021%. For VS, they performed a docking calculation of 235,000 commercially available molecules using DOCK 3.5. After docking, 1000 top-scoring molecules were considered for further evaluation of which 889 molecules were actually available. After visual inspection, 365 molecules were chosen for testing, of which only 127 molecules were found to be active with IC_{50} less than 100 µM, corresponding to a hit rate of 34.8%. Their work showed that structure-based docking enriched the hit rate by 1700-fold over random screening. Further, they found that the hits from VS and HTS were very different from each other, which implies that combination of VS and HTS may be more helpful for lead discovery.

In another study by Minini *et.al.* (Minini, Álvarez, González, Cerecettob, & Merlino, 2015), the binding modes of 1,2,4-thiadiazol, phenazine and 1,2,6-thiadiazine derivatives to *Tc*TIM have been investigated using molecular docking studies combined with molecular dynamics (MD) simulations. Their results indicated that two of the derivatives showed dimer-disrupting properties towards *Tc*TIM and had allosteric effects in the conformation of the active site. They also showed that one derivative caused significant decrease in enzyme mobility in both monomers and loss of conformational flexibility. This suggested that this inhibitor could prevent essential motions of the enzyme required for optimal activity.

Rajkhowa et.al. (Rajkhowa, Jha, & Deka, 2015), in their study have used a combination of different computational techniques to virtually screen out four potential drug candidates from a list of possible ones. They employed molecular docking studies to screen molecules from 62 compounds. From this screening they found 32 compounds having good binding modes and top re-rank scores. A comparative study on the chemical properties have been employed using DFT calculations. DFT based global reactivity descriptors (GRD), such as hardness, chemical potential, chemical softness, electrophilicity index, Fukui function, and local philicity calculated at the optimized geometries were used to investigate the uscfulness of these descriptors for understanding the reactive nature and sites of the molecules. QSAR equations were built using these descriptors considering these 32 compounds. Four common compounds showing the best correlation and the best docking scores were considered for the ADMET property calculations and their dynamical movements have been studied using molecular dynamics simulations. Their results showed that these four compounds were chemically more active than triclosan and have the potential to inhibit the *Mycobacterium tuberculosis* enoyl acyl carrier protein reductase. Their results showed that combined studies of DFT based QSAR along with docking and MD simulations may provide a better insight towards drug discovery with a higher success rate.

5. CONCLUSION

An extensive summary of the types of docking methods has been presented in this chapter. The receptor flexibility and movement of several key secondary elements of the receptor involving ligand binding and the catalysts, is still a major problem in docking studies. Some of the methods to deal with side chain flexibility have been proven to be effective and adequate in certain areas. An ensemble of proteins is considered as a popular solution for conformer selection with respect to global flexibility. It requires an efficient way to obtain and select reliable protein structures used for docking. Therefore, those structures in which the ligand can fit in should be included in the ensemble. In this regard, Local Move Monte Carlo (LMMC) can be used as an appropriate method for sampling a ligand within loop-containing active sites. This is because the loop tends to be more flexible and hard to model using existing approaches (Meng, et al., 2011).

Many docking programs exist, but no single program has yet been emerged that has outperformed all others in all cases. Generally, the programs do an adequate job of searching conformational space and generate correct ligand poses (binding modes), however, the scoring functions need improvement. This is evidenced by the fact that the correlation between calculated and observed binding affinities is often low and that separate scoring functions are frequently needed for pose evaluation and affinity prediction. Scoring function is a very important component in docking and therefore, further improvement is needed. Accurate and low computational cost scoring functions may bring docking application to a new stage.

Docking is very useful in reducing a collection of virtual compounds down to a manageable number to be synthesized and in selecting compounds from an external collection. VS is faster than HTS and inexpensive by any standard. It is often recommended to use docking studies in parallel with other techniques to dock to multiple conformations of the target, and to use the docking results to select as many compounds as possible for experimental confirmation.

Considering the known successful applications, the progress made in this field, and the on-going developments, it can be comprehended that the importance and impact of docking will continue to increase significantly.

ACKNOWLEDGMENT

SR acknowledges Department of Science and Technology (DST), New Delhi for providing the DST-INSPIRE Fellowship (Fellow Code: IF110258).

REFERENCES

Abagyan, R., Totrov, M., & Kuznetsov, D. (1994). ICM-A new method for protein modeling and design: Applications to docking and structure prediction from the distorted native conformation. *Journal of Computational Chemistry*, *15*(5), 488–506. doi:10.1002/jcc.540150503

Bailey, D., & Brown, D. (2001). High-throughput chemistry and structure-based design: Survival of the smartest. *Drug Discovery Today*, *6*(2), 57–59. doi:10.1016/S1359-6446(00)01596-8 PMID:11166243

Bajorath, J. (2002). Integration of virtual and high-throughput screening. *Nature Reviews. Drug Discovery*, *1*(11), 882–894. doi:10.1038/nrd941 PMID:12415248

Berendsen, H. J. C., van der Spoel, D., & van Drunen, R. (1995). GROMACS: A message-passing parallel molecular dynamics implementation. *Computer Physics Communications*, *91*(1-3), 43–56. doi:10.1016/0010-4655(95)00042-E

Boehr, D. D., Nussinov, R., & Wright, P. E. (2009). The role of dynamic conformational ensembles in biomolecular recognition. *Nature Chemical Biology*, *5*(11), 789–796. doi:10.1038/nchembio.232 PMID:19841628

Bohm, H. J. (1992). LUDI: Rule-based automatic design of new substituents for enzyme inhibitor leads. *Journal of Computer-Aided Molecular Design*, *6*, 593–606. doi:10.1007/BF00126217 PMID:1291628

Bohm, H. J. (1994). The development of a simple empirical scoring function to estimate the binding constant for a protein-ligand complex of known three-dimensional structure. *Journal of Computer-Aided Molecular Design*, *8*, 243–256. doi:10.1007/BF00126743 PMID:7964925

Brady, G. P. Jr, & Stouten, P. F. (2000). Fast prediction and visualization of protein binding pockets with PASS. *Journal of Computer-Aided Molecular Design*, *14*(4), 383–401. doi:10.1023/A:1008124202956 PMID:10815774

Brint, A. T., & Willett, P. (1987). Algorithms for the Identification of Three-Dimensional Maximal Common Substructures. *Journal of Chemical Information and Computer Sciences*, *27*(4), 152–158. doi:10.1021/ci00056a002

Bron, C., & Kerbosch, J. (1973). Algorithm 457: Finding All Cliques of an Undirected Graph. *Communications of the ACM*, *16*(9), 575–576. doi:10.1145/362342.362367

Brooijmans, N., & Kuntz, I. D. (2003). Molecular recognition and docking algorithms. *Annual Review of Biophysics and Biomolecular Structure*, *32*(1), 335–373. doi:10.1146/annurev.biophys.32.110601.142532 PMID:12574069

Brooks, B. R., Bruccoleri, R. E., Olafson, B. D., States, D. J., Swaminathan, S., & Karplus, M. (1983). CHARMM: A program for macromolecular energy, minimization, and dynamics calculations. *Journal of Computational Chemistry*, *4*(2), 187–217. doi:10.1002/jcc.540040211

Burkhard, P., Taylor, P., & Walkinshaw, M. D. (1998). An example of a protein ligand found by database mining: Description of the docking method and its verification by a 2.3 A X-ray structure of a thrombin-ligand complex. *Journal of Molecular Biology*, *277*(2), 449–466. doi:10.1006/jmbi.1997.1608 PMID:9514757

Cavasotto, C. N., & Abagyan, R. A. (2004). Protein flexibility in ligand docking and virtual screening to protein kinases. *Journal of Molecular Biology*, *337*(1), 209–225. doi:10.1016/j.jmb.2004.01.003 PMID:15001363

Clark, K. P., & Ajay, . (1995). Flexible ligand docking without parameter adjustment across four ligandreceptor complexes. *Journal of Computational Chemistry*, *16*(10), 1210–1226. doi:10.1002/jcc.540161004

Claussen, H., Buning, C., Rarey, M., & Lengauer, T. (2001). FlexE: Efficient molecular docking considering protein structure variations. *Journal of Molecular Biology*, *308*(2), 377–395. doi:10.1006/jmbi.2001.4551 PMID:11327774

Cornell, W. D., Cieplak, P., Bayly, C. I., Gould, I. R., Merz, K. M., Ferguson, D. M., & Kollman, P. A. et al. (1995). A second generation force field for the simulation of proteins, nucleic acids, and organic molecules. *Journal of the American Chemical Society*, *117*(19), 5179–5197. doi:10.1021/ja00124a002

Coupez, B., & Lewis, R. A. (2006). Docking and scoring--theoretically easy, practically impossible? *Current Medicinal Chemistry*, *13*(25), 2995–3003. doi:10.2174/092986706778521797 PMID:17073642

Cross, J. B., Thompson, D. C., Rai, B. K., Baber, J. C., Fan, K. Y., Hu, Y., & Humblet, C. (2009). Comparison of several molecular docking programs: Pose prediction and virtual screening accuracy. *Journal of Chemical Information and Modeling*, *49*(6), 1455–1474. doi:10.1021/ci900056c PMID:19476350

DesJarlais, R. L., Sheridan, R. P., Dixon, J. S., Kuntz, I. D., & Venkataraghavan, R. (1986). Docking flexible ligands to macromolecular receptors by molecular shape. *Journal of Medicinal Chemistry*, *29*(11), 2149–2153. doi:10.1021/jm00161a004 PMID:3783576

Desmet, J., Maeyer, M. D., Hazes, B., & Lasters, I. (1992). The dead-end elimination theorem and its use in protein sidechain positioning. *Nature*, *356*(6369), 539–542. doi:10.1038/356539a0 PMID:21488406

Diller, D. J., & Merz, K. M. Jr. (2001). High throughput docking for library design and library prioritization. *Proteins*, *43*(2), 113–124. doi:10.1002/1097-0134(20010501)43:2<113::AID-PROT1023>3.0.CO;2-T PMID:11276081

Doman, T. N., McGovern, S. L., Witherbee, B. J., Kasten, T. P., Kurumbail, R., Stallings, W. C., & Shoichet, B. K. et al. (2002). Molecular docking and high-throughput screening for novel inhibitors of protein tyrosine phosphatase-1B. *Journal of Medicinal Chemistry*, *45*(11), 2213–2221. doi:10.1021/jm010548w PMID:12014959

Ewing, T. J., Makino, S., Skillman, A. G., & Kuntz, I. D. (2001). DOCK 4.0: Search strategies for automated molecular docking of flexible molecule databases. *Journal of Computer-Aided Molecular Design*, *15*(5), 411–428. doi:10.1023/A:1011115820450 PMID:11394736

Fischer, D., Norel, R., Wolfson, H., & Nussinov, R. (1993). Surface motifs by a computer vision technique: Searches, detection, and implications for protein-ligand recognition. *Proteins*, *16*(3), 278–292. doi:10.1002/prot.340160306 PMID:8394000

Fischer, E. (1894). Einfluss der configuration auf die wirkungderenzyme. *Berichte der Deutschen Chemischen Gesellschaft*, *27*(3), 2985–2993. doi:10.1002/cber.18940270364

Frauenfelder, H., Sligar, S. G., & Wolynes, P. G. (1991). The energy landscapes and motions of proteins. *Science*, *254*(5038), 1598–1603. doi:10.1126/science.1749933 PMID:1749933

Friesner, R. A., Banks, J. L., Murphy, R. B., Halgren, T. A., Klicic, J. J., Mainz, D. T., & Shenkin, P. S. et al. (2004). Glide: A new approach for rapid, accurate docking and scoring. 1 Method and assessment of docking accuracy. *Journal of Medicinal Chemistry*, *47*(7), 1739–1749. doi:10.1021/jm0306430 PMID:15027865

Frishman, D., & Argos, P. (1995). Knowledge-based protein secondary structure assignment. *Proteins*, *23*(4), 566–579. doi:10.1002/prot.340230412 PMID:8749853

Gabb, H. A., Jackson, R. M., & Sternberg, M. J. (1997). Modelling protein docking using shape complementarity, electrostatics and biochemical information. *Journal of Molecular Biology*, *272*(1), 106–120. doi:10.1006/jmbi.1997.1203 PMID:9299341

Gilad, Y., & Senderowitz, H. (2014). Docking Studies on DNA Intercalators. *Journal of Chemical Information and Modeling*, *54*(1), 96–107. doi:10.1021/ci400352t PMID:24303988

Glaser, F., Morris, R. J., Najmanovich, R. J., Laskowski, R. A., & Thornton, J. M. (2006). A method for localizing ligand binding pockets in protein structures. *Proteins*, *62*(2), 479–488. doi:10.1002/prot.20769 PMID:16304646

Gohlke, H., & Klebe, G. (2002). Approaches to the description and prediction of the binding affinity of smallmolecule ligands to macromolecular receptors. *Angewandte Chemie International Edition in English*, *41*(15), 2644–2676. doi:10.1002/1521-3773(20020802)41:15<2644::AID-ANIE2644>3.0.CO;2-O PMID:12203463

Goodford, P. J. (1985). A computational procedure for determining energetically favorable binding sites on biologically important macromolecules. *Journal of Medicinal Chemistry*, *28*(7), 849–857. doi:10.1021/jm00145a002 PMID:3892003

Goodsell, D. S., Lauble, H., Stout, C. D., & Olson, A. J. (1993). Automated docking in crystallography: Analysis of the substrates of aconitase. *Proteins*, *17*(1), 1–10. doi:10.1002/prot.340170104 PMID:8234239

Goodsell, D. S., & Olson, A. J. (1990). Automated docking of substrates to proteins by simulated annealing. *Proteins*, *8*(3), 195–202. doi:10.1002/prot.340080302 PMID:2281083

Halperin, I., Ma, B., Wolfson, H., & Nussinov, R. (2002). Principles of docking: An overview of search algorithms and a guide to scoring functions. *Proteins*, *47*, 09–443.

Hart, T. N., & Read, R. J. (1992). A multiple-start Monte Carlo docking method. *Proteins*, *13*(3), 206–222. doi:10.1002/prot.340130304 PMID:1603810

Istyastono, E. P., Kooistra, A. J., Vischer, H. F., Kuijer, M., Roumen, L., Nijmeijer, S., & de Graaf, C. et al. (2015). Structure-based virtual screening for fragment-like ligands of the G protein-coupled histamine H4 receptor. *MedChemComm*, *6*(6), 1003–1017. doi:10.1039/C5MD00022J

Jones, G., Willett, P., Glen, R. C., Leach, A. R., & Taylor, R. (1997). Development and validation of a genetic algorithm for flexible docking. *Journal of Molecular Biology*, *267*(3), 727–748. doi:10.1006/jmbi.1996.0897 PMID:9126849

Jorgensen, W. L. (2004). The many roles of computation in drug discovery. *Science*, *303*(5665), 1813–1818. doi:10.1126/science.1096361 PMID:15031495

Kastenholz, M. A., Pastor, M., Cruciani, G., Haaksma, E. E., & Fox, T. (2000). GRID/CPCA: A new computational tool to design selective ligands. *Journal of Medicinal Chemistry*, *43*(16), 3033–3044. doi:10.1021/jm000934y PMID:10956211

Kitao, A., Hayward, S., & Go, N. (1998). Energy landscape of a native protein: Jumping-among-minima model. *Proteins*, *33*(4), 496–517. doi:10.1002/(SICI)1097-0134(19981201)33:4<496::AID-PROT4>3.0.CO;2-1 PMID:9849935

Kitchen, D. B., Decornez, H., Furr, J. R., & Bajorath, J. (2004). Docking and scoring in virtual screening for drug discovery: Methods and applications. *Nature Reviews. Drug Discovery*, *3*(11), 935–949. doi:10.1038/nrd1549 PMID:15520816

Klebe, G. (2006). Virtual ligand screening: Strategies, perspectives and limitations. *Drug Discovery Today*, *11*(13-14), 580–594. doi:10.1016/j.drudis.2006.05.012 PMID:16793526

Knegtel, R. M., Kuntz, I. D., & Oshiro, C. M. (1997). Molecular docking to ensembles of protein structures. *Journal of Molecular Biology*, *266*(2), 424–440. doi:10.1006/jmbi.1996.0776 PMID:9047373

Kokh, D. B., Wade, R. C., & Wenzel, W. (2011). Receptor flexibility in small-molecule docking calculations. *WIREs Computational Molecular Science*, *1*(2), 298–314. doi:10.1002/wcms.29

Kollman, P. A. (1996). AMBER 5.0. San Francisco, CA: University of California.

Kontoyianni, M., Madhav, P., Suchanek, E., & Seibel, W. (2008). Theoretical and practical considerations in virtual screening: A beaten field? *Current Medicinal Chemistry*, *15*(2), 107–116. doi:10.2174/092986708783330566 PMID:18220766

Koshland, D. E. Jr. (1963). Correlation of Structure and Function in Enzyme Action. *Science*, *142*(3599), 1533–1541. doi:10.1126/science.142.3599.1533 PMID:14075684

Kramer, B., Rarey, M., & Lengauer, T. (1999). Evaluation of the FLEXX incremental construction algorithm for protein-ligand docking. *Proteins*, *37*(2), 228–241. doi:10.1002/(SICI)1097-0134(19991101)37:2<228::AID-PROT8>3.0.CO;2-8 PMID:10584068

Kroemer, R. T. (2007). Structure-based drug design: Docking and scoring. *Current Protein & Peptide Science*, *8*, 312–328. doi:10.2174/138920307781369382 PMID:17696866

Kubinyi, H. (2006). *Computer Applications in Pharmaceutical Research and Development*. New York: John Wiley.

Kuntz, I. D., Blaney, J. M., Oatley, S. J., Langridge, R., & Ferrin, T. E. (1982). A geometric approach to macromolecule-ligand interactions. *Journal of Molecular Biology*, *161*(2), 269–288. doi:10.1016/0022-2836(82)90153-X PMID:7154081

Kuntz, I. D., & Leach, A. R. (1992). Conformational analysis of flexible ligands in macromolecular receptor sites. *Journal of Computational Chemistry*, *13*(6), 730–748. doi:10.1002/jcc.540130608

Lahana, R. (1999). How many leads from HTS. *Drug Discovery Today*, *4*(10), 447–448. doi:10.1016/S1359-6446(99)01393-8 PMID:10481138

Langer, T., & Hoffmann, R. D. (2001). Virtual screening: An effective tool for lead structure discovery? *Current Pharmaceutical Design*, *7*(7), 509–527. doi:10.2174/1381612013397861 PMID:11375766

Langer, T., & Krovat, E. M. (2003). Chemical feature-based pharmacophores and virtual library screening for discovery of new leads. *Current Opinion in Drug Discovery & Development*, *6*, 370–376. PMID:12833670

Laskowski, R. A. (1995). SURFNET: A program for visualizing molecular surfaces, cavities, and intermolecular interactions. *Journal of Molecular Graphics*, *13*(5), 323–330. doi:10.1016/0263-7855(95)00073-9 PMID:8603061

Leach, A. R. (1994). Ligand docking to proteins with discrete side-chain flexibility. *Journal of Molecular Biology*, *235*(1), 345–356. doi:10.1016/S0022-2836(05)80038-5 PMID:8289255

Levitt, D. G., & Banaszak, L. J. (1992). POCKET: A computer graphics method for identifying and displaying protein cavities and their surrounding amino acids. *Journal of Molecular Graphics*, *10*(4), 229–234. doi:10.1016/0263-7855(92)80074-N PMID:1476996

Levy, R. M., & Karplus, M. (1979). Vibrational Approach to the Dynamics of an alpha-Helix. *Biopolymers*, *18*(10), 2465–2495. doi:10.1002/bip.1979.360181008

Li, X., Li, Y., Cheng, T., Liu, Z., & Wang, R. (2010). Evaluation of the performance of four molecular docking programs on a diverse set of protein-ligand complexes. *Journal of Computational Chemistry*, *31*(11), 2109–2125. doi:10.1002/jcc.21498 PMID:20127741

Lionta, E., Spyrou, G., Vassilatis, D. K., & Cournia, Z. (2014). Structure-based virtual screening for drug discovery: Principles, applications and recent advances. *Current Topics in Medicinal Chemistry*, *14*(16), 1923–1938. doi:10.2174/1568026614666140929124445 PMID:25262799

McConkey, B. J., Sobolev, V., & Edelman, M. (2002). The performance of current methods in ligand-protein docking. *Current Science*, *83*, 845–855.

McMartin, C., & Bohacek, R. S. (1997). QXP: Powerful, rapid computer algorithms for structure-based drug design. *Journal of Computer-Aided Molecular Design*, *11*(4), 333–344. doi:10.1023/A:1007907728892 PMID:9334900

Meng, X. Y., Zhang, H. X., Mezei, M., & Cui, M. (2011). Molecular docking: A powerful approach for structure-based drug discovery. *Current Computer-aided Drug Design*, *7*(2), 146–157. doi:10.2174/157340911795677602 PMID:21534921

Mezei, M. (2003). A new method for mapping macromolecular topography. *Journal of Molecular Graphics & Modelling*, *21*(5), 463–472. doi:10.1016/S1093-3263(02)00203-6 PMID:12543141

Miller, M. D., Kearsley, S. K., Underwood, D. J., & Sheridan, R. P. (1994). FLOG: A system to select 'quasi-flexible' ligands complementary to a receptor of known three-dimensional structure. *Journal of Computer-Aided Molecular Design*, *8*(2), 153–174. doi:10.1007/BF00119865 PMID:8064332

Minini, L., Álvarez, G., González, M., Cerecettob, H., & Merlino, A. (2015). Molecular docking and molecular dynamics simulation studies of *Trypanosoma cruzi* triosephosphate isomerase inhibitors. Insights into the inhibition mechanism and selectivity. *Journal of Molecular Graphics & Modelling*, *58*, 40–49. doi:10.1016/j.jmgm.2015.02.002 PMID:25829097

Moitessier, N., Englebienne, P., Lee, D., Lawandi, J., & Corbeil, C. R. (2008). Towards the development of universal, fast and highly accurate docking/scoring methods: A long way to go. *British Journal of Pharmacology*, *153*(S1), S7–S26. doi:10.1038/sj.bjp.0707515 PMID:18037925

Morris, G. M., Goodsell, D. S., Halliday, R. S., Huey, R., Hart, W. E., Belew, R. K., & Olson, A. J. (1998). Automated docking using a Lamarckian genetic algorithm and an empirical binding free energy function. *Journal of Computational Chemistry*, *19*(14), 1639–1662. doi:10.1002/(SICI)1096-987X(19981115)19:14<1639::AID-JCC10>3.0.CO;2-B

Morris, G. M., Huey, R., Lindstrom, W., Sanner, M. F., Belew, R. K., Goodsell, D. S., & Olson, A. J. (2009). AutoDock4 and AutoDockTools4: Automated docking with selective receptor flexibility. *Journal of Computational Chemistry*, *30*(16), 2785–2791. doi:10.1002/jcc.21256 PMID:19399780

Muralidharan, A. R., Selvaraj, C., Singh, S. K., Sheu, J. R., Thomas, P. A., & Geraldine, P. (2015). Structure-Based Virtual Screening and Biological Evaluation of a Calpain Inhibitor for Prevention of Selenite-Induced Cataractogenesis in an in Vitro System. *Journal of Chemical Information and Modeling*, *55*(8), 1686–1697. doi:10.1021/acs.jcim.5b00092 PMID:26270943

Noguti, T., & Go, N. (1989). Structural basis of hierarchical multiple substates of a protein. I: Introduction. *Proteins*, *5*(2), 97–103. doi:10.1002/prot.340050203 PMID:2748581

Norel, R., Fischer, D., Wolfson, H. J., & Nussinov, R. (1994). Molecular surface recognition by a computer vision-based technique. *Protein Engineering*, *7*(1), 39–46. doi:10.1093/protein/7.1.39 PMID:8140093

Oshiro, C. M., Kuntz, I. D., & Dixon, J. S. (1995). Flexible ligand docking using a genetic algorithm. *Journal of Computer-Aided Molecular Design*, *9*(2), 113–130. doi:10.1007/BF00124402 PMID:7608743

Patil, S. A. (2012). Role of Medicinal Chemist in the Modern Drug Discovery and Development. *Organic Chemistry: Current Research*, *1*, e110.

Plewczynski, D., Lazniewski, M., Augustyniak, R., & Ginalski, K. (2010). Can we trust docking results? Evaluation of seven commonly used programs on PDB bind database. *Journal of Computational Chemistry*, *32*(4), 742–755. doi:10.1002/jcc.21643 PMID:20812323

Rajkhowa, S., & Deka, R. C. (2014). DFT based QSAR/QSPR models in the development of novel anti-tuberculosis drugs targeting Mycobacterium tuberculosis. *Current Pharmaceutical Design*, *20*(27), 4455–4473. doi:10.2174/13816128196661311181658 24 PMID:24245759

Rajkhowa, S., Hussain, I., Hazarika, K. K., Sarmah, P., & Deka, R. C. (2013). Quantitative structure-activity relationships of the antimalarial agent artemisinin and some of its derivatives - a DFT approach. *Combinatorial Chemistry & High Throughput Screening*, *16*(8), 590–602. doi:10.2174/1386207311316080002 PMID:23597248

Rajkhowa, S., Jha, A. N., & Deka, R. C. (2015). Anti-tubercular drug development: Computational strategies to identify potential compounds. *Journal of Molecular Graphics & Modelling*, *62*, 56–68. doi:10.1016/j.jmgm.2015.09.007 PMID:26386453

Rarey, M., Kramer, B., & Lengauer, T. (1999). The particle concept: Placing discrete water molecules during protein-ligand docking predictions. *Proteins*, *34*(1), 17–28. doi:10.1002/(SICI)1097-0134(19990101)34:1<17::AID-PROT3>3.0.CO;2-1 PMID:10336380

Rarey, M., Kramer, B., Lengauer, T., & Klebe, G. (1996). A fast flexible docking method using an incremental construction algorithm. *Journal of Molecular Biology*, *261*(3), 470–489. doi:10.1006/jmbi.1996.0477 PMID:8780787

Reich, S., Thomsen, C., & Maultzsch, J. (2004). Carbon nanotubes: basic concepts and physical properties. Wiley-VCH.

Rich, D. H., Bursavich, M. G., & Estiarte, M. A. (2002). Discovery of nonpeptide, peptidomimetic peptidase inhibitors that target alternate enzyme active site conformations. *Biopolymers*, *66*(2), 115–125. doi:10.1002/bip.10231 PMID:12325161

Saikia, N., Rajkhowa, S., & Deka, R. C. (2013). Density functional and molecular docking studies towards investigating the role of single-wall carbon nanotubes as nanocarrier for loading and delivering of pyrazinamide antitubercular drug on to pncA protein. *Journal of Computer-Aided Molecular Design*, *27*(3), 57–276. doi:10.1007/s10822-013-9638-6

Sandak, B., Nussinov, R., & Wolfson, H. J. (1995). An automated computer vision and robotics-based technique for 3-D flexible biomolecular docking and matching. *Computer Applications in the Biosciences, 11*, 87–99. PMID:7796279

Schnecke, V., & Kuhn, L. A. (2000). Virtual Screening with Solvation and Ligand-Induced Complementarity. *Perspectives in Drug Discovery and Design, 20*(1), 171–190. doi:10.1023/A:1008737207775

Sherman, W., Day, T., Jacobson, M. P., Friesner, R. A., & Farid, R. (2006). Novel procedure for modeling ligand/receptor induced fit effects. *Journal of Medicinal Chemistry, 49*(2), 534–553. doi:10.1021/jm050540c PMID:16420040

Shoichet, B. K., McGovern, S. L., Wei, B., & Irwin, J. J. (2002) Hits, leads and artifacts from virtual and high throughput screening. In *Proceedings of the International Beilstein Workshop*. Molecular Informatics: Confronting Complexity.

Sindhu, T., & Srinivasan, P. (2015). Identification of potential dual agonists of FXR and TGR5 using e-pharmacophore based virtual screening. *Molecular BioSystems, 11*(5), 1305–1318. doi:10.1039/C5MB00137D PMID:25787676

Storn, R., & Price, K. (1995). Differential Evolution - A Simple and Efficient Adaptive Scheme for Global Optimization over Continuous Spaces. Technical Report. Berkley, CA: International Computer Science Institute.

Taylor, J. S., & Burnett, R. M. (2000). DARWIN: A program for docking flexible molecules. *Proteins, 41*(2), 173–191. doi:10.1002/1097-0134(20001101)41:2<173::AID-PROT30>3.0.CO;2-3 PMID:10966571

Taylor, R. D., Jewsbury, P. J., & Essex, J. W. (2002). A review of protein-small molecule docking methods. *Journal of Computer-Aided Molecular Design, 16*(3), 151–166. doi:10.1023/A:1020155510718 PMID:12363215

Teague, S. J. (2003). Implications of protein flexibility for drug discovery. *Nature Reviews. Drug Discovery, 2*(7), 527–541. doi:10.1038/nrd1129 PMID:12838268

ten Brink, T., & Exner, T. E. (2009). Influence of protonation, tautomeric, and stereoisomeric states on proteinligand docking results. *Journal of Chemical Information and Modeling, 49*(6), 1535–1546. doi:10.1021/ci800420z PMID:19453150

Teodoro, M. L., & Kavraki, L. E. (2003). Conformational flexibility models for the receptor in structure based drug design. *Current Pharmaceutical Design, 9*(20), 1635–1648. doi:10.2174/1381612033454595 PMID:12871062

Thomsen, R., & Christensen, M. H. (2006). MolDock: A new technique for high-accuracy molecular docking. *Journal of Medicinal Chemistry, 49*(11), 3315–3321. doi:10.1021/jm051197e PMID:16722650

Totrov, M., & Abagyan, R. (2008). Flexible ligand docking to multiple receptor conformations: A practical alternative. *Current Opinion in Structural Biology, 18*(2), 178–184. doi:10.1016/j.sbi.2008.01.004 PMID:18302984 van der Spoel, D., Lindahl, E., Hess, B., Groenhof, G., Mark, A. E., & Berendsen, H. J. C. (2005). GROMACS: Fast, Flexible and Free. *Journal of Computational Chemistry*, *26*(16), 1701–1718. doi:10.1002/jcc.20291 PMID:16211538

Verdonk, M. L., Cole, J. C., Hartshorn, M. J., Murray, C. W., & Taylor, R. D. (2003). Improved protein-ligand docking using GOLD. *Proteins*, *52*(4), 609–623. doi:10.1002/prot.10465 PMID:12910460

Vuorinen, A., Engeli, R., Meyer, A., Bachmann, F., Griesser, U. J., Schuster, D., & Odermatt, A. (2014). Ligand-based pharmacophore modeling and virtual screening for the discovery of novel 17β-hydroxysteroid dehydrogenase 2 inhibitors. *Journal of Medicinal Chemistry*, *57*(14), 5995–6007. doi:10.1021/jm5004914 PMID:24960438

Walters, W. P., Stahl, M. T., & Murcko, M. A. (1998). Virtual screening - an overview. *Drug Discovery Today*, *3*(4), 160–178. doi:10.1016/S1359-6446(97)01163-X

Weine, S. J., Kollman, P. A., Case, D. A., Singh, U. C., Ghio, C., Alagona, G., & Weiner, P. et al. (1984). New force field for molecular mechanical simulation of nucleic acids and proteins. *Journal of the American Chemical Society*, *106*(3), 765–784. doi:10.1021/ja00315a051

Welch, W., Ruppert, J., & Jain, A. N. (1996). Hammerhead: Fast, fully automated docking of flexible ligands to protein binding sites. *Chemistry & Biology*, *3*(6), 449–462. doi:10.1016/S1074-5521(96)90093-9 PMID:8807875

Yang, J. M., & Chen, C. C. (2004). GEMDOCK: A generic evolutionary method for molecular docking. *Proteins*, *55*(2), 288–304. doi:10.1002/prot.20035 PMID:15048822

Yu, R., Wang, J., Wang, R., Lin, Y., Hu, Y., Wang, Y., & Lin, Z. et al. (2015). Combined pharmacophore modeling, 3D-QSAR, homology modeling and docking studies on CYP11B1 inhibitors. *Molecules (Basel, Switzerland)*, *20*(1), 1014–1030. doi:10.3390/molecules20011014 PMID:25584832

Zsoldos, Z., Reid, D., Simon, A., Sadjad, B. S., & Johnson, A. P. (2006). eHiTS: An innovative approach to the docking and scoring function problems. *Current Protein & Peptide Science*, *7*(5), 421–435. doi:10.2174/138920306778559412 PMID:17073694

Chapter 9
Scoring Functions of Protein–Ligand Interactions

Zhiqiang Yan
Changchun Institute of Applied Chemistry, China

Jin Wang
Stony Brook University, USA

ABSTRACT

Scoring function of protein-ligand interactions is used to recognize the "native" binding pose of a ligand on the protein and to predict the binding affinity, so that the active small molecules can be discriminated from the non-active ones. Scoring function is widely used in computationally molecular docking and structure-based drug discovery. The development and improvement of scoring functions have broad implications in pharmaceutical industry and academic research. During the past three decades, much progress have been made in methodology and accuracy for scoring functions, and many successful cases have be witnessed in virtual database screening. In this chapter, the authors introduced the basic types of scoring functions and their derivations, the commonly-used evaluation methods and benchmarks, as well as the underlying challenges and current solutions. Finally, the authors discussed the promising directions to improve and develop scoring functions for future molecular docking-based drug discovery.

INTRODUCTION

The experiment-based discovery and development of novel drugs are highly complex, time-consuming and expensive (Dickson, 2004). As the number of determined three-dimensional protein structures grows and the power of high-performance computing improves, structure-based drug design and discovery is gaining prominence to identify lead compounds. This field has experienced a rapid growth in academia and pharmaceutical industry. Molecular docking is the most popular computational method employed in structure-based drug design and discovery (Kitchen, 2004; Meng, 2011; Bello, 2013).

Molecular docking mimics the recognition process in which a small molecule (ligand) translates, rotates, and twists thoroughly in the active site of a macromolecule (normally protein), with the goal of finding the most favorable binding mode measured by a scoring function of the binding interactions.

DOI: 10.4018/978-1-5225-0115-2.ch009

Thus, the protein-ligand docking method can be described as a combination of a search algorithm and a scoring function. The search algorithm generates a number of possible poses that fit the ligand into the binding pocket of the protein. The scoring function scores each pose and then ranks the different poses that are generated by the search algorithm.

In the past three decades, multiple search algorithms were employed in the molecular docking and different types of scoring functions were developed (Dias, 2008; Guo, 2014; Liu, 2015). Various docking programs based on different search algorithms and scoring functions were developed to perform molecular docking studies and applications in structure-based drug design and discovery (Plewczynski, 2011; Sousa, 2013). There have been a mounting number of successful stories with docking-based virtual screening (Tuccinardi, 2009; Villoutreix, 2009; Danishuddin, 2015). Those successes are the outcome of the concerted applications from sampling algorithms, scoring functions as well as filtering criteria. In this chapter, the authors would like to focus on the introduction of scoring functions.

BACKGROUND

Experimentally, the binding affinity (free energy) ($\Delta G_{Bind} = G_{Complex} - G_{Protein} - G_{Ligand}$) of a protein-ligand complex can be given by the determined dissociation constant (K_d) with the equation $\Delta G_{Bind} = RT\ln K_d$, where R is the ideal gas constant, and T is the temperature. Computationally, there are a wide variety of different methods capable of predicting the binding affinity of protein-ligand complex. Relatively accurate binding methods, such as free energy perturbation (FEP), thermodynamic integration (TI), linear interaction energy (LIE), Molecular mechanics Poisson-Boltzmann (or Generalized Born) surface area (MM-PB(GB)SA) continuum solvation models (Kollman, 1993; Aqvist, 1994; Gutiérrez-de, 2012; Kollman, 2000), require large computational cost, especially in the computational virtual screening of small molecule databases. Instead, the scoring functions adopt various assumptions and simplifications with approximate mathematical methods to estimate the binding free energy between the protein and the ligand; they are fast in calculating the protein-ligand interactions. Thus, the accuracy is the main task of developers in improving and developing scoring functions.

The accuracy of a scoring function can be evaluated in accordance with its applications in the structure-based drug design and discovery (Huang, 2010; Li, 2014). The first one refers to the ability to rank the computationally generated docking poses, and identify the top ranked one as the "native" binding pose which should be close to experimentally determined structure. The second one refers to the ability to estimate the binding affinities which should be proportional to the experimentally determined binding affinities. The third one refers to the ability to discriminate the true binders against inactive ones in the virtual database screening according to computational binding scores. With the appropriate evaluation methods, these three kinds of abilities can be validated by testing on the protein-ligand benchmarks or with blind assessment exercises.

Scoring functions are either implemented in docking programs or stand-alone in scoring and ranking the generated poses or ligands. So far, the scoring functions have grown into a large diverse family with more than one hundred, whereas, many scoring functions employ common methodologies. Therefore, the scoring functions can be roughly grouped into four main categories: force-field-based, empirical, knowledge-based, and machine-leaning-based scoring functions (Liu, 2015). In addition, to take the advantages and balance the deficiencies of different scoring functions, the consensus scoring technique has been introduced to improve the probability of finding correct solutions by combining the scores from multiple scoring functions.

Though the molecular docking is playing an in increasing role in structure-based drug design and discovery, important challenges are still existed, such as water molecules in binding site, solvation effect, protein flexibility, treatment of entropy, covalent docking, metal ions, and binding specificity (Yan, 2013; Yuriev, 2013; Brás, 2014). Some challenges lie in the side of scoring functions and common to all types of scoring functions. By addressing these obstacles, many attempts have been made to the scoring functions to increase the robustness of reliability and accuracy in their applications, which have paved the way for future directions of improving and developing new scoring functions.

SCORING FUNCTIONS

Classes of Scoring Functions

1. Force-Field-Based Scoring Function

Since the pioneer works by Martin Karplus and coworkers employed molecular mechanics force fields to model biological macromolecules (Karplus, 1976; Brooks, 1983), force-field-based scoring functions have been used for more than two decades to compute the binding affinity between the protein and the ligand (Meng, 1992; Kollman, 1993; Goodsell, 1996; Makino, 1997; Jones, 1997; Kramer, 1999; Yin, 2008). Force-field-based scoring functions generally quantify the sum of two energies, the protein-ligand interaction energy and the internal energy of the ligand. The functional form of the internal ligand energy is very similar to the protein-ligand interaction energy. The energies are normally composed of a van der Waals and an electrostatic energy terms. The general functional form is like

$$S = E_{PL} + E_{L} \tag{1}$$

where E_{PL} represents the energy between the protein and ligand and E_L represents the internal energy of the ligand. Different force-field-based scoring functions have similar energy forms with a few different extension energy terms as shown in Table 1. The van der Waals energy term can be given by the Lennard-Jones potential function, the parameters of the Lennard-Jones potential vary by using 12-6 or 8-4 Lennard-Jones potential to adjust the "hardness" of the potential. The electrostatic term is accounted for by a Coulombic formulation with a distance-dependent dielectric function. The effect of solvent is implicitly considered by introducing a simple distance dependent dielectric constant in the Coulombic formulation, which lessens the contribution from charge-charge interactions by mimicking the screening effect. The extensions of general force-field-based scoring functions include a torsional entropy term for ligands and the inclusion of explicit protein-ligand hydrogen-bonding terms. The torsional entropy term is to account for the entropy associated with the rotatable bonds of the ligands. For example, immobilization of a rotatable bond on binding leads to loss of its torsional (or rotational) entropy. The hydrogen-bonding term is modeled to increase the potential of specific ligand binding with noticeably different ways. For example, G-Score (Kramer, 1999) includes different hydrogen-bonding terms depending on the nature and geometry of the interaction while AutoDock represents all of the hydrogen bonds by a 12-10 Lennard-Jones potential with an angular weighting factor.

Table 1. Examples of force-field-based scoring functions

Name	Functional Form
AutoDock4.2 (Morris, 2009)	$S=W_{vdw}\sum_{i,j}\left(\frac{A_{ij}}{r_{ij}^{12}}-\frac{B_{ij}}{r_{ij}^{6}}\right)+W_{hb}\sum_{i,j}E(t)\left(\frac{C_{ij}}{r_{ij}^{12}}-\frac{D_{ij}}{r_{ij}^{10}}\right)+W_{ele}\sum_{i,j}\frac{q_i q_j}{\varepsilon(r_{ij})r_{ij}}+W_{sol}\sum_{i,j}(S_iV_j+S_jV_i)e^{(-r_{ij}^2/2\sigma^2)}$
Dock (Makino, 1997)	$S=\sum_{i,j}\left[\left(\frac{A_{ij}}{r_{ij}^{12}}-\frac{B_{ij}}{r_{ij}^{6}}\right)+332.0\frac{q_i q_j}{\varepsilon(r_{ij})r_{ij}}\right]$
GoldScore (Jones, 1997)	$S=\sum_{i,j}\left[\left(\frac{A_{ij}}{r_{ij}^{a}}-\frac{B_{ij}}{r_{ij}^{b}}\right)+332.0\frac{q_i q_j}{\varepsilon(r_{ij})r_{ij}}\right]+(E_{hb})_{optional}$
D-Score (Meng, 1992)	$S=\sum_{i,j}\left[\left(\frac{A_{ij}}{r_{ij}^{12}}-\frac{B_{ij}}{r_{ij}^{6}}\right)+332.0\frac{q_i q_j}{\varepsilon(r_{ij})r_{ij}}\right]$
G-Score (Jones, 1997)	$S=\sum_{i,j}\left[\left(\frac{A_{ij}}{r_{ij}^{a}}-\frac{B_{ij}}{r_{ij}^{b}}\right)+332.0\frac{q_i q_j}{\varepsilon(r_{ij})r_{ij}}\right]+\sum_{lig}\frac{1}{2}V\left[1+\frac{n}{\lvert n\rvert}\cos(\lvert n\rvert\omega)\right]+E_{hb}$

Atoms i and j are either from the protein and ligand respectively or both from the ligand; A, B, C and D are van der Waals parameters for given atoms, C and D are specialized for hydrogen bonds with oxygen and nitrogen; Deferent W are the weighting constants for the energy terms; E(t) is the angle weighting factor at angel t from ideal hydrogen bonding geometry; ε is the dielectric constant; E_{hb} is the hydrogen bonding term; a and b means that the protein-ligand interaction energy and the internal energy would take different Lenard-Jones parameters; n, w and V are the parameters for the torsional energy in the ligand. For more details, readers are suggested to look into the related citations.

Force-field-based scoring functions were primarily formulated to model enthalpy gas-phase contributions to the energetics. Thus, the major limitations of force-field-based scoring functions are the lack of solvation and entropic terms which are important contributors to the binding free energy between the protein and the ligand. This is in part corrected through the inclusion of additional terms. For example, G-Score includes a torsional entropy term to estimate the conformational entropy loss of ligand upon binding. A straightforward way to account for the solvent effect is to treat water molecules explicitly such as free energy perturbation (FEP) and thermodynamic integration (TI). Alternatively, implicit solvent models include Poisson-Boltzmann (PB) or Generalized Born (GB) continuum solvation models (Kollman, 2000), were developed for scoring use in molecular docking. These methods compute energies as ensemble averages in an attempt to incorporate more physics at the expense of significantly increased computational cost. In practice, these methods are often used in post-scoring of docking programs.

2. Empirical Scoring Function

The empirical scoring function was first proposed by Böhm in 1992 (Böhm, 1992). An empirical scoring function computes the fitness of protein-ligand binding by summing up the contributions of a number of individual terms. The functional form of empirical scoring functions is often simpler than force-field-based scoring functions, although many of the individual contributing terms have counterparts in the force-field molecular mechanics terms. Generally, the empirical scoring functions decompose the overall binding free energy into several energetic terms as

$$S = \sum_{i} W_i E_i \tag{2}$$

The coefficients W_i for various terms E_i are obtained from regression analysis using experimentally determined binding free energies and sometimes structural information. Representative examples of empirical scoring functions are given in Table 2. Though the basic expressions are similar between different empirical scoring functions, the energetic terms have rather different implementations. For

Table 2. Examples of empirical scoring functions

Name	Functional Form
LUDI (Böhm, 1992)	$S=\Delta G_{hb}\sum_{hb} f(\Delta R, \Delta\alpha) + \Delta G_{ion}\sum_{ion} f(\Delta R, \Delta\alpha) + \Delta G_{hp}\sum_{hp} \lvert A_{hp} \rvert + \Delta G_{rot} N_{rot} + \Delta G_0$
ChemScore (Eldridge, 1997)	$S=\Delta G_{hb}\sum_{hb} f(\Delta R, \Delta\alpha) + \Delta G_{met}\sum_{met} f(\Delta R, \Delta\alpha) +$ $\Delta G_{lipo}\sum_{lipo} f(\Delta R) + \Delta G_{rot}\sum_{rot} f(P_{nl}, P'_{nl}) + \Delta G_0$
F-Score (Rarey, 1996)	$S=\Delta G_{hb}\sum_{hb} f(\Delta R, \Delta\alpha) + \Delta G_{ion}\sum_{ion} f(\Delta R, \Delta\alpha) + \Delta G_{arom}\sum_{arom} f(\Delta R) + \Delta G_{con}\sum_{con} f(\Delta R, \Delta\alpha)$ $+ \Delta G_{rot} N_{rot} + \Delta G_0$
GlideScore (Wang, 2002)	$S=C_{lipo\text{-}lipo}\sum f(r_{lr}) + C_{hb-neut-neut}\sum g(\Delta r)h(\Delta\alpha) + C_{hb-neut-ch}\sum g(\Delta r)h(\Delta\alpha)$ $+ C_{hb-ch-ch}\sum g(\Delta r)h(\Delta\alpha) + C_{max\text{-}met\text{-}ion}\sum f(r_{lr}) + C_{rot}H_{rot} + C_{pol-hp}V_{pol-hp} + C_{coul}E_{coul}$ $+ C_{vdw}E_{vdw} + E_{sol}$

The scoring above uses energy terms including hydrogen bonding (hb), ionic (ion), hydrophobic (hp), ligand rotational entropy (rotor), contact (con), lipophilic (lipo), metal (met) Coulomb (coul), van der Waals (vdw) and solvation components. The scoring functions differ in energy terms and corresponding regression coefficients. Different ΔG are obtained by multiple linear regressions, ΔG_0 is a regression constant, and the energy term is calculated with a function f or g or h which depends on the angular and/or a distance. GlideScore modifies and extends the ChemScore by classifying hydrogen bonding term into several weighted components depending on whether the donor and the acceptor are neutral (neut) or charged (ch), and incorporating the Coulomb (coul) and van der Waals (vdw) as well as solvation (sol) component. For more details, the readers are suggested to look into the related citations.

instance, LUDI function computes hydrophobic contributions based on the molecular surface area; Glide XP and Fresno (Rognan, 1999) functions include the solvation term to account for the desolvation effect; F-Score function takes into account the interaction between aromatic groups.

Both force-field-based and empirical scoring functions decompose protein-ligand binding free energy into individual energy terms. The major difference between them is that force-field-based scoring function employs the theoretical framework, including the energy function and the associated parameters, from well-established models; whereas the empirical scoring function usually adopts a flexible functional form which is obtained from regression fit. Several force-field-based methods have introduced empirical parameters to reconcile the contributions of its energy terms just like an empirical scoring function. For example, AutoDock4 (Huey, 2007) calibrated the energy terms using a large set of diverse protein-ligand complexes (see Table 1).

The main drawback of the empirical scoring functions is their dependence on the experimental data set used to perform regression analyses and fitting. This means that based on the limited training set, it is rather difficult to establish a comprehensive and consistent description of all possible weighting factors in protein-ligand binding within the framework of an empirical scoring function. i.e. there is no guarantee that an empirical scoring functions is able to predict the binding affinity of the ligands structurally very different from those used in the training set. Nowadays, researchers are able to use several thousands of protein-ligand complexes at a better quality to calibrate or test their scoring functions. The doubt is largely removed while the remaining concern is that the public experimental data are usually measured under different conditions.

3. Knowledge-Based Scoring Function

Since the pioneer work by Tanaka and Scheraga (Tanaka, 1976), a large number of knowledge-based scoring functions have been developed and widely applied to protein structure prediction and protein-ligand studies. The knowledge-based scoring function is also known as statistical potentials. It employs energy potentials that are derived from the structural information embedded in experimentally determined atomic structures (Miyazawa, 1985; Sippl, 1990). The principle behind knowledge-based scoring functions is simple. The interactions of protein-ligand complexes are modeled using pairwise potentials. The pairwise potentials are directly obtained from the occurrence frequency of atom pairs in a database using the inverse Boltzmann relation. The potentials are computed by

$$S = \sum w(r) \tag{3}$$

$$w(r) = -k_B T \ln[g(r)] = -k_B T \ln\left[\frac{f(r)}{f^*(r)}\right] \tag{4}$$

where w(r) is the pairwise potential between protein and ligand, k_B is the Bolzmann constant, T is the absolute temperature of the system, f(r) is the occurrence frequency of the atom pair at distance r and f*(r) is the averaged occurrence frequency in the reference state. If an interatomic distance occurs more often than the average value in the reference state where interatomic interactions are zero, the pairwise potential is assumed as a favorable contact, and vice versa. A challenge in deriving knowledge-based scoring functions is how to achieve the reference state (Thomas, 1996).

The various knowledge-based scoring functions differ between each other in the training sets of protein-ligand complexes used to obtain the potentials, the form of the energy function, the definition of protein and ligand atom types, distance cutoff, and the approximation of the reference state. For example, SMoG uses 13 atom types, two distance intervals and a reference state determined by a self-consistent method (Ishchenko, 2002). SPA uses 22 atom types, 16 distance intervals and a reference state circumvented by Monte Carlo optimization (Yan, 2012). Some methods have combined knowledge-based potentials with solvation and entropy terms (Huang, 2010; Gohlke, 2000; Yan, 2015). In this sense, the boundary between knowledge-based and empirical scoring functions was blurred. The representative knowledge-based scoring functions are SMoG (DeWitte, 1996), PMF(Muegge, 1999), DrugScore (Gohlke, 2000), BLEEP (Mitchell, 1999), ITScore (Huang, 2006), SPA (Yan, 2012).

Unlike force-field-based and empirical scoring functions, knowledge-based potentials attempt to capture all of the energetic factors in protein-ligand interaction implicitly with the pairwise potentials. The major advantage of the knowledge-based scoring functions is their computational simplicity. The deduction of the potentials required for the scoring function only requires the knowledge of a set of protein-ligand complex structures, and such knowledge is relatively rich and still increasing due to the contributions from structural biologists. This overcomes the disadvantage of knowledge-based scoring function that its derivation is based on structural information implicitly encoded in the limited sets of protein-ligand complex structures. In addition, these scoring functions can be easily applied to systems that were not used for the setup of the scoring function, which makes them attractive to apply for very large compound databases.

4. Machine-Learning-Based Scoring Function

Machine-learning-based (ML-based) scoring function is relatively new compared to above three classes of scoring functions since it was first emerged in 2004 (Deng, 2004). These methods employ quantitative structure-activity relationship (QSAR) analysis into protein-ligand interaction evaluation. The machine-learning techniques are applied to derive statistical models that compute protein-ligand binding scores, once the properties of the ligand and the protein as well as their interaction patterns are coded with certain descriptors. The starting point of the method is generally a large pool of descriptors including the descriptors accounting for specific interactions (electrostatic interactions, hydrogen bonds, or aromatic stacking), geometrical descriptors (surface or shape properties), and conventional ligand-based descriptors (molecular weight, number of rotatable bonds, etc.). A variety of machine-learning algorithms, such as neural network, Bayesian classifiers, random forest, and support vector machine, are employed for variable selection. Similar to empirical scoring functions, the method also needs a training set of protein-ligand complexes with known structures and binding data to derive their final models. Typical examples include NNScore (Durrant, 2010), RF-Score (Ballester, 2010), SFCscore^{RF}(Zilian, 2013) and ID-Score (Li, 2013).

ML-based scoring functions have attracted attentions and discussions recently due to their good performance in predicting the binding affinity (Ballester, 2010; Kramer, 2010; Ballester, 2011; Zilian, 2013; Ballester, 2014). ML-based scoring functions achieved Pearson correlation coefficients over 0.700 for the testing benchmark, which is superior to other types of scoring functions (shown in Table 4 in the next section). However, Gabel et al. argued that the ML-based scoring functions perform very poorly in the docking test (Gabel, 2014). It means ML-based scoring functions are unable to identify the "native" pose out of the docking decoys. The reason could be that the ML-based functions are over-trained on descriptors that do not depict protein-ligand interactions but interaction-independent atom-pair counts which are insensitive to docking poses. Thus, appropriate improvement is needed to address the issue and extensive validations are necessary for the ML-based scoring functions.

5. Consensus Scoring Function.

Consensus scoring function in essence is not a type of scoring function but combines the scores from multiple scoring functions. It is first proposed by Charifson et al. (1999) and become a new trend in this field. The aim of consensus scoring is to balance errors in single scores and improve the probability of identifying "true" modes/binders. Generally, they are employed as a post-processing step after docking runs. The crucial implementation in consensus scoring is the design of an appropriate consensus scoring strategy of individual scores so that the "true" modes/binders can be discriminated from others (Wang, 2001; Clark, 2002). Commonly used consensus scoring strategies include vote-by-number, number-by-number, rank-by-number, average rank, linear combination, etc (Oda, 2006). Representative examples of consensus scoring are MultiScore (Terp, 2001), X-Cscore (Wang, 2002), GFscore (Betzi, 2006), SCS (Teramoto, 2007a), SeleX-CS (Bar-Haim, 2009).

Many studies have showed that employing consensus scoring approaches can improve the performance by compensating for the deficiencies of multiple individual scoring functions and tend to reduce the number of false positives that are identified by individual scoring functions (Bissantz, 2000; Feher, 2006). It is worth noting that the potential advantage of consensus scoring is limited when the predictions from different scoring functions are highly correlated.

Evaluation of Scoring Functions

1. Evaluation Methods

As introduced, there are three applications for the scoring functions, identifying the "native" pose out of the conformational decoys, predicting the binding affinity of the protein-ligand interactions and virtual database screening. In line with these three applications, the performance of a scoring function can be evaluated with three related but independent methods. Researchers have named them as docking power, scoring power and screening power (Li, 2014).

The docking power requires the structural comparisons of the docking decoys with the "native" conformation of the ligand. It determines the performance of a scoring function in identifying the best-scored decoy as the "native" binding mode. Normally, if RMSD value of the best-scored binding pose to the "native" pose is less than a predefined cutoff, it is considered as a successful identification of the "native" pose by the scoring function. However, this criterion could be problematic in some cases due to its simplicity. For example, small or nearly symmetrical ligands are likely to obtain relative small RMSD values even when they are randomly placed in a small active site. On the contrary, large flexible ligands are likely to obtain the large RMSD values. To overcome these limitations, several alternative methods have been proposed for pose evaluations, such as relative displacement error (RDE) (Abagyan, 1997), interaction-based accuracy classification (IBAC) (Kroemer, 2004), real space R-factor (RSR) (Yusuf, 2008), and generally applicable replacement for RMSD (GARD) (Baber, 2009).

The scoring power refers to the prediction accuracy of a scoring function on how closely it can estimate the binding affinities in comparison with the experimental measurements. The predicted scores with scaling by scoring functions usually reproduce the relative values of experimental binding affinities. Therefore, the correlations between the predicted binding scores and experimentally measured binding affinities were generally used to evaluate the accuracy of binding affinity prediction. The higher the correlation coefficient is, the better the scoring power is. The commonly-used correlation is the Pearson correlation (CP) which is computed as

$$C_P = \frac{\sum (S_i - < S >)(G_i - < G >)}{\sqrt{\sum (S_i - < S >)^2} \sqrt{\sum (G_i - < G >)^2}} \tag{5}$$

where S_i is the score of a protein-ligand complex computed by a certain scoring function, G_i is the experimental binding affinity. According to the comparative assessment of scoring functions (Li, 2014), the performance of scoring functions in binding affinity is limited with the correlations below 0.70. This could be resulted from two aspects. Firstly, current scoring functions are still in need of improvement as stated in this chapter. Second, the correlation coefficient is dependent on the accuracy and reliability of experimental affinities. Experimental affinities are collected from diverse sources where the binding constants are likely measured under different conditions; these differences can lead to inherent errors among the experimental affinities although carefully cultivated by the researchers (Wang, 2004; Block, 2006; Dunbar Jr, 2011). The inherent errors of the collected experimental binding constant limit the correlation, and make the improvement much harder when correlation coefficient is relatively high.

The screening power determines the ability of a scoring function in discriminating the true binders to a given protein against a large database of compounds. Virtual database screening tests whether or not a scoring function is able to rank the known binders/inhibitors above the inactive compounds in a database. The enrichment test is often taken to quantify the performance of a scoring function in virtual database screening. The enrichment is defined as the accumulated rate of active inhibitors/binders found above a certain percentile of the ranked database that includes the active binders and inactive ligands. A higher enrichment at a fixed percentage of the ranked database can be taken to indicate a better scoring function. Other measurements for virtual database screening are AUC (Area Under the Curve, such as AOC (receiver operating characteristic) curve) and KS (Kolmogorov-Smirno) test, which are more appropriate when the number of inactive ligands is comparable to the number of active binders.

Ideally, an accurate scoring function should be able to perform well above three evaluations. However, most of the existing scoring functions usually perform well on only one or two of the evaluations and fail on others (Li, 2014). In addition, the performances of scoring functions are sometimes sensitive to the testing set. Therefore, to fully evaluate the performance of a scoring function, the above three evaluations are suggested to be carried out with more than one test sets.

2. Evaluation Exercise and Benchmark

An objective and publicly available exercise or benchmark for assessing the performance of the scoring functions can provide practical guidance for the users to make appropriate choices. It can also elucidate the common weakness in the scoring functions for future improvements. The commonly-used exercises and benchmarks in the research community are CSAR (Community Structure-Activity Resource, http://www.csardock.org/) (Smith, 2011) and SAMPL (Statistical Assessment of the Modeling of Proteins and Ligands, http://www.eyesopen.com/SAMPL) (Nicholls, 2008), CASF (Comparative Assessment of Scoring Functions, http://www.pdbbind-cn.org/) (Cheng, 2009) and DUD (Directory of Useful Decoys, http://dud.docking.org/) (Huang, 2006; Irwin, 2008). CSAR and SAMPL exercise results are blind to participants before the exercises were finished. The exercises are normally carried out annually and the benchmarks are often updated and improved with new and higher-quality data. It is worth mentioning that scoring functions are often applied in combination with molecular docking methods to predict ligand binding poses and ligand binding affinities or to identify active compounds through virtual screening. Thus, the evaluation exercises or benchmarks are designed to assess not only the scoring functions but also the docking programs.

3. Comparative Evaluation

By applying the evaluation methods and using the benchmark, researchers have recently compared the performance of different scoring functions testing on the benchmark CASF-2013 (Li, 2014; Yan, 2015). CASF-2013 is composed of 195 protein-ligand complexes with known structures and experimental affinities. To evaluate the scoring functions, a set of binding poses were generated for each protein-ligand complex with RMSDs ranging from 0 to 10 angstroms. The binding poses were scored by the scoring functions including Force-field-based, empirical and knowledge-based scoring functions. The testing results of docking power and scoring power were listed in the Table 3 and Table 4.

It can be seen that all the scoring functions predict the affinities with correlation coefficients lower than 0.7. This suggests that all the testing scoring functions are still not accurate in predicting binding affinities. Compared to the scoring power, more than half scoring functions achieves success rate over 80% in the docking power when top three poses are considered, suggesting that the scoring functions have higher capability in identifying the "native'' conformation. Based on the comparative evaluations of different scoring functions, there is no clear advantage for a particular class of scoring function since the performance of the scoring functions in the same class is diversely ranked in the tables. It is worth noting that SPA-SE ranks best for both docking power and scoring power, as stated in the paper (Yan, 2015), this is because SPA-SE incorporates solvation effect in the scoring and binding specificity in the optimization, which are discussed next.

CHALLENGES AND SOLUTIONS

The molecular docking involves concerted performance of sampling and scoring the binding poses. To this end, there are some major challenges, such as protein flexibility in the sampling, implicit or explicit solvation effect of the binding site, entropy contribution to the binding free energy, covalent bonding between the protein and the ligand, metal ion coordination during binding, binding specificity of molecular recognition, structural water molecules in the binding site and so on. Most of the challenges are related to both sampling and scoring methods and a few are beyond them. Here, the authors introduced the major challenges on the side of the scoring function and recent solutions, such as incorporating

Table 3. Comparative evaluations of scoring functions in the docking power

Scoring Function	Classification	Success Rate (%)		
		Top One	Top two	Top Three
SPA-SE	Knowledge-based	83.1	89.2	92.8
SPA	Knowledge-based	82.1	88.2	90.8
ChemPLP@GOLD	Empirical	81.0	86.7	89.7
ChemScore@GOLD	Empirical	77.9	83.1	88.2
GlideScore-SP	Empirical	78.5	85.6	87.7
LigScore2@DS	Empirical	76.9	84.1	86.7
PLP1@DS	Empirical	76.9	84.1	86.2
Alpha-HB@MOE	Empirical	75.4	82.6	86.2
GlideScore-XP	Empirical	74.4	82.6	85.6
ASP@GOLD	Knowledge-based	71.8	81.5	87.2
GoldScore@GOLD	Force-field-based	71.3	81.0	85.6
LUDI1@DS	Empirical	59.0	75.4	83.1
Affinity-dG@MOE	Empirical	63.1	74.9	81.0
ChemScore@SYBYL	Empirical	59.5	69.2	75.4
London-dG@MOE	Empirical	59.5	73.8	78.5
X-Score^{HM}	Empirical	61.0	73.3	77.9
G-Score@SYBYL	Force-field-based	45.1	61.5	72.3
Jain@DS	Empirical	48.2	62.1	70.8
PMF@SYBYL	Knowledge-based	51.8	60.0	66.7
PMF04@DS	Knowledge-based	51.8	62.6	66.2
ASE@MOE	Empirical	51.3	60.0	63.6
D-Score@SYBYL	Force-field-based	18.5	29.7	42.6

The results of scoring function SPA-SE and SPA are obtained from the paper (Yan, 2015) and others are from the paper (Li, 2014). For more details, the readers are suggested to look into the citations.

the implicit or explicit solvation effect of the binding site, incorporating the entropy contribution to the binding free energy and taking account of binding specificity in the development strategy. Scoring functions accounting for such contributions are more accurate than those without these considerations.

Solvation Effect

It is believed that water molecules actively participate in and affect the process of protein-ligand binding in aqueous solvent (Dill, 2005; Levy, 2006; Berne, 2009; Yan, 2015). The hydration shell that forms around the protein and the ligand is of particular importance to their binding activity. During the binding process, both solvated protein and ligand will be partly dehydrated and the water layers will be gradually excluded into the bulk water, making the ligand diffuse into the active site of the protein in the solvent. The role of water molecules not only assists the formation of the protein-ligand interface but also prob-

Table 4. Comparative evaluations of scoring functions in the scoring power

Scoring Function	Classification	C_P	D_S	N_C
SPA-SE	Knowledge-based	0.662	1.69	195
SPA	Knowledge-based	0.656	1.70	195
X-Score^{HM}	Empirical	0.614	1.78	195
ChemScore@SYBYL	Empirical	0.592	1.82	195
ChemPLP@GOLD	Empirical	0.579	1.84	195
PLP1@DS	Empirical	0.568	1.86	195
G-Score@SYBYL	Force-field-based	0.558	1.87	195
ASP@GOLD	Knowledge-based	0.556	1.88	195
ASE@MOE	Empirical	0.544	1.89	195
ChemScore@SYBYL	Empirical	0.536	1.90	189
D-Score@SYBYL	Force-field-based	0.526	1.92	195
Alpha-HB@MOE	Empirical	0.511	1.94	195
LUDI3@DS	Empirical	0.487	1.97	195
GoldScore@GOLD	Force-field-based	0.483	1.97	189
Affinity-dG@MOE	Empirical	0.482	1.98	195
LigScore2@DS	Empirical	0.456	2.02	190
GlideScore-SP	Empirical	0.452	2.03	169
Jain@DS	Empirical	0.408	2.05	191
PMF@DS	Knowledge-based	0.364	2.11	194
GlideScore-XP	Empirical	0.277	2.18	164
London-dG@MOE	Empirical	0.242	2.19	195
PMF@SYBYL	Knowledge-based	0.221	2.20	191

The results of scoring function SPA-SE and SPA are obtained from the paper (Yan, 2015) and others are from the paper (Li, 2014). C_P is the Pearson correlation between experimental and predicted affinities computed by Equation 5; the D_S is the standard deviation in the linear correlation; The N_C is the number of complexes used in the Pearson correlation calculation. For more details, the readers are suggested to look into the citations.

ably makes the active site more adaptable for ligand binding than other possible binding sites, smoothing the binding energy landscape (Papoian, 2003; Levy, 2006). In this sense, the solvation effect is assumed to be a regulator to the energetics as well as the specificity of ligand binding.

To account for the solvation effect, much progress has been made in the development of both explicit and implicit solvent models (Feig, 2004; Purisima, 2013). Explicit model with inclusion of water molecules can provide detailed structural and energetic information; whereas it significantly increases computational cost (Levy, 1998). Especially, it is unaffordable to screen a large library of small molecules in the docking-based drug discovery. The charge-dependent implicit solvent models, such as PB/GBSA (Poisson-Boltzmann/Generalized Born Surface Area) models, are widely taken as post-processing methods to compute free energies based on molecular dynamics simulations or energy minimizations (Kollman, 2000; Lindstrom, 2011); whereas, these models are rarely incorporated into the scoring function for the process of protein-ligand docking (Zou, 1999). Alternatively, the charge-independent implicit solvent models have the advantage of less computational cost due to simple additive expression since the models are based on solvent accessible surface area (SASA) of each atom and the atomic solvation parameter (Eisenberg, 1986). However, the atomic solvation parameters are usually independently optimized (Wang, 2001; Hou, 2002; Guvench, 2004; Hou, 2005), which are not transferable to the scoring functions of protein-ligand interactions which are described with different classifications of atomic types compared to those used in the solvent models.

Given above difficulties, two knowledge-based scoring functions SPA-SE (Yan, 2015) and ITScore (Huang, 2010) have effectively combined statistical atomic-pair potentials and the atomic solvation energy. The atomic solvation energy is represented by the charge-independent implicit solvent model which accounts for the contribution of solvation energy to the protein-ligand binding based on the change of SASA (Solvent Accessible Surface Area) upon binding. SPA-SE optimized the energetic parameters by maximizing the affinity and specificity predictions of ligand binding, while ITScore employed an iterative method to derive the parameters until the predicted energy converges to the experimentally observed one. Both methods have showed improvements in predicting the binding affinity and identifying the "native" pose.

Entropy Contribution

Conformational entropy is an important component to the free energy change upon binding of a ligand to the protein (Finkelstein, 1989; Murray, 2002; Salaniwal, 2007). The entropy contribution to the binding free energy essentially is arisen from the reduction of the translational and rotational degrees of freedom in the ligand and from changes in the flexibility of the protein during binding, as well as the water molecules involved (Hermans, 1997; Amzel, 1997). The computation of the entropic contribution is very time-consuming (Baron, 2009). Therefore, incorporating such effects in an efficient scoring function is still an important challenge in the field of protein-ligand docking. Some attempts to incorporate entropy in scoring functions have been reported in the literatures which implemented them in the re-scoring schemes (Ruvinsky, 2005; Ruvinsky, 2007; Lee, 2008).

Ruvinsky and Kozintsev (2005) proposed a way to estimate contributions of the translational, rotational and torsional entropy to protein-ligand affinity. In their approach, A number of docking runs are carried out and the generated conformations are clustered by structural similarity. The size of each cluster is then measured to estimate the entropic contribution based on the assumption that large clusters of conformations are indicative of favorable entropic contributions of the local energy landscapes. With this approach, testing scoring functions have shown some improvements in identifying the "native" binding pose (Ruvinsky, 2007). Lee and Seok (2008) introduced a probability function to analyze the populations of diverse binding poses in the context of statistical mechanics. The approach estimates the contribution of the state represented by a sampled conformation of the configurational integral (Xiang, 2002). The approach in combination with 11 scoring functions has resulted in improved accuracy in binding mode prediction. Both approaches are relatively low computational cost and need the combination with other pre-existing scoring functions. However, only the entropy contribution resulted from the ligand were accounted for in these studies. Some other MD-based methods were proposed to estimate the conformational entropy of the protein (Andricioae, 2001; Baron, 2006; Baron, 2008), but they are challenging to be widely employed in the scoring functions due to highly computational demanding.

In addition to the entropic contributions of the ligand and protein, other key components still require our attention. For example, the entropic contribution of the solvent molecules, particularly those at the binding pocket is normally forgotten (Chiba, 2011; Breiten, 2013). The very limited movement of buried waters in the protein-ligand complex and the huge variation arising from desolvation make this an important issue.

Binding Specificity

Biomolecular recognition requires both stability and specificity to realize its function (Janin, 1995; Havranek, 2003; Wang, 2003; Wang, 2007; Kortemme, 2004; Bolon, 2005). The stability of the complex is determined by the binding affinity while the specificity is measured by the discrimination against competitive binding complexes. Most scoring functions of protein-ligand binding, whether force-field based, empirical, knowledge-based scoring functions or other quantitative description of protein-ligand interactions, mainly aim to predict the binding affinities. The strategy of developing these methods seeks to optimize the stability based on the combination of energetics and shape complementarity but without explicit consideration of binding specificity.

Practically, the quantification of binding specificity is a challenge since it requires comparison of all the possible binding complexes. To circumvent the challenge, the concept of intrinsic specificity was proposed to describe the binding specificity (Wang, 2003; Wang, 2007, Yan, 2012). According to the funnel-like energy landscape of protein-ligand binding (Wang, 2003; Wang, 2007; Bryngelson, 1995; Janin, 1996; Rejto, 1996; Miller, 2008; Tsai, 1999; Dominy, 2004; Z. Liu, 2004; Levy, 2004), the intrinsic specificity ratio (ISR) was quantified to measure the binding specificity. With the quantification of the binding specificity, Yan and Wang (Yan, 2012) developed a novel scoring function SPA by optimizing both specificity and affinity. The development strategy for SPA is to simultaneously optimize binding specificity and affinity of ligand binding. SPA was validated by testing the benchmark set and the performance was compared with 16 popular scoring functions. The test results showed that SPA outperformed the compared scoring functions. The work built up a new frame to develop and improve the scoring function of protein-ligand interactions by incorporating the binding specificity.

FUTURE RESEARCH DIRECTIONS

With the efforts in the past three decades, scoring functions of protein-ligand interactions have made much progress and have been used for many successful applications. Currently, there are large data sets of protein-ligand complexes publicly available with known structures and binding affinities to improve the quality. Also, quite a few benchmarks have been created so that different scoring functions can be compared in a more objective manner. This will be undoubtedly further enhancing the improvement of current scoring functions.

Despite the numerous successful applications of molecular docking-based drug discovery, a widely accepted fact is that the major weakness of current computational protein-ligand docking lies not in the sampling methods but in the scoring functions (Kitchen, 2004; Danishuddin, 2015). The benchmark testing suggests that current scoring functions can provide promising results in predicting the correct ligand binding pose, but their performance is relatively weak in predicting protein-ligand binding affinities. In virtual screening, it is still challenging for scoring functions to differentiate low- to medium-affinity binders from non-binders. Thus, considerable efforts are still needed to the development and improvement of the scoring function aiming to rapidly and accurately quantify the interaction between the protein and the ligand (Huang, 2010; Wang, 2013; Liu, 2015). As discussed, a current trend in the field is to focus on the inclusion of solvation effect, entropy contributions as well as binding specificity. Scoring functions accounting for those are more accurate but are also computationally more expensive. The fast and accurate prediction of a ligand tightly and specifically binding to a target protein will be the final aim of the scoring functions in the docking-based drug discovery and design.

So far, there are a large number of alternatives of the scoring functions, most of them are generic models derived from the large-scale experimental data of protein-ligand complexes and are presumably applicable to all kinds of target classes. However, it is still difficult to develop a scoring function which can perform equally well across many different protein families (Seifert, 2009; Scharfe, 2012; Cheng, 2012). Binding sites and recognition processes have unique features that ultimately render protein-ligand interactions specific, which in turn might require re-calibrating of scoring schemes on a case-by-case basis. Thus, scoring functions can be improved by tailoring them to a specific target site or to multiple related receptors. For example, the kinase family-specific potential of mean force (kinase-PMF) (Xue, 2010) is kinase-specific scoring function adjusted from the original PMF04 (Muegge, 2006). Target-specific scoring functions have shown improved performance and enhanced screening enrichment in the tests (Teramoto, 2007b; Pham, 2008; Seifert, 2008). It is a promising trend to develop target-specific scoring functions in the future since the improvement of generic scoring functions is often limited with the compromise between the accuracy and computational cost.

CONCLUSION

The discovery and development of novel drugs are highly complex, time-consuming and expensive. Docking-based drug discovery and design is gaining prominence as a method to filter the very large chemical space. Molecular docking-based virtual screening becomes routine in both pharmaceutical companies and academic groups for early-stage drug discovery. Scoring function is at the heart of molecular docking by assisting a docking program to efficiently explore the binding space of a ligand. It is also responsible for evaluating the binding affinity once the binding poses are generated. Therefore, the performance of scoring functions has a significant impact on the docking-based drug discovery. A multitude of scoring functions has been reported in the past three decades and new ones are still emerging. The current trend to improve the performance of scoring functions focusing on the inclusion of additional factors to account for solvation and entropic effects, as well as the strategy to quantify the binding specificity and to develop the target-specific scoring functions. As large data sets of protein-ligand complexes with high quality are available to the public, as well as the computation power increasingly improves, researchers will reach better tradeoff between the accuracy and computational cost for the scoring functions in the coming years. It will undoubtedly produce more intriguing and important applications through molecular docking-based drug design and discovery.

REFERENCES

Abagyan, R. A., & Totrov, M. M. (1997). Contact area difference (CAD): A robust measure to evaluate accuracy of protein models. *Journal of Molecular Biology*, *268*(3), 678–685. PMID:9171291

Amzel, L. M. (1997). Loss of translational entropy in binding, folding, and catalysis. *Proteins:* Structure Function and Genetics, *28*(2), 144–149. PMID:9188731

Andricioaei, I., & Karplus, M. (2001). On the calculation of entropy from covariance matrices of the atomic fluctuations. *The Journal of Chemical Physics*, *115*(14), 6289–6292.

Aqvist, J., Medina, C., & Samuelsson, J.-E. (1994). A new method for predicting binding affinity in computer-aided drug design. *Protein Engineering*, *7*(3), 385–391. PMID:8177887

Baber, J. C., Thompson, D. C., Cross, J. B., & Humblet, C. (2009). GARD: A generally applicable replacement for rmsd. *Journal of Chemical Information and Modeling*, *49*(8), 1889–1900. PMID:19618919

Ballester, P. J., & Mitchell, J. B. (2010). A machine learning approach to predicting protein–ligand binding affinity with applications to molecular docking. *Bioinformatics (Oxford, England)*, *26*(9), 1169–1175. PMID:20236947

Ballester, P. J., & Mitchell, J. B. (2011). Comments on leave-cluster-out cross-validation is appropriate for scoring functions derived from diverse protein data sets: Significance for the validation of scoring functions. *Journal of Chemical Information and Modeling*, *51*(8), 1739–1741. PMID:21591735

Ballester, P. J., Schreyer, A., & Blundell, T. L. (2014). Does a more precise chemical description of protein–ligand complexes lead to more accurate prediction of binding affinity? *Journal of Chemical Information and Modeling*, *54*(3), 944–955. PMID:24528282

Bar-Haim, S., Aharon, A., Ben-Moshe, T., Marantz, Y., & Senderowitz, H. (2009). Selex-CS: A new consensus scoring algorithm for hit discovery and lead optimization. *Journal of Chemical Information and Modeling*, *49*(3), 623–633. PMID:19231809

Baron, R., Hunenberger, P. H., & McCammon, J. A. (2009). Absolute single-molecule entropies from quasi-harmonic analysis of microsecond molecular dynamics: Correction terms and convergence properties. *Journal of Chemical Theory and Computation*, *5*(12), 3150–3160. PMID:20011626

Baron, R., & McCammon, J. A. (2008). (Thermo) dynamic role of receptor flexibility, entropy, and motional correlation in protein–ligand binding. *ChemPhysChem*, *9*(7), 983–988. PMID:18418822

Baron, R., van Gunsteren, W. F., & Hünenberger, P. H. (2006). Estimating the configurational entropy from molecular dynamics simulations: Anharmonicity and correlation corrections to the quasi-harmonic approximation. *Trends in Physical Chemistry*, *11*, 87–122.

Bello, M., Mart'ınez-Archundia, M., & Correa-Basurto, J. (2013). Automated docking for novel drug discovery. *Expert Opinion on Drug Discovery*, *8*(7), 821–834. PMID:23642085

Berne, B. J., Weeks, J. D., & Zhou, R. (2009). Dewetting and hydrophobic interaction in physical and biological systems. *Annual Review of Physical Chemistry*, *60*, 85. PMID:18928403

Betzi, S., Suhre, K., Chétrit, B., Guerlesquin, F., & Morelli, X. (2006). GFScore: A general nonlinear consensus scoring function for high-throughput docking. *Journal of Chemical Information and Modeling*, *46*(4), 1704–1712. PMID:16859302

Bissantz, C., Folkers, G., & Rognan, D. (2000). Protein-based virtual screening of chemical databases. 1. evaluation of different docking/scoring combinations. *Journal of Medicinal Chemistry*, *43*(25), 4759–4767. PMID:11123984

Block, P., Sotriffer, C. A., Dramburg, I., & Klebe, G. (2006). AffinDB: A freely accessible database of affinities for protein–ligand complexes from the pdb. *Nucleic Acids Research*, *34*(suppl 1), D522–D526. PMID:16381925

Böhm, H.-J. (1992). LUDI: Rule-based automatic design of new substituents for enzyme inhibitor leads. *Journal of Computer-Aided Molecular Design*, *6*(6), 593–606. PMID:1291628

Bolon, D. N., Grant, R. A., Baker, T. A., & Sauer, R. T. (2005). Specificity versus stability in computational protein design. *Proceedings of the National Academy of Sciences of the United States of America*, *102*(36), 12724–12729. PMID:16129838

Brás, N., Cerqueira, N., Sousa, S., Fernandes, P., & Ramos, M. (2014). Protein ligand docking in drug discovery. *Protein Modelling*, 249.

Breiten, B., Lockett, M. R., Sherman, W., Fujita, S., Al-Sayah, M., Lange, H., & Whitesides, G. M. et al. (2013). Water networks contribute to enthalpy/entropy compensation in protein–ligand binding. *Journal of the American Chemical Society*, *135*(41), 15579–15584.

Brooks, B. R., Bruccoleri, R. E., Olafson, B. D., States, D. J., Swaminathan, S., & Karplus, M. (1983). CHARMM: A program for macromolecular energy, minimization, and dynamics calculations. *Journal of Computational Chemistry*, *4*(2), 187–217.

Bryngelson, J. D., Onuchic, J. N., Socci, N. D., & Wolynes, P. G. (1995). Funnels, pathways, and the energy landscape of protein folding: A synthesis. *Proteins: Structure, Function, and Bioinformatics*, *21*(3), 167–195. PMID:7784423

Charifson, P. S., Corkery, J. J., Murcko, M. A., & Walters, W. P. (1999). Consensus scoring: A method for obtaining improved hit rates from docking databases of three-dimensional structures into proteins. *Journal of Medicinal Chemistry*, *42*(25), 5100–5109. PMID:10602695

Cheng, T., Li, Q., Zhou, Z., Wang, Y., & Bryant, S. H. (2012). Structure-based virtual screening for drug discovery: A problem-centric review. *The AAPS Journal*, *14*(1), 133–141. PMID:22281989

Cheng, T., Li, X., Li, Y., Liu, Z., & Wang, R. (2009). Comparative assessment of scoring functions on a diverse test set. *Journal of Chemical Information and Modeling*, *49*(4), 1079–1093. PMID:19358517

Chiba, S., Harano, Y., Roth, R., Kinoshita, M., & Sakurai, M. (2012). Evaluation of protein-ligand binding free energy focused on its entropic components. *Journal of Computational Chemistry*, *33*(5), 550–560.

Clark, R. D., Strizhev, A., Leonard, J. M., Blake, J. F., & Matthew, J. B. (2002). Consensus scoring for ligand/protein interactions. *Journal of Molecular Graphics & Modelling*, *20*(4), 281–295. PMID:11858637

Cosconati, S., Forli, S., Perryman, A. L., Harris, R., Goodsell, D. S., & Olson, A. J. (2010). Virtual screening with AutoDock: Theory and practice. *Expert Opinion on Drug Discovery*, *5*(6), 597–607. PMID:21532931

Danishuddin, M., & Khan, A. U. (2015). Structure based virtual screening to discover putative drug candidates: Necessary considerations and successful case studies. *Methods (San Diego, Calif.)*, *71*, 135–145. PMID:25448480

Deng, W., Breneman, C., & Embrechts, M. J. (2004). Predicting protein-ligand binding affinities using novel geometrical descriptors and machine-learning methods. *Journal of Chemical Information and Computer Sciences*, *44*(2), 699–703. PMID:15032552

DeWitte, R. S., & Shakhnovich, E. I. (1996). Smog: De novo design method based on simple, fast, and accurate free energy estimates. 1. Methodology and supporting evidence. *Journal of the American Chemical Society*, *118*(47), 11733–11744.

Dias, R., de Azevedo, J., & Walter, F. (2008). Molecular docking algorithms. *Current Drug Targets*, *9*(12), 1040–1047. PMID:19128213

Dickson, M., & Gagnon, J. P. (2004). Key factors in the rising cost of new drug discovery and development. *Nature Reviews. Drug Discovery*, *3*(5), 417–429. PMID:15136789

Dill, K. A., Truskett, T. M., Vlachy, V., & Hribar-Lee, B. (2005). Modeling water, the hydrophobic effect, and ion solvation. *Annual Review of Biophysics and Biomolecular Structure*, *34*, 173–199. PMID:15869376

Dominy, B. N., & Shakhnovich, E. I. (2004). Native atom types for knowledge-based potentials: Application to binding energy prediction. *Journal of Medicinal Chemistry*, *47*(18), 4538–4558. PMID:15317465

Dunbar, J. B. Jr, Smith, R. D., Yang, C.-Y., Ung, P. M.-U., Lexa, K. W., Khazanov, N. A., & Carlson, H. A. et al. (2011). CSAR benchmark exercise of 2010: Selection of the protein–ligand complexes. *Journal of Chemical Information and Modeling*, *51*(9), 2036–2046. PMID:21728306

Durrant, J. D., & McCammon, J. A. (2010). NNScore: A neural-network-based scoring function for the characterization of protein- ligand complexes. *Journal of Chemical Information and Modeling*, *50*(10), 1865–1871. PMID:20845954

Eisenberg, D., & Mclachlan, A. (1986). Solvation energy in protein folding and binding. *Nature*, *319*(6050), 199–203. PMID:3945310

Eldridge, M. D., Murray, C. W., Auton, T. R., Paolini, G. V., & Mee, R. P. (1997). Empirical scoring functions: I. the development of a fast empirical scoring function to estimate the binding affinity of ligands in receptor complexes. *Journal of Computer-Aided Molecular Design*, *11*(5), 425–445. PMID:9385547

Feher, M. (2006). Consensus scoring for protein–ligand interactions. *Drug Discovery Today*, *11*(9), 421–428. PMID:16635804

Feig, M., & Brooks,, C. L. III (2004). Recent advances in the development and application of implicit solvent models in biomolecule simulations. *Current Opinion in Structural Biology*, *14*(2), 217–224. PMID:15093837

Finkelstein, A. V., & Janin, J. l. (1989). The price of lost freedom: Entropy of bimolecular complex formation. *Protein Engineering*, *3*(1), 1–3. PMID:2813338

Friesner, R. A., Banks, J. L., Murphy, R. B., Halgren, T. A., Klicic, J. J., & Mainz, D. T. et al. others. (2004). Glide: A new approach for rapid, accurate docking and scoring. 1. method and assessment of docking accuracy. *Journal of Medicinal Chemistry*, *47*(7), 1739–1749. PMID:15027865

Friesner, R. A., Murphy, R. B., Repasky, M. P., Frye, L. L., Greenwood, J. R., Halgren, T. A., & Mainz, D. T. et al. (2006). Extra precision glide: Docking and scoring incorporating a model of hydrophobic enclosure for protein-ligand complexes. *Journal of Medicinal Chemistry*, *49*(21), 6177–6196. PMID:17034125

Gabel, J., Desaphy, J., & Rognan, D. (2014). Beware of machine learning-based scoring functionsr on the danger of developing black boxes. *Journal of Chemical Information and Modeling*, *54*(10), 2807–2815. PMID:25207678

Gohlke, H., Hendlich, M., & Klebe, G. (2000). Knowledge-based scoring function to predict protein-ligand interactions. *Journal of Molecular Biology*, *295*(2), 337–356. PMID:10623530

Goodsell, D. S., Morris, G. M., & Olson, A. J. (1996). Automated docking of flexible ligands: Applications of autodock. *Journal of Molecular Recognition*, *9*(1), 1–5. PMID:8723313

Guo, L., Yan, Z., Zheng, X., Hu, L., Yang, Y., & Wang, J. (2014). A comparison of various optimization algorithms of protein–ligand docking programs by fitness accuracy. *Journal of Molecular Modeling*, *20*(7), 1–10. PMID:24935106

Gutiérrez-de Terán, H., & Aqvist, J. (2012). Linear interaction energy: method and applications in drug design. In *Computational drug discovery and design* (pp. 305–323). Springer.

Guvench, O., & Brooks, C. L. (2004). Efficient approximate all-atom solvent accessible surface area method parameterized for folded and denatured protein conformations. *Journal of Computational Chemistry*, *25*(8), 1005–1014. PMID:15067676

Havranek, J. J., & Harbury, P. B. (2003). Automated design of specificity in molecular recognition. *Nature Structural Biology*, *10*(1), 45–52. PMID:12459719

Hermans, J., & Wang, L. (1997). Inclusion of loss of translational and rotational freedom in theoretical estimates of free energies of binding. application to a complex of benzene and mutant t4 lysozyme. *Journal of the American Chemical Society*, *119*(11), 2707–2714.

Hou, T., Qiao, X., Zhang, W., & Xu, X. (2002). Empirical aqueous solvation models based on accessible surface areas with implicit electrostatics. *The Journal of Physical Chemistry B*, *106*(43), 11295–11304.

Hou, T., Zhang, W., Huang, Q., & Xu, X. (2005). An extended aqueous solvation model based on atom-weighted solvent accessible surface areas: SAWSA v2. 0 model. *Journal of Molecular Modeling*, *11*(1), 26–40. PMID:15565273

Huang, N., Shoichet, B. K., & Irwin, J. J. (2006). Benchmarking sets for molecular docking. *Journal of Medicinal Chemistry*, *49*(23), 6789–6801. PMID:17154509

Huang, S.-Y., Grinter, S. Z., & Zou, X. (2010). Scoring functions and their evaluation methods for protein–ligand docking: Recent advances and future directions. *Physical Chemistry Chemical Physics*, *12*(40), 12899–12908. PMID:20730182

Huang, S.-Y., & Zou, X. (2006). An iterative knowledge-based scoring function to predict protein–ligand interactions: I. derivation of interaction potentials. *Journal of Computational Chemistry*, *27*(15), 1866–1875. PMID:16983673

Huang, S.-Y., & Zou, X. (2010). Inclusion of solvation and entropy in the knowledge-based scoring function for protein-ligand interactions. *Journal of Chemical Information and Modeling*, *50*(2), 262–273. PMID:20088605

Huey, R., Morris, G. M., Olson, A. J., & Goodsell, D. S. (2007). A semiempirical free energy force field with charge-based desolvation. *Journal of Computational Chemistry*, *28*(6), 1145–1152. PMID:17274016

Irwin, J. J. (2008). Community benchmarks for virtual screening. *Journal of Computer-Aided Molecular Design*, *22*(3-4), 193–199. PMID:18273555

Ishchenko, A. V., & Shakhnovich, E. I. (2002). Small molecule growth 2001 (SMoG2001): An improved knowledge-based scoring function for protein-ligand interactions. *Journal of Medicinal Chemistry*, *45*(13), 2770–2780. PMID:12061879

Janin, J. (1995). Principles of protein-protein recognition from structure to thermodynamics. *Biochimie*, *77*(7), 497–505. PMID:8589061

Janin, J. (1996). Quantifying biological specificity: The statistical mechanics of molecular recognition. *Proteins: Structure, Function, and Bioinformatics*, *25*(4), 438–445. PMID:8865339

Jones, G., Willett, P., Glen, R. C., Leach, A. R., & Taylor, R. (1997). Development and validation of a genetic algorithm for flexible docking. *Journal of Molecular Biology*, *267*(3), 727–748. PMID:9126849

Karplus, M., & Weaver, D. L. (1976). Protein-folding dynamics. *Nature*, *260*, 404–406. PMID:1256583

Kitchen, D. B., Decornez, H., Furr, J. R., & Bajorath, J. (2004). Docking and scoring in virtual screening for drug discovery: Methods and applications. *Nature Reviews. Drug Discovery*, *3*(11), 935–949. PMID:15520816

Kollman, P. (1993). Free energy calculations: Applications to chemical and biochemical phenomena. *Chemical Reviews*, *93*(7), 2395–2417.

Kollman, P. A., Massova, I., Reyes, C., Kuhn, B., Huo, S., & Chong, L. et al. others. (2000). Calculating structures and free energies of complex molecules: Combining molecular mechanics and continuum models. *Accounts of Chemical Research*, *33*(12), 889–897. PMID:11123888

Kortemme, T., Joachimiak, L. A., Bullock, A. N., Schuler, A. D., Stoddard, B. L., & Baker, D. (2004). Computational redesign of protein-protein interaction specificity. *Nature Structural & Molecular Biology*, *11*(4), 371–379. PMID:15034550

Kramer, B., Rarey, M., & Lengauer, T. (1999). Evaluation of the flexx incremental construction algorithm for protein–ligand docking. *Proteins: Structure, Function, and Bioinformatics*, *37*(2), 228–241. PMID:10584068

Kramer, C., & Gedeck, P. (2010). Leave-cluster-out cross-validation is appropriate for scoring functions derived from diverse protein data sets. *Journal of Chemical Information and Modeling*, *50*(11), 1961–1969. PMID:20936880

Kroemer, R. T., Vulpetti, A., McDonald, J. J., Rohrer, D. C., Trosset, J.-Y., Giordanetto, F., & Stouten, P. F. et al. (2004). Assessment of docking poses: Interactions-based accuracy classification (IBAC) versus crystal structure deviations. *Journal of Chemical Information and Computer Sciences*, *44*(3), 871–881. PMID:15154752

Lee, J., & Seok, C. (2008). A statistical rescoring scheme for protein–ligand docking: Consideration of entropic effect. *Proteins: Structure, Function, and Bioinformatics*, *70*(3), 1074–1083. PMID:18076034

Levy, R. M., & Gallicchio, E. (1998). Computer simulations with explicit solvent: Recent progress in the thermodynamic decomposition of free energies and in modeling electrostatic effects. *Annual Review of Physical Chemistry*, *49*(1), 531–567. PMID:9933909

Levy, Y., & Onuchic, J. N. (2006). Water mediation in protein folding and molecular recognition. *Annual Review of Biophysics and Biomolecular Structure*, *35*, 389–415. PMID:16689642

Levy, Y., Wolynes, P. G., & Onuchic, J. N. (2004). Protein topology determines binding mechanism. *Proceedings of the National Academy of Sciences of the United States of America*, *101*(2), 511–516. PMID:14694192

Li, G.-B., Yang, L.-L., Wang, W.-J., Li, L.-L., & Yang, S.-Y. (2013). Id-score: A new empirical scoring function based on a comprehensive set of descriptors related to protein–ligand interactions. *Journal of Chemical Information and Modeling*, *53*(3), 592–600. PMID:23394072

Li, Y., Han, L., Liu, Z., & Wang, R. (2014). Comparative assessment of scoring functions on an updated benchmark: II. Evaluation methods and general results. *Journal of Chemical Information and Modeling*, *54*, 1717–1736. PMID:24708446

Li, Y., Liu, Z., Li, J., Han, L., Liu, J., Zhao, Z.-X., & Wang, R. (2014). Comparative assessment of scoring functions on an updated benchmark: I. Compilation of the test set. *Journal of Chemical Information and Modeling*, *54*, 1700–1716. PMID:24716849

Lindstrom, A., Edvinsson, L., Johansson, A., Andersson, C. D., Andersson, I. E., Raubacher, F., & Linusson, A. (2011). Postprocessing of docked protein-ligand complexes using implicit solvation models. *Journal of Chemical Information and Modeling*, *51*(2), 267–282. PMID:21309544

Liu, J., & Wang, R. (2015). On classification of current scoring functions. *Journal of Chemical Information and Modeling*, *55*(3), 475–482. PMID:25647463

Liu, Z., Dominy, B. N., & Shakhnovich, E. I. (2004). Structural mining: Self-consistent design on flexible protein-peptide docking and transferable binding affinity potential. *Journal of the American Chemical Society*, *126*(27), 8515–8528. PMID:15238009

Makino, S., & Kuntz, I. D. (1997). Automated flexible ligand docking method and its application for database search. *Journal of Computational Chemistry*, *18*(14), 1812–1825.

Meng, E. C., Shoichet, B. K., & Kuntz, I. D. (1992). Automated docking with grid-based energy evaluation. *Journal of Computational Chemistry*, *13*(4), 505–524.

Meng, X.-Y., Zhang, H.-X., Mezei, M., & Cui, M. (2011). Molecular docking: A powerful approach for structure-based drug discovery. *Current Computer-aided Drug Design*, *7*(2), 146. PMID:21534921

Miller, D. W., & Dill, K. A. (2008). Ligand binding to proteins: The binding landscape model. *Protein Science*, *6*(10), 2166–2179. PMID:9336839

Mitchell, J. B., Laskowski, R. A., Alex, A., & Thornton, J. M. (1999). Bleeppotential of mean force describing protein–ligand interactions: I. generating potential. *Journal of Computational Chemistry*, *20*(11), 1165–1176.

Miyazawa, S., & Jernigan, R. L. (1985). Estimation of effective interresidue contact energies from protein crystal structures: Quasi-chemical approximation. *Macromolecules*, *18*(3), 534–552.

Morris, G. M., Huey, R., Lindstrom, W., Sanner, M. F., Belew, R. K., Goodsell, D. S., & Olson, A. J. (2009). AutoDock4 and autodocktools4: Automated docking with selective receptor flexibility. *Journal of Computational Chemistry*, *30*(16), 2785–2791. PMID:19399780

Muegge, I. (2006). PMF scoring revisited. *Journal of Medicinal Chemistry*, *49*(20), 5895–5902. PMID:17004705

Muegge, I., & Martin, Y. C. (1999). A general and fast scoring function for protein-ligand interactions: A simplified potential approach. *Journal of Medicinal Chemistry*, *42*(5), 791–804. PMID:10072678

Murray, C. W., & Verdonk, M. L. (2002). The consequences of translational and rotational entropy lost by small molecules on binding to proteins. *Journal of Computer-Aided Molecular Design*, *16*(10), 741–753. PMID:12650591

Nicholls, A., Mobley, D. L., Guthrie, J. P., Chodera, J. D., Bayly, C. I., Cooper, M. D., & Pande, V. S. (2008). Predicting small-molecule solvation free energies: An informal blind test for computational chemistry. *Journal of Medicinal Chemistry*, *51*(4), 769–779. PMID:18215013

Oda, A., Tsuchida, K., Takakura, T., Yamaotsu, N., & Hirono, S. (2006). Comparison of consensus scoring strategies for evaluating computational models of protein-ligand complexes. *Journal of Chemical Information and Modeling*, *46*(1), 380–391. PMID:16426072

Papoian, G. A., Ulander, J., & Wolynes, P. G. (2003). Role of water mediated interactions in protein-protein recognition landscapes. *Journal of the American Chemical Society*, *125*(30), 9170–9178. PMID:15369374

Pham, T. A., & Jain, A. N. (2008). Customizing scoring functions for docking. *Journal of Computer-Aided Molecular Design*, *22*(5), 269–286. PMID:18273558

Plewczynski, D., La'zniewski, M., Augustyniak, R., & Ginalski, K. (2011). Can we trust docking results? Evaluation of seven commonly used programs on pdbbind database. *Journal of Computational Chemistry*, *32*(4), 742–755. PMID:20812323

Purisima, E., & Sulea, T. (2013). Solvation models: Theory and validation. *Current Pharmaceutical Design*, *20*(20), 3266–3280. PMID:23947651

Rarey, M., Kramer, B., Lengauer, T., & Klebe, G. (1996). A fast flexible docking method using an incremental construction algorithm. *Journal of Molecular Biology*, *261*(3), 470–489. PMID:8780787

Rejto, P. A., & Verkhivker, G. M. (1996). Unraveling principles of lead discovery: From unfrustrated energy landscapes to novel molecular anchors. *Proceedings of the National Academy of Sciences of the United States of America*, *93*(17), 8945–8950. PMID:8799133

Rognan, D., Lauemøller, S. L., Holm, A., Buus, S., & Tschinke, V. (1999). Predicting binding affinities of protein ligands from three-dimensional models: Application to peptide binding to class i major histocompatibility proteins. *Journal of Medicinal Chemistry*, *42*(22), 4650–4658. PMID:10579827

Ruvinsky, A. M. (2007). Role of binding entropy in the refinement of protein–ligand docking predictions: Analysis based on the use of 11 scoring functions. *Journal of Computational Chemistry*, *28*(8), 1364–1372. PMID:17342720

Ruvinsky, A. M., & Kozintsev, A. (2005). New and fast statistical-thermodynamic method for computation of protein-ligand binding entropy substantially improves docking accuracy. *Journal of Computational Chemistry*, *26*(11), 1089–1095. PMID:15929088

Salaniwal, S., Manas, E. S., Alvarez, J. C., & Unwalla, R. J. (2007). Critical evaluation of methods to incorporate entropy loss upon binding in high-throughput docking. *Proteins: Structure, Function, and Bioinformatics*, *66*(2), 422–435. PMID:17068803

Scharfe, M., Pippel, M., & Sippl, W. (2012). Development of target-biased scoring functions for protein-ligand docking. *Journal of Cheminformatics*, 4(Suppl 1).

Seifert, M. H. (2008). Optimizing the signal-to-noise ratio of scoring functions for protein-ligand docking. *Journal of Chemical Information and Modeling*, *48*(3), 602–612. PMID:18293951

Seifert, M. H. (2009). Targeted scoring functions for virtual screening. *Drug Discovery Today*, *14*(11), 562–569. PMID:19508918

Shoichet, B. K. (2004). Virtual screening of chemical libraries. *Nature*, *432*(7019), 862–865. PMID:15602552

Sippl, M. J. (1990). Calculation of conformational ensembles from potentials of mena force: An approach to the knowledge-based prediction of local structures in globular proteins. *Journal of Molecular Biology*, *213*(4), 859–883. PMID:2359125

Smith, R. D., Dunbar, J. B. Jr, Ung, P. M.-U., Esposito, E. X., Yang, C.-Y., Wang, S., & Carlson, H. A. (2011). CSAR benchmark exercise of 2010: Combined evaluation across all submitted scoring functions. *Journal of Chemical Information and Modeling*, *51*(9), 2115–2131. PMID:21809884

Sousa, S., Cerqueira, N., Fernandes, P., & Ramos, M. (2010). Virtual screening in drug design and development. *Combinatorial Chemistry & High Throughput Screening*, *13*(5), 442–453. PMID:20236061

Sousa, S., Ribeiro, A., Coimbra, J., Neves, R., Martins, S., Moorthy, N., & Ramos, M. et al. (2013). Protein-ligand docking in the new millennium–a retrospective of 10 years in the field. *Current Medicinal Chemistry*, *20*(18), 2296–2314. PMID:23531220

Tanaka, S., & Scheraga, H. A. (1976). Medium-and long-range interaction parameters between amino acids for predicting three-dimensional structures of proteins. *Macromolecules*, *9*(6), 945–950. PMID:1004017

Teramoto, R., & Fukunishi, H. (2007a). Supervised consensus scoring for docking and virtual screening. *Journal of Chemical Information and Modeling*, *47*(2), 526–534. PMID:17295466

Teramoto, R., & Fukunishi, H. (2007b). Supervised scoring models with docked ligand conformations for structure-based virtual screening. *Journal of Chemical Information and Modeling*, *47*(5), 1858–1867. PMID:17685604

Terp, G. E., Johansen, B. N., Christensen, I. T., & Jørgensen, F. S. (2001). A new concept for multidimensional selection of ligand conformations (Multiselect) and multidimensional scoring (multiscore) of protein-ligand binding affinities. *Journal of Medicinal Chemistry*, *44*(14), 2333–2343. PMID:11428927

Thomas, P. D., & Dill, K. A. (1996). Statistical potentials extracted from protein structures: How accurate are they? *Journal of Molecular Biology*, *257*(2), 457–469. PMID:8609636

Tsai, C.-J., Kumar, S., Ma, B., & Nussinov, R. (1999). Folding funnels, binding funnels, and protein function. *Protein Science*, *8*(6), 1181–1190. PMID:10386868

Tuccinardi, T. (2009). Docking-based virtual screening: Recent developments. *Combinatorial Chemistry & High Throughput Screening*, *12*(3), 303–314. PMID:19275536

Villoutreix, B. O., Eudes, R., & Miteva, M. A. (2009). Structure-based virtual ligand screening: Recent success stories. *Combinatorial Chemistry & High Throughput Screening*, *12*(10), 1000–1016. PMID:20025565

Wang, J., & Verkhivker, G. M. (2003). Energy landscape theory, funnels, specificity, and optimal criterion of biomolecular binding. *Physical Review Letters*, *90*(18), 188101. PMID:12786043

Wang, J., Wang, W., Huo, S., Lee, M., & Kollman, P. A. (2001). Solvation model based on weighted solvent accessible surface area. *The Journal of Physical Chemistry B*, *105*(21), 5055–5067.

Wang, J., Zheng, X., Yang, Y., Drueckhammer, D., Yang, W., Verkhivker, G., & Wang, E. (2007). Quantifying intrinsic specificity: A potential complement to affinity in drug screening. *Physical Review Letters*, *99*(19), 198101. PMID:18233118

Wang, J.-C., & Lin, J.-H. (2013). Scoring functions for prediction of protein-ligand interactions. *Current Pharmaceutical Design*, *19*(12), 2174–2182. PMID:23016847

Wang, R., Fang, X., Lu, Y., & Wang, S. (2004). The pdbbind database: Collection of binding affinities for protein-ligand complexes with known three-dimensional structures. *Journal of Medicinal Chemistry*, *47*(12), 2977–2980. PMID:15163179

Wang, R., Lai, L., & Wang, S. (2002). Further development and validation of empirical scoring functions for structure-based binding affinity prediction. *Journal of Computer-Aided Molecular Design, 16*(1), 11–26. PMID:12197663

Wang, R., & Wang, S. (2001). How does consensus scoring work for virtual library screening? an idealized computer experiment. *Journal of Chemical Information and Computer Sciences, 41*(5), 1422–1426. PMID:11604043

Xiang, Z., Soto, C. S., & Honig, B. (2002). Evaluating conformational free energies: The colony energy and its application to the problem of loop prediction. *Proceedings of the National Academy of Sciences of the United States of America, 99*(11), 7432–7437. PMID:12032300

Xue, M., Zheng, M., Xiong, B., Li, Y., Jiang, H., & Shen, J. (2010). Knowledge-based scoring functions in drug design. 1. developing a target-specific method for kinase-ligand interactions. *Journal of Chemical Information and Modeling, 50*(8), 1378–1386. PMID:20681607

Yan, Z., & Wang, J. (2012). Specificity quantification of biomolecular recognition and its implication for drug discovery. *Scientific Reports, 2*, 309. PMID:22413060

Yan, Z., & Wang, J. (2015). Optimizing the affinity and specificity of ligand binding with the inclusion of solvation effect. *Proteins: Structure, Function, and Bioinformatics*, 83(9), 1632-1642.

Yan, Z., Zheng, X., Wang, E., & Wang, J. (2013). Thermodynamic and kinetic specificities of ligand binding. *Chemical Science, 4*(6), 2387–2395.

Yin, S., Biedermannova, L., Vondrasek, J., & Dokholyan, N. V. (2008). Medusascore: An accurate force field-based scoring function for virtual drug screening. *Journal of Chemical Information and Modeling, 48*(8), 1656–1662. PMID:18672869

Yuriev, E., & Ramsland, P. A. (2013). Latest developments in molecular docking: 2010–2011 in review. *Journal of Molecular Recognition, 26*(5), 215–239. PMID:23526775

Yusuf, D., Davis, A. M., Kleywegt, G. J., & Schmitt, S. (2008). An alternative method for the evaluation of docking performance: RSR vs RMSD. *Journal of Chemical Information and Modeling, 48*(7), 1411–1422. PMID:18598022

Zilian, D., & Sotriffer, C. A. (2013). Sfcscore rf: A random forest-based scoring function for improved affinity prediction of protein–ligand complexes. *Journal of Chemical Information and Modeling, 53*(8), 1923–1933. PMID:23705795

Zou, X., Sun, Y., & Kuntz, I. D. (1999). Inclusion of solvation in ligand binding free energy calculations using the generalized-born model. *Journal of the American Chemical Society, 121*(35), 8033–8043.

KEY TERMS AND DEFINITIONS

Benchmark: A set of diverse protein-ligand complexes used for testing the scoring functions or docking methods.

Conformational Space: The degrees of freedom involved in the process of placing a ligand onto the protein.

Docking: The process of searching the possible locations for a ligand onto the binding site of a protein.

Sampling: The exploration of the conformational space given by the movements of the docking partners.

Scoring: The measurement of the fitness for the interactions of a ligand with a protein.

Solvation: The process that the water molecules surround the protein and the ligand.

Specificity: The feature of a ligand to recognize its own binding partner and to discriminate against others.

Virtual Screening: Computationally search the library of small molecules to identify the lead compounds which most favorably bind with a target protein.

Chapter 10
Molecular Docking Technique to Understand Enzyme–Ligand Interactions

Kailas Dashrath Sonawane
Shivaji University, India

Maruti Jayram Dhanavade
Shivaji University, India

ABSTRACT

Molecular docking has advanced to such an extent that one can rapidly and accurately identify pharmaceutically useful lead compounds. It is being used routinely to understand molecular interactions between enzyme and ligand molecules. Several computational approaches are combined with experimental work to investigate molecular mechanisms in detail at the atomic level. Molecular docking method is also useful to investigate proper orientation and interactions between receptor and ligand. In this chapter we have discussed protein-protein approach to understand interactions between enzyme and amyloid beta (Aβ) peptide. The Aβ peptide is a causative agent of Alzheimer's disease. The Aβ peptides can be cleaved specifically by several enzymes. Their interactions with Aβ peptide and specific enzyme can be investigated using molecular docking. Thus, the molecular information obtained from docking studies might be useful to design new therapeutic approaches in treatment of Alzheimer's as well as several other diseases.

BACKGROUND

Large numbers of three-dimensional structures of biomolecules are being uploaded in the various structural databases. These 3-D structures along with molecular models may be used in the docking process to understand proper interactions between receptor and ligand molecules. Molecular docking method is useful to investigate proper orientation and interactions between protein-protein, protein-ligand and DNA-protein molecules. In this chapter we have discussed protein-protein approach to understand in-

DOI: 10.4018/978-1-5225-0115-2.ch010

teractions between human cathepsin B and *Aβ* peptide a causative agent of Alzheimer's disease with the help of AutoDock. Docking methodology and procedure has been discussed with proper figures.

INTRODUCTION

Molecular docking has become a useful technique in structure-based drug design (Sousa, 2006). We can predict the preferred orientation of receptor and ligand bound to each other to form a stable complex (Lengauer & Rarey, 2006). The molecular docking is a multistep process which starts with the representation followed by docking algorithms to identify best matched conformation to receptor structure. The docking algorithms complemented by scoring functions which is used for evaluation of interactions between ligand and receptor based on shape and electrostatic complementarities (Kitchen, et al., 2004). The two main features of molecular docking technique are conformation search algorithm and scoring function estimates the binding affinity between molecules. There are three types of docking such as protein-protein (Sonawane & Barage, 2004; Tseng, et al., 2007; Barage, et al., 2014), protein-ligand (Jalkute, et al., 2014; Jalkute, et al., 2015) and DNA-protein (Parulekar, et al., 2013) which may apply depending upon the receptors and ligands used in the docking experiments. There are two types of docking approaches used such as rigid docking and flexible docking. In rigid docking approach both the protein and ligand are considered as rigid bodies whereas in the flexible-ligand docking, receptor and ligand kept flexible. Most of the molecular docking programs treat protein as a rigid molecule and ligand as a flexible molecule. For molecular docking studies several docking programs such as AutoDock (Morris, et al., 2009), HEX (Macindoe, et al., 2010), PatchDock (Duhovny, et al., 2002; Schneidman-Duhovny, et al., 2005), Ludi (Böhm, et al., 1992), Glide (Halgren, et al., 2004), etc. are being used with appropriate docking algorithms and scoring functions.

In this chapter to understand the protein-protein docking we have taken an example of docking between cathepsin B from human with *Aβ* peptide (Musil, et al., 1991; Crescenzi, et al., 2002; Dhanavade, et al., 2013; Dhanavade, et al., 2014). Amyloid plaque formation is the pathological hallmark in the Alzheimer's disease (Hardy & Selkoe, 2002; Glenner & Wong, 1984). The Aβ peptides (1–42) are the main constituent of amyloid plaques in Alzheimer's disease (Gouras, et al., 1998; Masters et al., 1985; Tanzi, et al., 2005). There are various enzymes such as IDE, NEP, ECE, ACE, MMP-9, and plasmin are known to degrade Aβ peptides (Iwata, et al., 2001; Qui, et al., 1998; Tucker, et al., 2000; Yin, et al., 2006; Zou, et al., 2007; Cimerman, et al., 1999). Similarly, certain bacterial enzymes like aminopeptidase from *Streptomyces grieus* KK565 (SGAK), Nattokinase from *Bacillus subtilis* Natto, and Angiotensin Converting Enzyme (ACE) from *Stigmatella aurantiaca* has potential to degrade Aβ peptide (Jalkute, et al., 2015; Yoo, et al., 2010; Hsu, et al., 2009). The Aβ peptide degradation is well studied experimentally but their cleavage mechanism is yet to be understood properly. Hence, in such situation molecular docking studies between respective enzyme and ligand plays an important role to understand molecular mechanism in detail at the molecular level. The detailed procedure of docking is explained and then the results are analyzed to understand the degradation of Aβ peptide by cathepsin B from human. Finally the importance of molecular docking studies is discussed.

MATERIALS

Downloading AutoDock and AutoGrid

Download freely available version of AutoDock and AutoGrid software (Morris, et al., 2009) for Windows operating system from the AutoDock webpage (http://autodock.scripps.edu/downloads/autodock-registration/autodock-4-2-download-page/). Install the AutoDock and AutoGrid as per the installation instructions.

Downloading the AutoDock Tools (Version 1.5.4, Version 3)

The AutoDock Tools (ADT) (mgltools_win32_1.5.6) is a tool required for the preparation and visualization of molecules along with AutoDock and AutoGrid. This mgltools_win32_1.5.6 has been made for Windows which can be downloaded from link: http://mgltools.scripps.edu/downloads. After downloading it can be installed as per the installation instructions (Morris, et al., 2009).

Other useful software for molecular docking studies:

- **MODELLER 9v7:** This software is developed by Sali and Blundell in 1993 (Sali & Blundell, 1993) and being used routinely to predict three-dimensional structures of proteins using homology modeling technique.
- **AutoDock:** This is developed by Morris and coworkers in 2009 (Morris, et al., 2009) and being used in the docking studies.
- **RasMol:** This is useful to visualize and analyze the 3D structures. This software is developed by Roger Sayle and coworkers in the year 1995 (Sayle, & Milner-White, 1995).
- **Chimera:** The Chimera is developed by the Pettersen and coworkers in the year 2004 (Pettersen, et al., 2004) and is useful for the structure visualization, analysis of models as well as to prepare attractive images of docked complex.

The above discussed software's are freely available and may be used for the basic study of molecular docking but besides this various online servers are also available for homology modeling process.

METHODS

- **Homology Modeling:** In order to start molecular docking process, three dimensional structures of receptor and ligand must be downloaded from RCSB and PUBCHEM respectively. If the 3D structure of receptor is not available then it has to be modeled by using MODELLER 9V software. In this current chapter we have considered human Cathepsin B (Musil, et al., 1991) as a receptor. Its 3-D structure (2IPP.pdb) has been directly extracted from RCSB Protein Data Bank (Musil, et al., 1991). For protein-protein docking, the amyloid beta protein, 1IYT.pdb (Macindoe, et al., 2010) has been taken as a ligand and retrieved from RCSB. If ligand is a chemical compound then its structure may be extracted from NCBI PUBCHEM.

- **Virtual Screening:** If receptor is an enzyme and its 3-D structure data available in Protein Data Bank (PDB) then one can use structural information of active site residues of receptor to search suitable drug molecules from ZINC database (Irwin, et al., 2005) which can bind effectively to the receptor. Dockblaster program has a flexible-ligand sampling algorithm which superimposes atoms of the docked molecule as per the supplied binding site which represent positions to bind ligand atoms (Irwin, et al., 2009). Basically the selection of ligand was performed by bin size; distance tolerance up to 2.0 Å, each ligand configuration was scored for electrostatic, van der Waals interaction complementarity and corrected for ligand desolvation. The screening of ligand is done in ZINC database of commercially available molecules, which contains 3 million compounds available ready to dock (Irwin, et al., 2005).

Molecular Docking

Molecular docking study mainly aims to predict the intramolecular interactions formed between two or more molecules. Molecular docking has been described as a useful tool in structure-based drug design and discovery (Sousa, et al., 2006). Earlier, significant amount of work has been made in the molecular modeling and drug discovery (Blaney & Dixon, 1993; Jones & Willet, 1995; Lybrand, 1995; Rosenfeld, et al., 1995; Gschwend, et al., 1996; Lengauer & Rarey, 1996). Molecular docking is a method to predict preferred orientation of receptor and ligand interacting to each other to form a stable complex (Lengauer & Rarey, 1996). Docking method considers only translational and orientational degrees of freedom of the ligand with respect to receptor (Kuntz, et al., 1982). However, the molecule's geometry may change upon binding into the active site of an enzyme, so now most of the method considers ligand as a flexible molecule. Proteins are not rigid structures therefore ideally it should be flexible during the docking simulations. However, introducing flexibility in the macromolecular proteins during docking procedure is quite difficult because of the requirement of the huge computational power for such processes. Hence, docking studies are performed using a flexible ligand and a rigid receptor protein.

Docking is multistep process which starts with molecular representation followed by docking algorithms to identify best matched conformation of receptor and ligand complex. The docking algorithms based on shape and electrostatic complementarities with scoring functions are used to evaluate the interactions between receptor and ligand (Kitchen, et al., 2004). The AutoDock (Morris, et al., 2009) was used to predict protein-protein and protein-ligand interactions with an appropriate docking algorithms and scoring functions. The search algorithms are classified into three types such as systematic, stochastic and deterministic search algorithms (Brooijmans & Kuntz, 2003). The grid value for all degrees of freedom in molecule can be explored by systematic algorithms in a combinatorial way. The incremental construction and conformational search are carried out by some systematic search algorithms (Halperin, et al., 2002). The random changes to single ligand or to the population of ligand can be done by stochastic search algorithm operations. The algorithms such as Monte Carlo methods, genetic algorithms and tabu search are included in stochastic algorithms. The deterministic search algorithms started with known initial state up to the state of lesser energy at which the termination criteria is met (Halperin, et al., 2002). Monte Carlo methods, Genetic algorithms and Lamarckian genetic algorithms have been adopted in AutoDock, a flexible docking program (Morris, et al., 2009).

Molecular Docking Procedure

The molecular docking methodology can be divided into three steps: file preparation, docking, and analysis. Before preparing the receptor, the bound ligand and water molecules must be deleted from the crystal structure. Firstly, the addition of polar hydrogen's, computing the partial charges, and assigning of AutoDock atom types has to be performed. After that two files such as rigid.pdbqt and flexible.pdbqt are generated from the receptor file. The flexible.pdbqt file is generated by keeping the active site residues of receptor as flexible residues while remaining residues are taken as a rigid in rigid.pdbqt file. Then the parameter file must be prepared to run the AutoGrid. The docking grid should be specified to constrain the search area for the docking. The grid parameter file (gpf) will be generated to show the macromolecule location, size and atoms types to compute grid map. While running AutoGrid, the parameters will be taken from the gpf file and generated grid map will be used further in AutoDock. AutoDock will generate a docking parameter file (dpf) for the ligand and then modified with the parameters for Lamarckian genetic algorithm. When all the files such as rigid.pdbqt, flexible.pdbqt and ligand.pdbqt which are needed for the docking process are generated, then one can start actual docking process. The time required for docking depends on various factors like size and number of rotatable bonds in the ligand, the docking grid size, the genetic algorithm parameters, and the number of runs. After the completion of docking process, AutoDock will give output in docking log file (dlg) which contains docked structures and binding free energies. The pose of the ligand, structural coordinates, cluster size and binding free energies can be obtained after analyzing the dlg file. The docked complex can be further converted into pdb file format and analyzed by different 3D structure visualizing software such as RasMol (Sayle & Milner-White, 1995) and Chimera (Pettersen, et al., 2004). In order to understand structural stability of docked complex molecular dynamics (MD) simulations can be carried out using available packages such as GROMACS (Spoel, et al., 2005), CHARMM (Brooks, et al., 2010) and AMBER (Case, et al., 2005).

Preparing the Macromolecule

For the preparation of macromolecule the stepwise procedure is given below.

- **Start AutoDock Tools:** Open the macromolecule structure in ADT: Click on File-----Read Molecule and then get your interested molecule e.g. 2IPP.pdb in the working area of ADTviewer.

Add hydrogens to the receptor file: Click on Edit-----Hydrogens-----Add. This will add hydrogen atoms in your interested PDB file. Then save pdb file with hydrogen atoms in the same folder; File-----Save-----Write PDB-----Receptor.pdb.

- **Preparing the Rigid and Flexible Residue File:** In this part the rigid and flexible residue files are prepared which has to be saved in the PDBQT file format. The active site residues of receptor are kept as flexible residues while the other residues are considered as rigid; Click on Flexible Residues-----Input-----Choose Macromolecule. A window will appear then select Receptor.pdb-----Select Molecule (Figure 1). This will compute Gasteiger partial charges, assigning AutoDock4 atom types to receptor and merged nonpolar hydrogens (Figure 1).

Figure 1. Selecting the receptor molecule

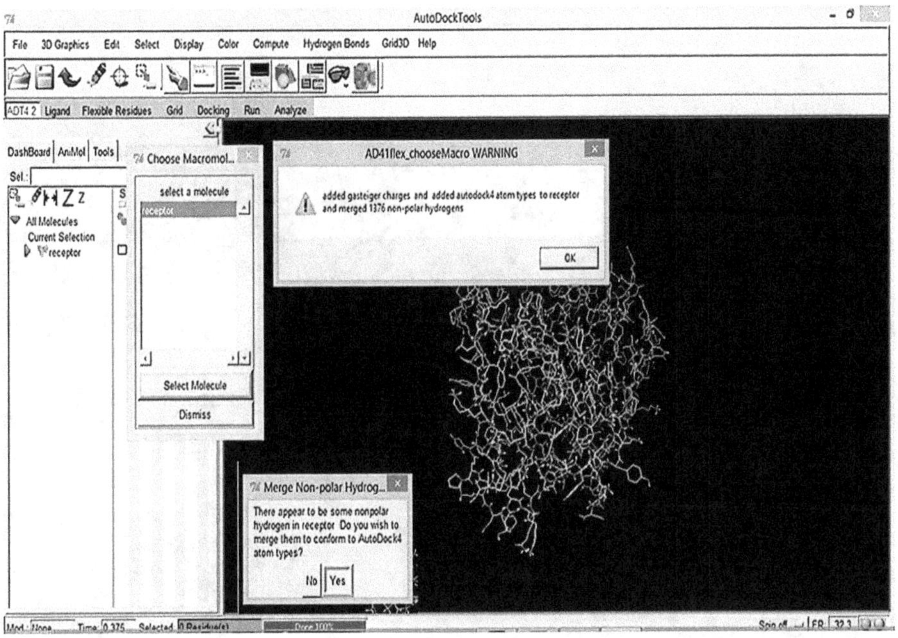

- **Selecting and Saving Flexible Residues:** The information about the active site residues are required to prepare the flexible residues file. The active site residues are considered as flexible residues.

Flexible Residues-----Input-----Choose Macromolecule. Then the receptor is selected. Then select flexible residues by clicking on Select-----Select From String-----Molecule List in which we have to select molecule (Receptor) then in the Residue column enter the names of active site residues (CYS29, HIS110, HIS111) then click on Add and then Dismiss (Figure 2). After that for setting torsions, click on Flexible Residues-----Choose Torsions in Currently Selected Residues. This will give the information about the torsion count in a separate window (Figure 3). Finally, the Flexible residue file (Flexible.pdbqt) and rigid residue file (Rigid.pdbqt) are saved as;

Flexible Residues-----Output-----Save Flexible PDBQT and

Flexible Residues-----Output-----Save Rigid PDBQT (Figure 4).

- **Preparing the Ligand File for AutoDock:** In AutoDock the ligand files are written in special keywords which can be recognized by AutoDock. The keywords such as ROOT, ENDROOT, BRANCH, and ENDBRANCH establish a "torsion tree" object or *torTree* which has a *root* and *branches*. The root is nothing but the rigid set of atoms, whereas the branches are rotatable groups of atoms connected to the rigid root. The keyword TORSDOF signals the number of torsional degrees of freedom in the ligand. The TORSDOF for a ligand is the total numbers of possible torsions in the ligand minus the number of torsions that only rotate hydrogens. TORSDOF is mainly used during calculation of the change in free energy caused due to the loss of torsional degrees of freedom upon binding. In the current example the ligand which is a patch of Aβ peptide (1IYT.pdb) is used for the docking. For the preparations of ligand.pdbqt following steps can be performed.

Figure 2. Selecting the active site residues

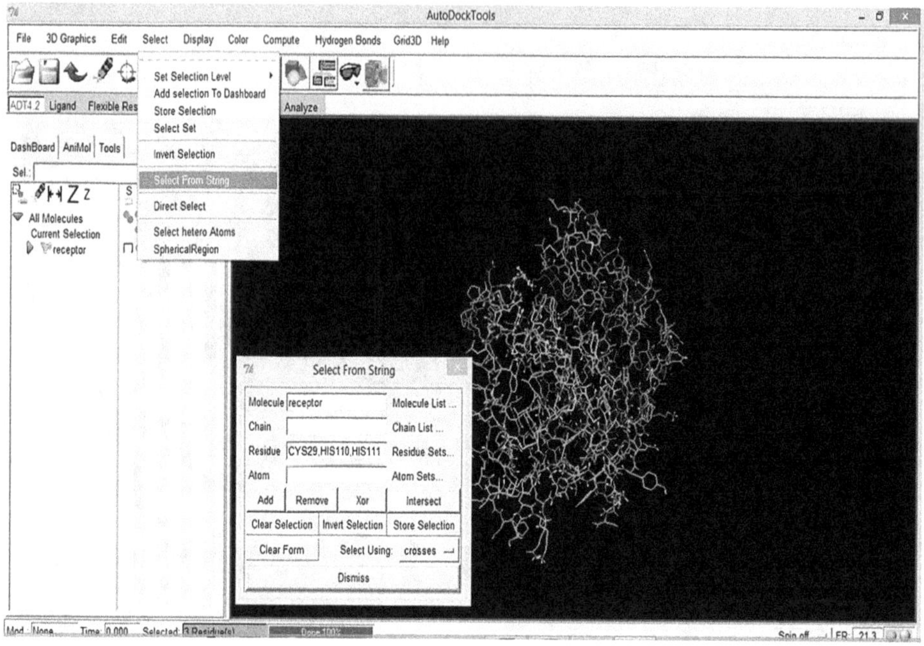

Figure 3. Choose torsions in active site residues

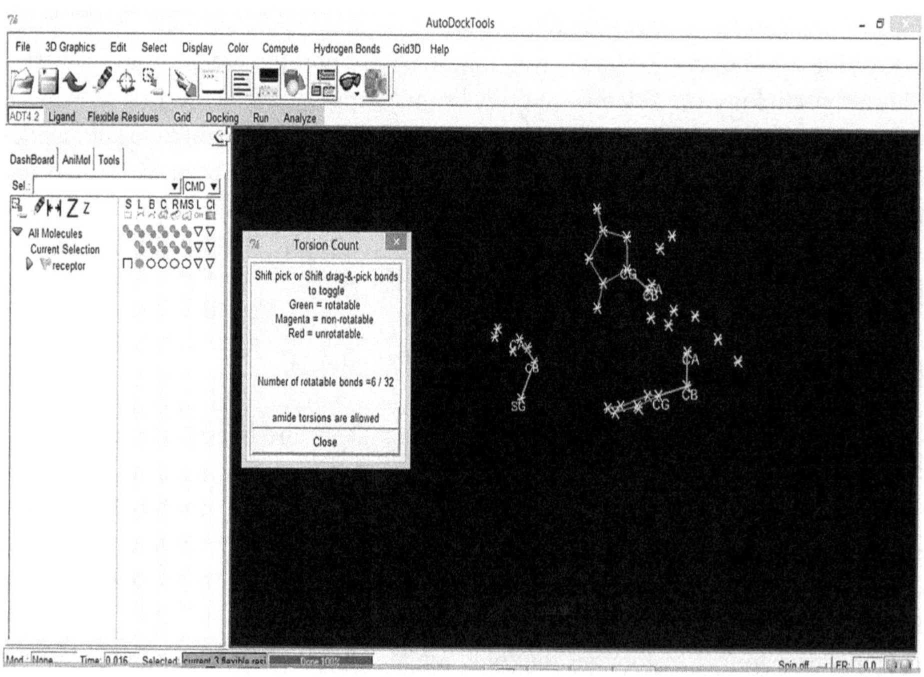

Figure 4. Save rigid and flexible files

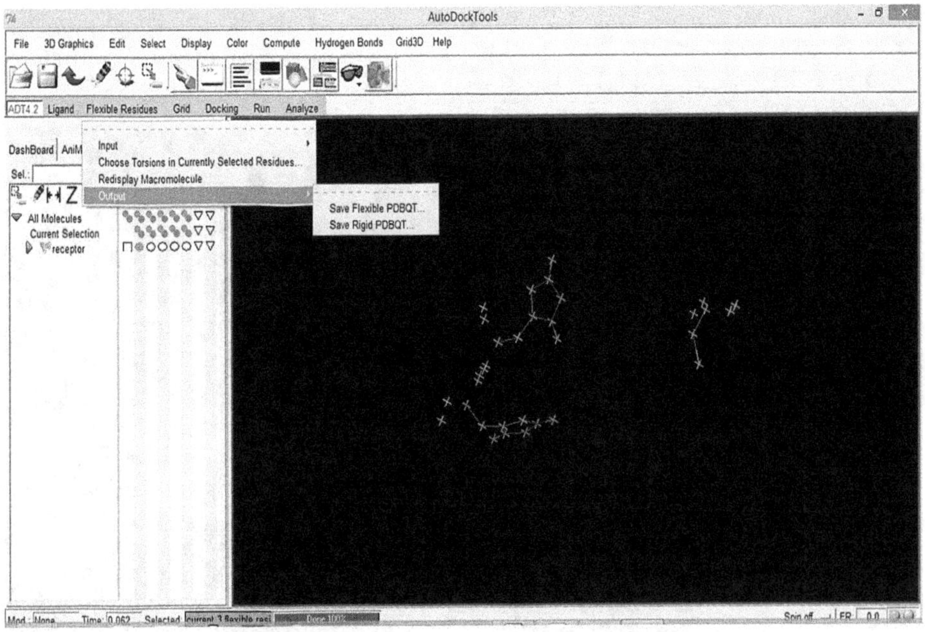

Click on Ligand-----Input-----Open to choose ligand PDB file i.e. IYT.PDB as shown in Figure 5. After choosing ligand.pdb click on Select Molecule for AutoDock4. Then the window will appear, in which the summary for ligand.pdb will be displayed with the ligand information as can be seen in Figure 5 and then click on OK.

Then following steps are carried out to detect root.

Ligand-----Torsion Tree-----Detect Root

Ligand-----Torsion Tree-----Hide Root Expansion

Ligand-----Torsion Tree-----Show/Hide Root Marker

ADT determines the atom which fits its idea of the best root and mark it with a green sphere. This best root is nothing but the atom in the ligand having small largest subtree.

Ligand-----Torsion Tree-----Choose Torsions

These operations will open the Torsion Count widget which displays the number of currently active bonds showing; Red-non rotated bonds, Purple-currently marked as inactive, Green- Currently active. Those bonds which are in cycles cannot be rotated. ADT determines the bonds which could be rotated ('possibleTors'). In ADT one can see rotatable bonds ('activeTors') by inactivating the other bonds. We can also toggle the activity of a bond or group of bonds by picking them in the viewer. Alternatively, buttons on this widget let you toggle the activity of a type of bonds like 'peptide bonds', 'amide bonds', 'bonds between selected atoms' or 'all rotatable bonds'. To do this we can click on Make all active bonds non-rotatable then click on Make all rotatable bonds rotatable. It also states that 11 out of 32 bonds on the ligand are rotatable. Click on Done in the Torsion Count window to exit (Figure 6).

Figure 5. Selecting the ligand Molecule

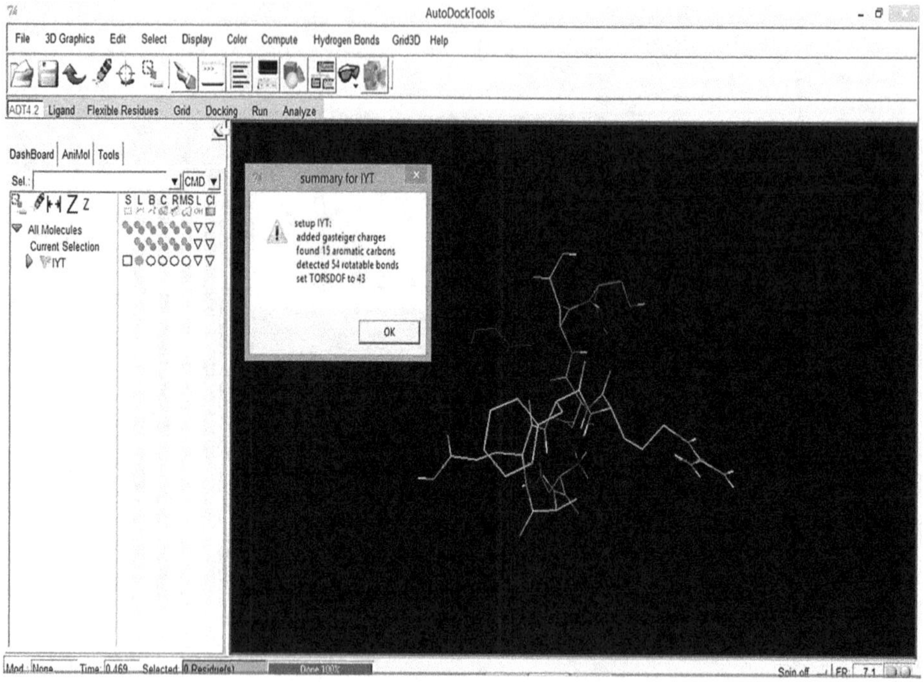

Figure 6. Choose Torsion count in ligands

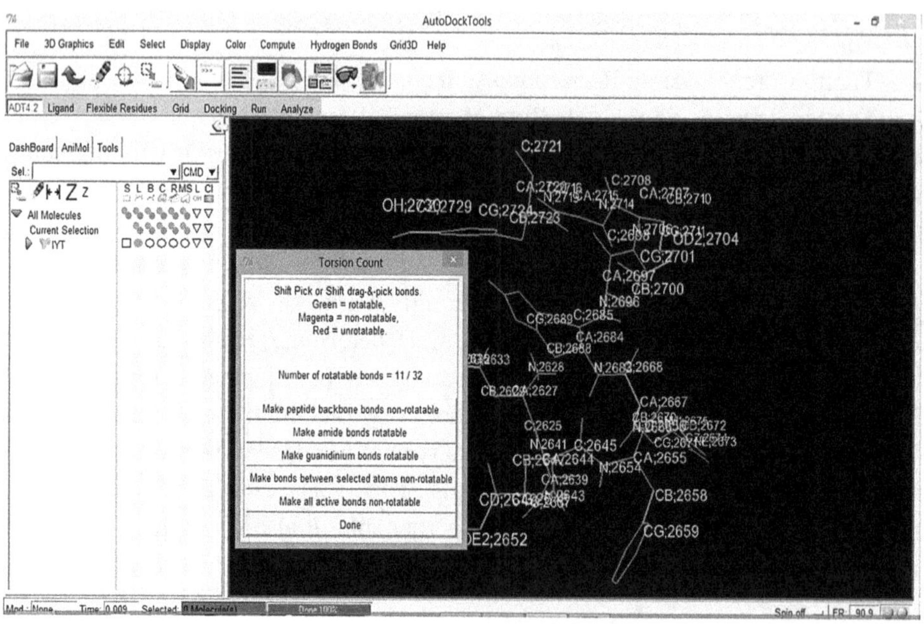

- **Saving the Ligand PDBQT (IYT.pdbqt):** To save ligand PDBQT file go to Ligand-----Output-----Save as PDBQT----- ligand.PDBQT. After these steps a ligand is prepared for the further docking procedure. This ligand.PDBQT file is then used for the actual docking process.
- **Preparing the Grid Parameter File:** Grid parameter file (gpf) is required to compute the types of maps, the location and extent of those maps as well as specific pair-wise potential energy parameters using AutoGrid. One map is calculated for each element present in the ligand with an electrostatic map. The following steps are performed for the preparation of gpf file.
- **Open the Macromolecule File RIGID.PDBQT in ADT:** Click on File-----Read Molecule. This will open a file browser in the directory where ADT was started. Select the file, RIGID.PDBQT, and click on Open. Then Click on Grid-----Choose Molecule-----Choose the RIGID.PDBQT file. After choosing the RIGID.PDBQT a window will appear asking to preserve input receptor charges instead of adding Gasteiger charges. Then Click on Yes to keep the charges. Also a WARNING box will then pop-up, showing the initialization of the pdbqt file, stating that the autodock_elements and gasteiger charges have already been added. Then click on OK.
- **Selecting the Ligand and Generate Maps:**

Click on Grid-----Set Map Types-----Open Ligand-----ligand.PDBQT.

Then Grid-----Set Map Types-----Choose Ligand. Choose Ligand window will appear. Click on ligand.PDBQT and click on Select Ligand (assuming the ligand file is already open). After selecting the ligand, the gpf will specify AutoGrid to generate maps for atom types present in the ligand.

- **Specify the Grid Box Dimensions, Spacing and Center:** The grid box can be constructed by using this step; Click on Grid-----Grid Box. The window will appear with Grid Options. Specify the grid box center where ligand is located and the grid Box size such as the active site residues should be in the box. In the Grid Options window, click on the Center tab and select Center on ligand. In the current window the grid box center is now located with dimensions such as -17.743, 66.133, -53.248. The size of the grid box, as shown in Fig 19, is 48 × 44 × 54 using a 0.375 Å resolution grid. For saving the modified grid option parameters, click on file and select close saving current (Figure 7).

Save the grid parameter file (grid.gpf): Click on-----Grid-----Output-----Save GPF. A file browser will appear in which file name like grid.gpf should be saved.

Edit the grid parameter file in a text editor. The Edit option is used to change required parameters of gpf file.

- **Running AutoGrid to Create Grid Maps:** To create the grid map for macromolecule run the AutoGrid: The AutoGrid must run in the same directory in which the ligand, macromolecule and grid parameter files are located. AutoGrid can be accessed by using ADT or command line. For running in ADT, click on

Run-----Run AutoGrid. After that a window will appear. The Macro Name and Host Name will specify your local machine by default. The location of AutoGrid executable must be specified by the program Pathname. The grid parameter file (gpf) specifies the parameter filename. The Log Filename is specifying the grid log file for the AutoGrid run. By default the name of the log file will be same as

Figure 7. Specifying the grid box dimensions

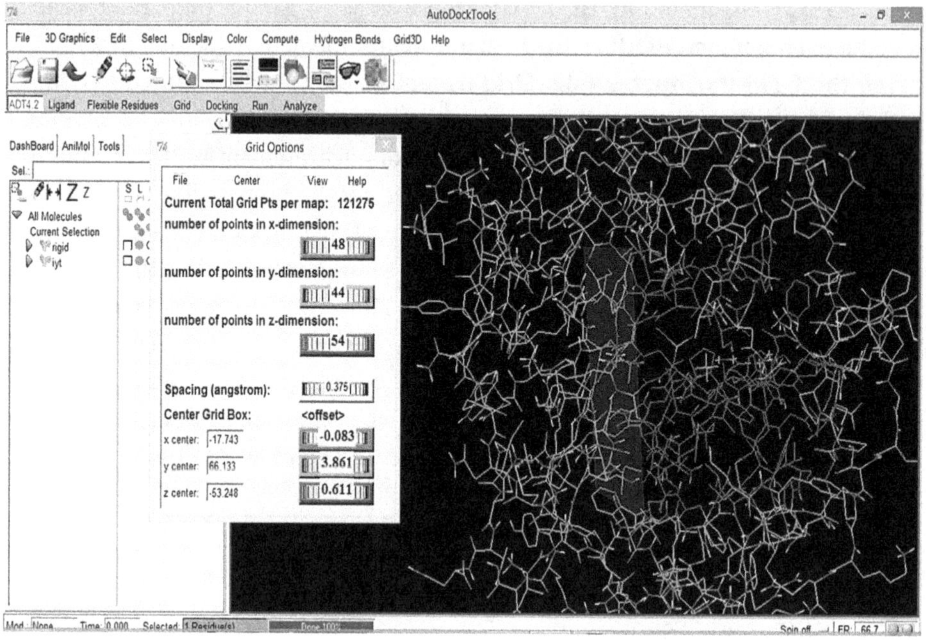

the grid parameter file, except with the extension.glg. The nice level for the remote jobs is also specified. The commands can be executed by running in cmd for windows. Click on Launch to start the job. We can see the status of job in the AutoDock Process Manager window which will appear after running the job. Alternatively, AutoGrid can be run via command line by typing: autogrid4 –p grid.gpf –l grid.glg& (Figure 8).

After completion of AutoGrid, the grid maps will be created for each specified atom type, electrostatic potential and desolvation energies. Also, other files created such as xyz file which specifies the grid box size, and the fld file which has other information about grid. The summary from the AutoGrid run is given in a grid log file (glg).

- **Creating the Docking Parameter File and Running AutoDock:** The docking parameter file (dpf) is necessary to run the final docking step. The files such as ligand.pdbqt, rigid.pdbqt and flexible.pdbqt are needed for docking should be placed in same directory. To prepare docking parameter file (dpf) firstly open all the files required for docking in ADT.

Click on-----Read Molecule-----Select RIGID.PDBQT-----Open.
Click on-----Read Molecule-----Select FLEXIBLE.PDBQT-----Open.
Then for selecting the rigid and flexible files for docking
Click on Docking-----Set Rigid Filename
Click on Docking-----Set Flexible Residue Filename
Finally open the ligand in the ADT
Click on Docking-----ligand-----Open
The search parameters are selected as;

Figure 8. Running AutoGrid command

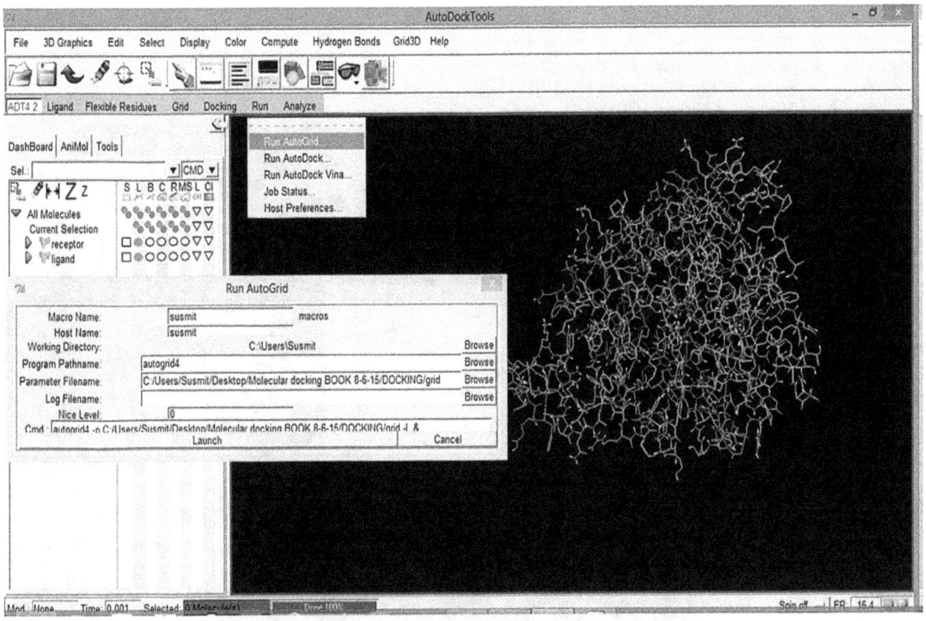

Docking-----Search Parameters-----Genetic Algorithm
The default docking parameters are used
Docking-----Docking Parameters-----Use Defaults-----Accept
Finally saving the docking parameter file (dpf)
Docking-----Output-----Lamarckian GA (4.2) -----dock.dpf
Save the docking parameter file (dock.dpf) in the same directory where all other files are generated.

- **Run the AutoDock:** AutoDock can be accessed by using ADT or command line. For running in ADT, click on Run-----Run AutoDock (Figure 9). Alternatively, AutoDock can also be run via command line by typing: autodock4 –p dock.gpf –l dock.dlg&
- **Analyzing Docking Results:** In AutoDock the docking results will be written in the docking log file (dlg).
  Firstly open the docking log file (dlg) in the ADT.
  Click on Analyze-----Docking-----Open-----dock.dlg. This will open the docking file.
  Open macromolecule in ADT
  Click on Analyze-----Macromolecule-----Open-----RIGID.PDBQT.
  Selection of lowest energy conformation

The lowest energy conformation must be selected from generated docked complexes. In this current procedure we have carried out 10 runs in which 10 different docked conformations were generated so we can see the energy of each conformation and then select the lowest energy conformation (Figure 10).

Figure 9. Running AutoDock command

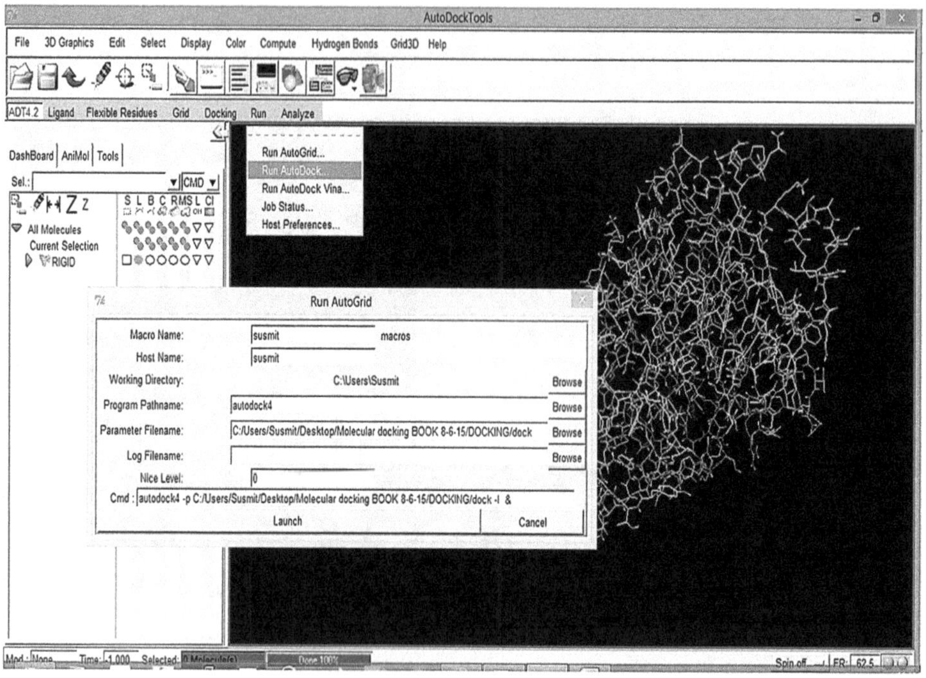

Figure 10. Selection of the lowest energy conformation

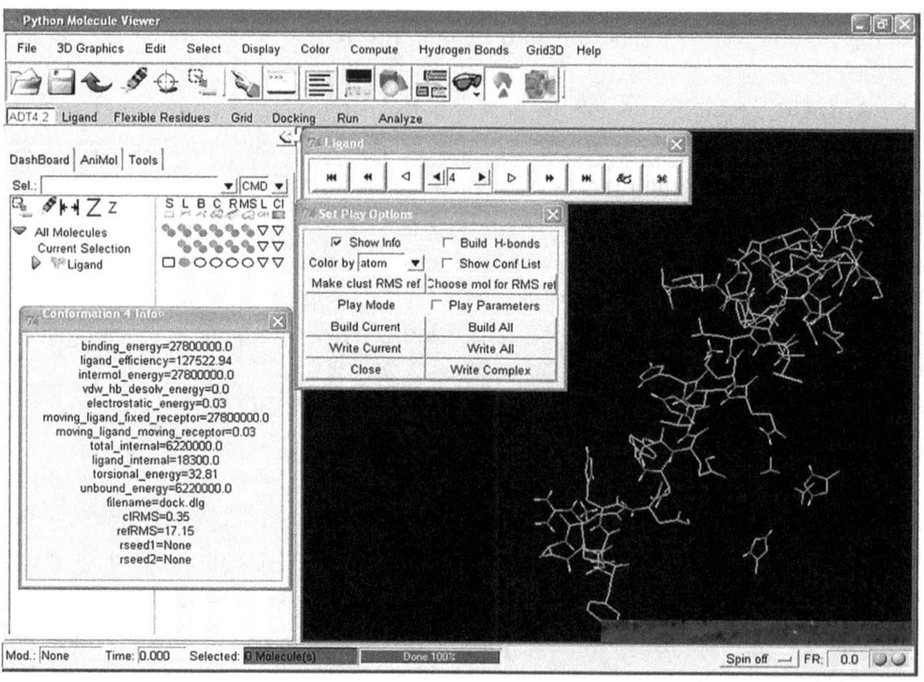

Click on Analyze-----Conformation-----Play

When this operation is performed a Conformation Player (CP) gets opened in which one can examine the docked conformations. The Conformation Player has a current conformations list (its sequence) and a current ID list. The sequence list consists of all the docked conformations, ordered by run. The ID list is [0, 1, 2, 3…10]. Then select the conformation on the basis of energy.

- **Saving the Docked Complex:** Finally, the docked complex with lowest energy can be saved by clicking on Write complex, and then the browser window will appear. After that select the location to save the docked complex and also change the file type and save the file in pdb file format as Docked.pdb (Figure 11). Similarly, other ligand types such as DNA and chemical compounds can be used for molecular docking studies as per the same procedure as explained above. The detailed methodologies for such docking studies have been given in (Sonawane, & Barage, 2014; Tseng, et al., 2007; Barage, et al., 2014; Jalkute, et al., 2014; Jalkute, et al., 2015; Parulekar, et al., 2013; Dhanavade, et al., 2013, Dhanavade, et al., 2014).
- **Visualizing the Docked Complex in Various Visualizing Software:** The docked complex can be visualized in the different visualizing software such as Chimera and RasMol. The images can be saved in the different image formats like TIFF, JPEG and PNG etc. The final docked complex between Cathepsin B and amyloid beta peptide which is a causative agent of Alzheimer's disease can be seen in the Figure 12. Then the docked complex can be taken further for MD simulation study to understand structural stability as explained in earlier studies (Dhanavade, et al., 2013).
- **Analyzing the Final Docked Complex:** For the analysis of docked complex various software having molecular graphics such as RasMol (Sayle & Milner-White, 1995), Chimera software (Pettersen, et al., 2004), etc. can be used. The interactions between the receptor and ligand can be investigated

Figure 11. Saving the docked PDB file

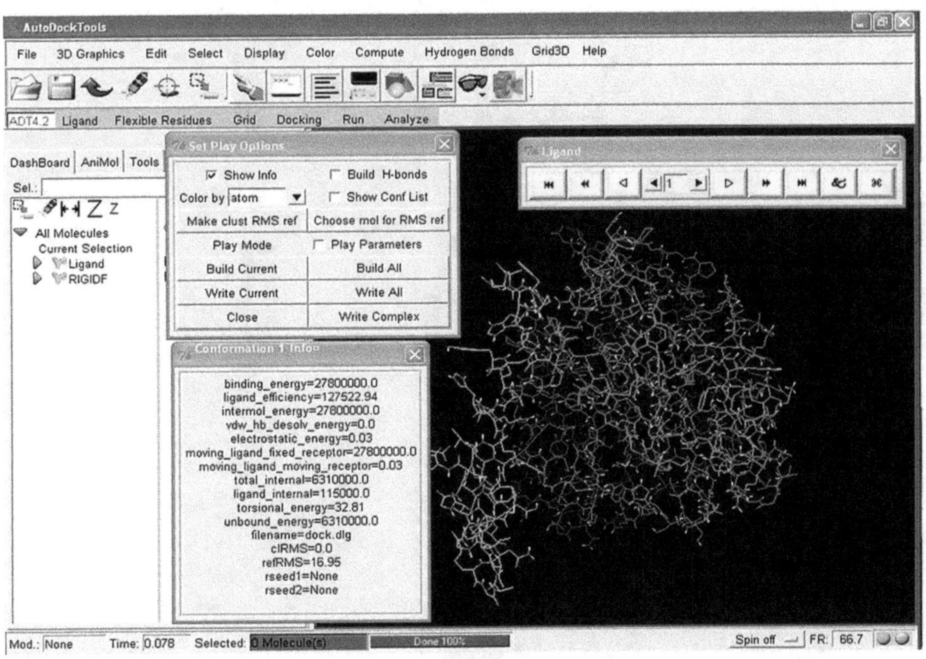

Figure 12. The docked complex between Cathepsin B and amyloid beta peptide

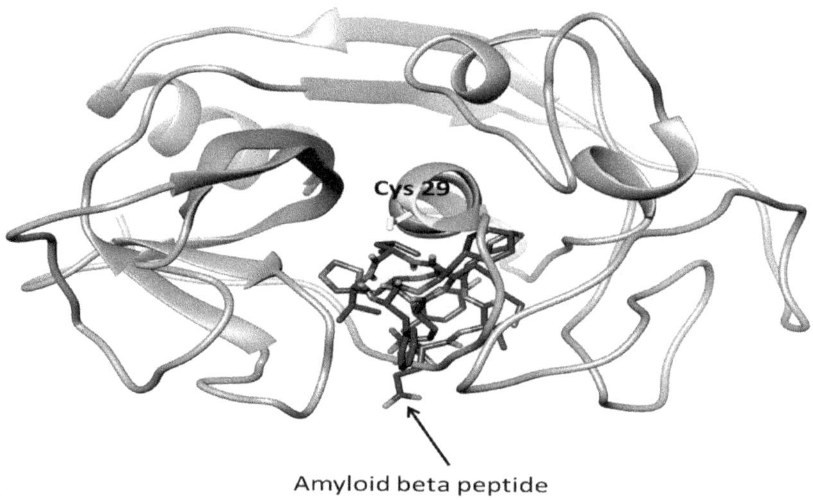

which would be helpful to make final conclusions. In this chapter we have considered a case study done by Dhanavade and coworkers in 2013 (Dhanavade, et al., 2013) in which docking has been carried out between cathepsin B (receptor) with *Aβ* peptide (ligand) as an example. The receptor cathepsin B (2IPP.pdb) which is a human enzyme has ability to degrade Aβ peptide. Hence, the molecular docking was carried out between the cathepsin B (receptor) with Aβ peptide (ligand) to find the active site residues of cathepsin B (receptor) involved in the interactions with the Aβ peptide (ligand).

Figure 13. Showing the hydrogen bonding between the cathepsin B (receptor) with Aβ peptide (ligand)

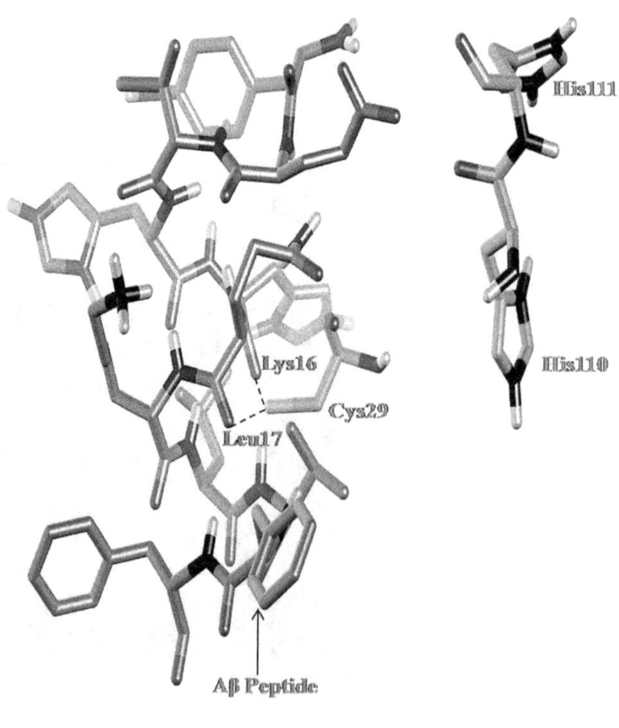

Finally, on the basis of interactions it is easy to find the active site residues of cathepsin B (receptor) which take part in the cleavage of Aβ peptide. The docked complex was analyzed by using Chimera (Pettersen, et al., 2004) and the results revealed that there is strong hydrogen bonding between the cathepsin B (receptor) with Aβ peptide (ligand) (Figure 13) (Dhanavade, et al., 2013). This hydrogen bonding between active site residue Cys 29 of cathepsin B with Lys 16 and Leu 17 of Aβ peptide indicate that cathepsin B can cleave Aβ peptide in between Lys 16-Leu 17 (Dhanavade, et al., 2013). There is experimental report available which confirm the degradation of Aβ peptide by human cathepsin B (Mueller-Steiner, et al., 2006). Hence, to study the exact cleavage mechanism of Aβ peptide (ligand) by cathepsin B the molecular docking technique has been found useful. Likewise, interactions between protein-protein, protein-ligand or drug, and protein-DNA can be studied using molecular docking technique.

CONCLUSION

The docking procedure and methodology discussed in this chapter would be useful for beginners to get into the field. The molecular modeling studies along with experimental findings could be useful to understand exact cause of several diseases in-detail at the molecular level. Such studies might be useful to discover new lead molecules to combat neurological and several other diseases.

Future Work

The docking results must consider as a good starting point for further experimental design. MD simulations studies are must to check the stability and dynamic behavior of docked complex. Then binding free energy calculations of docked complex might be useful to get exact idea about the interacting residues. The continuous increase in computational power can be helpful to expand the practical applications of molecular models. These results are then compared with experimental findings. In future, experimental studies coupled with docking and MD simulations would be more promising to discover new lead molecules.

ACKNOWLEDGMENT

KDS is thankful to University Grants Commission, New Delhi for providing financial support. KDS is also gratefully acknowledged to Department of Science and Technology, New Delhi for providing financial grants. MJD is thankful to DST for providing fellowship as research assistance under the scheme DST-PURSE. Authors are thankful to Computer Centre, Shivaji University, Kolhapur for providing the computational facility.

REFERENCES

Barage, S. H., Jalkute, C. B., Dhanavade, M. J., & Sonawane, K. D. (2014). Simulated interactions between endothelin converting enzyme and Aβ peptide: Insights into subsite recognition and cleavage mechanism. *International Journal of Peptide Research and Therapeutics*, *20*(4), 409–420. doi:10.1007/s10989-014-9403-2

Blaney, J. M., & Dixon, J. S. (1993). A Good Ligand is Hard to Find: Automatic Docking Methods. *Perspectives in Drug Discovery and Design*, *15*(2), 301–319. doi:10.1007/BF02174531

Böhm, H. J. (1992). The computer program LUDI: A new method for the de novo design of enzyme inhibitors. *Journal of Computer-Aided Molecular Design*, *6*, 61–78. doi:10.1007/BF00124387 PMID:1583540

Brooijmans, N., & Kuntz, I. D. (2003). Molecular recognition and docking algorithms. *Annual Review of Biophysics and Biomolecular Structure*, *32*(1), 335–373. doi:10.1146/annurev.biophys.32.110601.142532 PMID:12574069

Brooks, B. R., Brooks, C. L. III, MacKerell, A. D. Jr, Nilsson, L., Petrella, R. J., Roux, B., & Karplus, M. et al. (2010). CHARMM: The Biomolecular Simulation Program. *Journal of Computational Chemistry*, *30*(10), 1545–1614. doi:10.1002/jcc.21287 PMID:19444816

Case, D. A., Cheatham, T. E., Darden, T., Gohlke, H., Luo, R., Merz, K. M., & Woods, R. J. et al. (2005). The Amber biomolecular simulation programs. *Journal of Computational Chemistry*, *26*(16), 1668–1688. doi:10.1002/jcc.20290 PMID:16200636

Cimerman, N., Prebanda, M. T., Turk, B., Popovic, T., Dolenc, I., & Turk, V. (1999). Interaction of Cystatin C variants with papain and human cathepsins B, H and L. *Journal of Enzyme Inhibition*, *14*(2), 167–174. doi:10.3109/14756369909036552 PMID:10445041

Crescenzi, O., Tomaselli, S., Guerrini, R., Salvadori, S., D'Ursi, A. M., Temussi, P. A., & Picone, D. (2002). Solution structure of the Alzheimer amyloid beta-peptide (1–42) in an apolar microenvironment similarity with a virus fusion domain. *European Journal of Biochemistry*, *269*(22), 5642–5648. doi:10.1046/j.1432-1033.2002.03271.x PMID:12423364

Dhanavade, M. J., Jalkute, C. B., Barage, S. H., & Sonawane, K. D. (2013). Homology modeling, molecular docking and MD simulation studies to investigate role of cysteine protease from *Xanthomonas campestris* in degradation of Aβ peptide. *Computers in Biology and Medicine*, *43*(12), 2063–2070. doi:10.1016/j.compbiomed.2013.09.021 PMID:24290922

Dhanavade, M. J., & Sonawane, K. D. (2014). Insights into the molecular interactions between aminopeptidase and amyloid beta peptide using molecular modeling techniques. *Amino Acids*, *46*(8), 1853–1866. doi:10.1007/s00726-014-1740-0 PMID:24729013

Duhovny, D., Nussinov, R., & Wolfson, H. J. (2002). Efficient Unbound Docking of Rigid Molecules. In *Proceedings of the 2nd Workshop on Algorithms in Bioinformatics* (WABI) (LNCS), (vol. 2452, pp. 185-200). Springer Verlag. doi:10.1007/3-540-45784-4_14

Glenner, G. G., & Wong, C. W. (1984). Alzheimer's disease: Initial report of the purification and Characterization of a novel cerebrovascular Amyloid protein. *Biochemical and Biophysical Research Communications*, *120*(3), 885–890. doi:10.1016/S0006-291X(84)80190-4 PMID:6375662

Gouras, G. K., Xu, H., Jovanovic, J. N., Buxbaum, J. D., Wang, R., Greengard, P., & Gandy, S. et al. (1998). Generation and regulation of beta-amyloid peptide variants by neurons. *Journal of Neurochemistry*, *71*(5), 1920–1925. doi:10.1046/j.1471-4159.1998.71051920.x PMID:9798916

Gschwend, D. A., Good, A. C., & Kuntz, I. D. (1996). Molecular Docking Towards Drug Discovery. *Journal of Molecular Recognition*, *8*(2), 175–186. doi:10.1002/(SICI)1099-1352(199603)9:2<175::AID-JMR260>3.0.CO;2-D PMID:8877811

Halgren, T. A., Murphy, R. B., Friesner, R. A., Beard, H. S., Frye, L. L., Pollard, W. T., & Banks, J. L. (2004). Glide: A new approach for rapid, accurate docking and scoring. 2. Enrichment factors in database screening. *Journal of Medicinal Chemistry*, *47*(7), 1750–1759. doi:10.1021/jm030644s PMID:15027866

Halperin, I., Ma, B., Wolfson, H., & Nussinov, R. (2002). Principles of docking: An overview of search algorithms and a guide to scoring functions. *Proteins*, *47*(4), 409–443. doi:10.1002/prot.10115 PMID:12001221

Hardy, J., & Selkoe, D. J. (2002). The Amyloid hypothesis of Alzheimer's disease: Progress and problems on the road to therapeutics. *Science*, *297*(5580), 353–356. doi:10.1126/science.1072994 PMID:12130773

Hsu, R. L., Lee, K. T., Wang, J. H., Lily, Y., Lee, L., Rita, P., & Chen, Y. (2009). Amyloid-degrading ability of nattokinase from *Bacillus subtilis* Natto. *Journal of Agricultural and Food Chemistry*, *57*(2), 503–508. doi:10.1021/jf803072r PMID:19117402

Irwin, J. J., & Shoichet, B. K. (2005). ZINC--a free database of commercially available compounds for virtual screening. *Journal of Chemical Information and Modeling*, *45*(1), 177–182. doi:10.1021/ci049714+ PMID:15667143

Irwin, J. J., Shoichet, B. K., Mysinger, M. M., Huang, N., Colizzi, F., Wassam, P., & Cao, Y. (2009). Automated docking screens: A feasibility study. *Journal of Medicinal Chemistry*, *52*(18), 5712–5720. doi:10.1021/jm9006966 PMID:19719084

Iwata, N., Tsubuki, S., Takaki, Y., Shirotani, K., Lu, B., Gerard, N. P., & Saido, T. C. et al. (2001). Metabolic regulation of brain Aβ by neprilysin. *Science*, *292*(5521), 1550–1552. doi:10.1126/science.1059946 PMID:11375493

Jalkute, C. B., Barage, S. H., Dhanavade, M. J., & Sonawane, K. D. (2014). Identification of Angiotensin Converting Enzyme Inhibitor: An InSilico Perspective. *International Journal of Peptide Research and Therapeutics*, *21*(1), 107–115. doi:10.1007/s10989-014-9434-8

Jalkute, C. B., Barage, S. H., & Sonawane, K. D. (2015). Insight into molecular interactions of Aβ peptide and gelatinase from *Enterococcus faecalis*: A molecular modeling approach. *RSC Advances*, *5*(14), 10488–10496. doi:10.1039/C4RA09354B

Jalkute, C. B., & Sonawane, K. D. (2015). Evaluation of a possible role of *Stigmatella aurantiaca* ACE in Aβ peptide degradation: A molecular modeling approach. *Journal of Molecular Microbiology and Biotechnology*, *25*(1), 26–36. doi:10.1159/000370114 PMID:25677850

Jones, G., & Willet, P. (1995). Docking Small-Molecule Ligands into Active Sites. *Current Opinion in Biotechnology*, *6*(6), 652–656. doi:10.1016/0958-1669(95)80107-3 PMID:8527835

Kitchen, D. B., Decornez, H., Furr, J. R., & Bajorath, J. (2004). Docking and scoring in virtual screening for drug discovery: Methods and applications. *Nature Reviews. Drug Discovery*, *3*(11), 935–949. doi:10.1038/nrd1549 PMID:15520816

Kuntz, I. D., Blaney, J. M., Oatley, S. J., Langridge, R., & Ferrin, T. E. (1982). A Geometric Approach to Macromolecule-ligand interactions. *Journal of Molecular Biology*, *161*(2), 269–288. doi:10.1016/0022-2836(82)90153-X PMID:7154081

Lengauer, T., & Rarey, M. (1996). Computational Methods for Biomolecular Docking. *Current Opinion in Structural Biology*, *6*(3), 402–406. doi:10.1016/S0959-440X(96)80061-3 PMID:8804827

Lengauer, T., & Rarey, M. (1996). Computational Methods for Biomolecular Docking. *Current Opinion in Structural Biology*, *6*(3), 402–406. doi:10.1016/S0959-440X(96)80061-3 PMID:8804827

Lybrand, T. P. (1995). Ligand-Protein Docking and Rational Drug Design. *Current Opinion in Structural Biology*, *5*(2), 224–228. doi:10.1016/0959-440X(95)80080-8 PMID:7648325

Macindoe, G., Mavridis, L., Venkatraman, V., Devignes, M. D., & Ritchie, D. W. (2010). Hex Server: An FFT-based protein docking server powered by graphics processors. *Nucleic Acids Research*, *38*(Web Server), W445–W449. doi:10.1093/nar/gkq311 PMID:20444869

Masters, C. L., Simms, G., Weinman, N. A., Multhaup, G., McDonald, B. L., & Beyreuther, K. (1985). Amyloid plaque core protein in Alzheimer disease and Down syndrome. *Proceedings of the National Academy of Sciences of the United States of America*, *82*(12), 4245–4249. doi:10.1073/pnas.82.12.4245 PMID:3159021

Morris, G. M., Huey, R., Lindstrom, W., Sanner, M. F., Belew, R. K., Goodsell, D. S., & Olson, A. J. (2009). AutoDock4 and AutoDockTools4: Automated docking with selective receptor flexibility. *Journal of Computational Chemistry*, *30*(16), 2785–2791. doi:10.1002/jcc.21256 PMID:19399780

Mueller-Steiner, S., Zhou, Y., Arai, H., Roberson, E. D., Sun, B., Chen, J., & Gan, L. et al. (2006). Antiamyloidogenic and neuroprotective functions of cathepsin B: Implications for Alzheimer's disease. *Neuron*, *51*(6), 703–714. doi:10.1016/j.neuron.2006.07.027 PMID:16982417

Musil, D., Zucic, D., Turk, D., Engh, R. A., Mayr, I., Huber, R., & Bode, W. et al. (1991). The refined 2.15 Å X-ray crystal structure of human liver cathepsin B: The structural basis for its specificity. *The EMBO Journal*, *9*, 2321–2330. PMID:1868826

Parulekar, R. S., Barage, S. H., Jalkute, C. B., Dhanavade, M. J., & Sonawane, K. D. (2013). Homology Modeling, Molecular Docking and DNA Binding Studies of Nucleotide Excision Repair UvrC Protein from *M. tuberculosis*. *The Protein Journal*, *32*(6), 467–476. doi:10.1007/s10930-013-9506-1 PMID:23963890

Pettersen, E. F., Goddard, T. D., Huang, C. C., Couch, G. S., Greenblatt, D. M., Meng, E. C., & Ferrin, T. E. (2004). UCSF Chimera—a visualization system for exploratory research and analysis. *Journal of Computational Chemistry*, *25*(13), 1605–1612. doi:10.1002/jcc.20084 PMID:15264254

Qiu, W. Q., Walsh, D. M., Ye, Z., Vekrellis, K., Zhang, J., Podlisny, M. B., & Selkoe, D. J. et al. (1998). Insulin-degrading enzyme regulates extra-cellular levels of amyloid β-protein by degradation. *The Journal of Biological Chemistry*, *273*(49), 32730–32738. doi:10.1074/jbc.273.49.32730 PMID:9830016

Rosenfeld, R., Vajda, S., & Delisi, C. (1995). Flexible Docking and Design. *Annual Review of Biophysics*, *24*(1), 677–700. doi:10.1146/annurev.bb.24.060195.003333 PMID:7663131

Sali, A., & Blundell, T. L. (1993). Comparative protein modelling by satisfaction of spatial Restraints. *Journal of Molecular Biology*, *234*(3), 779–815. doi:10.1006/jmbi.1993.1626 PMID:8254673

Sayle, R. A., & Milner-White, E. J. (1995). RASMOL: Biomolecular graphics for all. *Trends in Biochemical Sciences*, *20*(9), 374–376. doi:10.1016/S0968-0004(00)89080-5 PMID:7482707

Schneidman-Duhovny, D., Inbar, Y., Nussinov, R., & Wolfson, H. J. (2005). PatchDock and Symm Dock: Servers for rigid and symmetric docking. *Nucleic Acids Research*, *33*(Web Server), W363-367. doi:10.1093/nar/gki481 PMID:15980490

Sonawane, K. D., & Barage, S. H. (2014). Structural analysis of membrane bound hECE1 dimer using molecular modeling techniques: Insights into conformational changes and Aβ1–42 peptide binding. *Amino Acids*, *47*(3), 543–559. doi:10.1007/s00726-014-1887-8 PMID:25501500

Sousa, S. F., Fernandes, P. A., & Ramos, M. J. (2006). Protein-ligand docking: Current status and future challenges. *Proteins*, *65*(1), 15–26. doi:10.1002/prot.21082 PMID:16862531

Spoel, V. D., Lindahl, E., Hess, B., Groenhof, G., Mark, A. E., & Berendsen, H. J. (2005). GROMACS: Fast, flexible, and free. *Journal of Computational Chemistry*, *26*(16), 1701–1718. doi:10.1002/jcc.20291 PMID:16211538

Tanzi, R. E., & Bertram, L. (2005). Twenty years of the Alzheimer's disease Amyloid hypothesis: A genetic perspective. *Cell*, *120*(4), 545–555. doi:10.1016/j.cell.2005.02.008 PMID:15734686

Tseng, G. N., Sonawane, K. D., Korolkova, Y. V., Zhang, M., Liu, J., Grishin, E. V., & Guy, H. R. (2007). Probing the outer mouth structure of the HERG channel with peptide toxin footprinting and molecular modeling. *Biophysical Journal*, *92*(10), 3524–3540. doi:10.1529/biophysj.106.097360 PMID:17293393

Tucker, H. M., Kihiko, M., Caldwell, J. N., Wright, S., Kawarabayashi, T., Price, D., & Estus, S. et al. (2000). The plasmin system is induced by and degrades amyloid-β aggregates. *The Journal of Neuroscience*, *20*, 3937–3946. PMID:10818128

Yin, K. J., Cirrito, J. R., Yan, P., Hu, X., Xiao, Q., Pan, X., & Turk, J. et al. (2006). Matrix metalloproteinases expressed by astrocytes mediate extracellular amyloid-beta peptide catabolism. *The Journal of Neuroscience*, *26*(43), 10939–10948. doi:10.1523/JNEUROSCI.2085-06.2006 PMID:17065436

Yoo, C., Ahn, K., Park, J. E., Kim, M. J., & Jo, S. A. (2010). An aminopeptidase from *Streptomyces sp.* KK565 degrades beta amyloid monomers, oligomers and fibrils. *FEBS Letters*, *584*(19), 4157–4162. doi:10.1016/j.febslet.2010.08.048 PMID:20828565

Zou, K., Yamaguchi, H., Akatsu, H., Sakamoto, T., Ko, M., Mizoguchi, K., & Kosaka, K. et al. (2007). Angiotensin converting enzyme converts Amyloid beta-protein1–42 (Aβ1–42) to Aβ1–40, and its inhibition enhances brain Abeta deposition. *The Journal of Neuroscience*, *27*, 8628–8635. doi:10.1523/JNEUROSCI.1549-07.2007 PMID:17687040

Chapter 11
Recent Advancements in Docking Methodologies

Vijay Kumar Srivastav
Shri Govindram Seksaria Institute of Technology and Science, India

Vineet Singh
Shri Govindram Seksaria Institute of Technology and Science, India

Meena Tiwari
Shri Govindram Seksaria Institute of Technology and Science, India

ABSTRACT

Nowadays molecular docking has become an important methodology in CADD (Computer-Aided Drug Design)-assisted drug discovery process. It is an important computational tool widely used to predict binding mode, binding affinity and binding free energy of a protein-ligand complex. The important factors responsible for accurate results in docking studies are correct binding site prediction, use of suitable small-molecule databases, consistent docking pose, high dock score with good MD (Molecular Dynamics), clarity whether the compound is an inhibitor or agonist, etc. However, still there are several limitations which make it difficult to obtain accurate results from docking studies. In this chapter, the main focus is on recent advancements in various aspects of molecular docking such as ligand sampling, protein flexibility, scoring functions, fragment docking, post-processing, docking into homology models and protein-protein docking.

INTRODUCTION

To accelerate the new drug discovery process, search of small molecules which bind to biological targets of interest, is rapidly increasing. Molecular docking is an important tool used in structure-based drug design which facilitates the drug discovery process with the aid of high speed computers. A typical docking program consists of two essential components viz. sampling and scoring. Sampling is generation of ligand orientations in the binding site of a protein, whereas scoring is prediction of its binding fitness. The top ranked orientation with lowest energy score is predicted as binding mode. Docking is

DOI: 10.4018/978-1-5225-0115-2.ch011

now routinely used in hit identification, lead optimization and virtual screening. With the development of protein structure prediction techniques and high speed computers, several advancements have been made in molecular docking methodologies. However, several limitations still exist, which make CADD approaches less reliable than experimental approaches. Several research articles can be found on docking studies but only few of them have succeeded in the development of drug candidate of interest.

This chapter covers general introduction to molecular docking process, its components, limitations and advancements in docking methodologies including ligand sampling, scoring functions, protein flexibility, fragment docking, post-processing, docking into homology models and protein-protein docking along with examples.

BACKGROUND

Enzyme-substrate, drug-protein, drug-nucleic acid, protein-protein and protein-nucleic acid interactions play an important role in many important cellular processes, such as signal transduction, cell regulation, gene expression, enzyme inhibition, antigen-antibody reaction, assembly of multi-domain proteins etc. These interactions form stable protein-protein or protein-ligand complexes which are essential for various biological functions. But, it is quite difficult and expensive to determine the complex structure of proteins by experimental methods i.e. X-ray crystallography and NMR (nuclear magnetic resonance). Thus, structure-based drug design approaches especially docking methodologies are now being considered as an important approach for understanding the protein-protein as well as protein-ligand interactions (Halperin, Ma, Wolfson, & Nussinov, 2002).

In the beginning, structure-based drug design was not easy due to low computational speed and non-availability of 3D (three dimensional) structure of proteins. One key methodology "docking" was pioneered during early 1980s, and became a thrust area of research. The first docking program DOCK was developed in 1982 by Irwin "Tack" Kuntz's group, which considered only geometric complementarities but not the ligand interactions (Kuntz, Blaney, Oatley, Langridge, & Ferrin, 1982). Subsequently, in 1985 Goodford developed a program GRID to understand and predict the ligand interactions in the binding site (Goodford, 1985). Further advancements in CADD resulted in development of various important computational approaches including high-throughput docking and virtual screening (VS). High-throughput docking is primarily used when only the structure of a target and its binding site is available while VS is used to identify a subset of chemical compounds from small-molecules database that contain relatively many hits against the target (Schneider & Böhm, 2002). Till date, several improvements have been made in the docking algorithm, and at the moment more than 50 docking programs are available. Now it is used as a tool not only for hit identification and lead optimization but also for understanding the drug metabolism as well as toxicity prediction (Halgren, 2007).

Development in structure determination techniques like high throughput X-ray crystallography, NMR, genomics and proteomics have increased the number of proteins with known 3D structure and are easily available as a repository of protein structures i.e. PDB (protein data bank). The 3D structure of a protein is an initial and prime requirement of docking program to estimate the binding mode and affinity of interacting ligands (Blundell, Jhoti, & Abell, 2002). Advancement in the structure-based drug design approaches has discovered a number of important drugs like inhibitors of HIV-1 protease, HIV-1 integrase, carbonic anhydrase, neuraminidase, protein kinases etc. Undoubtedly, docking methodologies are very important scientific advances to understand the nature of chemical molecules (i.e. drug and

Figure 1. Types of docking methodologies

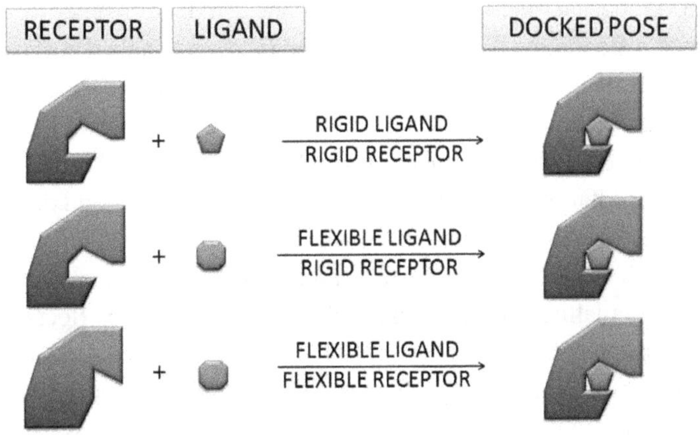

receptor), as noted particularly when three top computational scientists won the 2013 Nobel Prize in chemistry. Although, docking methodology has wide applications in drug discovery, there are various problems afflicting docking results and need to be resolved like correct selection of scoring functions, handling of water molecules and metals, covalent interactions and flexibility in the ligand and receptor.

MOLECULAR DOCKING

The complexity of docking methodologies can be represented in the following order: (i) rigid body docking (in which both the receptor and ligand are rigid); (ii) flexible ligand docking (where receptor is rigid, but the ligand is flexible); and (iii) flexible docking (in which both receptor and ligand are flexible). Rigid receptor/flexible ligand model is most commonly used docking methodology (Figure 1). Flexible docking may not be a good choice without a very powerful and high-speed computer to screen more than 10,000 compounds from a database. By contrast, to dock only few compounds in the specific protein binding site, at a specific pH and solvation, flexible docking program might be a good choice. Selection of docking programs, therefore, depends on type of hardware and size of database being screened (Hernández-Santoyo, Tenorio-Barajas, Altuzar, Vivanco-Cid, & Mendoza-Barrera, 2013; Teodoro, Phillips, & Kavraki, 2001).

Recognition of molecular features and their modifications, responsible for specific biological processes are very complex issues that are difficult to be simulated on a computer. A simple molecule contains many conformational degrees of freedom and sampling these degrees of freedom must be carried out with sufficient accuracy to identify the correct orientation that matches the structure of receptor. Ligand sampling and scoring are two important goals of a typical docking program. Sampling must be fast enough to assess thousands of molecules in a defined docking run (Kitchen, Decornez, Furr, & Bajorath, 2004). Calculation of the binding energy of ligand interactions is also one of the important aspects of molecular docking. Methods based on quantum mechanics and empirical energy functions are generally used to calculate this energy. The extensively used docking methods employ search algorithms based on Monte Carlo (MC), genetic algorithm (GA), MD and fragment-based approaches. DOCK, FlexX, GOLD, and ICM are used for high throughput docking of a large database of molecules (Dias & de Azevedo Jr, 2008).

In docking studies, energy estimation is usually carried out with various types of scoring functions. The two main features of a good scoring function are selectivity and efficiency (Teodoro, et al., 2001). Selectivity allows scoring function to differentiate between correct and incorrect docked poses while efficiency allows the docking program to run in a reasonable time. Generally, the available scoring functions are based on forcefield (i.e. force-field based scoring functions) and are used to simulate the receptor-ligand interaction energy and internal ligand energy (strain induced by binding). A forcefield is a function of expressing the energy of a system as a sum total of diverse bonded terms (bond angles, bond distances, torsional angles, etc.) and non-bonded terms [electrostatic and vdW (van der Walls) energy]. These terms are utilized for different types of atoms in the molecule by adjusting a series of empirical parameters. Most of the scoring functions are focused on calculating energetics than solvation and entropic effects. Incorporation of solvation and entropic effects in the scoring function is a challenging task, as it makes the process expensive and time consuming. More complex scoring schemes with electrostatic and vdW interactions along with solvation and entropic effects are now being used to further evaluate the pre-selected conformations with the use of high speed computers. Thus, the accuracy of the scoring function is directly related to its computational speed and expense. Now-a-days statistical scoring functions are also being used that are derived using experimental data. However, there are several problems associated with scoring functions which make docking methodologies complex. In addition, other complications like low resolution of crystallographic protein targets, protein flexibility, induced fit or other conformational changes, and the involvement of water molecules in protein-ligand interactions also exist (Taylor, Jewsbury, & Essex, 2002).

ADVANCEMENTS IN DOCKING METHODOLOGIES

The lock and key theory was used in earliest reported protein-ligand docking methodologies in which both ligand and protein were treated as rigid bodies. Further development introduced docking of flexible ligands to rigid proteins which remains the most popular docking methodology. Recent research is finding the ways of handling the situations where both ligand and protein are flexible. The failure or relative success of the early structure-based drug design methods urges in-depth analysis of docking methodologies to find out the ways of accurate prediction of binding mode and energy. Docking involves a compromise between the docking algorithm speed and its accuracy. In an attempt to achieve higher accuracy, one can employ more advanced scoring functions or more exhaustive sampling of the possible binding modes and flexibility, but these modifications usually add to the computational cost. This section summarizes the recent advancements in docking methodologies and related phenomena.

1. Molecular Representation

The representation of ligand and protein is very important to understand the docking process. Appropriate representation of molecules is now become crucial in analysis of small molecule databases. A molecule can be expressed by a pair of A (collection of atoms) and B (collection of bonds between pairs of atoms). Each atom has some standard properties, such as its vdW radius. There are three main properties associated with each bond: (i) bond length; (ii) bond angle; (iii) number of rotatable bonds (Figure 2). In common practice bond length and angles are considered as fixed because they do not change. Thus, degree of freedom of a molecule mainly depends upon rotatable bonds. Generally, ligands have 3-15 rotatable bonds, while receptors have 1,000-2,000 rotatable bonds (Teodoro, et al., 2001).

Figure 2. A drug molecule represented by ball and sticks model
(Teodoro, et al., 2001)

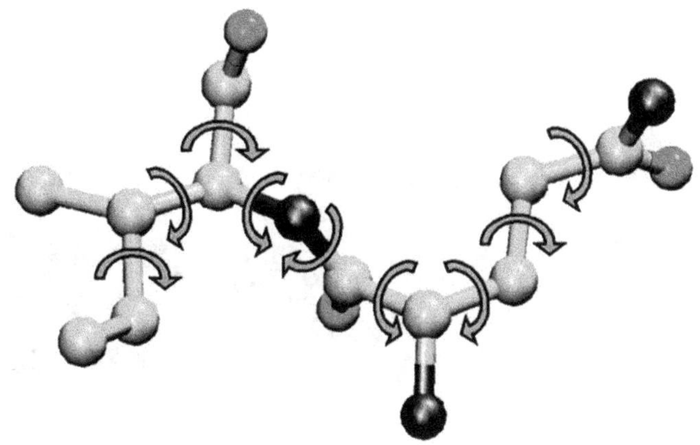

In commonly used docking programs, the protomeric and tautomeric states of ligands are user defined. The structure of a ligand can be modified as per the requirement, by adding or removing hydrogens and by adjusting pKa values. Mostly the structures at neutral pH are preferred. The correct atom typing is also very important otherwise wrong definition of acceptor and donor heteroatoms may lead to serious errors in docking results. Example: In an early experiment by Watson and Crick, helical model with purine-pyrimidine hydrogen bonded base pairs failed due to wrong assignment of tautomeric formula (enol forms) for the nucleic acid bases. As the correct tautomeric structures (keto forms) of the bases were assigned, all the key features of the three-dimensional structure of double helical DNA were readily attained (Mohan, Gibbs, Cummings, Jaeger, & DesJarlais, 2005). To describe the stereochemistry of the molecules, it is beneficial to generate all possible diastereoisomers and dock them individually to the receptor. Pipeline Pilot, Stergen and Stereoplex are example of programs available for generating all possible diastereoisomers of a given molecule.

The target protein can be represented by three ways: atomic representation, surface representation and Grid representation. Atomic representation is the most common representation used along with potential energy function and in final ranking procedures. Generally, higher the resolution of a crystal structure, better will be docking result. A docking experiment of reproducing the binding mode using ICM program showed that binding modes of ligands within 1 Å of the bound conformations were better than 2.0 Å. it is also very important that the B-factors of the atoms in the binding site should be reasonable, as high values imply that their coordinates are less reliable (Figure 3). Protein flexibility is described in detail in a later section.

2. Sampling Algorithms

Along with translational, rotational and conformational degrees of freedom, ligand and protein both have a large number of possible binding modes. Thus, it is very difficult to generate all possible conformations. At this stage sampling algorithm plays a very important role. It performs ligand sampling and generates putative ligand poses in the selected binding site. Subsequently the scoring functions select the correct pose based on ranking. In general practice, there are three types of sampling algorithms used in flexible protein-ligand docking: shape matching, systematic search, and stochastic algorithms.

Figure 3. Representation of 3D structure of the protein triose phosphate isomerase. Left: atom type representation. Middle: backbone conformation, colored by secondary structure. Right: surface representation (acidic residues red, basic residues blue, polar residues green, nonpolar residues white) (http://www.chempep.com/ChemPep-Protein-Overview.htm).

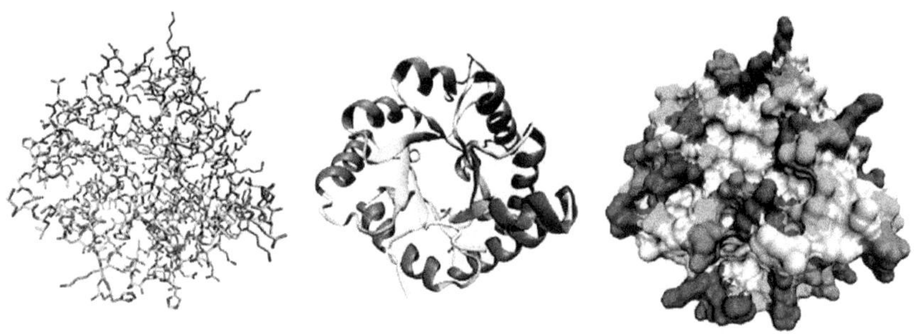

2.1. Shape Matching

The shape matching method places the ligand in the binding site with proper shape compatibility with the binding site of the protein. In this algorithm protein and ligand are considered as pharmacophores and each distance of the pharmacophore within the protein and ligand is calculated for a pair (Meng, Zhang, Mezei, & Cui, 2011). New ligand conformations are regulated by the distance matrix between the pharmacophore and the corresponding ligand atoms. Hydrogen-bond donor and acceptor properties of upcoming conformation are considered during match. Examples of docking programs based on shape matching algorithm include DOCK, EUDOC, FRED, FLOG, LigandFit, MS-DOCK, MDock and Surflex.

2.2. Systematic Search

Systematic search algorithms generally consider all degrees of freedom of ligand to generate all possible binding conformations in the binding site of protein. There are three types of systematic search methods: exhaustive search, fragmentation-based search and conformational ensemble search.

The exhaustive search algorithm operates in docking by rotating all possible rotatable bonds of the ligand in the binding site at a specified interval. Computationally these are too expensive as the search space is very large e.g. Glide and FRED (Kitchen, et al., 2004).

In fragmentation methods, the ligand is introduced in the binding site by dividing it into different fragments. The key fragment which interacts with target protein is selected to dock in the binding site. The remaining fragments are added one by one, to incrementally growing ligand, e.g. ADAM, DOCK, eHiTs, FlexX and LUDI. a new ranking based sampling algorithm have developed within DOCK 6 by softening the vdW interaction energy, improving the performance of the bump filter and by only ranking the layers of ligand growth to manage the sampling algorithm towards the correct pose.

In conformational ensemble methods, ligand flexibility is defined by docking an ensemble of ligand conformations generated programs such as OMEGA. Subsequently, binding modes of ligand from several docking runs are collected and ranked on the basis of their binding energy scores (Roy, Kar, & Das, 2015). Examples of programs based on conformational ensemble are DOCK3.5, FLOG, MS-DOCK, MDock, PhDOCK, and Q-Dock.

2.3. Stochastic Algorithms

Stochastic algorithms are based on the principle of randomness and solve global optimization problems in docking studies. It gives good solutions within reasonable time but do not guarantee the optimum solution, as the random change will be accepted or rejected according to a probabilistic criterion. There are four types of stochastic algorithms: Evolutionary algorithms (EAs), Monte Carlo (MC) methods, Swarm optimization (SO) methods and Tabu search methods (Table 1).

2.3.1. Evolutionary Algorithms

EAs searches for the correct binding mode based on evolutionary process in biological systems. The most popular type of EAs is the GA. It is based on Darwin's theory of evolution in which variables are considered as genes (Degrees of freedom of the ligand). These genes make up the 'chromosome' (pose of the ligand). GA searches for poses which show better fitness function through genetic operators like mutation and crossover. The process results in a pose with better fitness function (Michalewicz, 1996; Srivastav & Tiwari, 2015). An improved adaptive GA has been developed which incorporates a multi-population genetic strategy, entropy-based searching with self-adaptation, and the quasi-exact penalty. When this improved adaptive GA was applied to test a dataset of 134 complexes, good pose prediction (<2A°) was obtained (Kang, Li, Jiang, & Wang, 2009). A hybrid method developed by combining a

Table 1. Programs based on sampling algorithms

Random/Stochastic	Systematic	Simulation
ICM (MC)	DOCK (incremental)	DOCK
QXP (MC)	FlexX (incremental)	Glide
Prodock (MC)	Glide (incremental)	MOE-Dock
MCDOCK (MC)	Hammerhead (incremental)	Auto Dock
DockVision (MC)	FLOG (database)	Hammerhead
MOE-Dock (MC, Tabu Search)	FRED(exhaustive search)	
Auto Dock (EA, MC)	LUDI	
GOLD (EA)	ADAM	
DIVALI(EA)	eHiTs	
DARWIN (EA)	OMEGA	
MolDock (EA)	PhDOCK	
FLIPDock (EA)	MS-DOCK	
Lead finder (EA)	MDock	
EADock (EA)	Q-Dock	
PSI-DOCK (EA, Tabu Search)		
PRO_LEADS (Tabu Search)		
SODOCK (SO)		
Tribe-PSO (SO)		
PSO@Autodock (SO)		
PLANTS (SO)		

multi-deme Lamarckinan-GA (LGA) and a gradient-based local optimizer [a modified L-BFGS (Limited-memory-Broyden-Fletcher-Goldfarb-Shanno algorithm)-quasi-Newton method] along with the Gehlhaar scoring function. This hybrid method applied to test a diverse set for flexible ligand-protein docking. It was found faster and superior than other LGAs, which use stochastic optimization methods (Fuhrmann, Rurainski, Lenhof, & Neumann, 2010). GOLD, AutoDock, DIVALI, DARWIN, MolDock, PSI-DOCK, FLIPDock, Lead finder, and EADock are the programs based on EAs.

2.3.2. Monte Carlo Methods

The MC method searches the thermodynamically accessible states of a system (i.e. ligand binding orientation in the target) by generating small random changes in system configuration (i.e. ligand conformation), which can be either accepted or rejected according to the metropolis algorithm. A MC algorithm based method was successfully used to search possible binding modes of molecular fragments of methotrexate (considering translational and rotational degrees of freedom) at a specific site of dihydrofolate reductase from E. ***coli*** (Hart & Read, 1992). The docking programs based on MC methods include DockVision, ICM, QXP, Prodock, and MCDOCK.

2.3.3. Swarm Optimization

Particle Swarm Optimization (PSO) was introduced to solve the conformational sampling problem in flexible protein-ligand docking. It exploits a population of individuals to detect promising regions in the search space. In this context, the population is defined as a swarm, and the individuals are defined as particles (Liu et al., 2013). Examples of docking programs that use SO algorithms include SODOCK, Tribe-PSO, PSO@Autodock, and PLANTS.

2.3.4. Tabu Search

Already explored areas of the conformational space are utilized in tabu search algorithm. An RMSD (root mean square deviation) between the current binding conformation and previously recorded conformation is calculated and used as cutoff to select or reject the upcoming molecular conformation. Only those conformations are accepted which have RMSD value better that the cutoff, otherwise rejected (Westhead, Clark, & Murray, 1997). Examples of tabu search based docking programs are PRO- LEADS and PSI-DOCK.

2.4. Molecular Dynamics (MD)

Combining docking with MD simulation can handle the flexibility of both ligand and protein more effectively than other algorithms. In order to identify the ligand conformations, MD simulations efficiently use local optimization. An improved MD-based program DynaDock has been developed for docking peptides into flexible receptors (Antes, 2010). This method involves approximate ligand posing, followed by optimized potential MD, which uses soft-core potentials for protein–peptide interactions with a new optimization scheme. It showed better sampling results as compared with conventional MD. The only disadvantage is that it may lead to inadequate sampling when stepping over high energy barriers.

3. Protein Flexibility

A better description of protein-ligand interaction was described by induced-fit model, in which both protein and the ligand are considered as flexible (Koshland, 1958). In this model, to form a minimum energy perfect-fit, both protein and ligand structures change their conformations. Binding of ligand induces conformational changes in protein (induced fit) which ranges from rearrangement of side-chains to large domain motions. Due to the large size and many degrees of freedom of proteins, flexibility became most challenging issue in molecular docking. To overcome this issue several advancements have been done in the docking process such as soft docking, side-chain flexibility, molecular relaxation, and ensemble docking (Table 2). The inclusion of receptor flexibility in docking can result in more accurate binding pose prediction, better virtual screening enrichments, and provide more realistic description of protein-ligand binding interaction.

Table 2. Approaches to incorporate protein flexibility in docking

Approach	Merits	Limitations
Docking with soft potentials	• Low computational cost • Easy to interpret • Can detect subtle changes that other methods overlook	• Only subtle side-chain changes detected • Completely novel conformation cannot be found • Low level of ligand effects • No quantification of conformational changes • Biased by input conformations
Rotamer exploration	• Moderate to high computational cost • depending on size of rotamer library • Can find novel conformation if included in rotamer library • Moderate level of ligand effects • Ideal when active site shows only side-chain rotations	• Only side-chain conformational changes • Multiple solutions may lead to ambiguity in interpretation • May be biased by rotamer library • May fail if desired rotamer not included in library
Multiple protein structures (MPS) multiple dockings	• Side-chain and backbone changes but restricted by ensemble set used • Ligand effects included to moderate level • Multiple ways of generating multiple structures	• High computational cost if many protein structures • Novel conformation cannot be found • Ambiguity in interpretation of multiple equally good solutions • No quantification of conformational changes for specific ligand • Biased by input set of conformations • May fail if specific ligand induces novel conformation not included in MPS set
Multiple protein structures (MPS) ensemble docking	• Side-chain and backbone changes but restricted by ensemble set used • Easy to interpret • Computationally better than multiple structures for VS • Ligand effects included to moderate level	• Moderate computational cost • A small chance of finding novel conformation • No quantification of conformational changes • May fail if ensemble is biased • Average protein structure may not give a real conformation
Current hybrid methods	• Side-chain and backbone movements • Can find novel conformation to some extent • Moderate to high level of ligand effects • Low bias due to input protein conformation	• Moderate to high computational cost • Interpretation may be difficult if multiple equally good solutions found • May miss large conformational changes
MD simulations after docking	• Side-chain and backbone movements but restricted by simulation time • Can find novel conformation to limited extent • Moderate to high level of ligand effects • Quantification of conformational changes • Can handle solvent and ionic effects and membrane-bound proteins	• High computational cost • Analysis of huge output effort-intensive • Biased by starting structure and initial velocity • May fail owing to short times of simulations and may miss best solution

(B-Rao, Subramanian, & Sharma, 2009)

3.1. Soft Docking

For modeling the protein flexibility, soft docking was the earliest approach used, in which protein is kept rigid but soft scoring functions were used to evaluate the protein-ligand binding. Soft docking works by decreasing the vdW repulsion energy between the binding site and ligand. Decrease in vdW repulsion energy accounts for partial flexibility of protein. It affects minor side-chain movements but not the backbone movements and other structural rearrangements (Ferrari, Wei, Costantino, & Shoichet, 2004; Jiang & Kim, 1991).

3.2. Side-Chain Flexibility

Side-chain flexibility analysis gives valuable insights to improve docking algorithms. In this method the protein backbones are kept fixed and side-chain conformations are sampled to find out the best pose. Amino acid side-chain rotamer libraries are often used to model the conformational space by selecting the preferred side-chain conformations. A dead-end elimination algorithm is used to remove the side-chain conformations which do not contribute in the efficient binding (Desmet, Wilson, Joniau, De Maeyer, & Lasters, 1997; Leach, 1994).

3.3. Molecular Relaxation Process

In the molecular relaxation process first rigid-body docking is performed by placing the ligand into the binding site and minimization of protein-ligand complex is done by using MC, MD, and other methods. This method improves the docking outcome as it is capable of handling both backbone (up to a certain extent) and side-chain flexibility, and thus, preferred over side chain flexibility methods (Apostolakis, Plückthun, & Caflisch, 1998; Davis & Baker, 2009).

3.4. Multiple Protein Structures Ensemble Docking

Docking into multiple receptor conformations is known as ensemble docking and account for receptor flexibility. Instead of using a single rigid protein structure in docking as used in standard docking methods, it involves multiple conformations of the protein to dock the ligand. This ensemble of protein conformations mimics the conformational equilibrium which characterizes the native state of the target protein and provides a structural degree of freedom by which the conformation of the protein model may be matched to fit any particular ligand. Ensemble docking has advantage over other flexible-receptor docking methods like induced-fit docking and capable of handling any type of protein motion (Craig, Essex, & Spiegel, 2010; Knegtel, Kuntz, & Oshiro, 1997; Lang, Ray, Liu, Singh, & Cuny, 2015). FlexE and ICM are examples of ensemble docking programs (Table 3).

3.5. Other Advancements

To handle the protein flexibility, some programs have employed a hybrid approach by combining soft potentials, rotamer exploration and active-site mutation to simulate induced-fit effects, e.g. Schrodinger/ Glide (Sherman, Day, Jacobson, Friesner, & Farid, 2006). A knowledge based approach QSiCR (quantitative structure-induced conformation relationship) has been developed to handle protein flexibility.

Table 3. Recent programs for docking and virtual screening considering protein flexibility

Docking Programs	Novel Features
Autodock	Free open-source EA-based docking program, flexible ligand, and flexible protein side chains.
DOCK	Anchor-and-grow based docking program, flexible ligand, and flexible protein.
GOLD	GA-based docking program, flexible ligand, partial flexibility for protein.
Glide	Exhaustive search-based docking program exists in extra precision, standard precision and virtual high-throughput screening modes, ligand and protein flexible.
SCIGRESS	Desktop/server molecular modeling suite employing linear scaling semi-empirical quantum methods for protein optimization and ligand docking.
ICM	Docking program based on pseudo-brownian sampling and local minimization, ligand and protein flexible.
FlexX, FlexE	Incremental build-based docking program, flexible ligand, protein flexibility through ensemble of protein structure.
Fleksy	Program for flexible and induced fit docking using receptor ensemble (constructed using backbone-dependent rotamer library) to describe protein flexibility.
FITTED	Suite of programs to dock flexible ligands into flexible proteins, relies on a GA to account for flexibility of the two molecules and location of water molecules, and on a novel application of a switching function to retain or displace water molecules and to form potential covalent bonds (covalent docking) with the protein side chains.
Autodock Vina	MC-based docking program, flexible ligand and flexible protein side chains.
FlipDock	GA-based docking program using flextree data structures to represent a protein-ligand complex, flexible ligand, and flexible protein.

It is similar to QSAR (quantitative structure-activity relationship) but the only difference is, instead of predicting the biological activity it predicts the effect of ligand on structural conformation of key residues in the active site of protein (Subramanian, Sharma, & B-Rao, 2006, 2008). The relaxed complex scheme (RCS) has been developed as a computational methodology by combining the docking algorithm and MD. The merit of this method is that it considers flexibility of both receptor and the docked ligand. Several improvements have been made in RCS, including its extension to virtual screening, more rigorous characterization of local and global binding effects, and improvement in its computational efficiency (Amaro, Baron, & McCammon, 2008). A 4D (four dimensional)-docking protocol has been developed for Internal Coordinate Mechanics (ICM), using receptor conformation as a fourth dimension. In this protocol, multiple receptor conformations are represented by multiple grids and used as a variable in global optimization. The method was found to be more accurate as compared to methods using single grid (Abagyan & Totrov, 1994; Bottegoni, Kufareva, Totrov, & Abagyan, 2009).

4. Scoring Functions

The scoring function is an important aspect of protein-ligand docking algorithm, because it directly determines the accuracy and speed of the algorithm and differentiates the correct poses from incorrect poses. The most important issue to scientific community is how to improve the quality of scoring functions? The validation set of GOLD suggested that it is quite difficult to develop scoring functions that can be applied equally well for different protein families, irrespective of their complexity and sophistication. This is just because of uniqueness of binding sites and recognition processes of specific protein-ligand interactions. It was observed that when docking scores are plotted versus logarithm of binding constants, slopes and intercepts are different for structurally distinct binding sites. Thus, adjustment of individual scoring schemes is required on a case by case basis. Scoring functions can be grouped into following categories: force field-based, empirical, knowledge-based, and consensus (Meng, et al., 2011).

4.1. Force Field-Based Scoring Functions

Classical force field scoring functions calculate the binding energy by summing up the individual interaction terms such as vdW energies, electrostatic energies etc. The vdW energies can be described on the basis of Lennard-Jones potential function. The Lennard-Jones potential estimates the acceptability of contact between protein and ligand atoms. The electrostatic energies are assessed by Coulombic calculations. A distance-dependent dielectric function is generally used to describe the contribution of charge-charge interactions, as charge calculations are problematic in modeling the actual protein's environment. Hydrogen bonding, solvation and rotational entropy effects are now being considered in force field-based scoring functions. It has been found that scoring functions which include and explain the solvation and rotational entropy effects are more accurate than standard force-field scoring functions. However, force field-based scoring functions have a disadvantage of slow computational speed (Lyne, 2002).

Docking results with force-field based scoring functions can be refined by using linear interaction energy and free-energy perturbation methods (FEP), which improve the accuracy of binding energy predictions (Meng, et al., 2011). However, these methods together with their simplified approaches such as OWFEG and LIEPROFEC are computationally expensive. The Poisson-Boltzmann/surface area (PB/SA) models and the generalized-Born/surface area (GB/SA) models are typical examples of such implicit solvent models (Morris et al., 1998; Yuriev, Agostino, & Ramsland, 2011; Yuriev & Ramsland, 2013).

4.2. Empirical Scoring Functions

Empirical scoring functions assess the binding energy of a protein-ligand complex by summing up a set of empirical energy components such as hydrogen bonding, ionic interaction, hydrophobic effect, binding entropy term etc. Each component is multiplied by a coefficient and then summed up to give a final score. These Coefficients can be obtained from regression analysis of a test set of ligand-protein complexes with known binding affinities (Eldridge, Murray, Auton, Paolini, & Mee, 1997). GlideScore, PLP, SYBYL/F-Score, LigScore, LUDI, SCORE, X-Score, ChemScore, MedusaScore, AIScore, and SFCscore are examples of empirical scoring functions.

4.3. Knowledge-Based Scoring Functions

Knowledge-based scoring functions employ statistical analysis of structures of protein-ligand complexes to derive the potential parameters such as interatomic contact frequencies and distances between the protein and ligand. It follows the assumption that more favorable an interaction, greater will be the frequency of occurrence. These frequency distributions are then converted into pairwise atom-type potentials. The score is calculated by considering the preferable contacts and penalizing repulsive interactions between each atom in protein and ligand within a given cutoff. It can also be used to screen large chemical compounds databases. It can also deal with some uncommon aromatic-sulphur and cation-π interactions, which are poorly managed by empirical approaches. Knowledge-based scoring functions can be grouped into three categories: traditional atom-randomized reference state, corrected reference state, and circumventing the reference state (S.-Y. Huang & Zou, 2010). The main disadvantage of the atom-randomization approximation is that the effect of excluded volume and interatomic connectivity are overlooked. PMF and DFIRE methods have been introduced as correction terms for the reference state. To circumvent the long-standing reference state problem, ITScore scoring function was derived, which has been tested by prediction of binding mode, binding affinity and virtual screening (Gohlke, Hendlich, & Klebe, 2000). Examples of knowledge-based scoring functions are DrugScore, SMoG, BLEEP, GOLD/ASP, MScore, and KScore.

4.4. Consensus Scoring Functions

Consensus scoring is not a specific type of scoring function but it combines the scoring information from various other scoring functions to find out a correct pose and subsequently balancing the errors of the individual scoring functions. Example: X-CSCORE (which is a combination of ChemScore, DOCK-like, FlexX-like, GOLD-like, and PMF scoring functions). The consensus scoring function based programs work according to the following steps: (i) hierarchical clustering of all poses generated by a docking tool into families represented by a leader; (ii) definition of all consensus pairs from leaders generated by different docking programs; (iii) clustering of consensus pairs into classes, represented by a mean structure; and (iv) ranking the different means starting from the most populated class of consensus pairs (Paul & Rognan, 2002). Other development in consensus scoring method include MultiScore, VoteD-ock, VLifeDock (uses array of PLPscore, XCscore, steric and electrostatic scores), FLAP-Consensus Docking, ConsDock, PLSDA-DOCET, (Partial Least Squares Discriminant Analysis-Docking Opti-mized Combined Energetic Terms), a combination of quantitative structure-binding affinity relation-ship (knowledge based) and MedusaScore (force field-based score), combination of MCSS and GOLD docking with MM/GBSA rescoring, combination of AutoDock4 and AutoDock Vina, combination of DOCK4, FlexX, and PMF, and HarmonyDOCK, etc. (Avram, Pacureanu, Seclaman, Bora, & Kurunczi, 2011; Paul & Rognan, 2002; Plewczynski, Łażniewski, Grotthuss, Rychlewski, & Ginalski, 2011; Poli, Giuntini, Martinelli, & Tuccinardi, 2015).

4.5. Other Advancements in Scoring Functions

4.5.1. Weighted Score Function

Theoretically, a lower Gibbs free energy of a protein-ligand complex indicates its stability. It has been found that weighted-score function is more accurate than the consensus score (Chen, 2009). Consensus score permits the user to select several or all the scoring functions and evaluate the overall score. Each score function is of equal weight in the consensus scoring function. On the other hand, the weighted-scoring algorithm is based on regression coefficients. A higher coefficient shows a higher weight in this algorithm. Thus, the weighted score is more accurate than the consensus score (Feher, 2006).

$$WS = \sum_{i=1}^{n} Wi \text{ x } \frac{SFi}{SFc} = a \text{ x } pIC_{50} + \text{b} \tag{1}$$

where i is the number of scoring functions; pIC_{50} = $-logIC_{50}$; SFc is the value of the control scoring function; SFi is the value of the scoring function; Wi is the weight. SFc in Equation I was employed to normalize the weighted equation (Yang, Chang, Chen, & Chen, 2011).

4.5.2. Clustering and Entropy-Based Scoring Methods

To improve the performance of scoring functions, cluster-based scoring methods have been developed which include entropic terms. These methods work by dividing the ligand binding modes into different clusters and entropic effect in each cluster is evaluated by the configurational space of the ligand poses. Combination of these methods with other docking methods improved binding mode prediction (Chang, Belew, Carroll, Olson, & Goodsell, 2008; Ruvinsky, 2007).

4.5.3. Machine-Learning Scoring

Machine learning scoring functions are now being used for prediction of native or near-native pose of a ligand docked to a protein's binding site, and for structure prediction of proteins. In protein structure prediction, machine-learning scores utilize artificial neural network and support vector machine (SVM) to correctly rank the predicted structures of each target according to their quality. It combines multiple features such as physical characteristics and probability distribution based potentials, including different quality structures (Ashtawy & Mahapatra, 2015; L. Li, Wang, & Meroueh, 2011; Qiu, Sheffler, Baker, & Noble, 2008). Examples of machine learning scoring function include SVR (support vector regression)-score, SVR and SVM-KB and SVM-EP scoring functions, and SVM target-specific scoring function.

4.5.4. Combination of Scoring Functions with Experimental (Binding Affinity) Data

An empirical scoring function, PHOENIX, has been developed with 34 descriptors from molecular mechanics calculations, 7 volume-based and shape-based terms to capture entropic contributions, and various surface terms, hydrogen bond donor and acceptor counts, and LogP. The novelty of PHOENIX is that it utilizes experimental data from isothermal titration calorimetry. It results in accurate predictions of binding energy and average prediction of enthalpy and entropy terms.

4.5.5. MM/GBSA and MM/PBSA Calculations

(MM/PBSA) Molecular Mechanics/Poisson Boltzmann Surface Area and (MM/GBSA) Molecular Mechanics/Generalized Born Surface Area methods calculate binding free energies of macromolecules by combining molecular mechanics and continuum solvation models. Considering the computational efficiency, MM/GBSA can serve as a powerful tool in drug design, where correct ranking of inhibitors is often emphasized (Y. Li, Liu, & Wang, 2010). MM/GBSA and MM/PBSA calculations were applied on 24 diverse protein-ligand complexes, each containing NMR-derived conformations. Modest correlations between experimentally measured binding free energies and the MM/GBSA and MM/PBSA were obtained. A comparison with four selected scoring functions (X-Score, PLP, ChemScore, and DrugScore) showed that MM/GBSA and MM/PBSA are much more computationally demanding (Hou, Wang, Li, & Wang, 2011; Y. Li, et al., 2010).

PROTEIN-PROTEIN DOCKING

PPIs play an important role in many biological events within a cell such as DNA synthesis, gene- expression, signal transduction, immune responses etc. The knowledge of atomic structure of protein-protein complexes is vital for understanding their biological function, for predicting the effect of mutations and for rational design of new PPI-based therapeutic agents. Due to high cost and technical difficulties in structure determination by experimental methods, protein-protein docking is being used for determining the structures of protein-protein complex (Guo, Wisniewski, & Ji, 2014; Sable & Jois, 2015). Various docking programs with varying sampling algorithm and scoring functions have been developed over last decades. As less information about binding site is available in protein-protein docking, the possible binding solutions are obtained by performing global search considering six degrees of freedom (i.e. three translational and three rotational) without requiring theoretical information about the binding site. Based on search strategies, docking programs can be classified into three categories: fast Fourier transform (FFT) or spherical Fourier transform (SFT)-based correlation algorithms, local shape feature matching approaches, and randomized search algorithms (S. Huang, 2015). Examples of docking programs based on FFT correlation search algorithms are DOT, FTDock, GRAMM, MolFit, PIPER, ZDOCK 1.3, ZDOCK 2.1, ZDOCK 2.3.2, ZDOCK 3.0.2, etc. Docking programs ATTRACT, PatchDock, and SwarmDock are based on randomized search algorithm or local shape matching (Table 4).

More information on protein-protein docking programs can be found in several other research papers (Dominguez, Boelens, & Bonvin, 2003; Sable & Jois, 2015; Smith & Sternberg, 2002).

In global docking (or ab initio docking), large number of degrees of freedom in protein's backbone and side chains are treated as rigid bodies to make the search process feasible. To perform the fast evaluation of binding modes, early global protein-protein docking programs like GRAMM and MolFit employed only shape complementarity and hydrophobic interactions, while programs like PatchDock include geometric hashing for fast surface patch matching. To consider full shape complementarity, sophisticated pairwise method has also been included in ZDOCK 2.1. Electrostatic interactions were first considered

Table 4. Protein-protein docking programs

Program	Search Algorithm	Scoring Functions
ATTRACT	Randomized search	Lennard-Jones type effective potentials and electrostatics
DOT	FFT-based correlation	vdW and electrostatic energies
FRODOCK	SFT-based correlation	vdW, electrostatics, and desolvation
FTDock	FFT-based correlation	Shape complementarity and electrostatic interactions
GRAMM	FFT-based correlation	Shape and hydrophobic match
HEX	SFT-based correlation	Surface complementarity and electrostatics
MolFit	FFT-based correlation	Geometric complementarity, hydrophobic complementarity, and electrostatic interactions
PatchDock	Local shape match	Geometric shape complementarity
PIPER	FFT-based correlation	Shape complementarity, electrostatic interactions, and knowledge-based pair potentials
SDOCKb	FFT-based correlation	vdW attractive potential, geometric collision, electrostatic potential, and desolvation energy
ZDOCK 1.3	FFT-based correlation	Shape Complementarity, desolvation, and electrostatics
ZDOCK 2.1	FFT-based correlation	Pairwise shape complementarity
ZDOCK 2.3.2	FFT-based correlation	Pairwise shape complementarity, electrostatics, and desolvation
ZDOCK 3.0.2	FFT-based correlation	Shape complementarity, electrostatics, and knowledge-based pair potentials

(S. Huang, 2015)

in FTDock. Since then, electrostatics term is always used along with shape complementarity in upcoming programs like ATTRACT, DOT, HEX, MolFit etc. The desolvation effects have also been included in programs such as FRODOCK, SDOCK, ZDOCK 1.3, and ZDOCK 2.3. Pairwise knowledge-based potentials included in the FFT-based docking programs like PIPER and ZDOCK 3.0 which utilize the information of experimental structures. HADDOCK and RosettaDock programs are used for refinement or site-specific docking with available information about the binding site but the only drawback is high computational cost. CAPRI (Critical Assessment of PRedicted interactions) experiment has been founded to perform the blind predictions of protein-protein docking and scoring algorithms, the results of which contain various biological information (Janin, Wodak, Lensink, & Velankar, 2015). Among different search strategies, FFT (or SFT) correlation methods achieved a good balance between the computational efficiency and the exhaustive global search, in which sampling covers every position of the grid of six-dimensional space (de Vries, Schindler, Chauvot de Beauchêne, & Zacharias, 2015; Schindler, de Vries, & Zacharias, 2015). It is very difficult to study the interactions between membrane proteins that occur inside the membrane bilayers. Recently, a novel method, RosettaMPDock was developed to predict PPIs within the membranes (Leman, Alford, & Gray, 2015).

A Case Study of Hot Spot-Based Design of Protein-Protein Interaction Inhibitor

Identification of hot spots is the most important step in PPI inhibitors design. Docking algorithms are often used to find out the hot spots. In general small molecular probes such as organic solvents are placed on a dense grid around the protein. Using rigid body docking with fast Fourier transform correlation approach, each probe generates many bound poses on the protein surface. For each probe, billions of docked poses are sampled. With the help of RMSD and consensus clustering, the minimized poses are clustered, and hot spots are defined. The binding site having largest cluster is defined as one hot spot. PPI complexes such as Bcl-xL, IL-2IL2R, p53-MDM2, and TNFα-TNFR1 were tested by hot-spot method for druggability. (Kenneth Morrow & Zhang, 2012).

The VHL (Von Hippel-Lindau tumor suppressor) protein is a component of the E3 ligase which interacts with hypoxia-inducible factor 1a (HIF1a) and promotes its degradation by activating ubiquitin proteasomal degradation pathway. VHL/HIF1a inhibitors can be used to treat chronic anemia. 3-hydroxyl-L-proline residue of HIF1a was used as a hot spot for interacting with VHL. Following this hot spot as starting point, new inhibitors were designed with the help of de novo design program BOMB. Compound (1) was discovered which exhibited an IC_{50} value of 117 ± 10 μM (Figure 4). Isoxazole moiety of compound (1) was found to interact with water molecule in the VHL/HIF1a complex. Introduction of an oxazole ring in the para position of the benzyl group resulted in the discovery of compound (2) with an IC_{50} value of 4.1 ± 0.4 μM. Further replacement of oxazole ring with 4-methylthiazole ring led to the generation of compound (3) with an IC_{50} value of 0.90 ± 0.03 μM and better interactions with a hydrophobic pocket in VHL (Guo, et al., 2014).

Figure 4. Designing of hydroxyl-proline containing VHL/HIF1a PPI inhibitors
(Guo, et al., 2014)

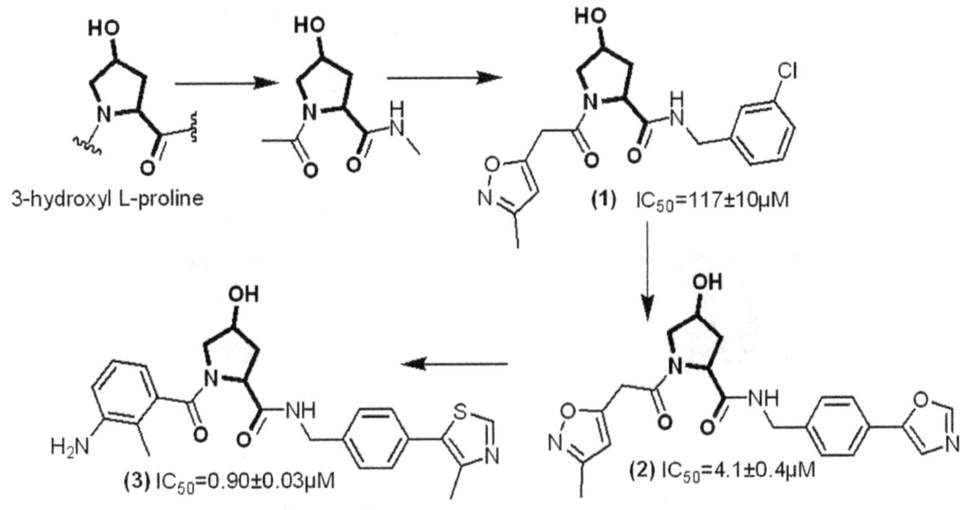

OTHER ADVANCEMENTS IN DOCKING METHODOLOGIES

1. Inverse Docking

Advancement in docking methodologies includes inverse docking, which has several applications in the field of drug research. First introduced in 2001, inverse docking has been used to identify and validate new potential target of known active compounds. It is also used to distinguish homology models of receptors, prediction of pharmacological profile of a new drug and helps to select less harmful treatment for a disease (Vasseur et al., 2015).

In a study, MDock program with scoring function (ITScore) was used to find out the targets of PRIMA 1. PRIMA 1 is a small molecule which restores mutant p53's tumor suppression function and leads to apoptosis of several cancer cells. Inverse docking involves a PDTD (potential drug target database) from which the target of known compound can be determined (Gao et al., 2008). A molecular surface was defined for each protein in the database, along with sphere points showing initial positions of ligand atoms. Ligand atoms matching the defined sphere points were sampled and ranked by the ITScore. The highest-score conformation was selected by performing rigid body docking of each pair of protein/PRIMA-1-conformations. Then each protein was ranked according to their ITScore. On the basis of sequence identity, the docked structures were then clustered into groups. BLAST search was run on PDTD proteins. Top scored human proteins were then identified and inhibitors were selected as anti-cancer agents for assay (Figure 5). As a result of study, oxidosqualene cyclase was identified as a potential target and Ro 48–8071 as a potent inhibitor of oxidosqualene cyclase and showed significant reduction in viability of cancer cells in bioassay (Grinter, Liang, Huang, Hyder, & Zou, 2011; S. Huang & Zou, 2006).

Figure 5. Inverse docking approach in target identification
(Grinter, et al., 2011)

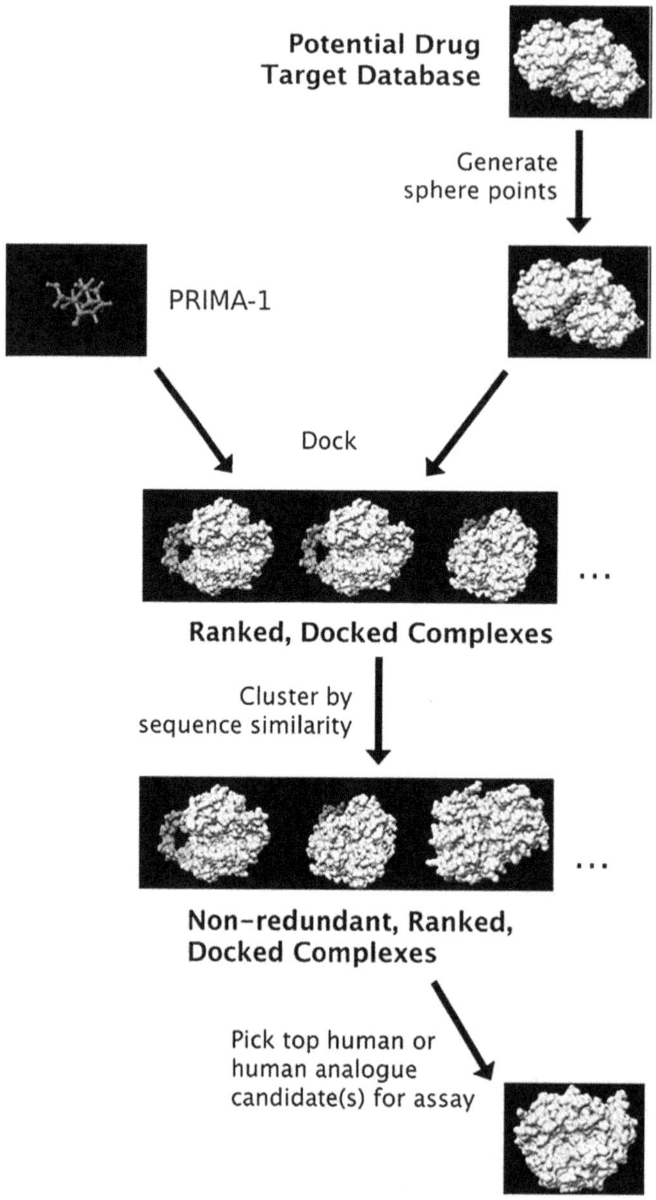

2. Fragment-Based Docking

Fragment-based drug design approaches are now being frequently used by computational chemists to discover drugs from small molecular fragments. To screen the chemical libraries, FBS (fragment-based screening) is now preferred over HTS (high throughput screening), as it produces higher hit rates, better ligand efficiency and has the ability to sample diverse set of molecules. In Figure 6 advantages of FBS over HTS is represented. It shows that HTS can identify a fully assembled, "drug-sized" ligands

Figure 6. Representation of drawback of HTS (left) and advantages of FBS (right)
(Erlanson & Jahnke, 2006)

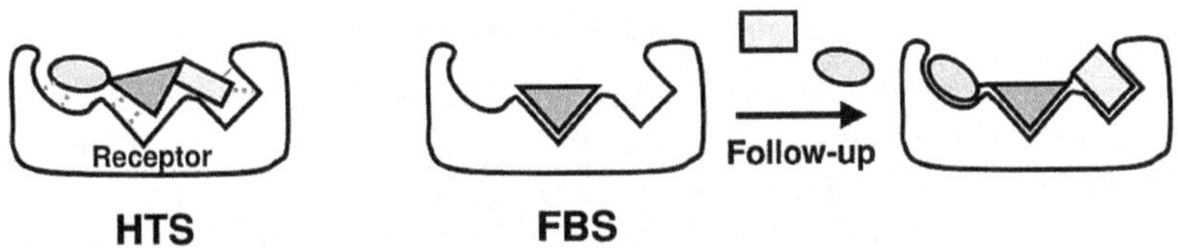

but with multiple compromised, non-optimal binding interactions whereas FBS can identify the ligands (with increase ligand affinity) for individual sub-pockets with good binding interactions (Erlanson & Jahnke, 2006).

In the same way, fragment-based docking is based on the fact that each molecular subunit involves at least one favorable interaction in the binding site. The fragment-based docking utilizes the following steps (1) finding optimal position of small molecular fragment in the binding site, (2) linking the small molecular fragments together to form complete molecule that is complementary to the target, and (3) estimating the binding affinity of the resulting molecules. Generally there are four phases of molecular decomposition i.e. ring identification, initial fragment definition, functional group merging, completion of valences. DAIM (Decomposition and Identification of Molecules)-SEED (Solvation Energy for Exhaustive Docking)-FFLD (Fast Flexible Ligand Docking) package and eHiTS are the examples of Fragment based docking programs (Rees, Congreve, Murray, & Carr, 2004).

Several advancements have been made in developing fragment-specific techniques. SERAPHiC (Selected Fragment-Protein Complexes), has been developed as a publicly available dataset of high quality protein-fragment complexes to use it as a validation tool in fragment docking (Favia, Bottegoni, Nobeli, Bisignano, & Cavalli, 2011).

3. Docking in Homology Models

Homology modeling or comparative modeling predicts the 3D-structure of an unknown protein by using homologous sequences of a known protein. The known structure serves as a template and the unknown structure is treated as a target. Homology modeling can be done in following steps: (1) template recognition and initial alignment, (2) alignment correction, (3) backbone generation, (4) loop modeling, (5) side-chain modeling, (6) model optimization, and (7) model validation (Krieger, Nabuurs, & Vriend, 2003).

A study described dopamine D1 receptor-agonist interactions by using homology model and mutagenesis. Dopamine D1 receptor is a G protein-coupled receptor which regulates intracellular signaling via agonist activation. Although the number of X-ray structures of G-protein coupled receptors has been steadily increasing, still no structure of the D1 receptor exists. A homology model was constructed for dopamine D1 receptor and effect of various catecholamines was studied. Several key interactions were found with conserved, trans-membrane residues and extra-cellular loop residues. MD simulations were used to examine the interactions. A mutagenesis of 12 orthosteric vicinity residues was used to examine the function of prototypical orthosteric D1 agonists and partial agonists. Mutation studies revealed that the ligands typically have weaker activity at many of the mutant receptors as compared to wild-type D1 receptor, reflecting the loss of interaction between the ligand and the receptor with the mutated side-chain (Mente, Guilmette, Salafia, & Gray, 2015).

Another homology model of CCR2 receptor was presented on the basis of the crystal structure of human beta-2 adrenoceptor (PDB ID: 2RH1). The reported model showed that 99.3% residues were in the core and allowed regions of the ramachandran plot but there was no residue present in the disallowed regions. The model also showed good RMSD value and alignment score with the template, human b2-adrenoceptor. The binding site is found within the transmembrane (TM) region and is large enough for docking the known CCR2 ligands. The docking results validated the homology model to meet various criteria that are necessary in molecular modeling studies (Singh & Elizabeth Sobhia, 2011). Promodel ("http://www.vlifesciences.com/products/BioPredicta/Product_BioPredicta.php," 2015), FoldX, RaptorX, MODELLER, Prime, LOMETS etc., are some examples of homology modeling programs (Chen, 2015).

4. Post-Processing

It is generally found that the top-ranked docking poses do not always represent the correct binding mode and misleading results were found after correlating docking score with biological data. Thus to improve the docking accuracy, post-processing approaches are now being frequently used. Rescoring by MM/PBSA and MM/GBSA techniques is an example of this approach. Using post-processing, binding free energy (MM/PBSA) of 222 Weel kinase inhibitors was successfully estimated and derived validated models were used for virtual screening. Prior to post processing, top-ranked docked poses showed wrong binding mode, whereas after post-processing they showed better binding mode. Subsequently, the docked poses giving the lowest binding free energy were selected and these values were used to establish a correlation with experimental data. A significant correlation between ΔG_{cal} and ΔG_{exp} ($r^2 = 0.58$) was obtained, where ΔG is binding energy. The study suggested that post-processing is a very useful approach in refining the docking outcome and can be used for correct prediction of biological activity of novel hits (Fukunishi, 2010; Wichapong, Lawson, Pianwanit, Kokpol, & Sippl, 2010).

5. Study of Absorption, Distribution, Metabolism, and Excretion (ADME) Properties

To reduce the use of experimental animals in clinical studies, research community is now being shifted towards the use of computational approaches which can correctly predict ADME properties of drugs. Now-a-days, docking methodologies are also being frequently used for *in silico* evaluation of ADME. Cytochrome P450 (CYP450) is an important enzyme which plays an important role in drug metabolism. Several cytochrome P450 isoforms are available which perform different metabolic functions. Various structure-based approaches have been developed which can predict the properties of compounds metabolized by or inhibit P450s (Park & Harris, 2003; Venhorst et al., 2003). Homology models have been developed for human P450 isoforms which serve as templates for docking to predict drug metabolism (de Groot, Ackland, Horne, Alex, & Jones, 1999). A crystal structure of human CYP2C9 complexed with warfarin was determined. It was found that the inhibitor binds to the iron-porphyrin system in the enzyme but do not interact directly with the cofactor. The binding mode of warfarin suggested that CYP2C9 may undergo an allosteric mechanism during its function. The newly discovered binding pocket also suggests that CYP2C9 may simultaneously accommodate multiple ligands during its biological function, and provides a possible molecular basis for understanding complex drug-drug interactions. Thus, metabolism of a compound may be predicted in silico with correct use of docking, MD, and density functional theoretical calculations (Williams et al., 2003).

FUTURE RESEARCH DIRECTIONS

Molecular docking has played a very important role in many successful drug discovery processes. At one hand various advancements have been done in various aspects of docking methodologies while on the other hand scientific community is also looking for refinement of docking results by addressing the limitations. Despite various advancements incorrect binding site prediction, use of undesirable small-molecule databases, inconsistent docking pose, high dock score, failure in MD simulations, lack of clarity whether the compound is an inhibitor or agonist, etc. are still not correctly answered by docking methodologies and urge further research. Even high scored conformations fail in the bioassays showing that there are shortcomings in docking methodologies which rank it lower than experimental approaches. It is still a question that how a correct forcefield can be applied to different set of compounds as well as proteins. Reliability of a protein structure available for docking study is still doubtful. Appropriate validation methods of protein structures are still a prime requirement for accurate docking e.g. Ramachandran plot. Handling large number of degrees of freedom with sufficient accuracy is very complex problem for a docking algorithm and this area still requires further advancement. Therefore, protein and ligand structure validation, protein flexibility, selection of best scoring function for different set of compounds and proteins, docking refinement, correct pose generation, Protein-protein docking, etc. are still an emerging area of future research.

CONCLUSION

The present chapter summarizes the key concepts of docking methodologies, including the recent advances to overcome the challenging issues like protein flexibility, ligand sampling, scoring function, protein-protein docking, etc. In addition, various future directions were also addressed in this chapter.

In the context of target discovery, an example has been discussed in which anti-cancer target identification has been done by inverse docking methodology. Homology modeling also plays an important role in discovery of important compounds for which 3D structure of protein targets is unavailable or difficult to be identified. An example of dopamine receptor and CCR2 receptor is presented here with the help of derived homology model.

From a thermodynamic point of view, scoring functions are still very primitive and need more attention. It has also been found that even high ranked conformations fail in bioassay, post-processing with the inclusion of correct pose sampling and rescoring can improve the docking results. Metabolism of a drug can be predicted with the help of docking studies including the MD and density function calculations, as exemplified by CYP450. As solvation and entropy calculations are costly and require high speed computers, now various scoring function incorporate these terms and give correct results.

Undoubtedly, molecular docking is an important methodology now being used by almost all major pharmaceutical companies and academic institutions. But it is also true that docking seems to have reached a performance plateau and is waiting for a real and important breakthrough.

REFERENCES

Abagyan, R., & Totrov, M. (1994). Biased Probability Monte Carlo Conformational Searches and Electrostatic Calculations for Peptides and Proteins. *Journal of Molecular Biology*, *235*(3), 983–1002. doi:10.1006/jmbi.1994.1052 PMID:8289329

Amaro, R., Baron, R., & McCammon, J. A. (2008). An improved relaxed complex scheme for receptor flexibility in computer-aided drug design. *Journal of Computer-Aided Molecular Design*, *22*(9), 693–705. doi:10.1007/s10822-007-9159-2 PMID:18196463

Antes, I. (2010). DynaDock: A new molecular dynamics-based algorithm for protein–peptide docking including receptor flexibility. *Proteins: Structure, Function, and Bioinformatics*, *78*(5), 1084–1104. doi:10.1002/prot.22629 PMID:20017216

Apostolakis, J., Plückthun, A., & Caflisch, A. (1998). Docking small ligands in flexible binding sites. *Journal of Computational Chemistry*, *19*(1), 21–37. doi:10.1002/(SICI)1096-987X(19980115)19:1<21::AID-JCC2>3.0.CO;2-0

Ashtawy, H. M., & Mahapatra, N. R. (2015). Machine-learning scoring functions for identifying native poses of ligands docked to known and novel proteins. *BMC Bioinformatics*, *16*(Suppl 6), S3. doi:10.1186/1471-2105-16-S6-S3 PMID:25916860

Avram, S., Pacureanu, L. M., Seclaman, E., Bora, A., & Kurunczi, L. (2011). PLS-DA - Docking Optimized Combined Energetic Terms (PLSDA-DOCET) Protocol: A Brief Evaluation. *Journal of Chemical Information and Modeling*, *51*(12), 3169–3179. doi:10.1021/ci2002268 PMID:22066983

B-Rao, C., Subramanian, J., & Sharma, S. D. (2009). Managing protein flexibility in docking and its applications. *Drug Discovery Today*, *14*(7–8), 394–400. PMID:19185058

Blundell, T. L., Jhoti, H., & Abell, C. (2002). High-throughput crystallography for lead discovery in drug design. *Nat Rev Drug Discov, 1*(1), 45-54.

Bottegoni, G., Kufareva, I., Totrov, M., & Abagyan, R. (2009). Four-Dimensional Docking: A Fast and Accurate Account of Discrete Receptor Flexibility in Ligand Docking. *Journal of Medicinal Chemistry*, *52*(2), 397–406. doi:10.1021/jm8009958 PMID:19090659

Chang, M. W., Belew, R. K., Carroll, K. S., Olson, A. J., & Goodsell, D. S. (2008). Empirical entropic contributions in computational docking: Evaluation in APS reductase complexes. *Journal of Computational Chemistry*, *29*(11), 1753–1761. doi:10.1002/jcc.20936 PMID:18351616

Chen, Y. (2009). Weighted Equation and Rules-A Novel Concept for Evaluating Protein-Ligand Interaction. *Journal of Biomolecular Structure & Dynamics*, *27*(3), 271–282. doi:10.1080/07391102.2009.10507315 PMID:19795911

Chen, Y. (2015). Beware of docking. *Trends in Pharmacological Sciences*, *36*(2), 78–95. doi:10.1016/j.tips.2014.12.001 PMID:25543280

Craig, I. R., Essex, J. W., & Spiegel, K. (2010). Ensemble Docking into Multiple Crystallographically Derived Protein Structures: An Evaluation Based on the Statistical Analysis of Enrichments. *Journal of Chemical Information and Modeling*, *50*(4), 511–524. doi:10.1021/ci900407c PMID:20222690

Davis, I. W., & Baker, D. (2009). RosettaLigand Docking with Full Ligand and Receptor Flexibility. *Journal of Molecular Biology*, *385*(2), 381–392. doi:10.1016/j.jmb.2008.11.010 PMID:19041878

de Groot, M. J., Ackland, M. J., Horne, V. A., Alex, A. A., & Jones, B. C. (1999). Novel Approach To Predicting P450-Mediated Drug Metabolism: Development of a Combined Protein and Pharmacophore Model for CYP2D6. *Journal of Medicinal Chemistry*, *42*(9), 1515–1524. doi:10.1021/jm981118h PMID:10229622

de Vries, S. J., Schindler, C. E. M., & de Beauchêne, C. (2015). A Web Interface for Easy Flexible Protein-Protein Docking with ATTRACT. *Biophysical Journal*, *108*(3), 462–465. doi:10.1016/j.bpj.2014.12.015 PMID:25650913

Desmet, J., Wilson, I. A., Joniau, M., De Maeyer, M., & Lasters, I. (1997). Computation of the binding of fully flexible peptides to proteins with flexible side chains. *The FASEB Journal*, *11*(2), 164–172. PMID:9039959

Dias, R., & de Azevedo, W. F. Jr. (2008). Molecular Docking Algorithms. *Current Drug Targets*, *9*(12), 1040–1047. doi:10.2174/138945008786949432 PMID:19128213

Dominguez, C., Boelens, R., & Bonvin, A. M. J. J. (2003). HADDOCK: A Protein–Protein Docking Approach Based on Biochemical or Biophysical Information. *Journal of the American Chemical Society*, *125*(7), 1731–1737. doi:10.1021/ja026939x PMID:12580598

Eldridge, M., Murray, C., Auton, T., Paolini, G., & Mee, R. (1997). Empirical scoring functions: I. The development of a fast empirical scoring function to estimate the binding affinity of ligands in receptor complexes. *Journal of Computer-Aided Molecular Design*, *11*(5), 425–445. doi:10.1023/A:1007996124545 PMID:9385547

Erlanson, D. A., & Jahnke, W. (2006). The Concept of Fragment-based Drug Discovery. *Fragment-based approaches in drug discovery*, 1-10.

Favia, A. D., Bottegoni, G., Nobeli, I., Bisignano, P., & Cavalli, A. (2011). SERAPhiC: A Benchmark for in Silico Fragment-Based Drug Design. *Journal of Chemical Information and Modeling*, *51*(11), 2882–2896. doi:10.1021/ci2003363 PMID:21936510

Feher, M. (2006). Consensus scoring for protein-ligand interactions. *Drug Discovery Today*, *11*(9–10), 421–428. doi:10.1016/j.drudis.2006.03.009 PMID:16635804

Ferrari, A. M., Wei, B. Q., Costantino, L., & Shoichet, B. K. (2004). Soft Docking and Multiple Receptor Conformations in Virtual Screening. *Journal of Medicinal Chemistry*, *47*(21), 5076–5084. doi:10.1021/jm049756p PMID:15456251

Fuhrmann, J., Rurainski, A., Lenhof, H.-P., & Neumann, D. (2010). A new Lamarckian genetic algorithm for flexible ligand-receptor docking. *Journal of Computational Chemistry*, *31*(9), 1911–1918. PMID:20082382

Fukunishi, Y. (2010). Post Processing of Protein-Compound Docking for Fragment-Based Drug Discovery (FBDD): In-Silico Structure-Based Drug Screening and Ligand-Binding Pose Prediction. *Current Topics in Medicinal Chemistry*, *10*(6), 680–694. doi:10.2174/156802610791111452 PMID:20337587

Gao, Z., Li, H., Zhang, H., Liu, X., Kang, L., Luo, X., & Jiang, H. et al. (2008). PDTD: A web-accessible protein database for drug target identification. *BMC Bioinformatics*, *9*(1), 104. doi:10.1186/1471-2105-9-104 PMID:18282303

Gohlke, H., Hendlich, M., & Klebe, G. (2000). Knowledge-based scoring function to predict protein-ligand interactions1. *Journal of Molecular Biology*, *295*(2), 337–356. doi:10.1006/jmbi.1999.3371 PMID:10623530

Goodford, P. J. (1985). A computational procedure for determining energetically favorable binding sites on biologically important macromolecules. *Journal of Medicinal Chemistry*, *28*(7), 849–857. doi:10.1021/jm00145a002 PMID:3892003

Grinter, S. Z., Liang, Y., Huang, S.-Y., Hyder, S. M., & Zou, X. (2011). An inverse docking approach for identifying new potential anti-cancer targets. *Journal of Molecular Graphics & Modelling*, *29*(6), 795–799. doi:10.1016/j.jmgm.2011.01.002 PMID:21315634

Guo, W., Wisniewski, J. A., & Ji, H. (2014). Hot spot-based design of small-molecule inhibitors for protein–protein interactions. *Bioorganic & Medicinal Chemistry Letters*, *24*(11), 2546–2554. doi:10.1016/j.bmcl.2014.03.095 PMID:24751445

Halgren, T. (2007). New Method for Fast and Accurate Binding-site Identification and Analysis. *Chemical Biology & Drug Design*, *69*(2), 146–148. doi:10.1111/j.1747-0285.2007.00483.x PMID:17381729

Halperin, I., Ma, B., Wolfson, H., & Nussinov, R. (2002). Principles of docking: An overview of search algorithms and a guide to scoring functions. *Proteins: Structure, Function, and Bioinformatics*, *47*(4), 409–443. doi:10.1002/prot.10115 PMID:12001221

Hart, T. N., & Read, R. J. (1992). A multiple-start Monte Carlo docking method. *Proteins: Structure, Function, and Bioinformatics*, *13*(3), 206–222. doi:10.1002/prot.340130304 PMID:1603810

Hernández-Santoyo, A., Tenorio-Barajas, A. Y., Altuzar, V., Vivanco-Cid, H., & Mendoza-Barrera, C. (2013). *Protein-Protein and Protein-Ligand Docking*. Academic Press.

Hou, T., Wang, J., Li, Y., & Wang, W. (2011). Assessing the Performance of the MM/PBSA and MM/GBSA Methods. 1. The Accuracy of Binding Free Energy Calculations Based on Molecular Dynamics Simulations. *Journal of Chemical Information and Modeling*, *51*(1), 69–82. doi:10.1021/ci100275a PMID:21117705

Huang, S. (2015). Exploring the potential of global protein-protein docking: An overview and critical assessment of current programs for automatic ab initio docking. *Drug Discovery Today*, *20*(8), 969–977. doi:10.1016/j.drudis.2015.03.007 PMID:25801181

Huang, S., & Zou, X. (2006). An iterative knowledge-based scoring function to predict protein-ligand interactions: I. Derivation of interaction potentials. *Journal of Computational Chemistry*, *27*(15), 1866–1875. doi:10.1002/jcc.20504 PMID:16983673

Huang, S.-Y., & Zou, X. (2010). Advances and Challenges in Protein-Ligand Docking. *International Journal of Molecular Sciences*, *11*(8), 3016–3034. doi:10.3390/ijms11083016 PMID:21152288

Janin, J., Wodak, S. J., Lensink, M. F., & Velankar, S. (2015). Assessing Structural Predictions of Protein–Protein Recognition: The CAPRI Experiment. *Reviews in Computational Chemistry*, *28*, 137–173.

Jiang, F., & Kim, S.-H. (1991). "Soft docking": Matching of molecular surface cubes. *Journal of Molecular Biology*, *219*(1), 79–102. doi:10.1016/0022-2836(91)90859-5 PMID:2023263

Kang, L., Li, H., Jiang, H., & Wang, X. (2009). An improved adaptive genetic algorithm for protein–ligand docking. *Journal of Computer-Aided Molecular Design*, *23*(1), 1–12. doi:10.1007/s10822-008-9232-5 PMID:18777161

Kenneth Morrow, J., & Zhang, S. (2012). Computational Prediction of Protein Hot Spot Residues. *Current Pharmaceutical Design*, *18*(9), 1255–1265. doi:10.2174/138161212799436412 PMID:22316154

Kitchen, D. B., Decornez, H., Furr, J. R., & Bajorath, J. (2004). Docking and scoring in virtual screening for drug discovery: methods and applications. *Nat Rev Drug Discov*, *3*(11), 935-949.

Knegtel, R. M. A., Kuntz, I. D., & Oshiro, C. M. (1997). Molecular docking to ensembles of protein structures1. *Journal of Molecular Biology*, *266*(2), 424–440. doi:10.1006/jmbi.1996.0776 PMID:9047373

Koshland, D. E. (1958). Application of a Theory of Enzyme Specificity to Protein Synthesis. *Proceedings of the National Academy of Sciences of the United States of America*, *44*(2), 98–104. doi:10.1073/pnas.44.2.98 PMID:16590179

Krieger, E., Nabuurs, S. B., & Vriend, G. (2003). Homology modeling. *Methods of Biochemical Analysis*, *44*, 509–524. PMID:12647402

Kuntz, I. D., Blaney, J. M., Oatley, S. J., Langridge, R., & Ferrin, T. E. (1982). A geometric approach to macromolecule-ligand interactions. *Journal of Molecular Biology*, *161*(2), 269–288. doi:10.1016/0022-2836(82)90153-X PMID:7154081

Lang, C. A., Ray, S. S., Liu, M., Singh, A. K., & Cuny, G. D. (2015). Discovery of LRRK2 inhibitors using sequential in silico joint pharmacophore space (JPS) and ensemble docking. *Bioorganic & Medicinal Chemistry Letters*, *25*(13), 2713–2719. doi:10.1016/j.bmcl.2015.04.027 PMID:25998502

Leach, A. R. (1994). Ligand docking to proteins with discrete side-chain flexibility. *Journal of Molecular Biology*, *235*(1), 345–356. doi:10.1016/S0022-2836(05)80038-5 PMID:8289255

Leman, J. K., Alford, R. F., & Gray, J. J. (2015). Rosetta-MPDock: A Novel Computational Tool for Protein-Protein Docking within the Membrane Bilayer. *Biophysical Journal*, *108*(2), 250a. doi:10.1016/j.bpj.2014.11.1382

Li, L., Wang, B., & Meroueh, S. O. (2011). Support Vector Regression Scoring of Receptor–Ligand Complexes for Rank-Ordering and Virtual Screening of Chemical Libraries. *Journal of Chemical Information and Modeling*, *51*(9), 2132–2138. doi:10.1021/ci200078f PMID:21728360

Li, Y., Liu, Z., & Wang, R. (2010). Test MM-PB/SA on True Conformational Ensembles of Protein-Ligand Complexes. *Journal of Chemical Information and Modeling*, *50*(9), 1682–1692. doi:10.1021/ci100036a PMID:20695488

Liu, Y., Zhao, L., Li, W., Zhao, D., Song, M., & Yang, Y. (2013). FIPSDock: A new molecular docking technique driven by fully informed swarm optimization algorithm. *Journal of Computational Chemistry*, *34*(1), 67–75. doi:10.1002/jcc.23108 PMID:22961860

Lyne, P. D. (2002). Structure-based virtual screening: An overview. *Drug Discovery Today*, *7*(20), 1047–1055. doi:10.1016/S1359-6446(02)02483-2 PMID:12546894

Meng, X.-Y., Zhang, H.-X., Mezei, M., & Cui, M. (2011). Molecular Docking: A Powerful Approach for Structure-Based Drug Discovery. *Current Computer-aided Drug Design*, *7*(2), 146–157. doi:10.2174/157340911795677602 PMID:21534921

Mente, S., Guilmette, E., Salafia, M., & Gray, D. (2015). Dopamine D1 receptor-agonist interactions: A mutagenesis and homology modeling study. *Bioorganic & Medicinal Chemistry Letters*, *25*(10), 2106–2111. doi:10.1016/j.bmcl.2015.03.079 PMID:25881819

Michalewicz, Z. (1996). *GAs: What Are They? Genetic Algorithms + Data Structures = Evolution Programs* (pp. 13–31). Springer Berlin Heidelberg. doi:10.1007/978-3-662-03315-9_2

Mohan, V., Gibbs, A. C., Cummings, M. D., Jaeger, E. P., & DesJarlais, R. L. (2005). Docking: Successes and Challenges. *Current Pharmaceutical Design*, *11*(3), 323–333. doi:10.2174/1381612053382106 PMID:15723628

Morris, G. M., Goodsell, D. S., Halliday, R. S., Huey, R., Hart, W. E., Belew, R. K., & Olson, A. J. (1998). Automated docking using a Lamarckian genetic algorithm and an empirical binding free energy function. *Journal of Computational Chemistry*, *19*(14), 1639–1662. doi:10.1002/(SICI)1096-987X(19981115)19:14<1639::AID-JCC10>3.0.CO;2-B

Park, J.-Y., & Harris, D. (2003). Construction and Assessment of Models of CYP2E1: Predictions of Metabolism from Docking, Molecular Dynamics, and Density Functional Theoretical Calculations. *Journal of Medicinal Chemistry*, *46*(9), 1645–1660. doi:10.1021/jm020538a PMID:12699383

Paul, N., & Rognan, D. (2002). ConsDock: A new program for the consensus analysis of protein–ligand interactions. *Proteins: Structure, Function, and Bioinformatics*, *47*(4), 521–533. doi:10.1002/prot.10119 PMID:12001231

Plewczynski, D., Łażniewski, M., Grotthuss, M. V., Rychlewski, L., & Ginalski, K. (2011). VoteDock: Consensus docking method for prediction of protein–ligand interactions. *Journal of Computational Chemistry*, *32*(4), 568–581. doi:10.1002/jcc.21642 PMID:20812324

Poli, G., Giuntini, N., Martinelli, A., & Tuccinardi, T. (2015). Application of a FLAP-Consensus Docking Mixed Strategy for the Identification of New Fatty Acid Amide Hydrolase Inhibitors. *Journal of Chemical Information and Modeling*, *55*(3), 667–675. doi:10.1021/ci5006806 PMID:25746133

Qiu, J., Sheffler, W., Baker, D., & Noble, W. S. (2008). Ranking predicted protein structures with support vector regression. *Proteins: Structure, Function, and Bioinformatics, 71*(3), 1175–1182. doi:10.1002/prot.21809 PMID:18004754

Rees, D. C., Congreve, M., Murray, C. W., & Carr, R. (2004). Fragment-based lead discovery. *Nat Rev Drug Discov, 3*(8), 660-672.

Roy, K., Kar, S., & Das, R. N. (2015). Other Related Techniques. In K. R. K. N. Das (Ed.), *Understanding the Basics of QSAR for Applications in Pharmaceutical Sciences and Risk Assessment* (pp. 357–425). Boston: Academic Press.

Ruvinsky, A. (2007). Calculations of protein-ligand binding entropy of relative and overall molecular motions. *Journal of Computer-Aided Molecular Design, 21*(7), 361–370. doi:10.1007/s10822-007-9116-0 PMID:17503189

Sable, R., & Jois, S. (2015). Surfing the Protein-Protein Interaction Surface Using Docking Methods: Application to the Design of PPI Inhibitors. *Molecules (Basel, Switzerland), 20*(6), 11569–11603. doi:10.3390/molecules200611569 PMID:26111183

Schindler, C. E. M., de Vries, S. J., & Zacharias, M. (2015). iATTRACT: Simultaneous global and local interface optimization for protein–protein docking refinement. *Proteins: Structure, Function, and Bioinformatics, 83*(2), 248–258. doi:10.1002/prot.24728 PMID:25402278

Schneider, G., & Böhm, H.-J. (2002). Virtual screening and fast automated docking methods. *Drug Discovery Today, 7*(1), 64–70. doi:10.1016/S1359-6446(01)02091-8 PMID:11790605

Sherman, W., Day, T., Jacobson, M. P., Friesner, R. A., & Farid, R. (2006). Novel Procedure for Modeling Ligand/Receptor Induced Fit Effects. *Journal of Medicinal Chemistry, 49*(2), 534–553. doi:10.1021/jm050540c PMID:16420040

Singh, R., & Elizabeth Sobhia, M. (2011). Homology modeling of human CCR2 receptor. *Medicinal Chemistry Research, 20*(9), 1704–1712. doi:10.1007/s00044-010-9497-9

Smith, G. R., & Sternberg, M. J. E. (2002). Prediction of protein–protein interactions by docking methods. *Current Opinion in Structural Biology, 12*(1), 28–35. doi:10.1016/S0959-440X(02)00285-3 PMID:11839486

Srivastav, V. K., & Tiwari, M. (2015). k-nearest neighbor molecular field analysis based 3D-QSAR and in silico ADME/T studies of cinnamoyl derivatives as HIV-1 integrase inhibitors. *Medicinal Chemistry Research, 24*(2), 684–700. doi:10.1007/s00044-014-1183-x

Subramanian, J., Sharma, S., & B-Rao, C. (2006). A Novel Computational Analysis of Ligand-Induced Conformational Changes in the ATP Binding Sites of Cyclin Dependent Kinases. *Journal of Medicinal Chemistry, 49*(18), 5434–5441. doi:10.1021/jm060172s PMID:16942017

Subramanian, J., Sharma, S., & B-Rao, C. (2008). Modeling and Selection of Flexible Proteins for Structure-Based Drug Design: Backbone and Side Chain Movements in p38 MAPK. *ChemMedChem, 3*(2), 336–344. doi:10.1002/cmdc.200700255 PMID:18081134

Taylor, R. D., Jewsbury, P. J., & Essex, J. W. (2002). A review of protein-small molecule docking methods. *Journal of Computer-Aided Molecular Design, 16*(3), 151–166. doi:10.1023/A:1020155510718 PMID:12363215

Teodoro, M. L., Phillips, G. N., & Kavraki, L. E. (2001). *Molecular docking: A problem with thousands of degrees of freedom.* Paper presented at the Robotics and Automation, 2001. doi:10.1109/ROBOT.2001.932674

Vasseur, R., Baud, S., Steffenel, L. A., Vigouroux, X., Martiny, L., Krajecki, M., & Dauchez, M. (2015). Inverse docking method for new proteins targets identification: A parallel approach. *Parallel Computing, 42*, 48–59. doi:10.1016/j.parco.2014.09.008

Venhorst, J., ter Laak, A. M., Commandeur, J. N. M., Funae, Y., Hiroi, T., & Vermeulen, N. P. E. (2003). Homology Modeling of Rat and Human Cytochrome P450 2D (CYP2D) Isoforms and Computational Rationalization of Experimental Ligand-Binding Specificities. *Journal of Medicinal Chemistry, 46*(1), 74–86. doi:10.1021/jm0209578 PMID:12502361

Westhead, D. R., Clark, D. E., & Murray, C. W. (1997). A comparison of heuristic search algorithms for molecular docking. *Journal of Computer-Aided Molecular Design, 11*(3), 209–228. doi:10.1023/A:1007934310264 PMID:9263849

Wichapong, K., Lawson, M., Pianwanit, S., Kokpol, S., & Sippl, W. (2010). Postprocessing of Protein–Ligand Docking Poses Using Linear Response MM-PB/SA: Application to Wee1 Kinase Inhibitors. *Journal of Chemical Information and Modeling, 50*(9), 1574–1588. doi:10.1021/ci1002153 PMID:20712342

Williams, P. A., Cosme, J., Ward, A., Angove, H. C., Matak Vinkovic, D., & Jhoti, H. (2003). Crystal structure of human cytochrome P450 2C9 with bound warfarin. *Nature, 424*(6947), 464-468.

Yang, S., Chang, S., Chen, H., & Chen, Y. (2011). Identification of Potent EGFR Inhibitors from TCM Database@Taiwan. *PLoS Computational Biology, 7*(10), e1002189. doi:10.1371/journal.pcbi.1002189 PMID:22022246

Yuriev, E., Agostino, M., & Ramsland, P. A. (2011). Challenges and advances in computational docking: 2009 in review. *Journal of Molecular Recognition, 24*(2), 149–164. doi:10.1002/jmr.1077 PMID:21360606

Yuriev, E., & Ramsland, P. A. (2013). Latest developments in molecular docking: 2010–2011 in review. *Journal of Molecular Recognition, 26*(5), 215–239. doi:10.1002/jmr.2266 PMID:23526775

Chapter 12
Docking Methodologies and Recent Advances

Ashwani Kumar
Guru Jambheshwar University of Science and Technology, India

Ruchika Goyal
Guru Jambheshwar University of Science and Technology, India

Sandeep Jain
Guru Jambheshwar University of Science and Technology, India

ABSTRACT

Docking, a molecular modelling method, has wide applications in identification and optimization in modern drug discovery. This chapter addresses the recent advances in the docking methodologies like fragment docking, covalent docking, inverse docking, post processing, hybrid techniques, homology modeling etc. and its protocol like searching and scoring functions. Advances in scoring functions for e.g. consensus scoring, quantum mechanics methods, clustering and entropy based methods, fingerprinting, etc. are used to overcome the limitations of the commonly used force-field, empirical and knowledge based scoring functions. It will cover crucial necessities and different algorithms of docking and scoring. Further different aspects like protein flexibility, ligand sampling and flexibility, and the performance of scoring function will be discussed.

INTRODUCTION

Molecular docking is an entrenched computational technique which can accurately predict the orientation of a ligand and its binding affinity within the constraints of a receptor binding site (Bachwani & Kumar, 2011). Docking of a ligand into the receptor pocket involves several degrees of freedom. Besides the six degrees of translational, rotational, as well as conformational degrees of freedom, some algorithm only uses the translation and orientation degrees of freedom. By increasing computer's performance and using various new algorithms we can explore the ligand conformational degree of freedoms. There are simple methods for docking of rigid ligand with rigid receptors and flexible ligand with rigid receptors,

DOI: 10.4018/978-1-5225-0115-2.ch012

but the general method of docking conformations of flexible ligand and receptors is problematic (Leach, Shoichet & Peishoff, 2006).

Besides protein ligand docking, there are protein-protein docking, ligand-nucleic acid docking and protein-nucleic acid docking. Protein-protein docking is important for knowing the structural and functional features of the cell. It is similar to protein ligand docking; scoring and searching algorithms are used. The softwares available for this kind of docking are ZDOCK, HADDOCK, GRAMM-X, Hex protein docking etc. (Ehrlich & Rebecca, 2001). Ligand-nucleic acid docking involves the docking of ligand with nucleic acid like DNA and RNA. Less attention has been made in regard to docking with nucleic acid. Cisplatin, actinomycin and netropsin are example of drugs that shows ligand-DNA docking but less effort has been done in designing of ligands that binds to RNA targets (Mohan, Gibbs, Cummings, Jaeger, & DesJarlais, 2005). Protein-nucleic acid docking requires a set of well defined test cases that forms a common ground for development, validation and comparison of docking method. Software that carried out this kind of docking is HADDOCK (Dijk and Bonvin, 2010).

Earlier 'soft-potential docking' and rotamer libraries were used for flexible receptor docking methods (B-Rao, Subramanian, & Sharma, 2009). Currently X-ray crystallography, NMR, Monte Carlo sampling, Normal Modes-based methods and Molecular Dynamics methods are being focused on constructing an ensemble of structures produced in flexible type docking (Feixas, Lindert, Sinko, & McCammon, 2014; Sousa et al., 2013; Durrant, & McCammon, 2010; Cavasotto, Orry, & Abagyan, 2005). Principle techniques for docking methods use various searching and scoring algorithms like molecular dynamics, Monte Carlo methods, genetic algorithms, fragment-based methods (Mohan, Gibbs, Cummings, Jaeger, & DesJarlais, 2005), point complementary methods, distance geometry methods, tabu searches, evolutionary algorithm, particle swarm optimization and ant colony optimization which are mentioned in detail in later part of this chapter (Taylor, Jewsbury, & Essex, 2002).

FUNDAMENTAL NECESSITIES

Molecular docking program emphasize on the following basic requirements (Mahajan, & Gill, 2014; Krovat, Steindl, & Langer, 2005):

1. A target protein structure with or without a bound ligand is detected by various experimental techniques like NMR or X-Ray crystallography, but if protein structure is not present then protein prediction is done by any technique like threading modelling, homology modelling.
2. Database containing existing or virtual compounds for the docking process
3. Sampling and scoring method, desired scoring and searching algorithms require a computational framework for its efficient working
4. The three-dimensional structure of the protein ligand complex has to be studied in depth of atomic resolution.

MOLECULAR DOCKING MODELS

1. **Lock and Key Model:** Emil Fischer, in 1890, proposed this model which explains the functioning of biological systems. Ligand fits into the active site of the macromolecule as depicts in Figure 1.

Figure 1. Lock and key model

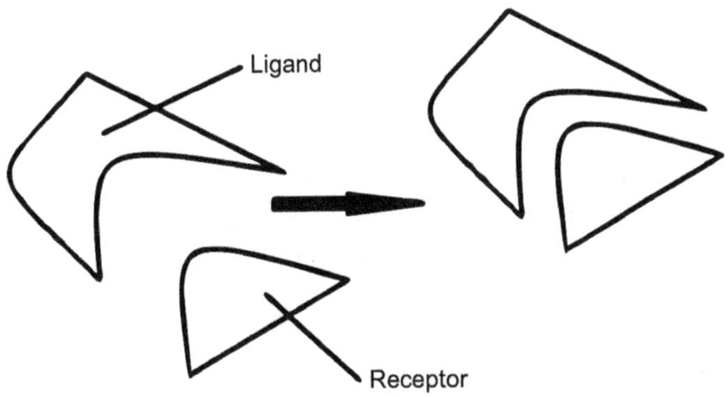

2. **Induced-Fit Model:** Daniel Koshland (1958) introduced the theory of induced-fit. In this, both ligand and receptor mutually interact with each other through small conformational change until an optimal complex is formed (Figure 2).
3. **Conformation Ensemble Model:** Buyong Ma et al., (2003) gave this theory. In this ligand flexibility is achieved by rigid docking of an ensemble of pregenerated ligand conformations with other docking programs.

DOCKING METHODOLOGIES

Docking has been divided in three parts based on the order of increasing complexity (Sousa et al., 2013; Meng, Zhang, Mezei, & Cui, 2011; Mohan et al., 2005):

1. **Rigid Body Docking:** Various docking suites use this type of docking method as first step. In this both ligand and receptor are considered as rigid, which limits the search space. It excludes any kind of flexibility and considers only three types of rotational and three translational degrees of freedom. Examples of software used for this type of docking are R-Dock, Z-Dock, MultiDock, FTDock.

Figure 2. Induced fit model

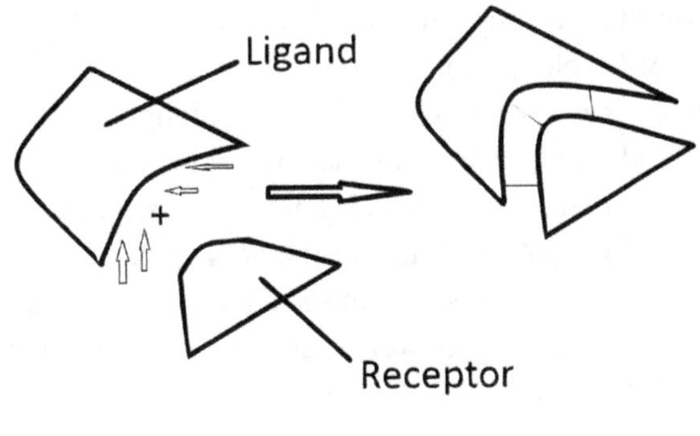

2. **Flexible Ligand Docking/Quasi Docking:** In this type of docking the receptor is kept rigid and ligand is considered as flexible. Gold, FlexX, Glide, AutoDock, ICM, Confort, Omega, Catalyst, Fred are the types of software using this method.
3. **Flexible Docking:** Both receptor and ligand are treated as flexible. Receptor flexibility is still a challenge in this type of docking. It comes under the category of Induced-fit paradigm. In this, both ligand and protein changes their conformations to set into minimum energy fit complex. Also, the high computational cost hurdles its use in screening of large chemical database. Affinity, FlexX-Ensemble, Forecaster and Slide are the software for this category.

OTHER METHODOLOGICAL ADVANCES

- **Improving Pose Prediction:** Docking algorithms produce a number of docked poses and ranked them based on scoring function, but the dependence only on scoring function sometimes results in false data. So, there is a need of some advances in methodology of scoring function which can improve the prediction of poses for better ranking (Yuriev, Agostino, & Ramsland, 2011).
- **Fragment Based Docking Method:** This method will be discussed in algorithm section of this chapter.
- **Inverse Docking:** Small molecule of interest (ligand) is docked into a library of receptor. It is used for identification and validation of potential targets of various active compounds. This method helps in prediction of various undesirable side-effects like toxicity (Yuriev, & Ramsland, 2013).
- **Blind Docking:** Blind docking is carried out when the ligand binding site is unknown. No major docking software are available that carried out this kind of docking. AutoDock is one of the software that carried out blind docking if structure of ligand, receptor is known without defining *a priori* ligand binding site (Grosdidier, Zoete, & Michielin, 2009)
- **Covalent Docking:** Covalent docking is a tool to study the covalent interactions between ligand and the receptor. It has much more stronger binding affinity that prolongs the duration of biological effect. Amoxicillin, Cefaclor, Isoniazid, Azacitidine, PenicillinV, Warfarin, Orlistat and various other drugs form covalent bond with the target proteins (Kumalo, Bhakat, & Soliman, 2015; Singh et al., 2011). CovalentDock package is used for covalent bond forming ligands. This is used for the accurate and large scale accurate covalent virtual screening. Key features of this package is that for the estimation of covalent binding, it has empirically calibrated energy model and it automatically detect and process structural patterns for the determination of prominent covalent linkage. Beside the advantage of its user friendliness, a limitation of this kind of package is that those who lack in experience face difficulties in compiling, installing and in using this software. To solve this issue, nowadays a fully functional web server the CovalenDock cloud is available (Ouyang, Zhou, Ge, Li & Kwoh, 2013).
- **Surflex Dock:** Surflex dock is used in virtual high throughput screening combined with state of the art speed, accuracy and usability. It requires an updated and re-parameterized scoring function called empirical scoring function, a searching engine and negative training data. It combines a unique scoring function with a patented state of the art search engine. This combination gives excellent docking accuracy and superior screening results. It provides accurate scoring, good speed, easier protein preparation, easier ligand preparation, protomol guided docking, ring flexing, and parallelization (Tripos, L.P., 2010).

- **Docking in Combination with Molecular Dynamics:** Docking only considers the binding affinity of a complex while molecular dynamics (MD) works on the complex shape which changes with time. Sometimes, in spite of the good docking score, it fails in MD simulations. So, it is necessary to validate the docking results with molecular dynamics. Docking is used in combination with molecular dynamics to study the dynamic behaviour of the complexes (Chen, 2015). Lafleur et al., 2009, studied the kinase-ligand binding mode determined by fragment-based docking to validate the MD. It is used as a screening tool for development and optimization of new drug delivery system (Yuriev et al., 2011).
- **Post Processing:** Various procedures are applied to improve the accuracy of the docking and virtual screening performance after taking pose generated by the existing docking and scoring protocols. Example of this kind of procedure is rescoring by MM/PBSA (Molecular Mechanics with Poisson-Boltzmann Surface Area) and MM/GBSA (Molecular Mechanics with Generalized Born Surface Area). Combination of rescoring techniques and multiple docked poses leads to better improvement in the docking methodologies (Meng et al., 2011).
- **Docking into Homology Models:** Docking is a computational calculation based on a range of assumptions and approximations that sometimes leads to some uncertain outcomes. Therefore, if adding another uncertain dimension, that is, a homology model which itself is a result of assumptions and approximations, increases the risk. When experimental data is not easily obtained and the information about complex structures or affinities is required (e.g., for GPCRs), then docking or VS is carried out against homology models and such protocols should be systematically evaluated for their robustness. Docking into various receptor proteins like neuronal nicotinic receptor (Mahasenan et al., 2011), *trypanosomabrucei* phosphodiesterases (Bland et al., 2011), prostaglandin E synthase (He,& Lai, 2011), aryl hydrocarbon receptor transcription factors (Motto, Bordogna, Soshilov, Denison & Bonati, 2011), as well as VS against the s1 receptor (Laurini et al., 2011), calpain (Stuart et al., 2011), kinases (Brylinski, & Skolnick, 2010), and GPCRs (Wada, Kanamori, Nakamura, & Fukunishi, 2011) are the examples of homology models (Yuriev et al., 2013).

As shown in Figure 3, the docking protocol is a combination of two functions: Searching function and scoring function (Hernández-Santoyo, Tenorio-Barajas, Altuzar, Vivano-Cid, & Mendoza-Barrera, 2013; Taylor et al. 2002).

SEARCHING FUNCTION

This function is used to represent the algorithms used for the flexible docking as it considers protein flexibility as well as ligand flexibility. It is used to explore the conformational space of the ligand that binds to the receptor. Various types of docking algorithms will be discussed in later parts of the chapter.

Protein Flexibility in Docking

There are many simple methods for docking of rigid ligands with rigid receptors but with flexible receptors is somehow, a formidable challenge in molecular docking studies. It is due to the large size of protein and many degrees of freedom, a small change in receptor conformation leads to increase in the

Figure 3. Docking protocol

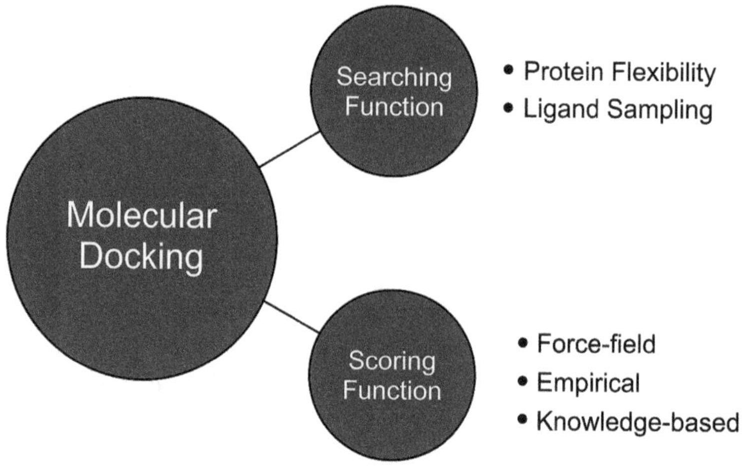

number of receptor conformations (Huang et al., 2010). At the time of docking, a powerful exploration of accessible points for a fully flexible active site is not feasible. Some approaches have been incorporated in order to restrict the number of degrees of freedom like soft docking, side-chain flexibility, molecular relaxation and protein ensemble docking (Huang, & Zou, 2010). In the last few years, advances are made in the field of protein flexibility with the improvement in experimental techniques like NMR, molecular dynamics simulations and sampling techniques. X-Ray crystallography and NMR techniques are basically used for structure determination of proteins (Huang, & Zou, 2007; Legge, Budi, Treutlein, & Yarovsky, 2006). But, some proteins cannot be crystallized then NMR techniques are used for those structures. NMR technique is also used to study the dynamic nature of proteins. But these methods are quite expensive. An alternative of this is taking different conformations snapshot from MD simulations. Some uses Monte Carlo minimization procedure e.g. IFREDA is used for generating ensemble of receptor conformations (Krovat et al., 2005). Cavasotto et al., 2005, gave a new algorithm to generate an ensemble of representative receptor back bone through a combination of normal mode analysis.

Soft Docking

It accounts for only small conformational changes and works by softening the van der Waals interactions and leads to a small degree of overlap between the ligand and receptor. This is computationally efficient method in which only scoring parameters is changed and other parameters remains the same as in case of rigid docking. It can detect only subtle changes (Sousa et al., 2013; Huang et al., 2010).

Side-Chain Flexibility

Discrete side-chain flexibility occurs due to the exploration of rotamer libraries. This method fails to fully simulate ligand-induced flexibility as backbone conformations are fixed only side chain conformations are flexible (Sousa et al., 2013; Huang et al., 2010). Zavodsky and Kuhn (2005) proposed a 'minimal rotation hypothesis' and used 'SLIDE' as a docking algorithm. They made several attempts to sort out ligand–receptor steric clashes by a minimizing the number of side-chain conformations (Totrov, & Abagyan, 2008).

Figure 4. Representation of molecular relaxation

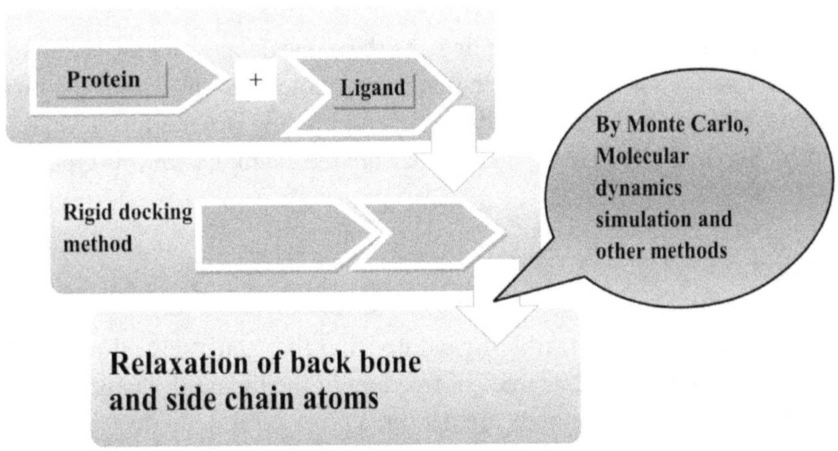

Molecular Relaxation

Despite being time consuming, this type of method is more demanding than side-chain flexibility method as it involves both side chain as well as back bone conformations. It first involves the method of rigid docking to fit the ligand into the receptor. After that the various relaxation procedures are used to relax the backbone and side-chain atoms like Monte Carlo, Molecular Dynamics simulations (Huang et al., 2010) and others as illustrated in Figure 4.

Protein Ensemble Docking

It is the most widely used protein flexibility method which utilizes multiple protein structures to represent various possible conformations (Huang et al., 2010). In this method, side-chain and back bone conformation changes in restricted manner. Multiple protein structure generates an average receptor grid by seeing the flexibility of multiple structures which therefore, significantly minimizes the cost. Ensemble of docking is generated by experimental procedure like NMR and others such as Monte Carlo simulations and Molecular Dynamics simulations (Huang et al., 2010).

Ligand Sampling and Flexibility

Ligand sampling is the most prominent area in the field of molecular docking by exploring all possible ligand degrees of freedom to get the experimental binding orientation. It requires sampling algorithms to explore various ligand conformations and their binding with the proteins. Binding site can be active site, a dimer interface or other site of interest (Huang et al., 2010). Ligand sampling is done by two approaches, *viz,* whole molecule approach and fragment based sampling approach. To sample the pose space whole molecule method is used and it produces various pre-calculated ligand conformations by Monte Carlo (MC), Molecular Dynamic simulations (MD), Genetic algorithms (GA), simulated annealing (SA) and fragment-based techniques. A rough idea of three types of sampling algorithms (shape matching, systematic search and stochastic algorithms) is given below:

Shape Matching

It is the simplest method used for ligand sampling. As the name depicts, it places the ligand in the position that its molecular surface is in complement with the molecular surface of the binding site of the receptor. There are six degrees of freedom for the ligand binding orientation. DOCK, FRED, FLOG, EUDOC, LigandFit, Surflex, MS-DOCK, and MDock are the examples for this type of method (Huang et al., 2010).

Systematic Searches

This method is commonly used for flexible-ligand docking that generates all possible ligand binding conformations after going through all degrees of freedom of the ligand. Exhaustive search, fragmentation and conformational ensemble methods are the three types of systematic search algorithms. It systematically endures all possible conformations and demonstrates the brute force solution to the docking problem. All molecules are supposed to be rigid and evaluation of their interaction energy from a force field model is done. Glide and FRED are two exhaustive examples of systematic search (Huang et al., 2010). Systematic search methods are of three types as depicted in Figure 5.

Figure 5. Types of systematic search methods

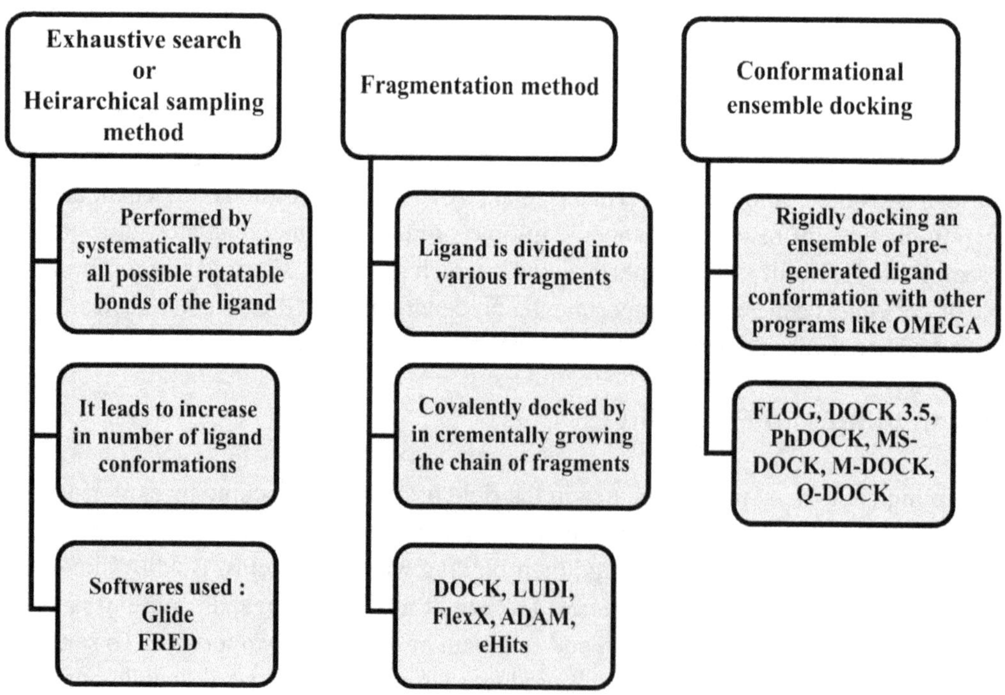

Stochastic Algorithms

Sampling of ligand binding orientations and conformations is done by making random changes to either a single ligand or population of ligands (Stark, & Powers, 2012). Four popular random approaches are Monte Carlo (MC) methods, evolutionary algorithms, Tabu search methods, Particle swarm optimization method, Ant colony optimization (Huang et al., 2010).

SCORING FUNCTIONS

Implication of an acceptable solution to the sampling problem is very challenging in finding out the correct binding mode. The rationale behind the use of the scoring procedure is to identify the correct binding pose by its lowest energy value and also to rank the protein-ligand complexes in accordance with their binding affinity which is calculated as a summation of independent terms. Scoring function depends upon the size of the molecule which is in proportionality with score, its entropy, and the solvation and desolvation effects. Scoring functions are subdivided into force-field, empirical, and knowledge based approaches (Schulz-Gasch, & Stahl, 2004).

Force-field based approach is used in protein-ligand docking as well as in molecular dynamics simulations. Binding affinity is calculated by adding van der Waals interactions, electrostatic interactions, bond stretching/ bending and torsional forces. The two shortcomings of this method are steric clash that cannot be avoided and the exaggerating polar interactions which can be compensated by using desolvation energy. AMBER and CHARMM (Wong, 2008) are the two examples of this category. DOCK is the typical force-field scoring function whose energy parameters are taken from AMBER force fields.

$$E = \sum_i \sum_j \left(\frac{A_{ij}}{r_{ij}^{12}} - \frac{B_{ij}}{r_{ij}^{6}} + \frac{q_i q_j}{\in\left(r_{ij}\right) r_{ij}} \right)$$

where, r_{ij} stands for the distance between protein atom i and ligand atom j, A_{ij} and B_{ij} are the VDW parameters, and q_i and q_j are the atomic charges. ε (r_{ij}) is usually set to 4r, reflecting the screening effect of water on electrostatic interactions. The effect of solvent is determined by using distance-dependent dielectric constant but it does not account for desolvation effect (Huang et al., 2010).

Empirical based approach is based on hydrophilic and hydrophobic contacts. In spiteof the key issues like the proper modelling of repulsive forces as well as the correct balance between hydrophilic and hydrophobic contacts, it is very successful in modeling complex molecular recognition phenomena with a very simplistic interaction model. Examples include LUDI (Wong, 2008), ChemScore, and X-SCORE.

Estimation of binding affinity of a complex is based on a set of weighed energy terms.

$$\Delta G = \sum_i W_i . \Delta G_i$$

where, ΔG_i represents individual empirical energy terms such as VDW energy, electrostatics, hydrogen bond, desolvation, entropy, hydrophobicity, etc. (Huang et al., 2010). The corresponding coefficients W_i are determined by reproducing the binding affinity data of training set of protein-ligand complexes with known three-dimensional structures, using least squares fitting.

It is faster than force field in binding energy calculations as it uses the simple energy terms.

Knowledge based approach is also known as statistical potential-based scoring function. As the name indicates, it is based on statistical analysis of experimentally determined complex structures. In this, the potential parameters are derived based on the structural information of the protein-ligand complex and after that sum of interactions of all protein-ligand atom pairs present in the complex are taken. The method shows a good balance between speed and accuracy as compared to the above two mentioned approaches. Some problems persist with this approach like excluded volume, reference state and sparse data problem. DFIRE, DrugScore, ITScore, PMF-score, and SMOG (Wong, 2008) are some of the examples based on the knowledge based scoring functions. The potential is calculated by using inverse Boltzmann relation:

where, $$w(r) = -\kappa_B T \ln\left[\frac{\rho(r)}{\rho*(r)}\right]$$

κ_B is the Boltzmann constant, T is the absolute temperature of the system, ρ(r) is the number density of the protein–ligand atom pair at distance r, and ρ*(r) is the pair density in a reference state where the inter-atomic interactions are zero (Huang et al., 2010).

OTHER ADVANCEMENT IN SCORING FUNCTION

Quantum Mechanics Method (QM Method)

Nowadays, Quantum mechanics method is boomed in the area of computational drug design and development due to its high accuracy for the estimation of binding affinities. For low to medium throughput *in-silico* screening, linear scaling QM was used for efficient approximations. QM based procedures are being used for the estimation of protonation states of ionisable groups, evaluating energies and molecular structure optimization. In HTS, it is employed to derive robust quantitative structure activity relationship (Zhou, Huang, & Caflisch, 2010). Raha and Merz gave this approach to overcome the inability of the above three approaches for a set of 18 carbonic anhydrase and five carboxy peptidase inhibitors (Krovat et al., 2005).

Consensus Scoring

Recent advances are made by researchers to increase the performance of scoring functions in spite of their limitations. This can be achieved by combining several scoring functions by using consensus scoring (Yuriev, Agostino, & Ramsland, 2011). It is used to lessen the false positive results of virtual screening and also to improve the structure prediction by improving the probability of finding a correct solution in hope to balance out the errors of the individual scoring functions. Examples of consensus scoring are MultiScore and X-Cscore (Huang et al., 2010).

Clustering and Entropy Based Scoring Methods

This method is also an advancement in improving the performance of scoring function which includes the entropic effects by dividing generated ligand binding modes into different clusters. The contribution of entropy is measured by the configurational space covered by the number of ligand poses in the cluster. Limitation of this method is that this approach is docking program dependent i.e. its performance depends upon the ligand sampling protocol used. If this method is used in combination with AutoDock, it improves the binding mode prediction (Yuriev, Agostino, & Ramsland, 2011).

Fingerprinting

This method identifies various interaction patterns that can guide and improve predictions. Interaction fingerprints (Ifs) and profile based methods have been used in scoring and virtual screening (Yuriev, Agostino, & Ramsland, 2011). For example Glide XP descriptors for locating key pharmacophoric features of docked fragments (Perez-Nueno, Rabal, Borrell, & Teixido, 2009) and weighted-Structural Interaction Fingerprints (w-SIFt) are used to determine inhibitor potency (Nandigam, Kim, Singh, & Chuaqui, 2013).

Combination with Experimental Data

Gonzalez-Ruiz, & Gohlke, 2009, combined drug score function with experimental data, specifically NMR amide proton chemical shift perturbations (CSP) which lead to improvement in pose prediction.

Target Specific Scoring Functions

It leads to improved robustness of the scoring function parameters that refrain to multiple targets of a target class (Yuriev et al., 2013). Target specific target system were developed for Kinase-PMF to assess the binding of ATP-competitive kinase inhibitors, HIV protease, carbonic anhydrase, and trypsin; enoyl acyl carrier protein reductase from Mycobacterium tuberculosis (Seifert, 2009).

PERFORMANCE OF SCORING FUNCTION

The three related and independent criteria are binding mode identification, binding affinity prediction and virtual screening database which are used to assess the performance of a scoring function. Firstly, the determination of binding mode is one of the essential features for the scoring function. The structure is ranked by their calculated binding scores. The rmsd value should be ≤ 2.0 Å between the top ligand conformation and the experimental value for the best result. However, this theory sometimes fails in case of nearly symmetrical ligands, large flexible ligand, solvent group and unimportant group which sometimes hide the performance of docking program. To solve this issue various other parameters like relative displacement error (RDE), interaction-based accuracy classification (IBAC), real space R-factor (RSR), and Generally Applicable Replacement for rmsD (GARD) are used. The tightness of the ligand binding to the receptor represents the binding affinity. It is difficult to achieve a score scale related to experimental binding data. To determine the correlation between theoretical and experimental data, the Pearson correlation coefficient and Spearman correlation coefficient are calculated. Prediction of the binding affinity is more challenging than binding mode because of various experimental errors and different experimental conditions. To assess the performance of scoring function, the enrichment test and the area under the receiver operating characteristic curve (AUC) of virtual data base screening are done. For a perfect scoring function, it is necessary to perform equally well on all the three criteria of any test set. However, due to the innate restrictions, most of the existing scoring functions executes well with only one or two of the criteria and fail on others (Huang, Grinter, & Zou, 2010).

DIFFERENT ALGORITHMS USED FOR DOCKING

Docking algorithms are used to obtain a fast and accurate method for the discovery of various novel lead compounds or to validate the experimental data (Dias, & de Azevedo, 2008). Here are some of the different algorithms used for molecular docking studies:

- **Molecular Dynamics:** Molecular dynamics (MD) is extensively used as a powerful simulation technique in the field of molecular modeling. The calculation of this method is based on Newton's equation of motions. It depicts the flexibility of both the ligand and protein more efficiently than any other algorithms. However, the disadvantage of MD simulations is that they progress petitely and thus have difficulties in reaching global energy minimum, which may results in improper sampling (Taylor et al., 2002). It is a very important technique as docking only considers binding affinity but MD accentuates the conformational change of the complex with time, thus, reducing the calculation capacity (Ian et al., 2013). So, MD simulation is considered as a validation before concluding any result of docking.
- **Monte-Carlo Simulation Method:** Monte Carlo (MC) methods produce various ligand poses during bond rotation, rigid-body translation or rotation. MC method is superior to MD to explore the search space by crossing the high energy barrier in potential energy space. It searches ligand conformational pose by random torsional sampling of rotatable bonds. Glide, ICM and Rosetta ligand methods are used to generate the pre-calculated ligand (Huang, 2014).
- **Genetic Algorithm Method:** Genetic algorithm (GA) is a well-known stochastic method. Idea of its stem comes from Darwin's theory of evolution. The ligand translational, rotational, and torsional degrees of freedom are considered for this type of algorithm. Genes are encoded as binary strings for degrees of freedom that make up the'chromosome' which represents the pose of the ligand. The genetic operators i.e. mutation and crossover affects the gene which results in a new ligand structure. Gold, Divali, Darwin and AutoDock programs are the softwares employed for genetic algorithms (Taylor et al., 2002).
- **Fragment Based Method:** It dissects the ligand molecule into a base fragment and various flexible fragments. Firstly, base fragment is docked in optimal orientation to establish favourable complementary interactions. After that, the addition of flexible fragments in a stepwise manner to that protein, explores all possible bond rotational conformations. This method can further be extended to de novo design programs (Taylor et al., 2002). FlexX, Hammerhead, and DOCK programs use fragment based ligand sampling methods (Chen, & Pohlhaus, 2010).

- **Point Complementary Methods:** Ligand can be docked into the binding site of the protein using point complementary method.This method is based upon the shape and chemical similarity between the interacting molecules. Perhaps, fragment based docking algorithms could also be included in this category, for that a division should be made between algorithms used for point complimentary and fragment based method. Examples of programs using point complementary methods are FTDOCK, SANDOCK, FLOG and the soft docking algorithm (Taylor et al., 2002).
- **Distance Geometry Methods:** It determines the binding mode between ligand and protein and considers only hydrogen bonding. It enables fast sampling of conformational space and thus helps in identifying palusible binding modes and is often used to express embedding algorithm. A recent example of a program using distance geometry in docking is DockIt. The resulting structures are then scored using PMF or PLP scoring functions (Taylor et al., 2002).
- **Tabu Searches:** This method is based on stochastic process in which n numbers of small random changes are made to the current conformation; the new solutions obtained are then scored and ranked in ascending order. Accept the best modification having a lower value than any other accepted modification even it is in the tabu list or not (Kitchen, Decornez, Furr, & Bajorath, 2004). An example of docking algorithm using tabu search is PRO LEADS (Taylor et al., 2002).
- **Evolutionary Algorithms:** This is the most commonly used stochastic method for flexible ligand docking. It is used to sample most favourable energetic conformation among the large search space of protein-ligand complexes (Thomsen R. 2003).
- **Ant Colony Optimization (ACO):** This method of optimization is inspired by the behaviour of real ants. Deposition of pheromone enables real ants to find the shortest route between the food and their nests without visual information. Similarly, in this method artificial ants can be deployed to solve complex optimization problems by using artificial pheromone deposition. Protein-Ligand ANT System is based on this type of algorithm used in protein-ligand docking. The artificial ants are used to obtain a minimum energy conformation of the ligand in the binding site with pheromone trails. This software is highly used for virtual screening of a large number of databases and lead optimization (Kumar et al., 2014). This software can be used for rigid body docking of multiple conformations of ligands into rigid as well as flexible receptors (Korb, Stutzle, & Exner, 2006).
- **Particle Swarm Optimization Method (PSO Method):** Dr. Eberhart and Dr. Kennedy, 1995, developed a population based stochastic optimization technique. They were inspired by the social behaviour of flying birds. As the name swarm depicts the behaviour of a large group of insects or birds, in this method a particle fly in the search space by following the current optimum particles. Movement of a ligand through search space is based on the concept of changing velocity of each particle towards its *p*best and *l*best locations.

Algorithm involved:

PSO is based on the behaviour of birds flocking. Suppose a group of birds in the sky are searching for food in an area and there is only one food in that area. Here, the birds are called as solutions or particle and food is optima by updating generations. All birds don't know the exact location of food but they know how far the food is from them in each iteration. So, to find the food they follow the bird which is nearest to the food. PSO used this phenomenon to solve the optimization problems by following the current optimum particles. All particles are having their optimized fitness function for the evaluation of fitness value and velocities to direct the flying of the particles. In every iteration, each particle is updated by following two best values:

1. The best solution, which is achieved, gives best fitness value is called as *p*best
2. Another best value which is tracked with PSO by any particle in the population is called as global best (*g*best). *l*best when particle takes all the particles as its topological neighbours.

After calculating the two best values, calculation of its velocity and position is done with the following equations (a) and (b).

$$v[\ ] = v[\ ] + c1 * rand(\ ) * (pbest[\ ] - present[\ ]) + c2 * rand(\ ) * (gbest[\ ] - present[\ ]) \quad \text{(a)}$$

$$present[\ ] = present[\ ] + v[\ ] \quad \text{(b)}$$

v $v[\ ]$ is the particle velocity, present [ ] is the current particle (solution). *p*best [ ] and *g*best [ ] are already defined, rand () is a random number between (0,1). c1, c2 are learning factors, usually c1 = c2 = 2.

Comparison of PSO with GA

PSO is similar to GA as it involves a population of random solutions and searches for optima by updating generations but no evolution operators such as crossover and mutation are involved. PSO can be used for a wide range of applications or to specific application determined for a specific requirement. It is a faster and cheaper technique as compared to other methods. Only few parameters need to be adjusted in PSO and that are easy to implement (Hu, Eberhart, & Shi, 2004).

A GLIMPSE OF VIRTUAL SCREENING

Structural based virtual screening (VS) begins with the use of computing system to identify ligand-binding site on the target molecule from a large database. It is a complementary approach to experimental screening methods. There are two types of VS, one is structure-based virtual screening (docking) and other is ligand-based virtual screening (Reddy, Pati, Kumar, Pradeep, & Sastry, 2007). The basic goal of VS is to decrease the cost of experimental screening and to identify various lead compounds. The whole process depends upon information available e.g. if 3D-structure is available then molecular docking method is used or if 3D-structure is not available then pharmacophore model is derived based upon various properties like molecular weight, lipophilicity, ADME properties. Recent advances have been done in the field of virtual screening based on molecular docking studies like improvement of old docking programs for example DOCK, FlexX (Tripos, 2005), AutoDock, Gold (Cambridge Crystallographic Data Centre, 1965-2015), ICM (MolSof, 2015). Several new docking programs which have been released in past decade, including EUDOC, LigandFit and eHiTS are used for VS (Hou, & Xu, 2004).

APPLICATIONS OF MOLECULAR DOCKING

Several research groups focus on evaluating the performance of various docking programs or on making improvements to the scoring functions when experimental testing has already been done. Such efforts could give meaningful guidance to choose the methodology for a particular target system and has a wide variety of applications. Docking and VS both are important for the identification of various hits, lead optimization for binding pose prediction, bioremediation. It results in a number of success stories, it is not a separate technique but it is entrenched in between the *in-silico* and experimental technique (Eweas, Maghrabi, & Namarneh, 2014). Some novel docking softwares along with their features and scoring details are given in Table 1.

CASE STUDY

Nolatrexed used for the treatment of unresectable hepatocellular carcinoma (HCC) found to have high inhibitory potential of thymidylate synthase (TS). Its discovery involves different structure-based molecular modelling methods and X-ray crystal structure data on TS of *Escherichia coli* which is highly homologous to human TS which led to the design of a potent quinazoline derivative CB3717. Various modifications in the compound CB3717 were carried out at position 2, 5, and 6 of ring and some were kept constant like 4-ketoquinazoline ring that accounted for hydrogen bonding and pi_pi stacking interaction and gave compound Nolatrexed (Figure 6). It completed the Phase III clinical trial (Roa, Kar, & Das, 2015).

Table 1. Some novel docking programs along with their key features and scoring functions

Programs	Key Features	Scoring Function
AutoDock Vina	• Automation of input and output • No limit on variables (number of atoms, rotatable bonds,grid map size, etc.) • Parallelism and multithreading on multicore machines	Vina (ML)
BetaDock	• Prioritizing shape complementarity, based on the theory of b-complex and the Voronoi diagram • Rigid bodies (both receptor and ligand)	AMBER + AutoDock (FF)
bhDock	• Identification of low-resolution binding sites • Ligand posing via simulated annealing MD-based global optimization	
EADock DSS	Efficient tree-based dihedral space sampling (DSS), based on hybrid sampling engine and multiobjective scoring	CHARMM22 (FF)
H-DOCK	Docking by hydrogen bond matching and shape complementarity	H-DOCK (HBM and SSC terms)
LigDockCSA	• Powerful global optimizer • Conformational space annealing (CSA)	AutoDock (FF) + PLP torsional energy term
MedusaDock	Modeling both ligand and receptor flexibility simultaneously with sets of discrete rotamers	MedusaScore (FF)
PD-DOCK	Parameter optimization for docking scores	PD-DOCK (K)

Figure 6. Case study of Nolatrexed

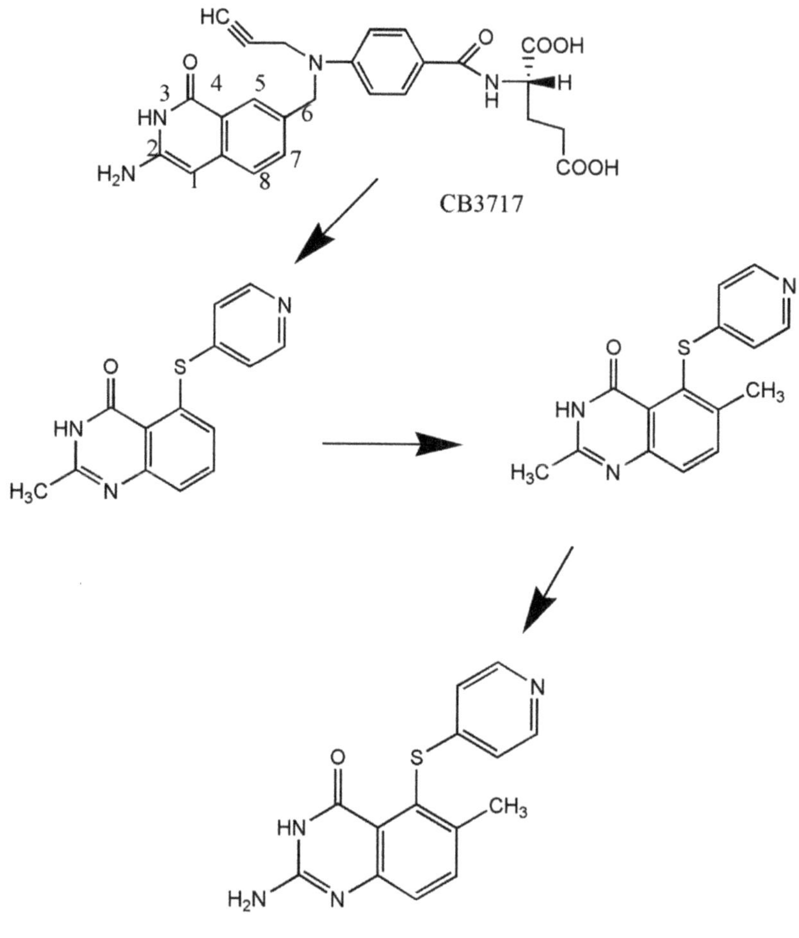

Nolatrexed

Boceprevir is used for the treatment of hepatitis C. It inhibits NS3-NS4A serine protease in hepatitis C virus (HCV), leading to the prevention of the replication of virus particles. Earlier high-throughput screening (HTS) analysis attempts for the development of HCV failed, that led to a structure-based design approach for potential molecules. Initially, it was observed that the inhibition of NS3-NS4A protease is facilitated by compounds that release N-terminal peptides. Later, it was found that Ser 139 in the viral protease was a target for electrophilic reaction with groups like aldehyde and ketones. Schering Plough Research Institute designed a series of α-ketoamide inhibitors. This drug was approved by FDA in 2011 (Roa, Kar, & Das, 2015).

Raltegravir, is a well-known drug approved by FDA in 2007 for the treatment of AIDS. It is a prominent HIV-1 integrase inhibitor. It is the first FDA-approved inhibitor. Binding site of this receptor was used in the molecular dynamics trajectory. Software used for the discovery was AutoDock along with the relaxed complex method. Initial molecules that showed inhibitory potential of HIV-integrase-catalyzed transfer of DNA were two classes of chemicals (namely 5, 6-dihydroxypyrimidine-4-carboxamides and N-methyl-4-hydroxypyrimidinone-carboxamides) and then through a series of structural changes and optimization, Raltegravir was discovered (Roa, Kar, & Das, 2015) as shown in Figure 7.

Figure 7. Case study of Raltegravir

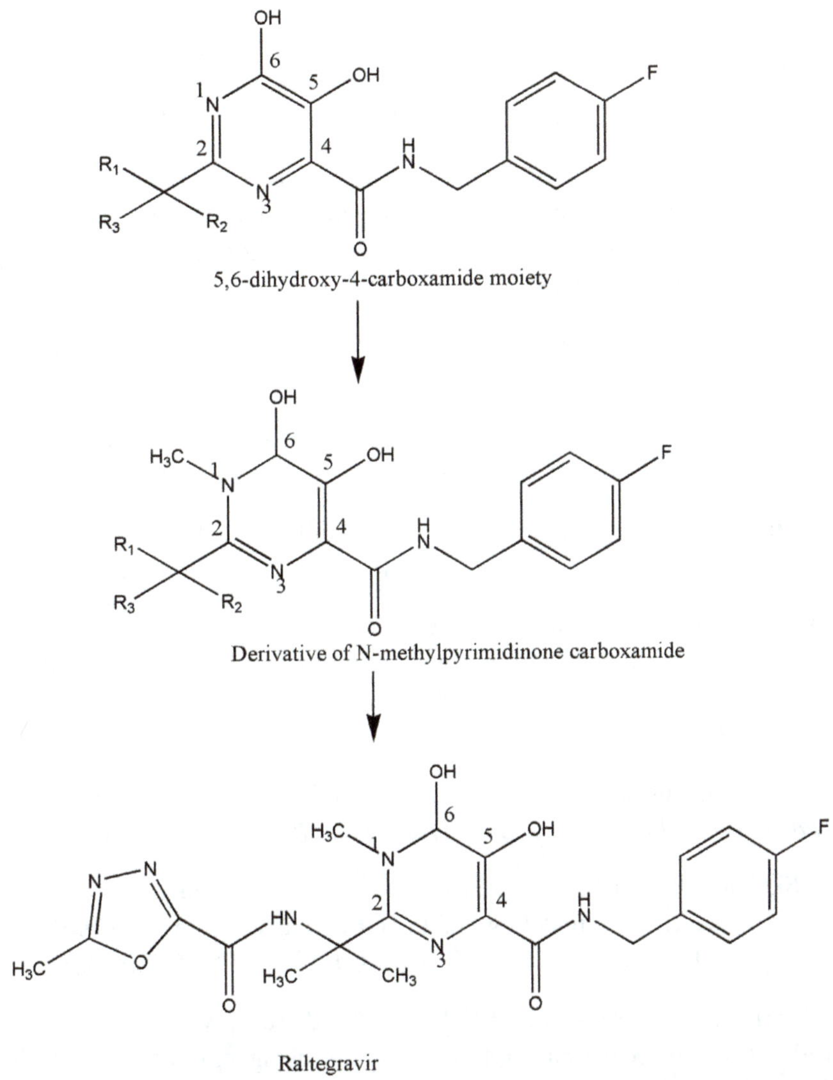

CONCLUSION

Molecular docking exhibit potential effect on identification and optimization in modern drug discovery. It helps us to rank the various lead compounds, predict the binding sites and affinities and elucidate the binding mechanism. Docking based virtual screening leads to much higher hit rate than traditional screening methods (HTS). It is not the replacement of experimental methods but both docking and experimental screening methods are complimentary and will reduce the cost of drug discovery to a large extent. Hopefully, innovative advancement in this field is expected in future which will ease the process of research on novel therapeutically active moieties.

REFERENCES

Abid, O. U. R., Babar, T. M., Ali, F. I., Ahmed, S., Wadood, A., Rama, N. M., & Choudhary, M. I. et al. (2010). Identification of novel urease inhibitors by high-throughput virtual and *in-vitro* screening. *ACS Medicinal Chemistry Letters*, *1*(4), 145–149. doi:10.1021/ml100068u PMID:24900188

Allenby, G., Bocquel, M.-T., Saunders, M., Kazmer, S., Speck, J., Rosenberger, M., & Chambon, P. et al. (1993). Retinoic acid receptors and retinoid X receptors: Interactions with endogenous retinoic acids. *Proceedings of the National Academy of Sciences of the United States of America*, *90*(1), 30–34. doi:10.1073/pnas.90.1.30 PMID:8380496

Bachwani, M., & Kumar, R. (2011). Molecular docking: A Review. *International Journal of Research in Ayurveda and Pharmacy*, *2*(6), 1746–1751.

Bailey, D., & Brown, D. (2001). High-throughput chemistry and structure-based design: Survival of the smartest. *Drug Discovery Today*, *6*(2), 57–59. doi:10.1016/S1359-6446(00)01596-8 PMID:11166243

Barril, X., & Fradera, X. (2006). Incorporating protein flexibility into docking and structure-based drug design. *Expert Opinion Drug Discovery*, *1*(4), 1–15. doi:10.1517/17460441.1.4.335 PMID:23495904

Bland, N. D., Wang, C., Tallman, C., Gustafson, A. E., Wang, Z., Ashton, T. D., & Pollastri, M. P. et al. (2011). Pharmacological validation of Trypanosoma brucei phosphodiesterases B1 and B2 as druggable targets for African sleeping sickness. *Journal of Medicinal Chemistry*, *54*(23), 8188–8194. doi:10.1021/jm201148s PMID:22023548

Borkotoky, S. (2012). Docking studies on HIV integrase inhibitors based on potential ligand binding sites. *International Journal of Bioscience. Biochemistry and Bioinformatics*, *2*(3), 21–29.

Brylinski, M., & Skolnick, J. (2010). Comprehensive structural and functional characterization of the human kinome by protein structure modeling and ligand virtual screening. *Journal of Chemical Information and Modeling*, *50*(10), 1839–1854. doi:10.1021/ci100235n PMID:20853887

Bustanji, Y., Al-Masri, I. M., Qasem, A., Al-Bakri, A. G., & Taha, M. O. (2009). *In silico* screening for non-nucleoside HIV-1 reverse transcriptase inhibitors using physicochemical filters and high-throughput docking followed by *in vitro* evaluation. *Chemical Biology & Drug Design*, *74*(3), 258–265. doi:10.1111/j.1747-0285.2009.00852.x PMID:19703027

Buzko, O. V., Bishop, A. C., & Shokat, K. M. (2002). Modified AutoDock for accurate docking of protein kinase inhibitors. *Journal of Computer-Aided Molecular Design*, *16*(2), 113–127. doi:10.1023/A:1016366013656 PMID:12188021

Capdeville, R., Buchdunger, E., Zimmermann, J., & Matter, A. (2002). Glivec (STI571, imatinib), a rationally developed, targeted anticancer drug. *Nature Reviews. Drug Discovery*, *1*(7), 493–502. doi:10.1038/nrd839 PMID:12120256

Carlson, H. A. (2002). Protein flexibility and drug design: How to hit a moving target. *Current Opinion in Chemical Biology*, *6*(4), 447–452. doi:10.1016/S1367-5931(02)00341-1 PMID:12133719

Cavasatto, C. N., Kovacs, J. A., & Abagyan, R. A. (2005). Representing receptor flexibility in ligand docking through relevant normal modes. *Journal of the American Chemical Society, 127*(26), 9632–9640. doi:10.1021/ja042260c PMID:15984891

Cavasotto, C. N., Orry, A. J. W., & Abagyan, R. A. (2005). The challenge of considering receptor flexibility in ligand docking and virtual screening. *Current Computer-aided Drug Design, 1*(4), 423–440. doi:10.2174/157340905774330291

Chen, Y. & Pohlhaus, D. T. (2010). In silico docking and scoring of fragments. *Drug Discovery Today: Technologies | Fragment based drug discovery, 7*(3), 149-156.

Chen, Y.-C. (2015). Beware of docking! *Trends in Pharmacological Sciences, 36*(2), 78–95. doi:10.1016/j.tips.2014.12.001 PMID:25543280

Cohen, P., & Alessi, D. R. (2013). Kinase drug discovery– What's next in the field? *ACS Chemical Biology, 18*(8), 96–104. doi:10.1021/cb300610s PMID:23276252

DesJarlais, R. L., Seibel, G. L., Kuntz, I. D., Furth, P. S., Alvarez, J. C., Babe, L. M., & Craik, Y., C. S. (1990). Structure-based design of nonpeptide inhibitors specific for the Human Immunodeficiency Virus 1 protease. *Proceedings of the National Academy of Sciences, 87*, 6644-6448. doi:10.1073/pnas.87.17.6644

Dias, R., & de Azevedo, W. F. Jr. (2008). Molecular Docking Algorithms. *Current Drug Targets, 9*(12), 1040–1047. doi:10.2174/138945008786949432 PMID:19128213

Dijk, M. V., & Bonvin, A. M. J. J. (2010). Pushing the limits of what is achievable in protein–DNA docking: Benchmarking HADDOCK's performance. *Nucleic Acids Research, 38*(17), 5634–5647. doi:10.1093/nar/gkq222 PMID:20466807

Durrant, J. D., & McCammon, J. A. (2010). Computer-aided drug-discovery techniques that account for receptor flexibility. *Current Opinion in Pharmacology, 10*(6), 77–0774. doi:10.1016/j.coph.2010.09.001 PMID:20888294

Ehrlich, L. P., & Rebecca, C. (2001). Protein-protein docking. *Reviews in Computational Chemistry, 17*, 61–98. doi:10.1002/0471224413.ch2

Eweas, A. F., Maghrabi, I. A., & Namarneh, A. I. (2014). Advances in molecular modeling and docking as a tool for modern drug discovery. *Der Pharma Chemica, 6*(6), 211–228.

Fabbro, D. (2015). 25 years of small molecular weight kinase inhibitors: Potentials and limitations. *Molecular Pharmacology, 87*(5), 766–775. doi:10.1124/mol.114.095489 PMID:25549667

Feixas, F., Lindert, S., Sinko, W., & McCammon, J. A. (2014). Exploring the role of receptor flexibility in structure-based drug discovery. *Biophysical Chemistry, 186*, 31–45. doi:10.1016/j.bpc.2013.10.007 PMID:24332165

Fraczek, T., Siwek, A., & Paneth, P. (2013). Assessing molecular docking tools for relative biological activity prediction: A case study of triazole HIV-1 NNRTIs. *Journal of Chemical Information and Modeling, 53*(12), 3326–3342. doi:10.1021/ci400427a PMID:24266618

Gane, P. J., & Dean, P. M. (2000). Recent advances in structure-based rational drug design. *Current Opinion in Structural Biology, 10*(4), 401–404. doi:10.1016/S0959-440X(00)00105-6 PMID:10981625

Ghanemi, A. (2015). Targeting G protein coupled receptor-related pathways as emerging molecular therapies. *Saudi Pharmaceutical Journal, 23*(2), 115–129. doi:10.1016/j.jsps.2013.07.007 PMID:25972730

Gonzalez-Ruiz, D., & Gohlke, H. (2009). Steering protein–ligand dockingwith quantitative NMR chemical shift perturbations. *Journal of Chemical Information and Modeling, 49*(10), 2260–2271. doi:10.1021/ci900188r PMID:19795907

Grinter, S. Z., & Zou, X. (2014). Challenges, Applications, and Recent Advances of Protein-Ligand Docking in Structure-Based Drug Design. *Molecules (Basel, Switzerland), 19*(7), 10150–10176. doi:10.3390/molecules190710150 PMID:25019558

Grosdidier, A., Zoete, V., & Michielin, O. (2009). Blind docking of 260 protein-ligand complexes with EADock 2.0. *Journal of Computational Chemistry, 00*, 1–10. PMID:19130502

Harris, R. A. (2013). Functional Validation of Virtual Screening for Novel Agents with General Anesthetic Action at Ligand-Gated Ion Channels. *Molecular Pharmacology, 84*(5), 670–678. doi:10.1124/mol.113.087692 PMID:23950219

He, S., & Lai, L. (2011). Molecular docking and competitive binding studydiscovered different binding modes of microsomal prostaglandin E synthase-1 inhibitors. *Journal of Chemical Information and Modeling, 51*(12), 3254–3261. doi:10.1021/ci200427k PMID:22077876

Hernández-Santoyo, A., Tenorio-Barajas, A. Y., Altuzar, V., Vivano-Cid, H. & Mendoza-Barrera, C. (2013). *Protein-Protein and Protein-Ligand Docking,* Croatia, Europe: INTECH

Heusser, S. A., Howard, R. J., Borghese, C. M., Cullins, M. A., Carlsson, J., Hou, T., & Xu, X. (2004). Recent Development and Application of Virtual Screening in Drug Discovery: An Overview. *Current Pharmaceutical Design, 10*(9), 1011–1033. doi:10.2174/1381612043452721 PMID:15078130

Hoi, P. M., Li, S., Vong, C. T., Tseng, H. H. L., Kwan, Y. W., & Lee, S. M.-Y. (2015). Recent advances in structure-based drug design and virtual screening of VEGFR tyrosine kinase inhibitors. *Methods (San Diego, Calif.), 71*, 85–91. doi:10.1016/j.ymeth.2014.09.004 PMID:25239735

Hu, X., Eberhart, R., & Shi, Y. (2004). *Recent advances in particle swarm.* IEEE Congress on Evolutionary Computation 2004, Portland, OR.

Huang, S.-Y. (2014). Search strategies and evaluation in protein–protein docking: Principles, advances and challenges. *Drug Discovery Today, 19*(8), 1081–1096. doi:10.1016/j.drudis.2014.02.005 PMID:24594385

Huang, S.-Y. (in press). Exploring the potential of global protein-protein docking: An overview and critical assessment of current programs for automatic *ab initio* docking. *Drug Discovery Today*. PMID:25801181

Huang, S.-Y., Grinter, S. Z., & Zou, X. (2010). Scoring functions and their evaluation methods for protein–ligand docking: Recent advances and future directions. *Physical Chemistry Chemical Physics, 12*(40), 12899–12908. doi:10.1039/c0cp00151a PMID:20730182

Huang, S.-Y., & Zou, X. (2007). Efficient molecular docking of NMR structures: Application to HIV-1 protease. *Protein Science*, *16*(1), 43–51. doi:10.1110/ps.062501507 PMID:17123961

Huang, S.-Y., & Zou, X. (2007). Ensemble Docking of Multiple Protein Structures: Considering Protein Structural Variations in Molecular Docking. *PROTEINS: Structure, Function, and Bioinformatics*, *66*(2), 399–421. doi:10.1002/prot.21214 PMID:17096427

Huang, S.-Y., & Zou, X. (2010). Advances and Challenges in Protein-Ligand Docking. *International Journal of Molecular Sciences*, *11*(8), 3016–3034. doi:10.3390/ijms11083016 PMID:21152288

Ian, I.-F., Pablo, C. V. J., Normande, C.-I., & Absalom, Z.-C. et al. (2013). Identification of pharmacological targets combining docking and molecular dynamics simulations. *American Journal of Agricultural and Biological Sciences*, *8*(1), 89–106. doi:10.3844/ajabssp.2013.89.106

Katritch, V., Jaakola, V.-P., Lane, J. R., Lin, J., IJzerman, A. P., Yeager, M., & Abagyan, R. et al. (2010). Structure-Based Discovery of Novel Chemotypes for Adenosine A_{2A}Receptor Antagonists. *Journal of Medicinal Chemistry*, *53*(4), 1799–1809. doi:10.1021/jm901647p PMID:20095623

Kitchen, D. B., Decornez, H., Furr, J. R., & Bajorath, J. (2004). Docking and scoring in virtual screening for drug discovery: Methods and applications. *Nature Reviews. Drug Discovery*, *3*(11), 935–949. doi:10.1038/nrd1549 PMID:15520816

Korb, O., Stutzle, T., & Exner, T. E. (2006). PLANTS: Application of Ant Colony optimization to Structure-Based Drug Design. *ANTS 2006. LNCS*, *4150*, 247–258.

Kroeger, M. B., Rizzo, R. C., Hose, B. M., Tirado-Rives, J., & Smith, R. H. (2003). Molecular modeling calculations of HIV-1 reverse transcriptase nonnucleoside inhibitors: Correlation of binding energy with biological activity for novel 2-aryl-substituted benzimidazole analogues. *Journal of Medicinal Chemistry*, *46*(10), 1940–1947. doi:10.1021/jm020271f PMID:12723956

Krovat, E. M., Steindl, T., & Langer, T. (2005). Recent advances in docking and scoring. *Current Computer-aided Drug Design*, *1*(1), 93–102. doi:10.2174/1573409052952314

Kumalo, H. M., Bhakat, S., & Soliman, M. E. S. (2015). Theory and applications of covalent docking in drug discovery: Merits and pitfalls. *Molecules (Basel, Switzerland)*, *20*(2), 1984–2000. doi:10.3390/molecules20021984 PMID:25633330

Kumar, A., Goyal, R., Kumar, S., Jain, S., Jain, N., & Kumar, P. (2015). Estrogenic and Anti-Alzheimer's studies of *Zingiber officinalis* as well as *Amomum subulatumRoxb*.: The success story of dry techniques. *Medicinal Chemistry Research*, *24*(3), 1089–1097. doi:10.1007/s00044-014-1166-y

Kumar, A., Kumar, S., Jain, S., Kumar, P., & Goyal, R. (2013). Study of binding of pyridoacridine alkaloids on Topoisomerase II using *In-Silico* tools. *Medicinal Chemistry Research*, *22*(11), 5431–5441. doi:10.1007/s00044-013-0496-5

Lafleur, K., Huang, D., Zhou, T., Caflisch, A., & Nevado, C. (2009). Structure based optimization of potent and selective inhibitors of the tyrosine kinase erythropoietin producing human hepatocellular carcinoma receptor B4 (EphB4). *Journal of Medicinal Chemistry*, *52*(20), 6433–6446. doi:10.1021/jm9009444 PMID:19788238

Laurini, E., Dal Col, V., Mamolo, M. G., Zampieri, D., & Pricl, S. et al. (2011). Homology model and docking-based virtual screening for ligands of the sigma (1) receptor. *ACS Medicinal Chemistry Letters*, *2*(11), 834–839. doi:10.1021/ml2001505 PMID:24900272

Leach, A. R., Shoichet, B. K., & Peishoff, C. E. (2006). Prediction of Protein-Ligand Interactions. Docking and Scoring: Successes and Gaps. *Journal of Medicinal Chemistry*, *49*(20), 5851–5855. doi:10.1021/jm060999m PMID:17004700

Legge, F. S., Budi, A., Treutlein, H., & Yarovsky, I. (2006). Protein flexibility: Multiple molecular dynamics simulations of insulin chain B. *Biophysical Chemistry*, *119*(2), 146–157. doi:10.1016/j.bpc.2005.08.002 PMID:16129550

Lyne, P. D., Lamb, M. L., & Saeh, J. C. (2006). Accurate prediction of the relative potencies of members of a series of kinase inhibitors using molecular docking and MM-GBSA scoring. *Journal of Medicinal Chemistry*, *49*(16), 4805–4808. doi:10.1021/jm060522a PMID:16884290

Ma, B., Shatsky, M., Wolfson, H. J., & Nussinov, R. (2002). Multiple diverse ligands binding at a single protein site: A matter of pre-existing populations. *Protein Science*, *11*(2), 184–197. doi:10.1110/ps.21302 PMID:11790828

Mahajan, A., & Gill, N. S. (2014). A review on molecular docking. *International Journal of Recent Adv Pharm Res.*, *4*(2), 1–7.

Mahasenan, K. V., & Li, C. (2012). Novel inhibitor discovery through virtual screening against multiple protein conformations generated via ligand-directed modeling: A maternal embryonic leucine zipper kinase example. *Journal of Chemical Information and Modeling*, *52*(5), 1345–1355. doi:10.1021/ci300040c PMID:22540736

McConkey, B. J., Soboley, V., & Edelmen, M. (2002). The performance of current methods in ligand–protein docking. *Current Science*, *83*(7), 845–956.

Mohan, V., Gibbs, A. C., Cummings, M. D., Jaeger, E. P., & DesJarlais, R. L. (2005). Docking: Successes and Challenges. *Current Pharmaceutical Design*, *11*(3), 323–333. doi:10.2174/1381612053382106 PMID:15723628

Motto, I., Bordogna, A., Soshilov, A. A., Denison, M. S., & Bonati, L. (2011). New aryl hydrocarbon receptor homology model targeted to improvedocking reliability. *Journal of Chemical Information and Modeling*, *51*(11), 2868–2881. doi:10.1021/ci2001617 PMID:21981577

Nandigam, R. K., Kim, S., Singh, J., & Chuaqui, C. (2009). Position specific interaction dependent scoring technique for virtual screening basedon weighted protein–ligand interaction fingerprint profiles. *Journal of Chemical Information and Modeling*, *49*(5), 1185–1192. doi:10.1021/ci800466n PMID:19415918

Nguyen, E. D., Norn, C., Frimurer, T. M., & Meiler, J. (2013). Assessment and Challenges of Ligand Docking into Comparative Models of G-Protein Coupled Receptors. *PLoS ONE*, *8*(7), e67302. PMID:23844000

Ogungbe, I. V., & Setzer, W. N. (2013). *In-silico* Leishmania target selectivity of antiparasitic terpenoids. *Molecules (Basel, Switzerland)*, *18*(7), 7761–7847. doi:10.3390/molecules18077761 PMID:23823876

Ouyang, X., Zhou, S., Ge, Z., Li, R., & Kwoh, C. K. (2013). CovalentDock Cloud: A web server for automated covalent docking. *Nucleic Acids Research*, 41. PMID:23677616

Perez-Nueno, V. I., Rabal, O., Borrell, J. I., & Teixido, J. (2009). APIF: A newinteractionfingerprint based on atom pairs and its application tovirtual screening. *Journal of Chemical Information and Modeling, 49*(5), 1245–1260. doi:10.1021/ci900043r PMID:19364101

Ranson, M., Mansoor, W., & Jayson, G. (2002). ZD1839 (IRESSA): A selective EGFR-TK inhibitor. *Expert Review of Anticancer Therapy*, *2*(2), 161–168. doi:10.1586/14737140.2.2.161 PMID:12113238

Rao, B. C., Subramanian, J., & Sharma, S. D. (2009). Managing protein flexibility in docking and its applications. *Drug Discovery Today*, *14*(7-8), 394–400. doi:10.1016/j.drudis.2009.01.003 PMID:19185058

Reddy, A. S., Pati, S. P., Kumar, P. P., Pradeep, H. N., & Sastry, G. N. (2007). Virtual Screening in Drug Discovery– A Computational Perspective. *Current Protein & Peptide Science*, *8*, 329–351. doi:10.2174/138920307781369427 PMID:17696867

Roy, K., Kar, S., & Das, R. N. (2015). *Understanding the basics of qsar for applications in pharmaceutical sciences and risk assessment*. Elsevier Inc.

Sali, A., & Blundell, T. L. (1993). Comparative protein modelling by satisfaction of spatial restraints. *Journal of Molecular Biology*, *234*(3), 779–815. doi:10.1006/jmbi.1993.1626 PMID:8254673

Schapira, M., Raaka, B. M., Samuels, H. H., & Abagyan, R. (2001). *In-silico* discovery of novel Retinoic Acid Receptor agonist structures. *BMC Structural Biology*, *1*(1), 1. doi:10.1186/1472-6807-1-1 PMID:11405897

Schulz-Gasch, T. & Stahl, M. (2004). Scoring functions for protein–ligand interactions: a critical perspective. *Drug discovery today: technologies | lead optimization, 1*(3), 231-239.

Seifert, M. H. (2009). Robust optimization of scoring functions for a target class. *Journal of Computer-Aided Molecular Design*, *23*(9), 633–644. doi:10.1007/s10822-009-9276-1 PMID:19471858

Singh, J., Petter, R. C., Baillie, T. A., & Whitty, A. (2011). The resurgence of covalent drugs. *Nature Reviews. Drug Discovery*, *10*(4), 307–317. doi:10.1038/nrd3410 PMID:21455239

Smith, J. (2005). Erlotinib: Small-molecule targeted therapy in the treatment of non-small-cell lung cancer. *Clinical Therapeutics*, *27*(10), 1513–1534. doi:10.1016/j.clinthera.2005.10.014 PMID:16330289

Sousa, S. F., Ribeiro, A. J. M., Coimbra, J. T. S., Neves, R. P. P., Martins, S. A., Moorthy, N. S. H. N., & Ramos, M. J. et al. (2013). Protein-ligand docking in the new millennium– a retrospective of 10years in the field. *Current Medicinal Chemistry*, *20*(18), 2296–2314. doi:10.2174/0929867311320180002 PMID:23531220

Stark, J. L., & Powers, R. (2012). Application of NMR and Molecular Docking in Structure-Based Drug Discovery. *Topics in Current Chemistry*, *326*, 1–34. doi:10.1007/128_2011_213 PMID:21915777

Stuart, B. G., Coxon, J. M., Morton, J. D., Abell, A. D., McDonald, D. Q., Aitken, S. G., & Bickerstaffe, R. et al. (2011). Molecular modeling: A search for a calpain inhibitor as a new treatment for cataractogenesis. *Journal of Medicinal Chemistry*, *54*(21), 7503–7522. doi:10.1021/jm200471r PMID:21955158

Tahir, R. A., Sehgal, S. A., Khattak, N. A., Khan Khattak, J. Z., & Mir, A. (2013). Tumor necrosis factor receptor superfamily 10B (TNFRSF10B): An insight from structure modeling to virtual screening for designing drug against head and neck cancer. *Theoretical Biology & Medical Modelling*, *10*(1), 38. doi:10.1186/1742-4682-10-38 PMID:23724937

Tautermann, C. S. (2014). GPCR structures in drug design, emerging opportunities with new structures. *Bioorganic & Medicinal Chemistry Letters*, *24*(17), 4073–4079. doi:10.1016/j.bmcl.2014.07.009 PMID:25086683

Taylor, R. D., Jewsbury, P. J., & Essex, J. W. (2002). A review of protein-small molecule docking methods. *Journal of Computer-Aided Molecular Design*, *16*(3), 151–166. doi:10.1023/A:1020155510718 PMID:12363215

Thomsen, R. (2003). Flexible ligand docking using evolutionary algorithms: Investigating the effects of variation operators and local search hybrids. *Bio Systems*, *72*(1-2), 57–73. doi:10.1016/S0303-2647(03)00135-7 PMID:14642659

Totrov, M., & Abagyan, R. (2008). Flexible ligand docking to multiple receptor conformations: A practical alternative. *Current Opinion in Structural Biology*, *18*(2), 178–184. doi:10.1016/j.sbi.2008.01.004 PMID:18302984

Tripathi, S. K., Muttineni, R., & Singh, S. K. (2013). Extra precision docking, free energy calculation and molecular dynamics simulation studies of CDK2 inhibitors. *Journal of Theoretical Biology*, *334*, 87–100. doi:10.1016/j.jtbi.2013.05.014 PMID:23727278

Tuccinardi, T., Botta, M., Giordano, A., & Martinelli, A. (2010). Protein kinases: Docking and homology modeling reliability. *Journal of Chemical Information and Modeling*, *50*(8), 1432–1441. doi:10.1021/ci100161z PMID:20726600

Vass, M., Schmidt, E., Horti, F., & Keseru, G. M. (2014). Virtual fragment screening on GPCRS: A case study on dopamine D3 and histamine H4 receptors. *European Journal of Medicinal Chemistry*, *77*, 38–46. doi:10.1016/j.ejmech.2014.02.034 PMID:24607587

Vyas, V., Jain, A., Jain, A., & Gupta, A. (2008). Virtual screening: A fast tool for drug design. *Scientia Pharmaceutica*, *76*(3), 333–360. doi:10.3797/scipharm.0803-03

Wada, M., Kanamori, E., Nakamura, H., & Fukunishi, Y. (2011). Selection of *in-silico* drug screening results for G-protein-coupled receptors by using universal active probes. *Journal of Chemical Information and Modeling*, *51*(9), 2398–2407. doi:10.1021/ci200236x PMID:21848279

Wlodawer, A. (2002). Rational approaches to AIDS drug design through structural biology. *Annual Review of Medicine*, *53*(1), 595–614. doi:10.1146/annurev.med.53.052901.131947 PMID:11818491

Wong, C. F. (2008). Flexible ligand–flexible protein docking in protein kinase systems. *Biochimica et Biophysica Acta*, *1784*(1), 244–251. doi:10.1016/j.bbapap.2007.10.005 PMID:17996742

Yuriev, E., Agostino, M., & Ramsland, P. A. (2011). Challenges and advances in computational docking: 2009 in review. *Journal of Molecular Recognition*, *24*(2), 149–164. doi:10.1002/jmr.1077 PMID:21360606

Yuriev, E., & Ramsland, P. A. (2013). Latest developments in molecular docking: 2010–2011 in review. *Journal of Molecular Recognition*, *26*(5), 215–239. doi:10.1002/jmr.2266 PMID:23526775

Zhang, X., Wang, H., Li, Y., Cao, R., Zhong, W., Zheng, Z., & Li, S. et al. (2013). Novel substituted heteroaromatic piperazine and piperidine derivatives as inhibitors of human enterovirus 71 and coxsackievirus a16. *Molecules (Basel, Switzerland)*, *18*(5), 5059–5071. doi:10.3390/molecules18055059 PMID:23629759

Zhou, T., Huang, D., & Caflisch, A. (2010). Quantum Mechanical Methods for Drug Design. *Current Topics in Medicinal Chemistry*, *10*(1), 33–45. doi:10.2174/156802610790232242 PMID:19929831

Chapter 13
Current Trends in Docking Methodologies

Shubhandra Tripathi
CSIR-Central Institute of Medicinal and Aromatic Plants, India

Amandeep Kaur Kahlon
International Centre for Genetic Engineering and Biotechnology, India

Akhil Kumar
CSIR-Central Institute of Medicinal and Aromatic Plants, India

Ashok Sharma
CSIR-Central Institute of Medicinal and Aromatic Plants, India

ABSTRACT

Molecular docking was earlier considered to predict the binding affinity of the receptor and ligand molecules. With the progress in computational power and developing approaches, new horizons are now opening for accurate prediction of molecular binding affinity. In the current book chapter, recent strategies for Computer-Aided Drug Designing (CADD) including virtual screening and molecular docking, encompassing molecular dynamics simulations, and binding free energy calculation methods are discussed. Brief overview of different binding free energy methods MMPBSA, MMGBSA, LIE and TI have also been given along with the recent Relaxed Complex Scheme protocol.

1. INTRODUCTION

The potential of computer aided drug design was recognized earlier. Fortune magazine in 1981 published a cover article titled "Next Industrial Revolution: Designing Drugs by Computer at Merck" (Van Drie, 2007). Initially, high throughput screening (HTS) was used to find novel therapeutic molecules in drug discovery. It is the process of testing a large number of diverse chemical structures against disease targets to identify 'hits'(Liu, Li, & Hu, 2004). Later, *in silico* techniques have acquired the competitive edge to filter out the lead molecules. Virtual screening is used for filtering molecules of interest from large chemical libraries. Screened molecules obtained as results are further evaluated using molecular docking. Moreover, the binding free energy methods viz. MMPBSA/MMGBSA, LIE, TI and FEP are used for stringent filtering from the molecular docking results. Lead molecules obtained from the *in silico* screening are used for biological evaluations (Figure 1).

DOI: 10.4018/978-1-5225-0115-2.ch013

Figure 1. Flowchart showing various sections in the book chapter

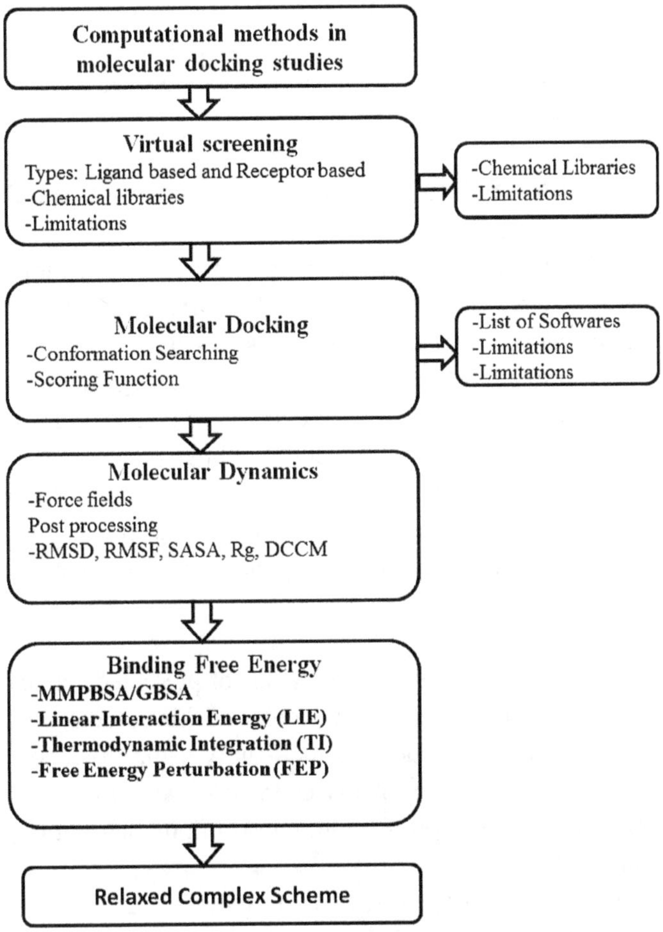

2. HIGH THROUGHPUT SCREENING (HTS)

HTS is characterized by taking the ligand-target interactions as the principle, leading to a higher information harvest (Liu et al., 2004). HTS lists contains a large percentage of false positives, making follow-up assays necessary to distinguish active from inactive substances (Jenkins, Kao, & Shapiro, 2003). HTS hit rate can be increased by using advance computational methods. Computational methods have several advantages over HTS as they are less time consuming, cost effective and require minimal compound design or prior knowledge. For example, researchers have used computational methods along with HTS to screen inhibitors of tyrosine phosphatase-1B, an enzyme implicated in diabetes. Virtual screening yielded nearly 35% hit rate as compared to 0.021% hit rate produced with HTS, demonstrating the power of computer aided drug designing (CADD) in drug discovery (Doman et al., 2002). Traditionally, HTS produces poor hit rate but in combination with CADD, its efficiency is greatly enhanced due to removal of false positives in less time and at lower computational cost.

One of the most remarkable uses of CADD in drug discovery process is the discovery of new targets for already existing drugs, a vice-versa approach. A few examples of approved drugs that owe their discovery to CADD includes: carbonic anhydrase inhibitor dorzolamide, approved in 1995 (Vijayakrishnan, 2009); the angiotensin-converting enzyme (ACE) inhibitor captopril, approved in 1981 as an antihypertensive drug (Talele, Khedkar, & Rigby, 2010).

2.1 Virtual Screening: Filter Efficient Compound from Large Chemical Space

The term "virtual screening" appeared for the first time in a peer-reviewed publication dating 1997. Virtual screening is the filtering of large chemical compound libraries and is extensively used in computer aided drug design. The main concept of virtual screening is to sort active molecules from a large chemical library which may bind to the target protein with reduced off target effects (Dror, Shulman-Peleg, Nussinov, &Wolfson, 2004). Nowadays, CADD is playing a crucial role in research and development of novel drugs. Different steps includes: virtual screening of large database, binding affinity estimation between protein ligand through molecular docking, activity predication of ligands, dynamical interaction of ligand with the protein via molecular dynamics and binding free energy calculations. ADMET is also an important part of CADD analysis. These components may be combined into three major steps in drug discovery:

1. Screening of large chemical libraries into smaller sets of predicted active compounds that can be tested in real time i.e. virtual screening and molecular docking studies.
2. Optimization of lead compounds, to increase its affinity using in silico methods viz. molecular dynamics simulations and binding free energy estimations and drug metabolism and pharmacokinetics (DMPK) properties estimation including absorption, distribution, metabolism, excretion and toxicity potential (ADMET).
3. Design novel compounds, either by "growing" starting molecule's functional group at a time or by combining together fragments into novel chemotypes.

Due to limitation of performing biological assay for vast number of commercially available compounds, compound selection based on some reasonable criteria is required. The process must be fast enough to screen large databases containing millions of compounds along with maintaining accuracy. Virtual screening fulfills that requirement and may also be called as fast docking process. The basic difference is that docking calculations used for drug designing are for maximum accuracy as compared to docking calculations in virtual screening having fast calculations with limited accuracy. In virtual screening docking algorithms is designed to be extremely fast at the cost of accuracy because a large quantity of data is being searched. It is necessary to use techniques taking very small time in analyzing each molecule.

Similarity search is a tool used in virtual screening to find out structurally similar compounds to the known inhibitors against the target receptor. Similarity search tends to find only those compounds which are structurally very similar to the known active compounds. They don't search compounds with different backbones but focus only on those compounds having same or very similar backbone and functional groups. A substructure search can also be used to find compounds with an identical structural motif.

Virtual screening can be classified as ligand based virtual screening (LBVS) and structure based virtual screening (SBVS)(Dror et al., 2004). LBVS basically uses the information from the experimentally known active set of molecules (Jahn, Hinselmann, Fechner, & Zell, 2009) and SBVS calculates the interaction between the target receptor protein and ligands. LBVS can be performed by quantitative structure activity relationships (QSAR), similarity and substructure searching and 3D QSAR pharmacophore (Villoutreix et al., 2007). SBVS uses the scoring and energy function to rank the docked pose into the active site to identify best hits.

First step of VS is to identify the biological therapeutic drug targets. The molecular structure can be obtained from the protein database (Berman, 2000) or can be generated using computational methods such as homology modeling. An unexplored aspect of SBVS is the flexibility of target protein structure (Evers, Hessler, Matter, & Klabunde, 2005).To deal with flexibility, all the conformational structures of target protein are clustered and all molecules are screened against representative structures of each cluster. Soft docking approach is another method to deal with target flexibility. Soft docking implements the softening of van der Waals potentials allowing small overlaps between the ligand and receptor without large steric penalties (Jiang & Kim, 1991). However, this may increase the rate of false positives because more diverse structures are allowed to bind (Lavecchia & Di Giovanni, 2013). One of the recent methods focusing on receptor flexibility is Relaxed Complex Scheme (RCS). In this approach, receptor ensemble is generated from structures extracted from molecular dynamics (MD) simulations of receptor molecule and these ensemble structures are screened against ligand libraries (Amaro, Baron, & McCammon, 2008). The next step is to screen chemical libraries in databases against particular drug target.

2.2 Chemical Libraries

Chemical libraries used for screening are broadly classified into 3 classes:

1. General libraries that can be used to screen against any target.
2. Focused libraries that are designed for a family of related targets.
3. Targeted libraries that are specifically designed for a single target.

Several target-specific libraries are available in Cambridge Structure Database viz., kinase biased, GPCR-biased and ion channel-biased sets. To screen these libraries, molecules as the ligand are to be prepared in 2D string format so that we can convert it into 3D structure. Also, ligand must be checked for incorrect bond length/distance, angle and order. Ligand structures requires preparation via 3D coordinate generation, geometry optimization, conformational sampling, assigning proper isometric and protonation state, and minimized molecular flexibility (Cavasotto & Phatak, 2011;Anderson, 2012). Chemical compounds can also be narrowed down by applying drug like or rule of five i.e. Lipinski filter but it may remove important scaffolds having higher binding affinity. Some of the important public databases are listed in Table1.

Besides, some other natural molecules database like Interbioscreen database, Duke phytochemical database, and French national chemical library are also popular having chemical libraries.

Once the chemical library is prepared, in SBVS we need a computational method to screen the top hit from the database. To achieve this goal several algorithms and scoring function are used. Mathematical algorithms are used to evaluate the binding of the ligand into the active site of target/ receptor. Scoring function and other criteria are used for ranking ligands. Scoring functions are very important in the SBVS and can be classified as force filed, knowledge based and empirical (Krovat, Steindl, & Langer, 2005; Jain, 2006). Scoring functions are described in detail in molecular docking section. Best option

Table 1. List of databases having chemical libraries

Database	Size	Type
ZINC	35 million	Annotated commercially available compounds
ChemDB	5 million	Annotated commercially available molecules
ChemSpider	34 million	Chemical structure database
Pubchem	30 million	Biologic activities of small molecules
DrugBank	4,800 drugs; 2,500 targets	Detailed drug data with comprehensive drug target Information
ChemBank	1,2 million	Biomedically relevant data and tools. Small molecules and small-molecule screens, and resources for studying the data
NCI Open Database	265,000	Compound libraries
ChEMBL	1 million	Chemicals extracted from literature, including target and bioactivity information
CoCoCo	7 million	Molecular structural information of commercial compounds from various chemical vendors by providing it in a ready-to-use format.

is to use combination of scoring function but it demands higher time and computational cost. In VS one has to screen millions of compounds, so docking and scoring function are selected with optimal accuracy and fast screening.

LBVS method uses set of active ligand and extract information from ligands necessary to bind. Ligands align itself to compute their shared pharmacophore. A pharmacophore is defined as the 3D arrangement of features that is crucial for a ligand molecule in order to interact with a target receptor in a specific binding site. Molecular docking interaction study is performed for lead molecules obtained from virtual screening. Utilizing molecular docking lead molecules are further filtered and those showing good binding affinity are processed further.

2.3 Limitations

Virtual screening plays a vital role in drug discovery. But it suffers from various limitations. The major limitations of the virtual screening and molecular docking are mentioned below.

1. The main and primarily goal of VS is to identify as many as possible diverse hits and not the most potent molecules. It may therefore, predict wrong binding pose due to biased algorithm and scoring function.
2. Currently, the role of solvent is not considered in virtual screening, while the water molecule have prominent role in hydrogen bond interaction.
3. Criteria for selecting the compounds after VS is ambiguous. Further, positive controls of negative data sets are rarely used in virtual screening.
4. Fundamental error may occur in converting molecular structure to another format. Various softwares have different protocols which are non-standardized. Serious concerns are regarding alternation in atomic coordinates, chirality, hybridization and partial charges.
5. One of the major challenges is generating an optimal set of conformations. Good coverage of conformational space is also important. Similarly, we have to remove high energy conformations, if they are not removed they will be present in the positive hits.

6. Flexibility of ligand is considered in VS but target flexibility is probably the most unexplored aspect of VS. It should be considered while doing SBVS.

VS require careful preparation of the database, judicious parameter choices, and sensible compromises between the different goals one attempts to obtain. A series of government-funded initiatives which mandate deposition of screening data in public repositories and promote the use of open source software has resulted into availability of vast datasets including more bioactivity classes. This addresses many of the biases inherent in earlier benchmark collections. While VS will always remain a probabilistic game, the odds of success can be greatly enhanced through careful planning and attention to details.

3. MOLECULAR DOCKING

Molecular docking is the interaction study of protein and ligand, the interaction is defined in terms of binding affinity. Kuntz et al., 1982 developed first rigid body docking software, based on geometric complementarity of interacting molecules. Docking has now improved to greater extent having efficient scoring functions and accurate predictions.

Molecules in the nature interact with each other to form various interactions. This is a vital process in nature to form various molecular associations. Using in silico molecular docking methods, various factors can be analyzed for molecular interactions. The fundamental approach used in molecular docking comprises, conformational space search of the three dimensional space of the protein with the ligand molecule (docking) and predicting the affinity of these poses using scoring function (scoring)(Kitchen, Decornez, Furr, & Bajorath, 2004).

3.1 Conformational Searching

Several degrees of freedom are computed for the ligand molecule in molecular docking. There are total six translational and rotational degrees of freedom that are calculated for the conformations of ligand molecule relative to receptor. Various methods includes genetic algorithm based on evolutionary approach (e.g. Autodock, GOLD), incremental construction method as in FlexX or methods based on molecular mechanics force field (e.g. Glide). A ligand molecule docked in the protein active site is shown in figure 2 (Tripathi, Kumar, Kumar, Negi, & Sharma, 2015).

3.2 Scoring Function

Scoring functions are the mathematical value used to calculate the binding of the ligand molecule with receptor protein. Scoring functions are majorly of three types: force field scoring function, empirical scoring function and knowledge based scoring functions (Bortolato, Fanton, Mason, & Moro, 2013). Molecular mechanics force field based scoring function utilizes various energy terms to calculate the binding free energy. The widely used force fields are AMBER(Wang, Wolf, Caldwell, Kollman, & Case, 2004) and OPLS (Jorgensen &Tirado-Rives, 1988). Empirical scoring function uses regression method to predict the binding free energy. Coefficients are calculated from ligands with known experimental

Figure 2. Protein-ligand docked complex image; ligand bound to the protein active site

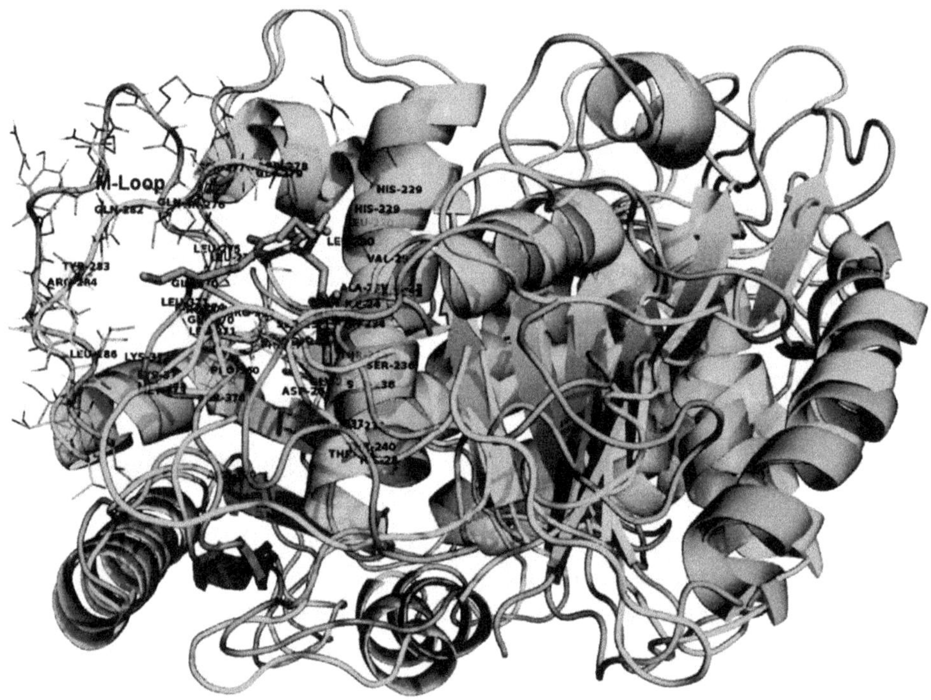

binding free energy. Similarly, in knowledge based scoring function, linear regression analysis or other methods like machine learning techniques are used to predict the binding free energy. Ligand conformations are ranked according to scoring function, and the best conformations are used for further analysis.

In real conditions, protein adopts different conformations upon ligand binding as compared to their unbound form. These changes in receptor conformations have significant impact in calculating the binding free energy. Therefore, the ligand binding predictions using methods like docking and scoring are not very accurate. There are two widely used methods to overcome this limitation:

1. **Conformational Selection:** In this method all the conformations of the unbound receptor available are utilized in docking and scoring (Monod, Wyman, & Changeux, 1965;Ma, Kumar, Tsai, & Nussinov, 1999). Receptor ensemble is generated using multiple X-ray, NMR structure of the receptor protein or through multiple snapshots from MD simulations (Lin, Perryman, Schames, & McCammon, 2002).
2. **Induced Fit:** The configuration of receptor is shifted to a different configuration upon ligand binding interaction. The study of receptor flexibility is performed using various techniques including Molecular Dynamics (MD), Monte Carlo, Umbrella Sampling, Metadynamics, and accelerated MD.

Table 2. List of molecular docking softwares

Software	Free for Academia	Details	Scoring Function
Autodock	Yes	Uses an interaction grid to account for receptor conformations and simulated annealing to account for ligand conformations. Available free of charge	Force field
DOCK	Yes	Docks small molecules, includes solvent effects, uses incremental construction for ligands. Available to academic users without charge	Force field or contact score
AutodockVina	Yes	AutoDockVina significantly improves the average accuracy of the binding mode predictions compared to AutoDock.	Force field
Molegro	Yes(30 days)	MolDock is based on a new heuristic search algorithm that combines differential evolution with a cavity prediction algorithm. The docking scoring function of MolDock is an extension of the piecewise linear potential (PLP) including new hydrogen bonding and electrostatic terms	Re-ranking scoring function
ZDOCK	Yes (Server)	Pair wise shape complementarity with desolvation and electrostatics	Pair wise shape complementarity base scoring function

Molecular docking is broadly classified into three categories on the basis of receptor flexibility while performing docking experiment-

1. Rigid docking, where docking of a rigid receptor molecule is performed with rigid ligand molecule.
2. Flexible docking, where receptor protein molecule behaves as a rigid entity while the ligand molecule has complete flexibility, its all rotational bonds can change internal torsions.
3. Full flexible docking, where the ligands have torsion angle flexibility while the selected amino acid residues of the protein active side chains also remain flexible.

Various docking softwares are available with different algorithm and scoring functions (Lavecchia & Di Giovanni, 2013). Some freely available softwares are listed in table2.

For the structure-based drug designing 3D structure of the target protein is required. 3D structure can be derived by X-ray, NMR or one can model from the primary sequence. X-ray high-resolution crystallography data of target protein is more preferred than NMR data in the structural based drug design. In virtual screening structure of both protein and ligand are used to calculate interaction energies for all compounds that are tested. Ligand based CADD is generally preferred when no or little structural information is available, often for membrane protein targets (Sliwoski, Kothiwale, Meiler, & Lowe, 2013).

The docking software provides a ranked list of ligands. Using computer graphics, these are visually evaluated in complex with the target to assess goodness of fit, formation of key hydrogen bonds or hydrophobic interactions, surface complementarity and stability of the bound conformation of the ligand compared to the free conformation. Several top-scoring ligands are then purchased or synthesized and evaluated in the laboratory using *in vitro* assays with the target. Compounds that succeed as hits in the laboratory assays are then subjected to further rounds of *in silico* screening using more focused libraries of potential analogs. The structure of the complex comprising the target and hit molecules is often experimentally determined in order to validate the binding mode. To further investigate the binding affinity and stability of the receptor-ligand interaction molecular dynamics simulations and binding free energy calculations are performed. Molecular dynamics investigates the conformational stability of the complex and binding free energy estimates stringent calculation of binding affinity.

3.3 Limitations

To predict accurate binding affinities, it will be necessary to go beyond predicting a single "dominant" conformation of the ligand docked with the protein. Major limitations of molecular docking are-

1. Binding free energy is not driven by a single conformation, but rather by the free energy landscape.
2. It is the shape of the energy landscape that is crucial, the shape and width of the minima influences entropies.
3. Entropies are key contributors to binding thermodynamics and are not observable in single bound structures.
4. Multiple ligand poses, protein conformations, strain energies, changes in water structure, and solvation and protonation all play roles in binding affinity. And none of these are observable in single structures.
5. Computational tools can help provide insight into the unseen landscape, so those doing crystallographic studies may want to complement their work by using computational tools to explore this landscape. And those relying on crystallographic data (e.g., in a drug design context) should be aware that there are various binding possibilities that might not be captured in a single crystal structure.

4. MOLECULAR DYNAMICS (MD) SIMULATION

Molecular dynamics simulation provides an important tool for validating the molecular docking results. Molecular dynamics simulation is to imitate the real conditions including solvent molecules, charge neutralizing by counter ions addition and then production MD to generate MD trajectory, resembling real time conformations of the system. Molecular dynamics imbibes the flexibility of the receptor molecules as well as the ligand molecule and helps in probing the interaction between them. In contrast with flexible docking where the protein flexibility remains limited, in MD simulations complete flexibility is maintained for the system. In MD of protein-ligand docked complex, ligand binding may cause subtle changes in the side chains rotation or may lead to large deviation in the backbone structure of the protein (Sinko, Lindert, & McCammon, 2013).Various conformational snapshots of different poses during the production run are saved in MD trajectories. All conformational changes in the protein and ligand can be assessed using the molecular dynamics trajectories. These MD trajectories are usually generated in nanosecond scale (usually 10-100 ns), containing multiple snapshots used for conformational investigations and binding free energy calculations. Molecular Dynamics simulations were first proposed based on classical law of motion (F= mxa) to describe atomic motions (McCammon, Gelin, & Karplus, 1977). Earlier complex quantum-mechanical calculations were utilized to define atomic motions.

In MD simulations all the forces acting on each atom of a biomolecular system are calculated using Newtonian physics. The energy contribution for every atom includes bonded interactions and non-bonded interactions(Cornell et al., 1995). The non-bonded interactions includes electrostatic interactions calculated using Columb's law, and vander Waals interactions using Lennard Jones Potentials (L-J potential)(Jones, 1924). The bonded interaction incorporates bond distances, bond angles, dihedral angles and improper angles, which are calculated using classical physical equations (Cornell et al., 1995). The chemical bonds and bond angles are represented as virtual springs, and dihedral angles are modeled as sinusoidal functions. All energy terms are optimized to fit the quantum-mechanical calculations to reproduce the real behavior of atomic motions of molecules (Durrant & McCammon, 2011).

4.1 Force Fields

All the optimized energy terms cumulatively corresponds to the "force field', describing the contribution of various atomics forces that govern molecular dynamics. Various widely used force fields for molecular dynamics simulation include Assisted Model Building with Energy Refinement (AMBER) (Wang et al., 2004), GROningen MAchine for Chemical Simulations (GROMACS) (Berendsen, van der Spoel, & van Drunen, 1995), Chemistry at HARvard Molecular Mechanics (CHARMM) (Brooks et al., 1983; Vanommeslaeghe et al., 2009) and Optimized Potential for Liquid Simulations (OPLS) (Jorgensen & Tirado-Rives, 1988). These force fields have different parameters and are optimized for different biomolecules viz. protein, nucleic acids, and small chemical molecules.

All the forces acting on each of the atom of system are calculated and then the position of atoms are moved according to direction of net applied force using Newton's law of motion. The time step in which the relative positions of atom are determined is usually 2 femtosecond (10-15 second). Then the relative positions of atoms of biomolecule are estimated in further steps and the generated snapshots are saved. These snapshots in a long timescale lead to formation of MD trajectories. MD trajectories are further used for analyzing various parameters like RMSD, RMSF etc.

Most commonly used softwares for molecular dynamics simulation are AMBER (Case et al., 2005; Wang et al., 2004), CHARMM (Brooks et al., 1983), GROMACS (Van Der Spoel et al., 2005; Hess, Kutzner, van der Spoel, & Lindahl, 2008) and NAMD (Kalé et al., 1999; Phillips et al., 2005).

4.2 MD Analysis

From MD simulation conformation trajectory is generated containing multiple snapshots of the system. Generated MD trajectory is utilized for various analysis to study the stability of the system. The secondary structure component or any other specific fluctuation in the 3-dimensional structure are analysed using various statistical methods. Methods used for analyzing the stability and structural changes are discussed below.

4.2.1 Root Mean Square Deviation (RMSD)

Root Mean Square Deviation (RMSD) of the atoms in a molecule is calculated with respect to a reference structure. The atoms in the molecule are fitted to the reference structure by roto-translational least square fitting (Karpen, Tobias, & Brooks, 1993). RMSD is calculated for the snapshot of protein molecule at two time point t1 and t2. Protein conformation at t1 is the reference structure and the RMSD of conformation at t2 is to be calculated by taking the average of the square of their atomic coordinate's difference. An example RMSD plot is shown in Figure 3 (Tripathi et al., 2015).

Figure 3. Plot showing RMSD graph, receptor protein (TUB), receptor protein with known ligand (TUB + Ctrl), receptor protein with other ligands (TUB + lig1 and lig2)

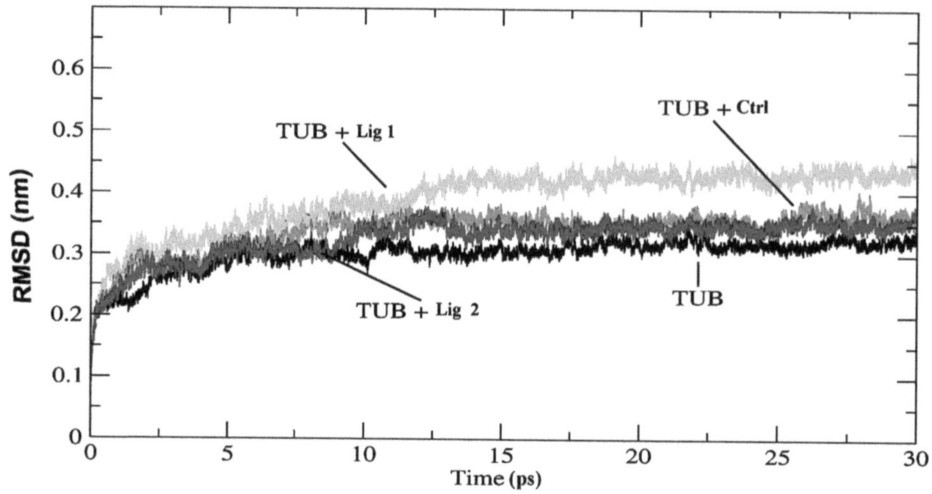

4.2.2 Root Mean Square Fluctuations (RMSF)

Root Mean Square Fluctuation (RMSF) is the standard deviation of the atomic positions in the MD trajectory fitting to a reference structure. Atomic fluctuations of individual amino acid in the protein are depicted by RMSF. RMSF calculations are used to investigate the higher flexibility shown in the loop regions of the receptor proteins. A RMSF plot is depicted in Figure 4 (Tripathi et al., 2015).

Figure 4. Plot showing RMSF graph, receptor protein unbound (TUB), receptor protein with known ligand (TUB + Ctrl), receptor protein with other ligands (TUB + lig1 and lig2)

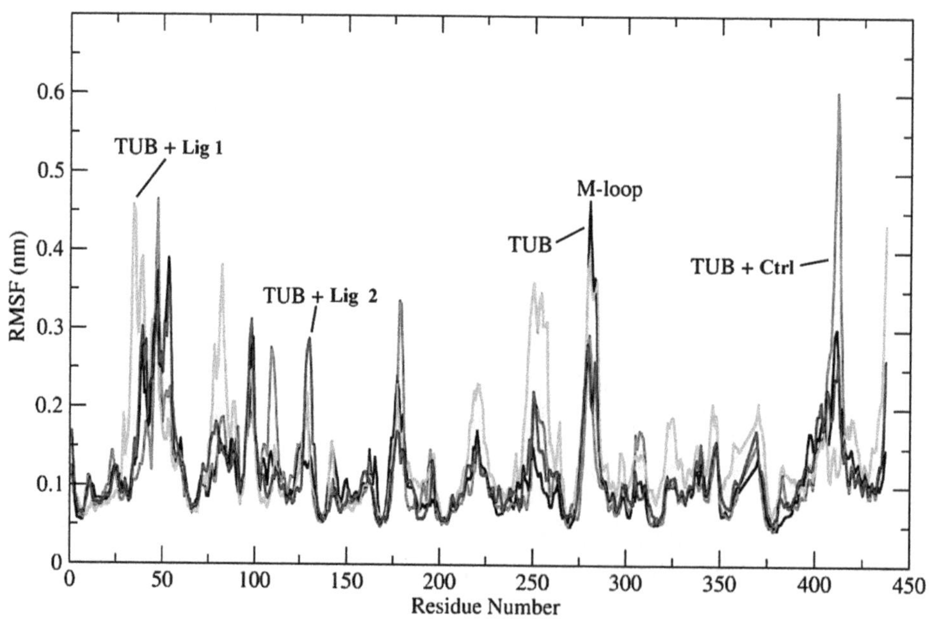

4.2.3 Solvent Accessible Surface Area (SASA)

Solvent Accessible Surface Area (SASA) is calculated from the MD trajectories. SASA includes hydrophobic, hydrophilic and total solvent accessible surface area of the protein molecule. The calculated surface area is the canonical surface area (Eisenhaber, Lijnzaad, Argos, Sander, & Scharf, 1995). The extent to which the amino acids interact with the solvent and the protein core is proportional to the surface area exposed to the solvent (Durham, Dorr, Woetzel, Staritzbichler, & Meiler, 2009).

4.2.4 Radius of Gyration (Rg)

Radius of gyration (Rg) represents the compactness of the protein structure. The Rg of a protein is derived in two steps, deriving the center of mass (R_c) of the 3-D coordinates and then the Rg was calculated for the estimation of the compactness of protein. Rg helps to probe the folding of the protein, a higher value of Rg suggest loose packing of various conformations. Stable Rg value from MD trajectories presents that unique stable spatial structure of protein is retained without any structural unfolding of protein (Lobanov, Bogatyreva, & Galzitskaya, 2008).

4.2.5 Hydrogen Bonding

The number of hydrogen bond interactions between protein and ligand, and the time span of individual hydrogen bond i.e. H-bond occupancy can also be calculated from the MD trajectories.

4.2.6 Domain Cross Correlation Map (DCCP)

DCCP analysis is performed to identify the protein domains from the generated MD trajectories (Arnold & Ornstein, 1997). It has a prominent role for the identification of domain in the protein structure. The collective domain motions in the protein structure and cross correlation between domains are analyzed along with the involvement of the active site in such motions (Luo & Bruice, 2002).

4.2.7 Essential Dynamics

All the collective motions of protein can be classified into two subspaces essential component containing few motions comprising of the most atomic fluctuations and the remaining component having very limited motions i.e. "physically constrained"(Amadei, Linssen & Bexerendsen, 1993). These essential components of the protein motions from the MD trajectories are responsible for overall fluctuations in the protein. Normal mode analysis (NMA) and Principle component analysis (PCA) are the two invaluable tools for studying conformational changes in proteins (Skjaerven, Martinez, & Reuter, 2011).

5. BINDING FREE ENERGY

Compared with molecular docking, MD simulations of protein-ligand docked complex have both protein and ligand in flexible manner. Earlier in docking programs priority was given to speed rather than accuracy while scoring but more accurate techniques are now available for the estimation of the binding free energy from MD trajectories. These advanced techniques includes molecular mechanics/Poisson-Boltzmann and surface area (MM/PB-SA) (P. A. Kollman et al., 2000), thermodynamic integration (TI) (P. Kollman, 1993; Adcock & McCammon, 2006), free energy perturbation(FEP) (P. Kollman, 1993; Kim et al., 2006), and linear interaction energy (LIE) (Johan Aqvist, Luzhkov, & Brandsdal, 2002). These are the sophisticated techniques to calculate the binding affinity between the protein and the ligand. TI and FEP are the most rigorous techniques but they are computationally too expensive. MMPBSA/GBSA methods are now days widely used for estimation of binding free energy with lower computational costs than TI and FEP. A brief account of each technique is given below.

5.1 MMPBSA and MMGBSA Methods

Molecular Mechanics Poisson Boltzman's Surface Area (MMPBSA) or Generalised Born Suface Area (MMGBSA) methods are utilized to obtain the relative free energies of binding, mostly used for comparing ligands binding affinity. This is the post-processing step after MD simulation, the representative snapshots from the stable trajectories are used to calculate the free energy change. It has three components, molecular mechanics energy calculations, Poisson Boltzman or Generalised Born method and estimation of surface area. The molecular mechanics component is used to calculate the energy difference in the gas phase. The solvation free energy component including both polar and non-polar solvation is calculated using the implicit solvent. Poisson Boltzman and Generalised Born are the two methods, used for calculating the polar solvation free energy component. The non-polar free energy component is calculated by estimating the surface area component. The entropy contribution to the total free energy was estimated further by the normal mode analysis or quasiharmonic approximation method. Ensemble based screening utilizes various receptor conformations for docking. Conformations can be determined from X-ray or NMR structures, Monte Carlo simulation, and molecular dynamics simulation or enhanced sampling methods.

5.2 Linear Interaction Energy

Linear interaction energy method estimates changes in the electrostatic and van der Waals interaction energy between the ligand and the surrounding environment, for the receptor bound ligand and the ligand in solution based on MD simulation (J. Aqvist, Medina, & Samuelsson, 1994). The binding free energy is estimated by calculating difference in electrostatic and van der Waals energies in a linear free energy equation. Difference in the solvent accessible surface area between bound and unbound ligand is also considered in estimating the free energy term (van Lipzig et al., 2004). These calculations are performed from the snapshots retrieved from the equilibrated MD trajectories.

5.3 Free Energy Perturbation (FEP) and Thermodynamics Integration (TI)

Free energy perturbation and Thermodynamics integration are rigorous and computationally expensive methods requiring higher computational cost. It's due to two main reasons, first the explicit solvent is used and second is to calculate the difference in free energy of two states on the basis of molecular dynamics simulations carried out at intermediate points along the transition path from one state to another state. Therefore, multiple molecular dynamics simulations are performed. The computational cost is reduced by using the implicit solvent approach to the end point states.

The advent of novel parallel computing architecture Compute Unified Device Architechture (CUDA) which utilizes Graphical Processing Units (GPUs), enhanced the sampling of the molecular modeling softwares. Future improvements and developments in the computational power will enhance the sampling of molecular docking, long time molecular dynamics simulations and stringent binding free estimation methods.

6. RELAXED COMPLEX SCHEME

Lin et al., 2002 proposed a novel computational methodology, accommodating the idea of receptor flexibility, for drug designing called Relaxed Complex Scheme (RCS). Later Amaro et al., 2008 improved the RCS method by including virtual screening, increasing computational efficiency and more stringent binding calculations for computer aided drug designing.

RCS is a benchmarking computational protocol for drug designing, including docking algorithms with dynamical structural information obtained through molecular dynamics (MD) simulations, accounting for flexibility of ligand as well as receptor under study. The conformational flexibility for both receptor and ligand are taken into consideration in receptor ensemble and ligand ensemble then flexible docking studies were performed followed by molecular dynamics (MD) simulations. Binding free energy are calculated using implicit solvent methods viz. MMPBSA (Damm & Carlson, 2007; Homeyer & Gohlke, 2012), MMGBSA or coupling-parameter approaches viz. thermodynamics integration (TI) and free energy perturbation (FEP) including extensive sampling of receptor, ligand, and solvents.

For a novel protein target, virtual screening approaches are utilized to identify novel potential inhibitors. In previous RCS version receptor ensemble were retrieved from the snapshots at regular time intervals from MD trajectories, resulting in thousands of receptor structures, most of them are conformational redundant. While in advanced RCS clustering method based on pair-wise RMSD of the aligned structures was utilized. For receptors having known binding region the clustering algorithm focuses on the subset of residues involved in receptor binding site. The receptor ensemble can also be derived from collection of independent X-ray or NMR experimental structures (Damm & Carlson, 2007; Huang & Zou, 2007). In this way, 5-10 non-redundant conformations representing the receptor ensembles were generated, which were further used for docking studies and binding free energy estimates. Ligand ensemble will be taken from the various databases, also all the rotations in the ligand bonds will be permitted while performing the docking studies.

Molecular docking is performed for the receptor ensemble and the ligand database with Autodock open-source software (Goodsell, Morris, & Olson, 1996). In the Post-Processing step binding free energy was further evaluated using MMPBSA, single step Perturbation, Linear Interaction Energy (LIE), Free Energy Perturbation (FEP) and Thermodynamics Integration (TI) methods.

REFERENCES

Adcock, S. A., & McCammon, J. A. (2006). Molecular dynamics: Survey of methods for simulating the activity of proteins. *Chemical Reviews*, *106*(5), 1589–1615. doi:10.1021/cr040426m PMID:16683746

Amadei, A., Linssen, A. B., & Berendsen, H. J. (1993). Essential dynamics of proteins. *Proteins*, *17*(4), 412–425. doi:10.1002/prot.340170408 PMID:8108382

Amaro, R. E., Baron, R., & McCammon, J. A. (2008). An improved relaxed complex scheme for receptor flexibility in computer-aided drug design. *Journal of Computer-Aided Molecular Design*, *22*(9), 693–705. doi:10.1007/s10822-007-9159-2 PMID:18196463

Anderson, A. C. (2012). Structure-Based Functional Design of Drugs: From Target to Lead Compound. In V. Espina & L. A. Liotta (Eds.), *Molecular Profiling* (Vol. 823, pp. 359–366). Totowa, NJ: Humana Press. doi:10.1007/978-1-60327-216-2_23

Aqvist, J., Luzhkov, V. B., & Brandsdal, B. O. (2002). Ligand binding affinities from MD simulations. *Accounts of Chemical Research*, *35*(6), 358–365. doi:10.1021/ar010014p PMID:12069620

Aqvist, J., Medina, C., & Samuelsson, J. E. (1994). A new method for predicting binding affinity in computer-aided drug design. *Protein Engineering*, *7*(3), 385–391. doi:10.1093/protein/7.3.385 PMID:8177887

Arnold, G. E., & Ornstein, R. L. (1997). Molecular dynamics study of time-correlated protein domain motions and molecular flexibility: Cytochrome P450BM-3. *Biophysical Journal*, *73*(3), 1147–1159. doi:10.1016/S0006-3495(97)78147-5 PMID:9284282

Berendsen, H. J. C., van der Spoel, D., & van Drunen, R. (1995). GROMACS: A message-passing parallel molecular dynamics implementation. *Computer Physics Communications*, *91*(1-3), 43–56. doi:10.1016/0010-4655(95)00042-E

Berman, H. M. (2000). The Protein Data Bank. *Nucleic Acids Research*, *28*(1), 235–242. doi:10.1093/nar/28.1.235 PMID:10592235

Bortolato, A., Fanton, M., Mason, J. S., & Moro, S. (2013). Molecular Docking Methodologies. In L. Monticelli & E. Salonen (Eds.), *Biomolecular Simulations* (Vol. 924, pp. 339–360). Totowa, NJ: Humana Press. doi:10.1007/978-1-62703-017-5_13

Brooks, B. R., Bruccoleri, R. E., Olafson, B. D., States, D. J., Swaminathan, S., & Karplus, M. (1983). CHARMM: A program for macromolecular energy, minimization, and dynamics calculations. *Journal of Computational Chemistry*, *4*(2), 187–217. doi:10.1002/jcc.540040211

Case, D. A., Cheatham, T. E., Darden, T., Gohlke, H., Luo, R., Merz, K. M., & Woods, R. J. (2005). The Amber biomolecular simulation programs. *Journal of Computational Chemistry*, *26*(16), 1668–1688. doi:10.1002/jcc.20290 PMID:16200636

Cavasotto, C. N., & Phatak, S. S. (2011). Docking methods for structure-based library design. *Methods in Molecular Biology (Clifton, N.J.)*, *685*, 155–174. doi:10.1007/978-1-60761-931-4_8 PMID:20981523

Cornell, W. D., Cieplak, P., Bayly, C. I., Gould, I. R., Merz, K. M., Ferguson, D. M., & Kollman, P. A. (1995). A Second Generation Force Field for the Simulation of Proteins, Nucleic Acids, and Organic Molecules. *Journal of the American Chemical Society*, *117*(19), 5179–5197. doi:10.1021/ja00124a002

Damm, K. L., & Carlson, H. A. (2007). Exploring experimental sources of multiple protein conformations in structure-based drug design. *Journal of the American Chemical Society*, *129*(26), 8225–8235. doi:10.1021/ja0709728 PMID:17555316

Doman, T. N., McGovern, S. L., Witherbee, B. J., Kasten, T. P., Kurumbail, R., Stallings, W. C., & Shoichet, B. K. (2002). Molecular docking and high-throughput screening for novel inhibitors of protein tyrosine phosphatase-1B. *Journal of Medicinal Chemistry*, *45*(11), 2213–2221. doi:10.1021/jm010548w PMID:12014959

Dror, O., Shulman-Peleg, A., Nussinov, R., & Wolfson, H. J. (2004). Predicting molecular interactions in silico: I. A guide to pharmacophore identification and its applications to drug design. *Current Medicinal Chemistry*, *11*(1), 71–90. doi:10.2174/0929867043456287 PMID:14754427

Durham, E., Dorr, B., Woetzel, N., Staritzbichler, R., & Meiler, J. (2009). Solvent accessible surface area approximations for rapid and accurate protein structure prediction. *Journal of Molecular Modeling*, *15*(9), 1093–1108. doi:10.1007/s00894-009-0454-9 PMID:19234730

Durrant, J. D., & McCammon, J. A. (2011). Molecular dynamics simulations and drug discovery. *BMC Biology*, *9*(1), 71. doi:10.1186/1741-7007-9-71 PMID:22035460

Eisenhaber, F., Lijnzaad, P., Argos, P., Sander, C., & Scharf, M. (1995). The double cubic lattice method: Efficient approaches to numerical integration of surface area and volume and to dot surface contouring of molecular assemblies. *Journal of Computational Chemistry*, *16*(3), 273–284. doi:10.1002/jcc.540160303

Evers, A., Hessler, G., Matter, H., & Klabunde, T. (2005). Virtual screening of biogenic amine-binding G-protein coupled receptors: Comparative evaluation of protein- and ligand-based virtual screening protocols. *Journal of Medicinal Chemistry*, *48*(17), 5448–5465. doi:10.1021/jm050090o PMID:16107144

Goodsell, D. S., Morris, G. M., & Olson, A. J. (1996). Automated docking of flexible ligands: Applications of AutoDock. *Journal of Molecular Recognition: JMR*, *9*(1), 1–5. doi:10.1002/(SICI)1099-1352(199601)9:1<1::AID-JMR241>3.0.CO;2-6 PMID:8723313

Hess, B., Kutzner, C., van der Spoel, D., & Lindahl, E. (2008). GROMACS 4: Algorithms for Highly Efficient, Load-Balanced, and Scalable Molecular Simulation. *Journal of Chemical Theory and Computation*, *4*(3), 435–447. doi:10.1021/ct700301q PMID:26620784

Homeyer, N., & Gohlke, H. (2012). Free Energy Calculations by the Molecular Mechanics Poisson–Boltzmann Surface Area Method. *Molecular Informatics*, *31*(2), 114–122. doi:10.1002/minf.201100135

Huang, S.-Y., & Zou, X. (2007). Ensemble docking of multiple protein structures: Considering protein structural variations in molecular docking. *Proteins*, *66*(2), 399–421. doi:10.1002/prot.21214 PMID:17096427

Jahn, A., Hinselmann, G., Fechner, N., & Zell, A. (2009). Optimal assignment methods for ligand-based virtual screening. *Journal of Cheminformatics*, *1*(1), 14. doi:10.1186/1758-2946-1-14 PMID:20150995

Jain, A. N. (2006). Scoring functions for protein-ligand docking. *Current Protein & Peptide Science*, *7*(5), 407–420. doi:10.2174/138920306778559395 PMID:17073693

Jenkins, J. L., Kao, R. Y. T., & Shapiro, R. (2003). Virtual screening to enrich hit lists from high-throughput screening: A case study on small-molecule inhibitors of angiogenin. *Proteins*, *50*(1), 81–93. doi:10.1002/prot.10270 PMID:12471601

Jiang, F., & Kim, S. H. (1991). "Soft docking": Matching of molecular surface cubes. *Journal of Molecular Biology*, *219*(1), 79–102. doi:10.1016/0022-2836(91)90859-5 PMID:2023263

Jones, J. E. (1924). On the Determination of Molecular Fields. II. From the Equation of State of a Gas. *Proceedings - Royal Society. Mathematical, Physical and Engineering Sciences*, *106*(738), 463–477. doi:10.1098/rspa.1924.0082

Jorgensen, W. L., & Tirado-Rives, J. (1988). The OPLS [optimized potentials for liquid simulations] potential functions for proteins, energy minimizations for crystals of cyclic peptides and crambin. *Journal of the American Chemical Society*, *110*(6), 1657–1666. doi:10.1021/ja00214a001

Kalé, L., Skeel, R., Bhandarkar, M., Brunner, R., Gursoy, A., Krawetz, N., & Schulten, K. (1999). NAMD2: Greater Scalability for Parallel Molecular Dynamics. *Journal of Computational Physics*, *151*(1), 283–312. doi:10.1006/jcph.1999.6201

Karpen, M. E., Tobias, D. J., & Brooks, C. L. (1993). Statistical clustering techniques for the analysis of long molecular dynamics trajectories: Analysis of 2.2-ns trajectories of YPGDV. *Biochemistry*, *32*(2), 412–420. doi:10.1021/bi00053a005 PMID:8422350

Kim, J. T., Hamilton, A. D., Bailey, C. M., Domaoal, R. A., Domoal, R. A., Wang, L., & Jorgensen, W. L. (2006). FEP-guided selection of bicyclic heterocycles in lead optimization for non-nucleoside inhibitors of HIV-1 reverse transcriptase. *Journal of the American Chemical Society*, *128*(48), 15372–15373. doi:10.1021/ja066472g PMID:17131993

Kitchen, D. B., Decornez, H., Furr, J. R., & Bajorath, J. (2004). Docking and scoring in virtual screening for drug discovery: Methods and applications. *Nature Reviews. Drug Discovery*, *3*(11), 935–949. doi:10.1038/nrd1549 PMID:15520816

Kollman, P. (1993). Free energy calculations: Applications to chemical and biochemical phenomena. *Chemical Reviews*, *93*(7), 2395–2417. doi:10.1021/cr00023a004

Kollman, P. A., Massova, I., Reyes, C., Kuhn, B., Huo, S., Chong, L., & Cheatham, T. E. (2000). Calculating Structures and Free Energies of Complex Molecules: Combining Molecular Mechanics and Continuum Models. *Accounts of Chemical Research*, *33*(12), 889–897. doi:10.1021/ar000033j PMID:11123888

Krovat, E., Steindl, T., & Langer, T. (2005). Recent Advances in Docking and Scoring. *Current Computer-aided Drug Design*, *1*(1), 93–102. doi:10.2174/1573409052952314

Kuntz, I. D., Blaney, J. M., Oatley, S. J., Langridge, R., & Ferrin, T. E. (1982). A geometric approach to macromolecule-ligand interactions. *Journal of Molecular Biology*, *161*(2), 269–288. doi:10.1016/0022-2836(82)90153-X PMID:7154081

Lavecchia, A., & Di Giovanni, C. (2013). Virtual screening strategies in drug discovery: A critical review. *Current Medicinal Chemistry*, *20*(23), 2839–2860. doi:10.2174/09298673113209990001 PMID:23651302

Lin, J.-H., Perryman, A. L., Schames, J. R., & McCammon, J. A. (2002). Computational drug design accommodating receptor flexibility: The relaxed complex scheme. *Journal of the American Chemical Society*, *124*(20), 5632–5633. doi:10.1021/ja0260162 PMID:12010024

Liu, B., Li, S., & Hu, J. (2004). Technological advances in high-throughput screening. *American Journal of Pharmacogenomics: Genomics-Related Research in Drug Development and Clinical Practice*, *4*(4), 263–276. doi:10.2165/00129785-200404040-00006 PMID:15287820

Lobanov, M. Y., Bogatyreva, N. S., & Galzitskaya, O. V. (2008). Radius of gyration as an indicator of protein structure compactness. *Molecular Biology*, *42*(4), 623–628. doi:10.1134/S0026893308040195 PMID:18856071

Luo, J., & Bruice, T. C. (2002). Ten-nanosecond molecular dynamics simulation of the motions of the horse liver alcohol dehydrogenase*PhCH2O- complex. *Proceedings of the National Academy of Sciences of the United States of America*, *99*(26), 16597–16600. doi:10.1073/pnas.262667599 PMID:12481026

Ma, B., Kumar, S., Tsai, C. J., & Nussinov, R. (1999). Folding funnels and binding mechanisms. *Protein Engineering*, *12*(9), 713–720. doi:10.1093/protein/12.9.713 PMID:10506280

McCammon, J. A., Gelin, B. R., & Karplus, M. (1977). Dynamics of folded proteins. *Nature*, *267*(5612), 585–590. doi:10.1038/267585a0 PMID:301613

Monod, J., Wyman, J., & Changeux, J. P. (1965). On the nature of allosteric transitions: A plausible model. *Journal of Molecular Biology*, *12*(1), 88–118. doi:10.1016/S0022-2836(65)80285-6 PMID:14343300

Phillips, J. C., Braun, R., Wang, W., Gumbart, J., Tajkhorshid, E., Villa, E., & Schulten, K. (2005). Scalable molecular dynamics with NAMD. *Journal of Computational Chemistry*, *26*(16), 1781–1802. doi:10.1002/jcc.20289 PMID:16222654

Sinko, W., Lindert, S., & McCammon, J. A. (2013). Accounting for Receptor Flexibility and Enhanced Sampling Methods in Computer-Aided Drug Design: Accounting for Receptor Flexibility. *Chemical Biology & Drug Design*, *81*(1), 41–49. doi:10.1111/cbdd.12051 PMID:23253130

Skjaerven, L., Martinez, A., & Reuter, N. (2011). Principal component and normal mode analysis of proteins; a quantitative comparison using the GroEL subunit. *Proteins*, *79*(1), 232–243. doi:10.1002/prot.22875 PMID:21058295

Sliwoski, G., Kothiwale, S., Meiler, J., & Lowe, E. W. (2013). Computational Methods in Drug Discovery. *Pharmacological Reviews*, *66*(1), 334–395. doi:10.1124/pr.112.007336 PMID:24381236

Talele, T. T., Khedkar, S. A., & Rigby, A. C. (2010). Successful applications of computer aided drug discovery: Moving drugs from concept to the clinic. *Current Topics in Medicinal Chemistry*, *10*(1), 127–141. doi:10.2174/156802610790232251 PMID:19929824

Tripathi, S., Kumar, A., Kumar, B. S., Negi, A. S., & Sharma, A. (2015). Structural investigations into the binding mode of novel neolignans Cmp10 and Cmp19 microtubule stabilizers by in silico molecular docking, molecular dynamics, and binding free energy calculations. *Journal of Biomolecular Structure & Dynamics*, 1–9. doi:10.1080/07391102.2015.1074941 PMID:26212016

Van Der Spoel, D., Lindahl, E., Hess, B., Groenhof, G., Mark, A. E., & Berendsen, H. J. C. (2005). GROMACS: Fast, flexible, and free. *Journal of Computational Chemistry*, *26*(16), 1701–1718. doi:10.1002/jcc.20291 PMID:16211538

Van Drie, J. H. (2007). Computer-aided drug design: The next 20 years. *Journal of Computer-Aided Molecular Design*, *21*(10-11), 591–601. doi:10.1007/s10822-007-9142-y PMID:17989929

Van Lipzig, M. M. H., terLaak, A. M., Jongejan, A., Vermeulen, N. P. E., Wamelink, M., Geerke, D., & Meerman, J. H. N. (2004). Prediction of ligand binding affinity and orientation of xenoestrogens to the estrogen receptor by molecular dynamics simulations and the linear interaction energy method. *Journal of Medicinal Chemistry*, *47*(4), 1018–1030. doi:10.1021/jm0309607 PMID:14761204

Vanommeslaeghe, K., Hatcher, E., Acharya, C., Kundu, S., Zhong, S., Shim, J., Mackerell, A. D. (2009). CHARMM general force field: A force field for drug-like molecules compatible with the CHARMM all-atom additive biological force fields. *Journal of Computational Chemistry*. http://doi.org/<ALIGNMENT.qj></ALIGNMENT>10.1002/jcc.21367

Vijayakrishnan, R. (2009). Structure-based drug design and modern medicine. *Journal of Postgraduate Medicine*, *55*(4), 301–304. doi:10.4103/0022-3859.58943 PMID:20083886

Villoutreix, B. O., Renault, N., Lagorce, D., Sperandio, O., Montes, M., & Miteva, M. A. (2007). Free resources to assist structure-based virtual ligand screening experiments. *Current Protein & Peptide Science*, *8*(4), 381–411. doi:10.2174/138920307781369391 PMID:17696871

Wang, J., Wolf, R. M., Caldwell, J. W., Kollman, P. A., & Case, D. A. (2004). Development and testing of a general amber force field. *Journal of Computational Chemistry*, *25*(9), 1157–1174. doi:10.1002/jcc.20035 PMID:15116359

Chapter 14
Protein Structure Prediction using Homology Modeling:
Methods and Tools

Akanksha Gupta
Netaji Subhas Institute of Technology, India

Pallavi Mohanty
Netaji Subhas Institute of Technology, India

Sonika Bhatnagar
Netaji Subhas Institute of Technology, India

ABSTRACT

Sequence-structure deficit marks one of the critical problems in today's scenario where high-throughput sequencing has resulted in large datasets of protein sequences but their corresponding 3D structures still needs to be determined. Homology modeling, also termed as Comparative modeling refers to modeling of 3D structure of a protein by exploiting structural information from other known protein structures with good sequence similarity. Homology models contain sufficient information about the spatial arrangement of important residues in the protein and are often used in drug design for screening of large libraries by molecular docking techniques. This chapter provides a brief description about protein tertiary structure prediction and Homology modeling. The authors provide a description of the steps involved in homology modeling protocols and provide information on the various resources available for the same.

INTRODUCTION

Knowledge of the protein structure is an important step towards rational structure based drug design and virtual screening of large molecular libraries. Sequence-structure deficit marks one of the critical problems in today's scenario where high-throughput sequencing has resulted in large datasets of protein sequences, but their corresponding 3D structures still need to be determined. Currently there are about 2 million protein sequences in Swissprot and TrEMBL, among them about 30,000 proteins have had

DOI: 10.4018/978-1-5225-0115-2.ch014

their structures solved experimentally. Although the rate of experimental structure determination will continue to increase, the number of newly discovered sequences grows much faster than the number of structures solved. Since experimental structures can only be determined for a small fraction of proteins, computational methods for protein structure modeling play an increasingly important role yielding models suitable for a wide spectrum of applications.

The term "homology modeling", also called comparative modeling refers to modeling of 3D structure of a protein by exploiting structural information from the known configurations of similar proteins. The method is based on the fact that structural conformation of a protein is more highly conserved than its amino acid sequence, and that small or medium changes in sequence normally result in little variation in the 3D structure. The necessary condition for successful homology modeling is a sufficient similarity (>30%) between the protein sequences. The importance and applicability of homology modeling is steadily increasing with the increase in the number of known structures in the protein structure database. Homology models contain sufficient information about the spatial arrangement of important residues in the protein, enabling us to study the binding site and design/dock drugs suitable for binding to the molecule. Stable and reliable repositories have been developed to give access to these annotated and evaluated models.

With the advent of new modeling softwares and algorithms, homology modeling is rapidly becoming the first choice for obtaining the 3D structure coordinates as well as other valuable structural and functional insights. The recent advancements in homology modeling, particularly in detecting and aligning sequences with Template structures, loop and side-chain modeling, model validation and evaluation have largely contributed to the prediction of more accurate models of protein structure, which was not possible several years ago. Improvement in sequence search/analysis, scoring systems and tertiary structure prediction methods allows development of models robust, error free models with high statistical significance. Publicly available, easy to use modeling servers and structure validation tools have further assisted the development of this technique.

Homology modeling is a multistep process, that can be summarized in following steps 1) Template identification and alignment, 2) alignment correction, 3) model generation, that includes backbone, loop and side chain generation 4) model optimization, and ; 5) validation. The first and foremost step is to fetch the Target protein sequence from various freely available databases. This is followed by Template identification, where, a protein Template whose 3D coordinates is known and that shares the maximum similarity with the Target sequence is selected. Similarity check between the Target and Template protein sequence is done by sequence alignment. The most widely used algorithm to run alignment is BLAST (Basic Local Alignment Search Tool). For a good alignment between both Target and Template sequence, the Template should not possess an E-value larger than 1. Sometimes it may be difficult to align two sequences in a region where the percentage sequence identity is very low. To resolve this issue, multiple Templates homologous to the Target protein are used, and an alignment correction is performed manually. After obtaining a good alignment, model generation starts. Quality of the model generated largely depends upon the correctness of the Template identified and accuracy of the alignment generated. If the alignment is accurate, there are less chances of the model being inaccurate. In Model generation, if the sequence alignment is good, the coordinates of those Template residues that show up in the alignment with the model sequence can be copied. If two aligned residues differ, only the backbone coordinates (N, Cα, C and O) are copied. The Target-Template alignment may contain gaps due to insertions and deletions of amino acids or bases. These loops are modeled using commercially available or manually generated loop libraries. There are generally two methods of loop modeling: 1) by finding similar loops in similar proteins and 2) by generating the segment *de novo*. Finding loops from similar proteins is done by taking some residues before and after the insertion as "anchor" residues and performing the loop search against databases like PDB with similar anchor residues. The best fitting loop is copied to the model. This has an advantage that all the loops are guaranteed to have reasonable internal geometries and conformations but again, they may not fit properly into the given model protein's framework. Loop selection must be done by carefully examining it for steric overlaps, and by checking loop atoms against rest of the protein's atoms. Structurally similar proteins have similar torsion angles about Cα-Cβ bond (psi angle) when comparing the side chain conformations. Therefore, in this step, the torsion angle between Cα and Cβ of the other residues can also be copied to the model because these rotamers tend to be conserved in similar proteins and many backbone configurations strongly prefer a specific rotamer. The modifications in the side chain can affect the backbone causing steric hindrance between atoms. Thus, optimization or idealization of the modeled structure using suitable a force field must be performed in order to obtain a refined model. Every homology model contains errors. Once the full scale 3D model is generated, model is validated for its authenticity, the overall correctness of the fold/structure, stereochemical parameters like bond lengths, torsion angles and bond angles, bump check etc. Moreover Ramachandran plot to visualize backbone dihedral angles (Phi)ψ against (Psi)φ of amino acid residues in protein structure have to be analyzed that whether the torsion angles fall into the allowed and the generously allowed region of the Ramachandran plot or not. Both the internal and external validation procedures need to be performed to verify that the generated model is precise and accurate. In spite of the tremendous growth in homology modeling procedures and algorithms, still there are several challenges in the technique of homology modeling which require further elucidation.

This chapter provides a brief description about the different approaches undertaken for protein tertiary structure prediction in order to lend statistical robustness to the problem of structure prediction. It provides a detailed overview of protein structure prediction, need of prediction methods and an introduction to various structure prediction approaches along with the challenges in protein structure prediction. Insights into the general features of the protein family, degree of conservation, the consensus sequence, Template identification and know-how of a correct alignment are also covered in the chapter. Recent developments in the field of protein structure prediction e.g. Comparative modeling and Threading algorithms have been discussed. Comparative modeling involves sequence-profile alignments while threading algorithms involves profile-profile alignments for identifying Templates against the query protein. Empirical techniques like *ab initio* structure prediction are also used when sufficient similarity of the Target protein to known models is not obtained. *Ab initio* strategy includes predicting the protein structure from its sequence from first principles. In different methods, this may entail calculating the physicochemical properties of proteins, evolutionary information etc. The information obtained is further used for generating spatial restraints or for identifying local structural building blocks.

Recent community-wide blind test like Critical Assessment of Protein Structure Prediction(CASP) have demonstrated significant advantages of composite (hybrid) approaches in protein structure prediction, which combine various techniques i.e Threading, *ab initio* methods, homology modeling and atomic level structure refinement. A variety of software tools, servers and meta-servers are now available for performing homology modelling. In spite of tremendous developments in structure prediction approaches, several challenges like modeling of proteins with very low similarities like GPCRs and better optimization methods still need further elucidation. The ongoing efforts in solving protein structures will continue to spur the development of novel algorithms and computational approaches to solve the problem of sequence-structure.

BACKGROUND

Functional classification requires structural information which relies on the primary sequence of the protein. Owing to the commendable efforts in genome sequencing, there has been an overwhelming flood of sequential information. Approximately 2000 nucleotide sequences are deposited per day in publicly available databases (Kulikova et al., 2004). However, while the structural genomics efforts succeeded in determining a large number of new protein structures, there are only about 109822 protein structures deposited in the Protein Data Bank (till June 2015), the repository of protein 3D structures. The thumb rule says that more the protein structures are determined experimentally, more would be the probability to find close structure analogues in the databases of known structures for almost any protein sequence in the near future. Due to the inability of experimental procedures for determining structural coordinates for most of the proteins, a major issue of sequence-structure deficit has arisen. This huge gap between the number of available sequences and experimentally solved protein structures can only be resolved by computational prediction methods.

Structure prediction methods can be divided into two broad categories: *ab initio* method (Lesk, 1997) and homology modeling(Kolinski, Rotkiewicz, Ilkowski, & Skolnick, 1999; Rost, Fariselli, & Casadio, 1996). The first approach has prediction from physical chemistry principles while the second method predicts the three-dimensional structure of the Target sequence based on its similarity to one or more proteins of known structures. In the early 1990s, fold recognition was considered as a separate field until recently CASPs (Critical Assessment of Structure Prediction) demonstrated that Fold recognition or threading is a combination approach combining the two extremes, in which remote homologues are detected by sampling the known protein PDB conformations with scoring energies(Venclovas, Zemla, Fidelis, & Moult, 2003; Xu, Xu, & Uberbacher, 1998).

Homology modeling, also called comparative modeling or Template based modeling refers to the construction of an unknown atomic resolution model of the 'Target' protein from the amino acid sequence of the Template protein. Homology modeling approach serves well if the sequence identity between the Target and the Template sequence is sufficiently high (>30%). The quality of the homology model generated depends largely on the extent of sequence conservation between the Target and the Template sequence (Chakravarty, Wang, & Sanchez, 2005; Marti-Renom et al., 2000). This type of modeling is based on the concept that protein's structural information is more conserved than the amino acid sequence.

Besides the increasing importance of homology modeling in the recent years, there still exists a cascade of challenges in front of the scientific community. The inaccuracy of force field makes conformation sampling much more difficult, since it could guide energy minimization to the conformation of global energy minimum that is totally different from the true conformation. Homology modeling is a technique whose stringent requirement is a good and accurate force field and completeness of the conformation searching.

In an era when the rate of experimental structural determination is approximately 6000 proteins per year, homology modeling has already saved hundreds of years of human effort. Even if homology modeling produces less accurate models than experimental methods, it can be of prime importance in proposing and testing hypothesis in molecular biology, function annotation, and drug design (Francoijs, Klomp, & Knegtel, 2000; Ginalski, Rychlewski, Baker, & Grishin, 2004; Takeda-Shitaka, Takaya, Chiba, Tanaka, & Umeyama, 2004; Zhou & Johnson, 1999). It can also provide good starting models for solving structures by experimental methods (Ceulemans & Russell, 2004; Talukdar & Wilson, 1999). Structural models generated by homology modeling can be of direct medical and biological relevance. Homology-based structural models are widely used in predicting the effects of single nucleotide polymorphisms, *in silico* drug designing. Structural models can also be used to design mutations that lead to specific changes in the function or stability of the modeled protein in the biological system. It is approximated that within the next 10 years, structural genomics might discover all distinct protein folds in nature, making homology modeling applicable to almost any protein sequence(Vitkup, Melamud, Moult, & Sander, 2001).

MAIN FOCUS OF THE CHAPTER

After studying this chapter, you shall be able to

1. **Know:** The basics of protein structure prediction, need of prediction methods and an introduction to various structure prediction approaches along with the challenges in protein structure prediction.
2. **Learn:** The method of comparative modeling in a detailed step by step manner.
3. **Identify:** The various checkpoints in comparative modeling where more emphasis has to be laid on, along with the type of errors that can emerge in the building of the model.
4. **Evaluate:** The modeled protein for its accuracy.
5. **Analyze:** The modeled protein for its structure-function relationships and annotate it.

Issues, Controversies, Problems

The biological role of a protein is determined by its function, which is in turn is determined from its 3D fold. Thus, there is enormous benefit in knowing the three dimensional structures of all the proteins. Although more and more structures are determined experimentally at an accelerated rate, but the gap is still huge. As more and more protein sequences are determined, there is pressing need for predicting protein structures computationally.

1. The main reason for modeling of protein structure is the sequence-structure deficit. While the sequence is available for many proteins, the structure has been determined for very few. This is largely due to the fact that structure determination techniques are time and resource intensive exercise. Also, experimental difficulties in purification and crystallization of proteins make the structure determination a lengthy process.
2. Additionally, homology modeling is not possible for proteins that do not share sufficient sequence-structure similarity with any of the available proteins in the public databases like PDB.
3. Even if the structural information has been obtained by experimental techniques, there are gaps in the structure due to experimental limitations. Especially, disordered structures may be refractive to X-ray crystallographic structure determination. On the other hand, NMR techniques can not handle very large structures.
4. A third issue is that of movement leading to conformational transitions. It is well known that a protein can occur in both global and local conformations, global conformations are undoubtedly the stable conformation but it's very likely that the local conformations are biologically relevant and plays an important role in disease prognosis. Thus, it is vital to gather the structural information from the literature, mutagenesis studies and binding assays to give us more insights to elucidate a more rational model that can be taken up for further drug designing.

5. Accurate protein tertiary structure prediction is essential for many bioinformatics applications, including studying catalytic mechanisms of the protein, mechanism of activation, and ligand docking in drug design.
6. Protein structure prediction also assumes a critically important role for exploration of mutations and mutagenesis on the function of the molecule.

Protein Tertiary Structure Prediction

There are three methods for protein structure prediction namely 1) *ab initio* methods 2) Template based Homology modeling 3) Fold recognition or threading. As the name suggests, *ab initio* methodology calculates the protein structure from its sequence from first principles or from the beginning. A range of *ab initio* methods rely on accurate energy functions and conformational search strategies to identify the most native-like model. However, *ab initio* methods require vast computational resources and have only proved successful for small proteins. Homology Modeling and Fold recognition, on the other hand, require the identification of homologues with known structure for the protein to be modeled (Target protein).

Homology modeling is one of the easiest ways to predict the structure of a protein and is based on the observation that during evolution, the structure of a protein changes much more slowly as compared to the sequence (Chothia C, 1986; Sander & Schneider, 1991). Thus, proteins with similar sequences adopt almost identical structures while proteins with lower sequence similarity still fold into overall similar structures. It was shown that as long as the identity between two sequences and the length of the overlap fall in the "safe" region (Figure 1), two sequences will almost certainly adopt the same fold (Rost, 1999). Typically, given that sufficient number of residues between two sequences aligns, a homolog with sequence identity above 40% can be used for homology modeling of the Target protein.

If the sequence similarity is less than 40%, the prediction can be done thorough fold recognition or threading. These methods differ from that of comparative modeling only at the level of Template selection and prediction. Here, sensitive sequence searching methods are used to align or thread the Target sequence to that of a protein with the same fold. The alignment can then be evaluated by statistical methods as well as manual inspection. If no suitable homolog is found in the database then the 3D structure is predicted through *ab initio* predictions. Typically, two or more Templates and methods of recognition are used to lend confidence to the modeling based on fold prediction.

1. **Template-Based Predictions:** Template based protein structure prediction (either Homology/ Comparative modeling OR threading based methods) uses knowledge of solved structures to model a protein sequence's native fold. First, sequence information is fetched for protein of interest (i.e. Target protein) from Expasy. BLAST (Basic Local Alignment Search Tool) search of Target protein against PDB will give us a rough idea of all the homologous protein with known 3D coordinates (Templates). Best hits may be selected from a BLAST run on the basis of low e-values (denoting statistically significant sequence similarity) or identified automatically by Homology modeling servers. Once the Template structure is found, the Target protein 3D structure is built by mapping the Target sequence onto the Template structure. This putative structure is refined using a combination of backbone moves, side-chain packing, and loop modeling. Template based protein structure prediction has always held great promise to produce atomically accurate models close to the native conformation. Comparative modeling techniques are extremely popular, reliable and fast where sequence homologues exist in the database. With increasing structural information, these

Figure 1. Safe zone: schematic representation of relationship between sequence identity, overlap, and confidence in structure
Source: Authors.

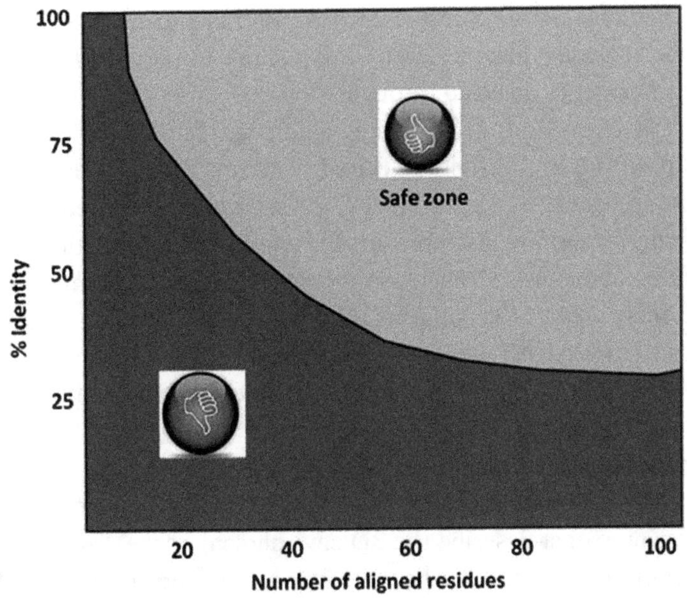

techniques should prove more useful. But the major limitation with Template based selections is that it is highly dependent on the sequence identity between query and Templates. A sequence identity above 50% leads to accurate models. The Model obtained is error prone if the identity range is less than 40%. Also, at times the side chains and loops may require remodeling in order to improve the modeled protein.

2. **Fold Recognition Methods:**
   a. **Template Structure Identification Using Threading Programs:** Threading is carried out when there is not sufficient sequence similarity between the Target and Template to carry out Comparative modeling. The recognition of the Template is a problem by itself and hence it is also called fold recognition. Threading involves steps similar to Comparative modeling. It differs in the fold identification and fold fitting step. There are many approaches, but the main theme is to find "folds' from a library of folds of known protein structures (Lu, Lu, & Skolnick, 2002; Skolnick, Kihara, & Zhang, 2004). Threading algorithm for protein structure prediction uses a database of known three-dimensional structures to match sequences without known structure with protein folds. A library of different protein folds is derived from the database of protein structures. Threading methods compare a Target sequence against a library of structural Templates. The query sequence is then optimally fitted to each library fold, allowing for relative insertions and deletions in loop regions. A scoring function is assigned to assess the fit of a sequence to a given fold. These functions are database driven and generally include a pairwise atom contact and solvation terms. The 'energy' of each possible fit is calculated by summation of the proposed pairwise interactions and solvation energy. The fold library is then ranked in the order of total energy, with the lowest energy fold being taken as the most suitable match. The best score is assumed to be the one adopted by the query sequence. The

basic assumption of the threading algorithm is that the tertiary structure is more conserved than the secondary structure and secondary structure is more conserved than the sequence. This method takes the advantage of knowledge of existing structures in the database and the physical properties they are stabilized with. Fold assignment and alignment are achieved by threading the sequence through each 3D structure in the fold library. Therefore structural modeling can be better detected through 2D or 3D structural homology instead of sequence homology. Two main kinds of Threading algorithms are available, namely:

i. **2D Threading or Prediction Based Methods:** These methods involve prediction of secondary structure (SS) or ASA (Accessible Surface Area) of the query protein. 2D Threading Algorithm also converts the coordinate information from PDB to a database containing sequence, SS and ASA information. Next, dynamic programming alignment of the query against the database is performed. These alignments are then evaluated on the basis of known SS and/or ASA matches with the Template structure. Lastly, it ranks the alignments and selects the most probable fold. Examples of 2D Threading servers are PredictProtein, 123D and GenThreader (Alexandrov & Shindyalov, 2003; Jones, 1999; Yachdav et al., 2014)

ii. **3D Profile Method:** This technique involves the alignment of the sequence to a string of descriptors that describe the 3D environment of the Target structure e.g. whether it is polar or non-polar. Other parameters could include α-helix, β-sheet or a turn conformation. Information whether a residue is fully buried, exposed or partially buried can also be considered. Here, each known fold is represented as a 2D matrix of inter-residue distance. An energy potential describes these distance dependent pair wise interactions for all possible combinations of amino acids. These are later compared with the 2D matrix of known protein structures. The trained set of contact potentials are then used to align an unknown protein sequence against a group of folds. The top scoring alignments are considered as possible Templates. In order to gain confidence and to increase the accuracy of the results, it is recommended to use more than one Template. Some of programs that predict through fold recognition are Loopp (version 2), 3DPSSM, PROSPECT, ProSa etc (Kelley, MacCallum, & Sternberg, 2000; Kim, Xu, Guo, Ellrott, & Xu, 2003; Teodorescu, Galor, Pillardy, & Elber, 2004; Wiederstein & Sippl, 2007). The widely used Loopp program models the protein through a library of folds derived from CATH and SCOP. It aligns by optimizing the interaction partners for each pair of residues in contact with the structure. It uses secondary structure predictions tools like Psipred to constrain the threading. ProSa uses contact potentials while 3DPSSM is used for position specific scoring matrix to align the Target with the Template. 3DPSSM is a hybrid method that combines optimization of alignment to a family with optimization of threading energy (Hong et al., 2011; Wang et al., 2014).

3. **Ab Initio Prediction:** *Ab initio* prediction is carried out when there is no suitable homolog found in the database. It is based on Anfinsen's hypothesis that the native state of the protein represents the global free energy minimum. *Ab initio* methods try to find these global minima of the protein. Finding the correct native-like conformation requires an efficient search method to explore the conformational space in the free energy landscape to get the global minima. To simplify the computation and to reduce the search space, the following models are used:

a. **Lattice Model:** This represents peptide chain as lattices. However, it fails to represent geometric attributes like strand twist. Also, its backbone prediction is error prone.
b. **Discrete State Off-Lattice Model:** This is an improved version of the lattice model. Off-lattice method are versatile than lattice method. It is able to represent subtle geometric considerations (strand twist, secondary structure propensities and packing) and can reproduce the backbone with greater accuracies (Park & Levitt, 1995).
c. **Using Local Structure Prediction:** Local structures can be excised from proteins that can fold independently from rest of the protein; this implies that there is a strong correlation between local structural information (biases) for short sequence segments. The strength and multiplicity of these local biases are highly sequence dependent.

The fitness of the model generated is calculated using a scoring function. There are two types of scoring functions in use, namely; knowledge based scoring function and force field based function. Former is an energy function derived from an analysis of known protein structures in the Protein Data Bank. This methodology relies on basic principle of free energy i.e. most stable conformation of a protein will be the lowest energy conformation. Force field based scoring function uses physics-based all-atom scoring function for protein structure prediction. This is analyzed with several widely used all-atom decoy sets. Currently there is not a single reliable scoring function or search method as both the scoring function are unable to explore the entire conformational space.

There is a new fully automated *ab initio* prediction method in which the Monte Carlo fragment insertion method (ROSETTA) and HMMSTR server are available, but with a limitation to model only small protein (>1000 residues) (Bystroff & Shao, 2002). The input sequence, after filtering out the low complexity region, is submitted to PSI-BLAST, and then converted to a sequence profile. The sequence profile is compared in a sliding window with each of the scoring matrices. Rosetta searches the conformational protein space using fragment insertion moves and by applying the Monte-Carlo acceptance criteria. The point where the fragment is inserted is selected at random and then a fragment of length 3 or 9 residues is selected at random from the move set. The backbone angles are changed to that of the fragment and the coordinates are then calculated. The move is accepted or rejected based on Monte Carlo criteria. The energy function is a structure based Bayesian conditional probability drawn form same PDB select database. For HMMSTR, the profile was submitted to each of the HMMSTR models namely HMMSTR-r for prediction of backbone angles, HMMSTR-d for the prediction of secondary structure and lastly HMMSTR-c for the prediction of super secondary structure.

Critical Assessment of Structural Prediction (CASP)

The biannual CASP (Moult, Fidelis, Kryshtafovych, Schwede, & Tramontano, 2014) experiments are carried out to test the different prediction methods/tools, which enables a direct comparison of a protein model to its X-ray structure. CASP is held every two years and the next one, CASP12, is to be held in mid 2016. In 2014 CASP Target list was of ~ 200 sequences with unknown 3D structures. Contestants are to predict the structure of protein using different algorithms and different methods - homology, fold recognition and *ab initio*. Subsequently, an assessment of the accuracy of the predictions is carried out. CASP concludes with a meeting to discuss the results that are published in CASP proceedings.

Table 1. Various modeling software tools and servers

Modeling Software	URL
SWISS Model	http://www.expasy.org/swissmod/SWISS-MODEL.html (Arnold, et al., 2006)
The CPHModels Server	http://www.cbs.dtu.dk/services/CPHmodels(Nielsen, Lundegaard, Lund, & Petersen, 2010)
EsyPred3D	http://www.fundp.ac.be/urbm/bioinfo/esypred (Lambert, Leonard, De Bolle, & Depiereux, 2002)
I-TASSER	http://zhanglab.ccmb.med.umich.edu/I-TASSER/(Yang et al., 2015; Zhang, 2008)
3D-JIGSAW	http://www.bmm.icnet.uk/~3djigsaw (Bates, Kelley, MacCallum, & Sternberg, 2001)
Modeller	http://salilab.org/modeler (Sali, 1995)
Structure Validation Resources	**URL**
WHAT IF	http://www.cmbi.kun.nl/swift/servers/ (Vriend, 1990)
PROCHECK	http://www.biochem.ucl.ac.uk (Laskowski, MacArthur, Moss, & Thornton, 1993)
VERIFY3D	http://www.doe-mbi.ucla.edu/Services(Eisenberg, Luthy, & Bowie, 1997)

Servers

There are various servers available for modeling as summarized in Table 1. This actually depends on the Templates chosen, the alignment of the query protein with the Template and also the algorithm or the method used in their prediction. It is a continuous benchmarking program and weekly evaluates the sensitivity and specificity of the different servers. Some of the servers that did well in both comparative and fold recognition modeling in CASP10 and CASP11 were 3DPSSM and Mgenthreader. Servers that outperformed in Template based modeling were I-Tasser and SwissModeller (Arnold, Bordoli, Kopp, & Schwede, 2006; Zhang, 2007).

Meta servers are servers that make their predictions based on the results of two or more different methods. Some simply look for consensus prediction between several methods while others calculate their own scores based on the results from the server. Some of the meta servers are Pcons, Pmodeller etc. The latest Live Bench experiments have demonstrated that meta servers clearly outperform the individual, autonomous servers.

Homology Modeling

Steps for Comparative Modeling

With the advent of automated protein modeling servers, submitting a sequence and obtaining its model has become a black box technique. However, all modeling protocols share certain steps, the understanding of which allows us to determine the quality of the model and the inherent errors in the system.

1. **Template Identification and Alignment:** For prediction of protein structure with reliably good accuracy, two conditions need to be fulfilled. First, the Target sequence must have significant similarity to the Template sequence. Second, computation of an accurate alignment must be feasible between the Target sequence and the Template structure. Overall model accuracy can be predicted by the degree of sequence similarity between the Target and the Template. The more the similar-

Figure 2. Multiple sequence alignment of MoxT (target) from B. anthracis with YdcE (template) from B. subtilis and Kid toxin from E. coli; the annotation of the catalytic residue E78(*) was derived by similarity from Kid.
Source: Based on Agarwal S, Mishra NK, Bhatnagar S, Bhatnagar R (2010).

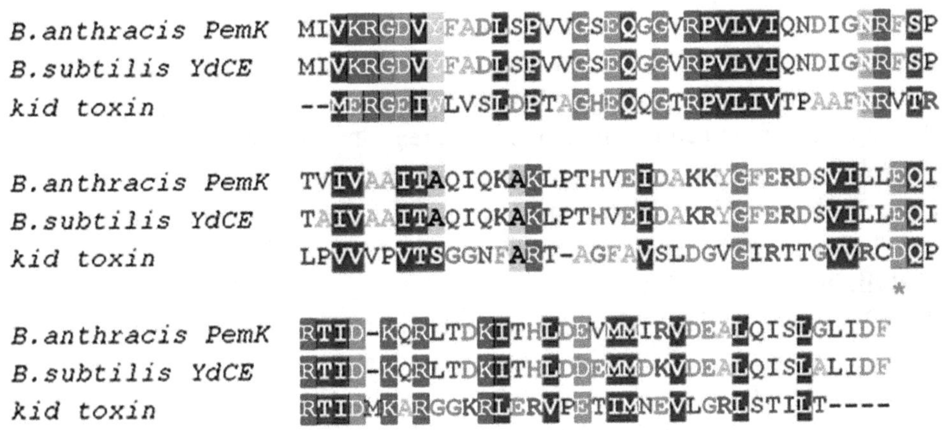

ity, more accurate is the model. However, while high accuracy models are more informative, low accuracy models may also provide coarse structural and functional annotation. Figure 2 shows a multiple sequence alignment of the Target (MoxT or the *B.anthracis* toxin) to the Template protein (*B. subtilis* YdcE) and *E.coli* Kid Toxin. The multiple sequence alignment was created using ClustalW(Thompson, Gibson, & Higgins, 2002) in Bioedit (Hall, 1999). The primary cause of errors in the final model is due to mistakes in the Target-Template alignment. Errors may also include incorrect loop building or side chain modeling. No modeling program can generate a good model from an incorrect starting alignment. Thus, good sequence-structure alignment must be ensured to improve the accuracy of the final model. Recent years mark the tremendous progress in the development of alignment methods e.g. HMMER, PSI-BLAST, HMM, SAM, HHsearch etc. Alignment can be pairwise or multiple alignments. Multiple alignments are usually heuristic, working on the principle of progressive alignment allowing large alignments of distantly related sequences eg ClustalW, ClustalX, T-Coffee. Probabilistic-based program like PROBCONS uses BAliBASE, which is a most accurate method available for multiple alignments. PROBCONS uses probabilities instead of the heuristic algorithms. HOMSTRAD is based on sequences with known 3D structures and PDB files.

2. **Alignment Correction:** An alignment is calculated by proper optimization of the scoring function. A scoring function includes gap penalty function and a substitution score matrix for matching residues in the Target and the Template sequence. Using a multiple sequence alignment program, structurally conserved region (SCRs) are searched. Insertions and deletions can be made in widely divergent parts of the molecule using multiple sequence alignment. Gaps have to be shifted to make them as small as possible. Gap is one or more empty spaces in one sequence aligned with letters in the other sequence. Gaps may or may not be treated as gap penalties. Gap penalty can be gap opening penalty or gap extension penalty. Gap penalties must be judiciously and carefully assigned as both the higher and the lower deviating from the optimum may lead to a poor alignment score. Precise importance has to be focused on residue-specific gap penalties which are responsible for

gaps in potential loop regions. Alignment score is a sum of gap penalties and the pair wise substitution scores which depend on the gap lengths and the residues matched. Despite inserting or deleting a gap and imposing gap penalties to correct the alignment, manual intervention may still be needed (Krieger, Nabuurs, & Vriend, 2003). In order to ensure a good alignment, it is recommended to choose to model only that part of the sequence which aligns without a large number of gaps in the alignment. Further, alignment errors are the main cause of deviations in comparative modeling even when the correct Template is chosen. In the alignment, gap positions should lie outside the secondary structures and in the loop areas. It is also recommended to integrate accessory information about secondary structures, conserved family residues, transmembrane helices, active site residues, etc. into the alignment in order to improve its accuracy. Last but not the least, no good alignment can be produced between sequences containing multiple repeats.

3. **Model Building:**
    a. **Backbone Generation:** When the alignment is correct, the backbone of the Target can be created. In this step, structurally conserved regions from the Template structure are used to generate the initial coordinates of the Target structure. For variable regions, the coordinates are copied from the Template structure only if the residue types are similar.
    b. **Loop Modeling:** The Target-Template alignment may contain gaps due to insertions and deletions of amino acids or bases. When the Target sequence contains a gap, the corresponding residue in the Template is deleted generating a hole. If there is an insertion in the Target, the Template will contain gap. Since, due to the gap in the Template, there will be no backbone coordinates known for these residues in the model, the Template backbone has to be cut to insert these residues. These large modifications are not feasible in secondary structural elements and therefore have to be placed in loops and strands. There are generally two methods of loop modeling: 1) by finding similar loops in similar proteins and 2) by generating the segment de novo (Li, 2013). Finding loops from similar proteins is done by taking some residues before and after the insertion as "anchor" residues and performing the loop search against databases like PDB with similar anchor residues. The best fitting loop is copied to the model. This has an advantage that all the loops are guaranteed to have reasonable internal geometries and conformations but again, they may not fit properly into the given model protein's framework. Loop selection must be done by carefully examining it for steric overlaps, and by checking loop atoms against rest of the protein's atoms.
    c. **Side Chain Modeling:** Structurally similar proteins have similar torsion angles about Cα-Cβ bond (psi angle) when comparing the side chain conformations. Therefore, in this step, the torsion angle between Cα and Cβ of the other residues can also be copied to the model because these rotamers tend to be conserved in similar proteins and many backbone configurations strongly prefer a specific rotamer. However, side chain conformations are also partially knowledge based. They are generated using rotamer libraries to select the best rotamer. To build a position-specific rotamer library, high-resolution protein structures are taken and all stretches of three to seven residues with a given amino acid at the center are collected. Prediction accuracy for side chain conformations is usually quite high for residues in the hydrophobic core (90%), while it is much lower for surface residues (50%).
    d. **Model Optimization:** The modifications in the side chain can affect the backbone causing steric hindrance between atoms. Errors in the prepared comparative model can be of different categories like: error in side chain modeling, distortions in the region correctly aligned with

the Template structure, drift in non-aligned regions, drift in the incorrectly aligned region, and structure misfolding due to incorrect Template. For the last case, a change of Template is required. In all other errors mentioned, optimization or idealization of the modeled structure using suitable a force field must be performed in order to obtain a refined model. A forcefield is a set of energy terms used to minimize the modeled structure. It is a collection of energy terms like energies of Van der Waals and electrostatic energy between all the atoms of the structure. For this, force fields used can be either quantum-based or self-parametrizing force fields. Bond-geometry correction and removal of bad contacts can be performed by forcefields like CHARMM (Brooks et al., 1983), AMBER (Cornell et al., 1995)or GROMOS (Scott et al., 1999). Refinement can be performed by using steepest descent, conjugate gradient or other energy minimization algorithms.

4. **Model Validation:** Homology models may contain errors. The reason may either be attributed to low percentage sequence identity among Target and Template or error in the Template. Therefore, the model must be evaluated for the overall correctness of the fold/structure, stereochemical parameters like bond lengths, torsion angles and bond angles, bump check etc. other properties like the distribution of polar/ nonpolar residues, bad contacts etc can also be compared with real structures. Models are routinely compared to Template structures to show the superposition (Figure 3) and the outcome is quantitatively expressed as root mean square deviation (RMSD). RMSD values up to 2.5Å between the Target and Template are indicative of good model quality. Validation can be of two types: Internal or External Validation.
    a. **Internal Validation:** Performed for self-consistency checks. It deals with the assessment of stereochemistry of the model like bond-lengths, bond-angles, dihedral angles, etc. it can be done by programs like WHATCHECK and PROCHECK.

Figure 3. Target template superposition; the figure highlights the superposition of the modeled MoxT (blue) upon the Template YdcE (green).
Source: Based on Agarwal S, Mishra NK, Bhatnagar S, Bhatnagar R (2010).

b. **External Validation:** Based on information that was not used in calculation of the model. External validation involves the check over whether the correct Template was used (based upon identity percent). This can be predicted by comparing the Z-score of the model and the Template structure. Z-score is the measure of the compatibility between a model's sequence and its structure. Moreover Ramachandran plot to visualize backbone dihedral angles φ and ψ of amino acid residues in protein structure have to be analyzed to determine whether the torsion angles fall into the allowed and the generously allowed region of the Ramachandran plot or not (Figure 4). Thus, validation forms an indispensable part of structure modeling and must be dealt with utmost precision.

SOLUTIONS AND RECOMMENDATIONS

1. The largest improvements will come from more experimentally determined structures. This will increase the chances of modeling assignments to recognize the fold correctly, thus increasing the accuracy of the model generated.
2. Further improvement in the sequence-structure alignment can also improve the accuracy of the structure model. A better aligned profile with fewer gaps improves the accuracy of the model. One possible way of improving sequence-structure alignment might be using threading based techniques to align the sequence to structures identified in comparative modeling.

Figure 4. Ramachandran plot
Source: Based on Agarwal S, Mishra NK, Bhatnagar S, Bhatnagar R (2010).

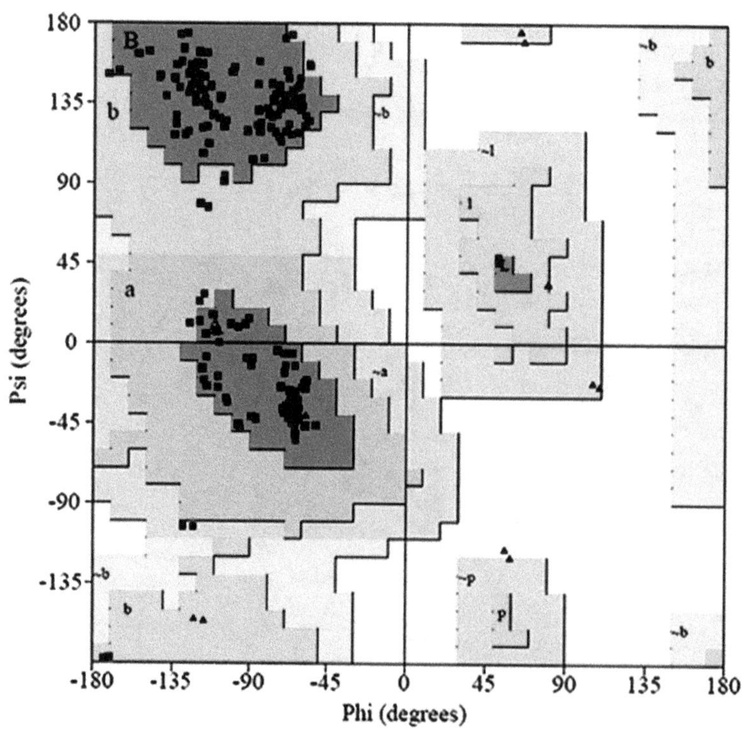

3. Protein tertiary structure prediction finds the solution to most of the issues and problems in the domain of deciphering protein structural coordinates. Multiple Template based homology modeling can be applied to the low sequence similarity situations where multiple protein structural Templates can be used to model the Target protein.
4. Refinements of the modeled structure generated from homology modeling, threading, or even *ab initio* methods can be done by running molecular dynamic simulations.
5. The model is supposed to be generated on the basis of background knowledge of the protein of interest; this can be achieved by knowledge from the literature on the protein, an in-depth knowledge of the functional relevant conformation and Active site residues information. The knowledge from the literature study can help us explaining the structural and functional attributes of the protein to be modeled.

CASE STUDY

Modeling of B.anthracis Toxin MoxT: This case study is based on "Agarwal S, Mishra NK, Bhatnagar S, Bhatnagar R (2010) PemK Toxin of *B.anthracis* Is a Ribonuclease: An insight into its active site, structure, and function. JBC 285:7254–7270 (Agarwal, Mishra, Bhatnagar, & Bhatnagar, 2010). "

B.anthracis toxin-antitoxin module (MoxXT) is a TA system wherein, PemK-like toxin (MoxT) functions as a ribonuclease and thus acts as a cellular endotoxin. Under normal conditions, it exists as a dimer in complex with its cognate antitoxin and is inactive. Modeling of this toxin was used for the purpose of mutating the possible catalytic site residues.

Methodology

Homology modeling of the MoxT dimer: Homology modeling of the MoxT dimer was performed using *B. subtilis* YdcE as Template (PDB ID: 1NE8) (Gogos, Mu, Bahna, Gomez, & Shapiro, 2003) using the Swiss-Model Workspace. Figure 2 shows a multiple sequence alignment of the Target (MoxT or the *B. anthracis* toxin) to the Template protein (*B. subtilis* YdcE) and *E.coli* Kid Toxin. The multiple sequence alignment was created using ClustalW (Thompson, et al., 2002) in Bioedit (Hall, 1999) (Figure 2). Homology modeling was done using the sequence alignment with YdcE while alignment with Kid was used to infer the catalytic site residues for further mutagenesis. The modeled structure superimposed on the Template with RMSD of 0.1 Å between 232 amino acids (Figure 3). The Ramachandran plot of the predicted model revealed that 92.8% of the residues lie in the most favoured regions, and 6.2% of the residues lie in the additionally allowed regions, whereas 1.0% of the residues lie in the generously allowed region (Figure 4). The Final model was validated using Procheck (Laskowski, et al., 1993).

The residue E78 is conserved in MoxT. The other catalytic residue of Kid (R73) is replaced by a non-polar Leucine in YdcE and MoxT. However, the ribonuclease activity essentially requires the presence of a negatively charged and a positively charged (or polar group) in close proximity. In order to find the second catalytic site residue, the polar groups in the neighborhood of the conserved negatively charged E78 were identified. Interestingly, the only positively charged residue located in the close proximity of E78 was found to be H59 (Figure 5). Site directed mutagenesis of E78 and the nearby H59 resulted in abrogation of the ribonuclease activity of MoxT.

Figure 5. Homology model of MoxT dimer; the monomer on the right shows the Cα trace of the monomer of the MoxT model (blue) superimposed upon the YdcE (black). A loop involved in binding of antitoxin is shown in orange. The catalytic base, Glu 78 and the catalytic acid, H59 are shown as magenta sticks. The monomer on the left is shown as a cartoon. The helices are shown in red, sheets in yellow, and loop regions in green. The figure rendering was accomplished using PyMOL (Delano Scientific).
Source: Delano (2002).

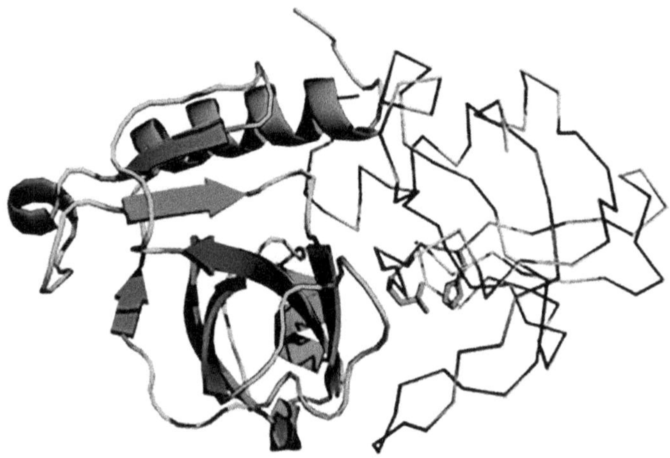

FUTURE RESEARCH DIRECTIONS

With the advancements in structural genomics projects, homology modeling methods is ought to play an attractive role in structure based drug discovery. Computational prediction methods provide a cost effective alternative in the absence of crystal structures. But the major drawback with prediction tools is to develop an accurate model. In this chapter, we report the current developments in this field, discuss *in silico* modeling limitations, and review the successful application of this technique to different stages of the drug discovery process. In-spite of the tremendous developments in the domain of protein tertiary structure prediction using homology modeling, there is still a requirement for better force fields and optimization functions. Better algorithmic approaches need to be developed to successfully incorporate divalent metal ions into the model

CONCLUSION

- Experimental macromolecular structure determination methods like X-ray and NMR methodology is expensive, exhaustive and cumbersome. Thus it cannot be always be undertaken.
- There are three basic approaches of structure prediction based upon the query-Template identity. If the identity is more than 30%, then the modeling lies in safe zone and comparative modeling approach is the preferred approach else threading and *ab initio* approach is used.
- If the query Template identity is less than 30%, then Threading is performed for protein structure prediction. This approach uses a database of known three-dimensional structures to match sequences without known structure with the protein folds.

- *Ab initio* methods of protein fold prediction use force-field-based energy minimization to simulate the process of protein folding.
- Comparative modeling refers to the prediction of the Target protein from the identified Template protein structure.
- In comparative modeling, first a suitable Template is identified based upon the Target-Template alignment. After getting a proper alignment, real model building starts which encompass backbone generation, side-chain modeling and loop modeling. This is followed by model optimization and model validation.
- The accuracy of the model largely depends upon the Template identified and its alignment with the Target sequence.

REFERENCES

Agarwal, S., Mishra, N. K., Bhatnagar, S., & Bhatnagar, R. (2010). PemK toxin of Bacillus anthracis is a ribonuclease: An insight into its active site, structure, and function. *The Journal of Biological Chemistry*, *285*(10), 7254–7270. doi:10.1074/jbc.M109.073387 PMID:20022964

Alexandrov, N., & Shindyalov, I. (2003). PDP: Protein domain parser. *Bioinformatics (Oxford, England)*, *19*(3), 429–430. doi:10.1093/bioinformatics/btg006 PMID:12584135

Arnold, K., Bordoli, L., Kopp, J., & Schwede, T. (2006). The SWISS-MODEL workspace: a web-based environment for protein structure homology modelling. *Bioinformatics, 22*(2), 195-201. doi: 10.1093/bioinformatics/bti770

Bates, P. A., Kelley, L. A., MacCallum, R. M., & Sternberg, M. J. (2001). Enhancement of protein modeling by human intervention in applying the automatic programs 3D-JIGSAW and 3D-PSSM. *Proteins*, *45*(S5), 39–46. doi:10.1002/prot.1168 PMID:11835480

Brooks, B. R., Bruccoleri, R. E., Olafson, B. D., States, D. J., Swaminathan, S., & Karplus, M. (1983). CHARMM: A program for macromolecular energy, minimization, and dynamics calculations. *Journal of Computational Chemistry*, *4*(2), 187–217. doi:10.1002/jcc.540040211

Bystroff, C., & Shao, Y. (2002). Fully automated ab initio protein structure prediction using I-SITES, HMMSTR and ROSETTA. *Bioinformatics (Oxford, England)*, *18*(Suppl 1), S54–S61. doi:10.1093/bioinformatics/18.suppl_1.S54 PMID:12169531

Ceulemans, H., & Russell, R. B. (2004). Fast fitting of atomic structures to low-resolution electron density maps by surface overlap maximization. *Journal of Molecular Biology*, *338*(4), 783–793. doi:10.1016/j.jmb.2004.02.066 PMID:15099745

Chakravarty, S., Wang, L., & Sanchez, R. (2005). Accuracy of structure-derived properties in simple comparative models of protein structures. *Nucleic Acids Res, 33*(1), 244-259. doi: 10.1093/nar/gki162

Chothia, C. L. A. (1986). The relation between the divergence of sequence and structure in proteins. *The EMBO Journal*, *5*, 823–836. PMID:3709526

Cornell, W. D., Cieplak, P., Bayly, C. I., Gould, I. R., Merz, K. M., Ferguson, D. M., & Kollman, P. A. et al. (1995). A Second Generation Force Field for the Simulation of Proteins, Nucleic Acids, and Organic Molecules. *Journal of the American Chemical Society, 117*(19), 5179–5197. doi:10.1021/ja00124a002

Delano, W. L. (2002). *The PyMOL Molecular Graphics System.* Retrieved from http://www.pymol.org

Eisenberg, D., Luthy, R., & Bowie, J. U. (1997). VERIFY3D: assessment of protein models with three-dimensional profiles. *Methods Enzymol, 277*, 396-404. doi: S0076-6879(97)77022-8

Francoijs, C. J., Klomp, J. P., & Knegtel, R. M. (2000). Sequence annotation of nuclear receptor ligand-binding domains by automated homology modeling. *Protein Engineering, 13*(6), 391–394. doi:10.1093/protein/13.6.391 PMID:10877848

Ginalski, K., Rychlewski, L., Baker, D., & Grishin, N. V. (2004). Protein structure prediction for the male-specific region of the human Y chromosome. *Proc Natl Acad Sci U S A, 101*(8), 2305-2310. doi: 101/8/2305

gkq535 [pii]

Gogos, A., Mu, H., Bahna, F., Gomez, C. A., & Shapiro, L. (2003). Crystal structure of YdcE protein from Bacillus subtilis. *Proteins, 53*(2), 320–322. doi:10.1002/prot.10457 PMID:14517982

Hall, T. A. (1999). BioEdit: A user-friendly biological sequence alignment editor and analysis program for Windows 95/98/NT. *Nucleic Acids Symposium Series, 41*, 95–98.

Hong, Y., Chintapalli, S. V., Ko, K. D., Bhardwaj, G., Zhang, Z., van Rossum, D., & Patterson, R. L. (2011). Predicting protein folds with fold-specific PSSM libraries. *PLoS ONE, 6*(6), e20557. doi:10.1371/journal.pone.0020557 PMID:21698189

jmbi.2000.3741 [pii]

Jones, D. T. (1999). GenTHREADER: an efficient and reliable protein fold recognition method for genomic sequences. *J Mol Biol, 287*(4), 797-815. doi: S0022-2836(99)92583-4

Kelley, L. A., MacCallum, R. M., & Sternberg, M. J. (2000). Enhanced genome annotation using structural profiles in the program 3D-PSSM. *Journal of Molecular Biology, 299*(2), 499–520. doi:10.1006/jmbi.2000.3741 PMID:10860755

Kim, D., Xu, D., Guo, J. T., Ellrott, K., & Xu, Y. (2003). PROSPECT II: Protein structure prediction program for genome-scale applications. *Protein Engineering, 16*(9), 641–650. doi:10.1093/protein/gzg081 PMID:14560049

Kolinski, A., Rotkiewicz, P., Ilkowski, B., & Skolnick, J. (1999). A method for the improvement of threading-based protein models. *Proteins, 37*(4), 592–610. doi:10.1002/(SICI)1097-0134(19991201)37:4<592::AID-PROT10>3.0.CO;2-2 PMID:10651275

Krieger, E., Nabuurs, S. B., & Vriend, G. (2003). Homology modeling. *Methods of Biochemical Analysis, 44*, 509–523. PMID:12647402

Kulikova, T., Aldebert, P., Althorpe, N., Baker, W., Bates, K., Browne, P., & Apweiler, R. et al. (2004). The EMBL Nucleotide Sequence Database. *Nucleic Acids Research*, *32*, D27–D30. doi:10.1093/nar/gkh120 PMID:14681351

Lambert, C., Leonard, N., De Bolle, X., & Depiereux, E. (2002). ESyPred3D: Prediction of proteins 3D structures. *Bioinformatics (Oxford, England)*, *18*(9), 1250–1256. doi:10.1093/bioinformatics/18.9.1250 PMID:12217917

Laskowski, R. A., MacArthur, M. W., Moss, D. S., & Thornton, J. M. (1993). PROCHECK: A program to check the stereochemical quality of protein structures. *Journal of Applied Crystallography*, *26*(2), 283–291. doi:10.1107/S0021889892009944

Lesk, A. M. (1997). CASP2: Report on ab initio predictions. *Proteins*, *29*(S1Suppl 1), 151–166. doi:10.1002/(SICI)1097-0134(1997)1+<151::AID-PROT20>3.0.CO;2-M PMID:9485507

Li, Y. (2013). Conformational sampling in template-free protein loop structure modeling: An overview. *Comput Struct Biotechnol J*, *5*(6), e201302003. doi:10.5936/csbj.201302003 PMID:24688696

Lu, L., Lu, H., & Skolnick, J. (2002). MULTIPROSPECTOR: An algorithm for the prediction of protein-protein interactions by multimeric threading. *Proteins*, *49*(3), 350–364. doi:10.1002/prot.10222 PMID:12360525

Marti-Renom, M. A., Stuart, A. C., Fiser, A., Sanchez, R., Melo, F., & Sali, A. (2000). Comparative protein structure modeling of genes and genomes. *Annu Rev Biophys Biomol Struct, 29*, 291-325.

Moult, J., Fidelis, K., Kryshtafovych, A., Schwede, T., & Tramontano, A. (2014). Critical assessment of methods of protein structure prediction (CASP)--round x. *Proteins*, *82*(Suppl 2), 1–6. doi:10.1002/prot.24452 PMID:24344053

Nielsen, M., Lundegaard, C., Lund, O., & Petersen, T. N. (2010). CPHmodels-3.0--remote homology modeling using structure-guided sequence profiles. *Nucleic Acids Res, 38*, W576-581. doi: 10.1093/nar/gkq535

Park, B. H., & Levitt, M. (1995). The complexity and accuracy of discrete state models of protein structure. *J Mol Biol, 249*(2), 493-507. doi: S0022-2836(85)70311-7

Rost, B. (1999). Twilight zone of protein sequence alignments. *Protein Engineering*, *12*(2), 85–94. doi:10.1093/protein/12.2.85 PMID:10195279

Rost, B., Fariselli, P., & Casadio, R. (1996). Topology prediction for helical transmembrane proteins at 86% accuracy. *Protein Science*, *5*(8), 1704–1718. doi:10.1002/pro.5560050824 PMID:8844859

Sali, A. (1995). Comparative protein modeling by satisfaction of spatial restraints. *Molecular Medicine Today*, *1*(6), 270–277. doi:10.1016/S1357-4310(95)91170-7 PMID:9415161

Sander, C., & Schneider, R. (1991). Database of homology-derived protein structures and the structural meaning of sequence alignment. *Proteins*, *9*(1), 56–68. doi:10.1002/prot.340090107 PMID:2017436

Scott, W. R. P., Hünenberger, P. H., Tironi, I. G., Mark, A. E., Billeter, S. R., Fennen, J., & van Gunsteren, W. F. et al. (1999). The GROMOS Biomolecular Simulation Program Package. *The Journal of Physical Chemistry A*, *103*(19), 3596–3607. doi:10.1021/jp984217f

Skolnick, J., Kihara, D., & Zhang, Y. (2004). Development and large scale benchmark testing of the PROSPECTOR_3 threading algorithm. *Proteins*, *56*(3), 502–518. doi:10.1002/prot.20106 PMID:15229883

Takeda-Shitaka, M., Takaya, D., Chiba, C., Tanaka, H., & Umeyama, H. (2004). Protein structure prediction in structure based drug design. *Current Medicinal Chemistry*, *11*(5), 551–558. doi:10.2174/0929867043455837 PMID:15032603

Talukdar, A. S., & Wilson, D. L. (1999). Modeling and optimization of rotational C-arm stereoscopic X-ray angiography. *IEEE Transactions on Medical Imaging*, *18*(7), 604–616. doi:10.1109/42.790460 PMID:10504094

Teodorescu, O., Galor, T., Pillardy, J., & Elber, R. (2004). Enriching the sequence substitution matrix by structural information. *Proteins*, *54*(1), 41–48. doi:10.1002/prot.10474 PMID:14705022

Thompson, J. D., Gibson, T. J., & Higgins, D. G. (2002). Multiple sequence alignment using ClustalW and ClustalX. In Curr Protoc Bioinformatics. doi:10.1002/0471250953.bi0203s00

Thompson, J. D., Higgins, D. G., & Gibson, T. J. (1994). CLUSTAL W: Improving the sensitivity of progressive multiple sequence alignment through sequence weighting, position-specific gap penalties and weight matrix choice. *Nucleic Acids Research*, *22*(22), 4673–4680. doi:10.1093/nar/22.22.4673 PMID:7984417

Venclovas, C., Zemla, A., Fidelis, K., & Moult, J. (2003). Assessment of progress over the CASP experiments. *Proteins*, *53*(S6), 585–595. doi:10.1002/prot.10530 PMID:14579350

Vitkup, D., Melamud, E., Moult, J., & Sander, C. (2001). Completeness in structural genomics. *Nature Structural Biology*, *8*(6), 559–566. doi:10.1038/88640 PMID:11373627

Vriend, G. (1990). WHAT IF: a molecular modeling and drug design program. *J Mol Graph, 8*(1), 52-56.

Wang, J., Wang, C., Cao, J., Liu, X., Yao, Y., & Dai, Q. (2014). Prediction of protein structural classes for low-similarity sequences using reduced PSSM and position-based secondary structural features. *Gene, 554*(2), 241-248. doi: S0378-1119(14)01197-4

Wiederstein, M., & Sippl, M. J. (2007). ProSA-web: interactive web service for the recognition of errors in three-dimensional structures of proteins. *Nucleic Acids Res, 35*, W407-410.

Xu, Y., Xu, D., & Uberbacher, E. C. (1998). An efficient computational method for globally optimal threading. *Journal of Computational Biology*, *5*(3), 597–614. doi:10.1089/cmb.1998.5.597 PMID:9773353

Yachdav, G., Kloppmann, E., Kajan, L., Hecht, M., Goldberg, T., Hamp, T., . . . Rost, B. (2014). PredictProtein--an open resource for online prediction of protein structural and functional features. *Nucleic Acids Res, 42*, W337-343. doi: 10.1093/nar/gku366

Yang, J., Yan, R., Roy, A., Xu, D., Poisson, J., & Zhang, Y. (2015). The I-TASSER Suite: Protein structure and function prediction. *Nature Methods*, *12*(1), 7–8. doi:10.1038/nmeth.3213 PMID:25549265

Zhang, Y. (2007). Template-based modeling and free modeling by I-TASSER in CASP7. *Proteins*, *69*(S8Suppl 8), 108–117. doi:10.1002/prot.21702 PMID:17894355

Zhang, Y. (2008). I-TASSER server for protein 3D structure prediction. *BMC Bioinformatics*, *9*(1), 40. doi:10.1186/1471-2105-9-40 PMID:18215316

Zhou, Y., & Johnson, M. E. (1999). Comparative molecular modeling analysis of-5-amidinoindole and benzamidine binding to thrombin and trypsin: Specific H-bond formation contributes to high 5-amidinoindole potency and selectivity for thrombin and factor Xa. *Journal of Molecular Recognition*, *12*(4), 235–241. doi:10.1002/(SICI)1099-1352(199907/08)12:4<235::AID-JMR460>3.0.CO;2-X PMID:10440994

KEY TERMS AND DEFINITIONS

Alignment: The process of lining up two or more sequences to achieve maximal levels of identity for the purpose of assessing the degree of similarity and the possibility of homology.

BLAST: Basic Local Alignment Search Tool. A sequence comparison algorithm used to search sequence databases for optimal local alignment to a query.

CASP: Critical assessment of Structure Prediction is a community-wide, worldwide experiment for protein structure prediction taking place every two years since 1994.

DDBJ: DNA Databank of Japan is a biological database that collects DNA sequences.

EMBL: European molecular biology laboratory is Europe's primary collection of all publicly available nucleotide sequences.

Fold: Arrangement of secondary structural elements like helix, sheets and coils of a protein in 3D space.

Gap: A position in an alignment that represents a deletion within one sequence relative to other at that position.

Gap Penalties: Penalty applied to a similarity score for the introduction of an insertion or deletion gap, gap extension, etc. gap penalty is the deduction from the cumulative score in alignment of two sequences.

Homology: Similarity in two sequences between individuals of the same or different species.

Modeling: Process where 3D structure of the macromolecule is generated (predicted), represented and manipulated to determine their molecular properties.

PDB: Protein Databank is an archive of experimentally determined 3D structures.

RMSD: Root mean square Deviation is the measure of the average distance between the atoms of superimposed proteins.

Chapter 15
Online Molecular Docking Resources

Adriana Isvoran
West University of Timisoara, Romania

ABSTRACT

This chapter aims to present the available online resources that are used for protein modeling with accent to online molecular docking resources. SwissDock, MTiAutoDock, and PatchDock online docking tools are described and a few illustrative examples concerning the molecular docking studies for the cytochrom P450 interactions with the fungicide difenoconazole. The results obtained using different servers based on distinct approaches are compared and the advantages and/or disadvantages of every server are illustrated.

INTRODUCTION

In the actual period we assist to the expansion of structural data concerning proteins and protein–ligand complexes with direct implications in the development of the computational strategies to enhance drug discovery. One of these strategies is the virtual screening (VS) technique that allows hits identification with lower costs compared with the traditional experimental high-throughput screening (Gohlke & Klebe, 2002). Virtual screening techniques can be classified into ligand-based and structure-based methods respectively.

The ligand-based method is used when a set of ligand molecules is known but the target's structure is unknown and it may be implemented using several approaches. When the information contained in the set of known ligands is used to build a model of the receptor molecule, the ligand-based method is implemented as pharmacophore modeling (Sun, 2008). The ligand-based approach may be implemented using the quantitative structure activity relationship (QSAR) method that takes into account the connection between chemical structures of some chemical compounds and their biological activities and predicts the activities of new chemicals (Nantasenamat et al, 2009). Another approach is the chemical similarity searching, a method based on the assumption that all compounds in a database that are similar to a query compound have similar biological activity (Willet et al., 1998). When applying this method, a

DOI: 10.4018/978-1-5225-0115-2.ch015

database of molecules is screened against one or more known active ligands structures in order to predict their biological activity. The ligand-based virtual screening may also be based on the shape similarity searching, molecules with shape similar to that of the active compound are considered to fit the target's binding site (Rush et al., 2005).

The structure-based drug design involves molecular docking as the most common method. It is based on docking of a candidate compound into the protein target and applying a scoring function to estimate the probability that the ligand will bind to the protein with high affinity (Kroemer, 2007). Computational programs based on diverse algorithms were developed to implement molecular docking studies, molecular docking being an increasingly important tool in actual pharmaceutical researches. From this point of view, it becomes important to evaluate the existing docking programs in order to choose the most suitable docking algorithm for a particular study.

Molecular docking studies may be implemented by using commercial software, free software designed for academic users and also using on-line specialized servers. There are many servers allowing online protein-protein and protein-ligand molecular docking. For example, Click2Drug server (http://www.click2drug.org/) contains a comprehensive list of docking web services, databases and computer-aided drug design tools.

Within this chapter, the use of some online resources designed to protein modeling and online servers allowing molecular docking studies based on different approaches is presented. The description is illustrated with examples of molecular docking computations concerning the prediction of the interactions of the cytochrome P450 with the fungicide difenoconazole (DFC). The results obtained using different servers are compared and discussed in terms of agreement or disagreement concerning the predictions of the receptor-ligand complexes obtained using the docking tools.

BACKGROUND

Online Databases

Last decades are characterized by the development of the internet and increase of the on-line resources. Another characteristic of the last decades is the massive accumulation of data concerning the molecules with therapeutic potential, pesticides and drugs, sequences and structures of the biological macromolecules. These data are now structured and deposited in databases, most of them being freely accessible for educational purposes and they can be used to build models for designing new drug molecules. Existing databases may be classified in databases containing information concerning the small chemical compounds structures and properties, databases comprising information about the sequences and structures of the biological macromolecules and databases relating the information about the chemical compounds with those concerning their molecular targets and pathways. Another more detailed classification of databases that are used for bioinformatics applications considers the following categories: nucleotide sequence databases, genome databases, protein sequence databases, protein structure databases, protein model databases, proteomics databases, nucleic acids databases, carbohydrate structure databases, protein-protein and other molecular interactions, signal transduction pathway databases, metabolic pathway and protein function databases, metabolomic databases and many other specialized databases.

Databases are indispensable resources for online-docking and/or other computational tools used in protein-protein and protein-ligand interactions analysis and drug discovery. A comprehensive view of the chemical databases is provided on the Computational Resources for Drug Discovery server (http://crdd.osdd.net/chemdatabase.php). A few examples of usual chemical libraries databases indispensable for molecular modeling studies are the following:

- ZINC database (http://zinc.docking.org/) provided by the Irwin and Shoichet Laboratories in the Department of Pharmaceutical Chemistry at the University of California and designed for structure-based virtual screening studies (Irwin and Stoichet, 2005; Irwin et al., 2012). It contains over 20 million ligands that may be downloaded in many common file formats (SMILES, mol2, ready-to-dock 3D formats compatible with most docking programs, etc.). The database also permits to extract information concerning the structure, biological activity, physical and chemical properties, name, catalog and Chemical Abstracts Service (CAS) numbers and vendor for every chemical compound. The chemical compound is prepared in multiple protonation states and multiple tautomeric forms and multiple conformations are available for the molecules. There also is the possibility for users to process their own molecules by uploading them to the server.
- PubChem compound database (http://www.ncbi.nlm.nih.gov/pccompound) provides substance information, compound structures and biological activities and is organized as three linked databases: PubChem Substance, PubChem Compound (Bolton et al., 2008) and PubChem BioAssay (Wang et al., 2014). PubChem Substance contains more than 180 million records, PubChem Compound covers more than 63 million structures and PubChem BioAssay comprises more than 1 million BioAssays. PubChem also provides a fast chemical structure similarity search tool.
- ChEMBL database (https://www.ebi.ac.uk/chembldb/) provides information about 1 million small drug-like molecules and 8200 drug targets.
- ChemDB database (http://cdb.ics.uci.edu/) includes information concerning the predicted or experimentally determined physicochemical properties and 3D structures for about 5 million chemicals (Chen et al., 2007).
- ChemSpider database (http://www.chemspider.com/) is owned by the Royal Society of Chemistry and provides access to structures, properties, and associated data integrated with publication information for over 35 million chemical molecules.
- Pesticide databases, such as Pesticide Action Network (PAN) database (http://www.pesticideinfo.org/) containing toxicity and regulatory information for pesticide and Pesticides Properties Database (PPDB, http://sitem.herts.ac.uk/aeru/ppdb/en/) comprising physicochemical, toxicological, ecotoxicological and other related data for about 1150 pesticides, 700 of their metabolites and 100 other related substances and
- Drugs databases: DrugBank database (http://www.drugbank.ca/), Therapeutic Target database (http://xin.cz3.nus.edu.sg/group/ttd/ttd.asp), PharmGKB database (https://www.pharmgkb.org/), Search Tool for Interactions of Chemicals (STITCH, http://stitch.embl.de/), Super Target Database (http://bioinf-apache.charite.de/supertarget_v2/).
- Universal Protein (UniProt) resource (http://www.uniprot.org/) is freely accessible and contains information concerning protein sequences and functions (Jain et al., 2009).The UniProt resource covers three linked databases: UniProt Knowledgebase (UniProtKB), the UniProt Reference Clusters (UniRef), and the UniProt Archive (UniParc).

- Protein Data Bank (PDB, http://www.rcsb.org/pdb/home/home.do) is the single worldwide online resource containing the three dimensional structures of the proteins and nucleic acids and of their complexes with other molecules (Bernstein et al., 1977). These structures are usually obtained through X-ray crystallography and/or NMR studies. PDB contains structures for biological molecules belonging to diverse organisms and it is updated weekly. On 11 of October 2015 this database contains 112 722 structures. The server housing the PDB resource also provides tools and additional information for researchers and students in structural biology, molecular biology, computational biology and other related fields.
- PDBsum (https://www.ebi.ac.uk/thornton-srv/databases/cgi-bin/pdbsum/GetPage.pl?pdbcode=index.html) is a pictorial database that offers a quick overview of the contents of each structural file deposited in the Protein Data Bank illustrating the molecule(s) contained in the structural file with a schematic diagram of their interactions (de Beer et al., 2014). In the case of proteins, for every PDB entry, the information deposited in the PDBSum is ranged on several pages and it contains: summary information (image of the structure and the molecules therein, enzyme reaction diagram), wiring diagram, topology diagram, residue conservation, description of ligands contained in the structural file, schematic representations of protein-protein interfaces and the residue-residue interactions made across them, top ten clefts in the surface of the protein classified by volume with the bound ligands shown and links to external databases for further information.

There also are other specialized online databases that are used for particular bioinformatics applications but it is not the aim of this chapter to present all available databases in the field.

Protein Modeling Resources

Due to advances in computer technology, the massive amount of available biological data and progresses in modeling methods development, protein modeling is playing a more and more important role in many fields of biological and biochemical sciences. Protein modeling tools refer to the prediction and/or analysis of the protein structure in correlation with its function and its underlying mechanism and may also provide insight to design experiments and suggest possible clues for drug design and discovery. There are online resources available for protein modeling and they usually refer to (Xu et al., 2000):

1. Protein sequence comparison including sequences alignment, sequence-based protein family classification, phylogenetic classification;
2. Protein sequence annotation including prediction of hydrophobic and/or flexibility profiles, of the secondary structures of proteins, of the transmembrane regions, active or signaling sites;
3. Three-dimensional structure prediction by homology modeling;
4. Protein structure analysis including structure visualization and geometry analysis, structure comparison and classification, surface analysis with pockets and cavities detection, electrostatics and hydrophobic surface coloring;
5. Molecular dynamics simulation and molecular docking aiming the protein-protein, protein-ligands and other protein interactions analysis and characterization.

Table 1. Free molecular visualization tools

Visualization Tool	URL	Reference
UCSF Chimera	http://www.cgl.ucsf.edu/chimera/	Pettersen et al., 2004
PyMol	https://www.pymol.org/	http://sourceforge.net/projects/pymol/
Swiss PDB Viewer	http://spdbv.vital-it.ch/	Guex and Peitsch, 1997
Jmol	http://jmol.sourceforge.net/	http://www.jmol.org/
Rasmol	https://www.umass.edu/microbio/rasmol/	https://www.umass.edu/microbio/rasmol/ distrib/rasman.htm
Visual molecular dynamics (VMD)	http://www.ks.uiuc.edu/Research/vmd/	Humphrey et al., 1996

Protein structure analysis, structures comparison and classification require molecular graphics visualization tools that enable the visualization of the structures that are encoded by the atomic coordinate files expressed in different formats and manipulation of these images to observe the molecules from various perspectives. There are several open source visualization tools that are freely download and used through a wide variety of operating systems (Table 1).

Proteins perform their biological functions by interacting with other molecules (substrates, antigens, nucleic acids, peptides, proteins, etc). These interactions occur in specific regions on the protein surface that are either concavities on the protein surface (pockets) either regions enveloped by the protein surface (cavities). Therefore, it is of interest to identify pockets and cavities in protein structures. Especially, the detection of pockets and cavities is a prerequisite for identification and characterization of pockets and/or cavities presents on the protein surface are important tasks. This information can provide insight into biological functions and reaction mechanisms of proteins and it is also a requirement for protein–ligand docking and structure-based drug design. There are a few online resources destined to the detection and characterization of pockets and/or cavities on the protein surface, such as:

- Fpocket (http://mobyle.rpbs.univ-paris-diderot.fr/cgi-bin/portal.py?form=fpocket2#forms::fpocket), a computational tool that enables the identification and characterization of the potent binding sites on the protein surface (Schmidtke et al. 2010).
- Computed Atlas of Surface Topography of proteins (CASTp, http://sts.bioe.uic.edu/castp/index.php) that allows the identification and measurements of surface accessible pockets and of interior inaccessible cavities for proteins and nucleic acids (Dundas et al., 2006).
- DEPTH (http://mspc.bii.a-star.edu.sg/tankp/intro.html) is a web server used for structural analyses of protein structures that permits to detect potential small-molecule ligand-binding cavities and compute their sizes and to predict the pKa of ionizable residues in proteins (Tan et al., 2013).
- Roll (http://altair.sci.hokudai.ac.jp/g6/Research/POCASA_e.html.) is an algorithm that can predict binding sites by detecting pockets and cavities of proteins with a rolling sphere (Yu et al., 2010). Roll is implemented in the program POCASA that can automatically predict the ligand binding sites of proteins. This program uses different probe spheres to determine the pockets and it makes the program adaptable for various ligands and proteins.

- *3D-SURFER* (http://dragon.bio.purdue.edu/3d-surfer) is a web-based tool destined to an interactive protein surface analysis, comparison and characterization of proteins at the global and local levels based on their surface shape (La et al., 2009). Concerning the protein surface analysis, the server also identifies and visualizes local features, such as pockets, protrusions, hollow spaces and flat regions. It also integrates identification of protein surface properties (electrostatic potentials, hydrophobicity and conservation) allowing detailed analysis designed to assist investigating function of proteins.

There also are resources concerning the protein spatial structure prediction that is one of the most important goals pursued by computational biochemistry especially for the drug design and biotechnological studies. The protein 3D structure prediction tools employ either homology based methods or *ab initio* methods. Homology based protein structure prediction requires the identification of homologs that can be used as templates for modeling with a significant amino acid sequence similarity of the model to the target sequence, this search involving the use of specialized computational tools (Wallner & Elofsson, 2005). Some online resources designed to the structure homology modeling of proteins are presented in Table 2.

Ab initio structure prediction methods are used when there are no available structural models for the investigated protein, the structure is predicted from sequence alone and they are usually designed to small proteins (Bonneau and Baker, 2001).

Molecular Docking Resources

Besides the great diversity of the bioinformatics tools, the molecular docking is one of the computational tools that help us to understand the interactions between biologically relevant molecules (proteins, nucleic acids, carbohydrates, lipids) at the atomic level, as well as to clarify fundamental biochemical processes, especially those based on the protein-protein and protein–ligands interactions.

Molecular docking is a computational technique focusing on prediction of the binding mode(s) (BMs) of a ligand with a protein (Morris & Lim-Wilby, 2008). It aims of finding an optimized conformation for both the protein and ligand and their relative orientations such that the free energy of the protein-ligand complex is minimum. There are two approaches used in molecular docking studies: one is the geometry-based approach that is centered on the surface complementarity of the ligand and protein (Shoichet et

Table 2. Free protein structure homology modeling tools

Structure Homology Tool	URL	Reference
MODELLER	http://salilab.org/modeller/	**Eswar et al, 2006**
SWISS-MODEL	http://swissmodel.expasy.org/interactive	Biasini et al, 2014
HHpred	http://toolkit.tuebingen.mpg.de/hhpred	Söding, 2005
Geno3D	http://geno3d-pbil.ibcp.fr/cgi-bin/geno3d_automat.pl?page=/GENO3D/geno3d_home.html	Combet et al., 2002
ESyPred3D	http://www.unamur.be/sciences/biologie/urbm/bioinfo/esypred	Lambert et al., 2002
PHYRE2	http://www.sbg.bio.ic.ac.uk/ phyre2	Kelley et al., 2015

al., 2004) and the other approach is energetic-based and consists on simulating the docking process and computing ligand-protein pairwise interactions energies (Feig et al., 2004). Every approach has both advantages and limitations (Meng et al., 2011).

Performing molecular docking using the surface complementarity approach means to obtain the molecular surfaces of both the protein and ligand (knowing their spatial structures) and to find the rigid transformations such as the tow objects have a good match, i.e. they share the maximum contact area between their molecular surfaces without overlapping regions (Norel et al., 1994). Usually, the entire molecular surfaces are utilized in the docking process, not only the location of the active center of the target protein, or some particular residues or atoms participating in the binding in either the protein or the ligand. There is a matching of every piece of surface of the ligand with every piece of surface of the target (Norel et al., 1999). Most docking algorithms use two kinds of molecular descriptors for the molecules involved, structural complementarity and binding complementarity descriptors (Oloff et al., 2006). Structural complementarity descriptors include solvent-accessible surface area, overall shape and geometric constraints between atoms in the protein and ligand and binding complementarity descriptors consider hydrogen bonds, hydrophobic contacts and van der Waals interactions. In order to limit the number of obtained binding modes, additional constraints could be used as additional filters, for example electrostatic interactions. The surface complementarity between the ligand and its target is implemented in two ways: grid-based (Gabb et al., 1997) and pairwise (Chen & Weng, 2003) shape complementarity respectively. Grid-based shape complementarity is computed by considering the grid points on at the binding interface (Gabb et al., 1997) and pairwise shape complementarity is computed by considering the total number of atom pairs between the protein and the ligand within a distance cutoff (Chen and Weng, 2003).

In the case of simulation of the docking process, the ligand and the protein are separated by a physical distance. This approach usually considers proteins as rigid bodies and the ligand is allowed to explore their conformational space. The ligand must find its position into the protein's active center using a certain number of changes in its conformational space (translations, rotations, torsions).The generated conformations of the ligand are docked consecutively into the protein, and a molecular dynamics (MD) simulation is performed. During the MD process, each of the ligand changes induces an energetic cost of the system and the system's total energy is calculated taking into account every change and the energies values determined are used for ranking the overall scoring (Halperin et al., 2002; Feig et al., 2004). Simulation of the docking process has the advantage that allows to easily consider the ligand flexibility, but it is computationally more expensive involving potentially hundreds of MD runs.

The implementation of docking algorithms implies that the three dimensional structures of both the ligand and its target are known and there are multiple formats for the files describing the three dimensional structures (Dalby et al., 1992). In the case of the target protein, the three dimensional structural files are obtained from the PDB. Structural data have a *PDB format*, a standard representation of the macromolecular structural data containing atomic coordinates. The PDB format (with the typical file extension .pdb) is a fixed-column-width format commonly used to express the proteins structures, but it may also be used for other types of molecules. This format is read and also written by many programs used to visualize and analyze the molecular structures. Some docking algorithms use as input for the target molecule its structural file in a *pdb* format, but others use the *mol2* format for this purpose. The *mol2* (with the typical file extension *.mol2*) format contains information about the atoms, bonds, connectivity and atoms coordinates, also being a standardized format that some modeling programs can read and write. Another type of file format is MDL Molfile with the typical extension *.mol.* It contains information concerning atoms coordinates, bonds and connectivity for a molecule and numerous computational applications are able to read it (Dalby et al., 1992).

In the case of small molecules (ligands), there are several formats offered by the chemical libraries. The common format used to encode small chemical compounds is the Simplified Molecular Input Line Entry Specification (SMILES) format. This format includes information concerning the connectivity but do not include 2D or 3D coordinates of the atoms and the hydrogen atoms are not represented (Weininger, 1988). Every atom is represented by its element symbol, a double bond is represented by the symbol "=", a triple bond by the symbol"#", branching is represented by "()" and rigs are indicated by pairs of digits. Another format widely used for the small chemical compounds is the Structure Data Format (SDF), a text file commonly used to represent multiple chemical structure records and associated data fields (Dalby et al., 1992).

Taking into account the diverse formats used to represent a molecule and that every computational tool used for viewing and editing molecular structures or for molecular modeling is capable to read and write only one (or a reduced number) of the file format types, format converters have been developed. OpenBabel (http://openbabel.org/wiki/Main_Page) is one of the commonly used online freely available tools specifically designed for converting between file formats (O'Boyle et al., 2011). It can read 88 formats and can write 89 formats. Another online format converter is PRODRG (http://davapc1.bioch.dundee.ac.uk/cgi-bin/prodrg). It reads many types of coordinates formats (pdb, mdl, mol2) or two-dimensional formats and generates a variety of topologies for use with molecular modeling tools, as well as energy-minimized coordinates in a variety of formats (Schüttelkopf & van Aalten, 2004).

There are multiple servers allowing online protein-ligand and protein-protein molecular docking. Table 3 reveals a short presentation of the most known online docking resources.

The use of PatchDock, SwissDock and MTiAutoDock servers is further illustrated.

SOLUTIONS AND RECOMMENDATIONS

When performing online molecular docking studies information concerning the ligand and protein structures is needed. The ligand structure and properties may be retrieved from ZINC (Irwin and Stoichet, 2005), PubChem (Bolton et al., 2008), ChEmbl, ChemDB (Chen et al., 2007), ChemSpider or other database. Depending on the format delivered by the chemical database and the format read by the docking algorithm, a format converter can be necessary. The structure of the target protein is retrieved from the Protein Data Bank (Bernstein et al., 1977). For molecular docking studies, the structures of protein-ligand complexes are considered if available and the ligand is removed when the receptor molecule is prepared for docking.

PatchDock server uses an algorithm that is based on finding the transformations that produce a local good geometric shape complementarity including the local individualities of the docked molecules with complementary features. The transformations are classified using a scoring function tacking into account both the geometric fit and atomic desolvation energy, the redundant solutions being rejected by applying a root mean square deviation clustering (RMSD). The output of PatchDock tool is a list of candidate complexes between the protein and the ligand molecule that were specified by user, the complexes being sorted according to the geometric shape complementarity score. For every obtained complex, the output contains the approximate interface area, the atomic contact energy (ACE) and the

Table 3. Online molecular docking servers

Molecular Docking Tool	URL	Short Description	Reference
DOCK Blaster	http://blaster.docking.org	DOCK Blaster is an online server used for the prediction of the ligand-protein complexes.	Irwin et al., 2009
Molecular Docking Server	http://www.dockingserver.com/web	DockingServer enables docking calculations and results evaluation. It can be used for docking of single ligands as well as for docking of ligand libraries to target proteins.	Bikadi and Hazai, 2009
ZDOCK	http://zdock.umassmed.edu	ZDOCK is a docking algorithm that searches all possible binding modes between the two proteins and evaluates each pose using an energy-based scoring function.	Pierce et al., 2014
Rosseta	http://rosie.rosettacommons.org/	Rosetta is an online tool used for docking small molecules into proteins.	Combs et al., 2013
PatchDock	http://bioinfo3d.cs.tau.ac.il/PatchDock/)	PatchDock is a webserver performing structure prediction for protein-small ligand complexes using a geometry based molecular docking algorithm .	Schneidman-Duhovny et al., 2005
SwissDock	http://www.swissdock.ch/	SwissDock accomplishes molecular docking predictions based on computing ligand-protein pairwise interactions energies.	(Grosdidier et al., 2011)
MTiAutoDock	http://mobyle.rpbs.univ-paris-diderot.fr/cgi-bin/portal.py#forms::MTiAutoDock	MtiAutoDock permits docking of small ligands into proteins cavities based on computing ligand-protein pairwise interactions energies.	Labbe et al., 2015
HexServer	http://hexserver.loria.fr/	HexServer is an interactive protein docking and molecular superposition program. It performs FFT-based rigid-body docking predictions.	Macindoe et al, 2010
HADDOCK (High Ambiguity Driven protein-protein DOCKing)	http://haddock.science.uu.nl/services/HADDOCK/	HADDOCK deals with modelling problems including protein-protein, protein-nucleic acids and protein-ligand complexes.	de Vries et al., 2010

3D transformations (3 rotational angles and 3 translational parameters) applied on the ligand molecule. To refine the PatchDock predictions, theFireDock webserver (http://bioinfo3d.cs.tau.ac.il/FireDock/) is used. It also provides the global energy of each protein-ligand complex predicted by PatchDock software (Andrusier et al., 2007). The user must upload the structural files of the protein (receptor) and ligand in the PDB format or provide their PDB codes entry. As crystal structures of proteins sometimes contains ions, water molecules and/or small ligands, preparation of the structures for docking studies keeping the functional molecules is usually recommended. The user must also provide an e-mail address to obtain the link to the docking results and the value of the RMSD used for the clustering, the recommended value being 1.5 Å for protein-small molecule docking. PatchDockserver uses different sets of parameters for distinct types of complexes (enzyme-inhibitor, antibody-antigen and protein-small ligand) and the user must specify the type of the complex. If the user does not know the type of the complex, the default complex type may be used. There also are some advanced options that are used when some information concerning the potential binding sites in the protein is known, a list of residues in an uploaded file may be specified.

The use of PatchDock server is illustrated here for the prediction of the interaction between the human cytochrome P450 subfamily 2C8 (CYP2C8) with the pesticide difenoconazole (DFC). The cytochrome P450 enzymes (CYPs) are proteins with critical roles in xenobiotic metabolism, and the CYP2C subfamily is involved in the metabolism of about 49% of widely used pesticides (Abass et al., 2011). Low levels of pesticides from residues in food and drinking water, respectively the occupational exposure may cause health problems in humans and the assessment of the pesticides interactions with human CYP2C8 is important as it is one of the main hepatic drug-metabolizing enzymes. There are 5 structural files in the PDB for the human CYP2C8 protein, their codes entry being: 1PQ2 for the complex of CYP2C8 with palmitic acid (Schoch et al, 2004), 2NNH, 2NNI, 2NNJ and 2VN0 for the complexes with 9-cis-retinoic acid, montelukast, felodipine, and troglitazone respectively (Schoch et al., 2008). In our study we have used the crystal structure of the CYP2C8 in complex with the inhibitor montelukast, code entry 2NNI. The inhibitor molecule, the palmitic acid molecule and the SO_4 ion have been removed, the heme group was kept and the resulting file has been prepared for docking using the UCSF Chimera computational tool. The output of the PatchDock server for the 20 best solutions is shown in the Table 4.

The solutions presented in Table 4 are ranked upon the shape complementarity score. The best solution predicted by PatchDock server (result.1.pdb) is illustrated in the Figure 1. Figure 1 reveals that DFC (presented as full surface) does not bind in the same place as montelukast (mesh surface), the known inhibitor of the protein.

Table 4. The 20 best solutions predicted by PatchDock for the DFC binding to the human CYP2C8

Solution No	Score	Interface Area (Å²)	ACE (kcal/mol)	Transformation	PDB File of the Complex
1	5694	664.50	-163.40	2.22 0.17 -2.97 52.79 13.42 -11.67	result.1.pdb
2	5540	678.40	-216.28	2.80 -0.22 -2.05 48.02 13.65 -19.35	result.2.pdb
3	5448	652.50	-174.57	-0.86 -0.04 0.11 46.88 12.71 -21.72	result.3.pdb
4	5430	626.50	-160.43	-1.17 0.03 1.43 49.62 10.23 -25.82	result.4.pdb
5	5372	619.20	-164.94	-0.92 0.16 0.37 48.34 11.88 -25.01	result.5.pdb
6	5244	691.50	-338.52	0.65 -0.13 -2.15 52.45 -0.47 -20.01	result.6.pdb
7	5242	663.10	-234.23	2.59 -0.15 -2.41 49.72 13.33 -19.86	result.7.pdb
8	5228	651.10	-257.81	1.37 0.01 0.55 47.03 11.80 -24.13	result.8.pdb
9	5222	599.10	-139.50	2.39 -0.21 -3.04 50.28 14.14 -12.68	result.9.pdb
10	5220	608.90	-171.76	-1.05 -0.04 -0.30 47.38 12.81 -23.67	result.10.pdb
11	5184	662.70	-245.84	2.27 0.17 -1.86 51.05 13.70 -20.15	result.11.pdb
12	5166	630.50	-199.85	1.84 -0.56 1.08 48.66 12.95 -18.09	result.12.pdb
13	5166	613.50	-228.08	-1.46 0.46 -2.90 48.01 16.30 -27.79	result.13.pdb
14	5158	627.50	-196.07	-1.16 0.28 0.09 48.16 11.22 -23.28	result.14.pdb
15	5156	648.10	-166.58	-1.36 -0.75 -1.35 47.28 16.26 -16.15	result.15.pdb
16	5144	621.70	-170.42	2.32 -0.02 2.92 51.05 12.82 -11.72	result.16.pdb
17	5134	636.10	-172.31	-1.15 -0.14 -1.10 46.94 15.56 -20.76	result.17.pdb
18	5128	617.40	-170.80	-0.96 0.61 0.43 50.64 12.64 -25.90	result.18.pdb
19	5124	649.60	-211.24	-1.40 0.77 0.18 50.34 10.92 -24.16	result.19.pdb
20	5122	643.40	-191.75	2.84 -0.54 -2.29 47.81 13.87 -16.78	result.20.pdb

Figure 1. The binding poses for the difenoconazole to the human CYP2C8 predicted using PatchDock server. The protein is shown as backbone, the difenoconazole in revealed as full surface, the inhibitor montelukast is shown as mesh surface.

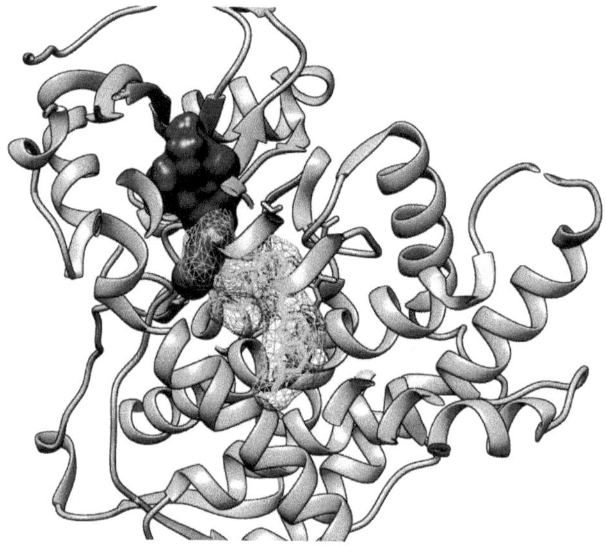

The best solutions obtained using PatchDock have been refined using FireDock on-line computational tool and the best 10 refined solutions are presented in Table 4. This table contains the new ranking after refining, the global energy, the attractive and repulsive van der Waals (VdW) energies, the atomic contact energy (ACE) and hydrogen bonds (HB). This table reflect that, after refining, the best solution is considered solution number 8 predicted by PatchDock, this solution being illustrated in the Figure 2. Table 5 reflects that solution number 8 has the highest values of the atomic contact energy.

Table 5. The 10 best refined solutions predicted by FireDock for the DFC binding to the human CYP2C8

Rank	Solution Number	Global Energy (kcal/mol)	Attractive VdW (kcal/mol)	Repulsive VdW (kcal/mol)	ACE (kcal/mol)	HB
1	8	-43.02	-17.47	7.62	-14.42	0.00
2	7	-35.57	-13.64	1.35	-10.23	0.00
3	3	-35.54	-19.93	11.40	-9.74	0.00
4	1	-35.52	-16.62	5.01	-9.43	0.00
5	2	-32.98	-14.52	4.22	-9.19	0.00
6	9	-31.36	-15.23	6.90	-8.97	0.00
7	5	-30.95	-13.65	4.05	-8.91	0.00
8	4	-30.24	-14.12	5.73	-8.96	0.00
9	10	-28.44	-14.98	9.11	-8.35	0.00
10	6	-2.49	-9.90	5.36	3.92	0.00

Figure 2. The binding poses for the difenoconazole to the human CYP2C8 predicted using PatchDock server and refined using FireDock server. The protein is shown in backbone, the difenoconazole in revealed as full surface, the inhibitor montelukast is shown in mesh surface.

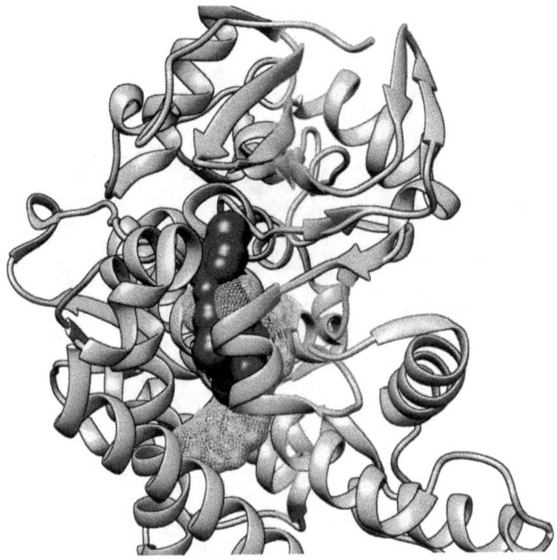

Figure 2 illustrates the importance of refining results when using molecular docking based on the shape complementarity of protein and ligand. After refining, the best solution reflects that DFC is situated almost in the same place as the known inhibitor and it suggests that DFC may also have an inhibitory effect on the human CYP2C8 protein.

SwissDock is a webserver (http://www.swissdock.ch/) performing molecular docking predictions based on computing ligand-protein pairwise interactions energies (Grosdidier et al., 2011). This server uses the EADock (Grosdidier et al., 2007) and EADockDSS (Grosdidier et al, 20111 b) algorithms to compute the interacting energies and outcomes the most favorable position and orientation of a ligand on a protein surface, i.e. the binding modes (BMs). The interaction energy is expressed as the FullFitness score computed by averaging the 30% most favorable effective energies of its elements. The effective energy considers the total energy of the system and the solvation energy. The obtained BMs are clustered and clusters are ranked by averaging the FullFitness scores of their elements. The protein structure can be specified by its code from the Protein Data Bank or by uploading the structural file. As mentioned above, it is recommended to upload a structural file of the protein prepared for docking studies. The ligand can be specified either by its identifier from the ZINC database or by uploading its structure files. There are three possibilities to select the docking type: very fast, fast and accurate. Three docking parameter presets can be selected: very fast, fast and accurate. They differ by the number of sampled BMs, the number of minimization steps and the number of predicted BMs. The user has the possibility to define or not the region of interest for docking. If there are not information concerning the binding site or the site of interest, the whole protein structure is considered during the docking process and the region of interest is not defined. If the binding pocket is known, the docking can be restricted to this region of the protein structure by defining the region of interest. The user may specify an email address

Figure 3. The binding poses for the difenoconazole to the human CYP2C8 predicted by SwissDock server. The protein is shown as backbone, the difenoconazole in revealed as full surface, the inhibitor montelukast is shown as mash surafce and the heme group in sticks.

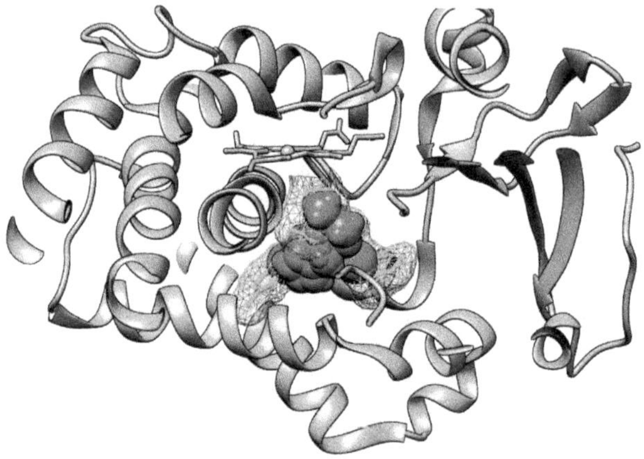

in the submission form and a link to the docking result web page is sent to this address when the docking process in finished. The predicted BMs may be visualized directly within the web browser or they may be downloaded for analyzed using Chimera software.

The SwissDock server is also used to predict the interaction between the CYP2C8 protein and the pesticide difenoconazole. In this purpose, the same structural files of the protein and ligand are used as previously explained for the PatchDock webserver and the region of interest is not defined. SwissDock webserver predicts 38 BMs, the best pose of the DFC binding to the CYP2C8 being presented in Figure 3. The fulfitness score for this pose is -2287.30 kcal/mol, the free energy of binding being -8.28kcal/mol.

The position of the predicted best pose (having the highest predicted fulfitness score and binding free energy) is compared to that of the known inhibitor of the CYP2C8, montelukast, in the crystallographic structure of the CYP2C8-montelukast complex. Figure 2 illustrates that DFC is superposed to the montelukast molecule.

MTiAutoDock (http://mobyle.rpbs.univ-paris-diderot.fr/cgi-bin/portal.py#forms::MTiAutoDock) is a recent designed webserver permitting docking of small ligands into proteins cavities and it is also based on computing ligand-protein pairwise interactions energies (Labbe et al., 2015). It uses the AutoDock4.2 algorithm to generate orientations/conformations of the compound (Morris et al, 2009). The webserver interface allows to the user to upload the ligand molecule in MOL2 format. For the protein, the user may provide the structure in PDB or MOL2 formats. When the protein is provided in PDB format, the structure is cleaned and all heteroatoms are removed and the hydrogen atoms are added. When the heteroatoms (or some of them) must be kept in the structure, the protein structure need to be prepared manually and uploaded in the MOL2 format. MTiAutoDock allows automated blind docking of ligands or ligands docking into a binding site defined by user and outputs the bound conformations and binding energies of ligands with their targets. Up to 10 ligands may be used for docking. The computation of the binding free energy is based on a semiempirical scoring function that allows incorporation of intramolecular energies and of charge based desolvation energy (Morris et al., 2009). When docking predictions are finished, the webserver allows the download and/or visualization in 3D of the 10 top scored poses for the docked ligand(s). MTiAutoDock also allows molecular docking studies with multiple ligands.

Figure 4. The predicted binding poses by MTiAutoDock server for the difenoconazole to the human CYP2C8 (A) and the pose having the highest free energy of binding (B). The protein is shown as backbone, the difenoconazole in revealed as mesh surface, the inhibitor montelukast is shown in as full surface and the heme group in sticks.

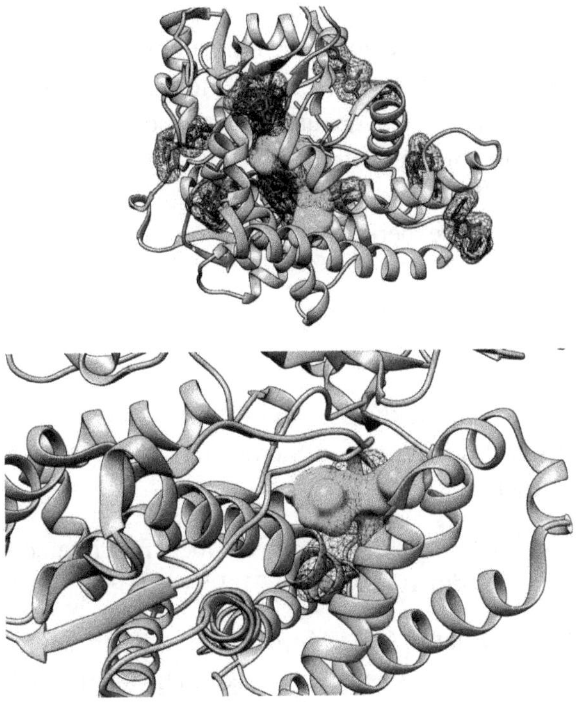

For the prediction of the interaction of CYP2C8 and DFC, MTiAutoDock server outcomes, for blind docking, 10 poses, but 6 of them are not predicted in the binding site as revealed by Figure 4A. The pose having the highest free energy of binding, -7.24 kcal/mol, is presented in Figure 4B.

There is a consensus for predicted binding region and free energy of binding values for the protein-difenoconazole complexes obtained using the three docking servers consolidating the prediction of the CYP2C8-fungicide interaction. All molecular docking servers predict that the fungicide may bind to the active site of the protein and that the binding region corresponds to that of the known inhibitor, montelukast, suggesting the inhibitory potential of the fungicide.

FUTURE RESEARCH DIRECTIONS

On-line facilities for performing bioinformatics analysis increase continuously. We assist during the last years to the considerable growth of the number of online resources and also of the accumulation of high-quality data describing structures and affinities of protein-drug ligand complexes in publicly available databases. These data are very important for development, validation, and benchmarking of drug docking algorithms.

CONCLUSION

This chapter presents online protein modeling and molecular docking facilities that can be successfully employed for distinct studies concerning drug design and biotechnology.

The online bioinformatics resources are important for research in all fields of life sciences and they are developing very fast. In the case of teaching, the online bioinformatics resources are important for both teachers and students because they develop interest to the students by facilitating understanding advanced biology studies and they promote quality human resource to undertake challenging research in the field of modern biology and biochemistry.

New online resources for in silico drug screening speed drug discovery and reduce drug development costs by allowing concentrating experimental efforts on the most favorable candidate compounds.

REFERENCES

Abass, K., Turpeinen, M., Rautio, A., Hakkola, J., & Pelkonen, O. (2011). Metabolism of Pesticides by Human Cytochrome P450 Enzymes In Vitro – A Survey. In Insecticides - Advances in Integrated Pest Management (pp 165-194). Rijeka: InTech.

Andrusier, N., Nussinov, R., & Wolfson, H. J. (2007). FireDock: Fast interaction refinement in molecular docking. *Proteins*, *69*(1), 139–159. doi:10.1002/prot.21495 PMID:17598144

Bernstein, F. C., Koetzle, T. F., Williams, G. J., Meyer, E. E. Jr, Brice, M. D., Rodgers, J. R., & Tasumi, M. et al. (1977). The Protein Data Bank: A Computer-based Archival File For Macromolecular Structures. *Journal of Molecular Biology*, *112*(3), 535–542. doi:10.1016/S0022-2836(77)80200-3 PMID:875032

Biasini, M., Bienert, T., Waterhouse, A., Arnold, K., Studer, G., Schmidt, T., & Schwede, T. et al. (2014). SWISS-MODEL: Modelling protein tertiary and quaternary structure using evolutionary information. *Nucleic Acids Research*, *42*(1), W252–W258. doi:10.1093/nar/gku340 PMID:24782522

Bikadi, Z., & Hazai, E. (2009). Application of the PM6 semi-empirical method to modeling proteins enhances docking accuracy of AutoDock. *Journal of Chemical Information*, *1*, 15. PMID:20150996

Bolton, E., Wang, Y., Thiessen, P. A., & Bryant, S. H. (2008). PubChem: Integrated Platform of Small Molecules and Biological Activities. In R. A. Wheeler & D. C. Spellmeyer (Eds.), *Annual Reports in Computational Chemistry* (pp. 217–241). Oxford, UK: Elsevier. doi:10.1016/S1574-1400(08)00012-1

Bonneau, R., & Baker, D. (2001). Ab initio protein structure prediction: Progress and prospects. *Annual Review of Biophysics and Biomolecular Structure*, *30*(1), 173–189. doi:10.1146/annurev.biophys.30.1.173 PMID:11340057

Chen, J. H., Linstead, E., Swamidass, S. J., Wang, D., & Baldi, P. (2007). ChemDB update-full-text search and virtual chemical space. *Bioinformatics (Oxford, England)*, *23*(17), 2348–2351. doi:10.1093/bioinformatics/btm341 PMID:17599932

Chen, R., & Weng, Z. (2003). A novel shape complementarity scoring function for protein-protein docking. *Proteins*, *51*(3), 397–408. doi:10.1002/prot.10334 PMID:12696051

Combet, C., Jambon, M., Deléage, G., & Geourjon, C. (2002). Geno3D: Automatic comparative molecular modelling of protein. *Bioinformatics (Oxford, England)*, *18*(1), 213–214. doi:10.1093/bioinformatics/18.1.213 PMID:11836238

Combs, S. A., Deluca, S. L., DeLuca, S. H., Lemmon, G. H., Nannemann, D. P., Nguyen, E. D., & Meiler, J. et al. (2013). Small-molecule ligand docking into comparative models with Rosetta. *Nature Protocols*, *8*(7), 1277–1298. doi:10.1038/nprot.2013.074 PMID:23744289

Dalby, A., Nourse, J. G., Hounshell, W. D., Gushurst, A. K. I., Grier, D. L., Leland, B. A., & Laufer, J. (1992). Description of several chemical structure file formats used by computer programs developed at Molecular Design Limited. *Journal of Chemical Information and Modeling*, *32*(3), 244–255. doi:10.1021/ci00007a012

de Beer, T. A. P., Berka, K., Thornton, J. M., & Laskowski, R. A. (2014). PDBsum additions. *Nucleic Acids Research*, *42*(D1), D292–D296. doi:10.1093/nar/gkt940 PMID:24153109

de Vries, S. J., van Dijk, M., & Bonvin, A. M. J. J. (2010). The HADDOCK web server for data-driven biomolecular docking. *Nature Protocols*, *5*(5), 883–897. doi:10.1038/nprot.2010.32 PMID:20431534

Dundas, J., Ouyang, Z., Tseng, J., Binkowski, A., Turpaz, Y., & Liang, J. (2006). CASTp: Computed atlas of surface topography of proteins with structural and topographical mapping of functionally annotated residues. *Nucleic Acids Research*, *34*(Web Server), W116–W118. doi:10.1093/nar/gkl282 PMID:16844972

Eswar, N., Marti-Renom, M.A., Webb, B., Madhusudhan, M.S., Eramian, D., Shen, M., Pieper, U., & Sali. A. (2006). Comparative Protein Structure Modeling With MODELLER. *Current Protocols in Bioinformatics*, *15*, 5.6.1-5.6.30.

Feig, M., Onufriev, A., Lee, M. S., Im, W., Case, D. A., & Brooks, C. L. (2004). Performance comparison of generalized born and Poisson methods in the calculation of electrostatic solvation energies for protein structures. *Journal of Computational Chemistry*, *25*(2), 265–284. doi:10.1002/jcc.10378 PMID:14648625

Gabb, H. A., Jackson, R. M., & Sternberg, M. J. (1997). Modelling protein docking using shape complementarity, electrostatics and biochemical information. *Journal of Molecular Biology*, *272*(1), 106–120. doi:10.1006/jmbi.1997.1203 PMID:9299341

Gohlke, H., & Klebe, G. (2002). Approaches to the description and prediction of the binding affinity of small-molecule ligands to macromolecular receptors. *Angewandte Chemie International Edition*, *41*(15), 2644–2676. doi:10.1002/1521-3773(20020802)41:15<2644::AID-ANIE2644>3.0.CO;2-O PMID:12203463

Grosdidier, A., Zoete, V., & Michielin, O. (2007). EADock: A multiobjective evolutionary framework for the docking of small molecules into protein active site. *Proteins: Structure, Function, and Bioinformatics*, *67*(4), 1010–1025. doi:10.1002/prot.21367

Grosdidier, A., Zoete, V., & Michielin, O. (2011). SwissDock - a protein-small molecule docking web service based on EADock DSS. *Nucleic Acids Research*, *39*(suppl), W270–W277. doi:10.1093/nar/gkr366 PMID:21624888

Grosdidier, A., Zoete, V., & Michielin, O. (2011). SwissParam: A fast force field generation tool for small organic molecule. *Journal of Computational Chemistry*, *32*(11), 2359–2368. doi:10.1002/jcc.21816 PMID:21541964

Guex, N., & Peitsch, M. C. (1997). SWISS-MODEL and the Swiss-PdbViewer: An environment for comparative protein modeling. *Electrophoresis*, *18*(15), 2714–2723. doi:10.1002/elps.1150181505 PMID:9504803

Halperin, I., Ma, B., Wolfson, H., & Nussinov, R. (2002). Principles of docking: An overview of search algorithms and a guide to scoring functions. *Proteins*, *47*(4), 409–443. doi:10.1002/prot.10115 PMID:12001221

Humphrey, W., Dalke, A., & Schulten, K. (1996). VMD - Visual Molecular Dynamics. *Journal of Molecular Graphics*, *14*(1), 33–38. doi:10.1016/0263-7855(96)00018-5 PMID:8744570

Irwin, J. J., & Shoichet, B. K. (2005). Zinc – a free database of commercially available compounds for virtual screening. *Journal of Chemical Information and Modeling*, *45*(1), 177–182. doi:10.1021/ci049714+ PMID:15667143

Irwin, J. J., Shoichet, B. K., Mysinger, M. M., Huang, N., Colizzi, F., Wassam, P., & Cao, Y. (2009). Automated Docking Screens: A Feasibility Study. *Journal of Medicinal Chemistry*, *52*(18), 5712–5720. doi:10.1021/jm9006966 PMID:19719084

Irwin, J. J., Sterling, T., Mysinger, M. M., Bolstad, E. S., & Coleman, R. G. (2012). ZINC: A Free Tool to Discover Chemistry for Biology. *Journal of Chemical Information and Modeling*, *52*(7), 1757–1768. doi:10.1021/ci3001277 PMID:22587354

Jain, E., Bairoch, A., Duvaud, S., Phan, I., Redaschi, N., Suzek, B. E., & Gasteiger, E. et al. (2009). Infrastructure for the life sciences: Design and implementation of the UniProt website. *BMC Bioinformatics*, *10*(1), 136. doi:10.1186/1471-2105-10-136 PMID:19426475

Kelley, L. A., Mezulis, S., Yates, C. M., Wass, M. N., & Sternberg, M. J. E. (2015). The Phyre2 web portal for protein modeling, prediction and analysis. *Nature Protocols*, *10*(6), 845–858. doi:10.1038/nprot.2015.053 PMID:25950237

Kroemer, R. T. (2007). Structure-based drug design: Docking and scoring. *Current Protein & Peptide Science*, *8*(4), 312–328. doi:10.2174/138920307781369382 PMID:17696866

La, D., Esquivel-Rodriguez, J., Venkatraman, V., Li, B., Sael, L., Ueng, S., & Kihara, D. et al. (2009). 3D-SURFER: Software for high throughput protein surface comparison and analysis. *Bioinformatics (Oxford, England)*, *25*(21), 2843–2844. doi:10.1093/bioinformatics/btp542 PMID:19759195

Labbé, C. M., Rey, J., Lagorce, D., Vavruša, M., Becot, J., Sperandio, O., & Miteva, M. A. et al. (2015). MTiOpenScreen: A web server for structure-based virtual screening. *Nucleic Acids Research*, *43*(W1), W448–W454. doi:10.1093/nar/gkv306 PMID:25855812

Lambert, C., Leonard, N., De Bolle, X., & Depiereux, E. (2002). ESyPred3D: Prediction of proteins 3D structures. *Bioinformatics (Oxford, England)*, *18*(9), 1250–1256. doi:10.1093/bioinformatics/18.9.1250 PMID:12217917

Macindoe, G., Mavridis, L., Venkatraman, V., Devignes, M.-D., & Ritchie, D. W. (2010). HexServer: An FFT-based protein docking server powered by graphics processors. *Nucleic Acids Research*, *38*(Web Serversuppl 2), W445–W449. doi:10.1093/nar/gkq311 PMID:20444869

Meng, X-Y., Zhang, H-X., Mezei, M, & Cui, M. (2011). Molecular Docking: A powerful approach for structure-based drug discovery. *Current Computational Aided Drug Design*, *7*(2), 146–157.

Morris, G. M., Huey, R., Lindstrom, W., Sanner, M. F., Belew, R. K., Goodsell, D. S., & Olson, A. J. (2009). AutoDock4 and AutoDockTools4: Automated docking with selective receptor flexibility. *Journal of Computational Chemistry*, *30*(16), 2785–2791. doi:10.1002/jcc.21256 PMID:19399780

Morris, G. M., & Lim-Wilby, M. (2008). Molecular docking. *Methods in Molecular Biology (Clifton, N.J.)*, *443*, 365–382. doi:10.1007/978-1-59745-177-2_19 PMID:18446297

Nantasenamat, C., Isarankura-Na-Ayudhya, C., Naenna, T., & Prachayasittikul, V. (2009). A practical overview of quantitative structure-activity relationship. *EXCLI Journal*, *8*, 74–88.

Norel, R., Fischer, D., Wolfson, H. J., & Nussinov, R. (1994). Molecular surface recognition by a computer vision-based technique. *Protein Engineering*, *7*(1), 39–46. doi:10.1093/protein/7.1.39 PMID:8140093

Norel, R., Petrey, D., Wolfson, H. J., & Nussinov, R. (1999). Examination of shape complementarity in docking of unbound proteins. *Proteins*, *36*(3), 307–317. doi:10.1002/(SICI)1097-0134(19990815)36:3<307::AID-PROT5>3.0.CO;2-R PMID:10409824

O'Boyle, N. M., Banck, M., James, C. A., Morley, C., Vandermeersch, T., & Hutchison, G. R. (2011). Open Babel: An open chemical toolbox. *Journal of Cheminformatics*, *3*(1), 33. doi:10.1186/1758-2946-3-33 PMID:21982300

Oloff, S., Zhang, S., Sukumar, N., Breneman, C., & Tropsha, A. (2006). Chemometric analysis of ligand receptor complementarity: Identifying complementary ligands based on receptor information (COLIBRI). *Journal of Chemical Information and Modeling*, *46*(2), 844–851. doi:10.1021/ci050065r PMID:16563016

Pettersen, E. F., Goddard, T. D., Huang, C. C., Couch, G. S., Greenblatt, D. M., Meng, E. C., & Ferrin, T. E. (2004). UCSF Chimera--a visualization system for exploratory research and analysis. *Journal of Computational Chemistry*, *25*(13), 1605–1612. doi:10.1002/jcc.20084 PMID:15264254

Pierce, B. G., Wiehe, K., Hwang, H., Kim, B. H., Vreven, T., & Weng, Z. (2014). ZDOCK server: Interactive docking prediction of protein-protein complexes and symmetric multimers. *Bioinformatics (Oxford, England)*, *30*(12), 1771–1773. doi:10.1093/bioinformatics/btu097 PMID:24532726

Rush, T. S., Grant, J. A., Mosyak, L., & Nicholls, A. (2005). A Shape-Based 3-D Scaffold Hopping Method and Its Application to a Bacterial Protein–Protein Interaction. *Journal of Medicinal Chemistry*, *48*(5), 1489–1495. doi:10.1021/jm040163o PMID:15743191

Schmidtke, P., Le Guilloux, V., Maupetit, J., & Tufféry, P. (2010). Fpocket: Online tools for protein ensemble pocket detection and tracking. *Nucleic Acids Research*, *38*(Web Server), W582–W589. doi:10.1093/nar/gkq383 PMID:20478829

Schneidman-Duhovny, D., Inbar, Y., Nussinov, R., & Wolfson, H. J. (2005). PatchDock and SymmDock: Servers for rigid and symmetric docking. *Nucleic Acids Research*, *33*(Web Server), W363-367. doi:10.1093/nar/gki481 PMID:15980490

Schoch, G. A., Yano, J. K., Sansen, S., Dansette, P. M., Stout, C. D., & Johnson, E. F. (2008). Determinants of cytochrome p4502c8 substrate binding - structures of complexes with montelukast, troglitazone, felodipine, and 9-cis-retinoic acid. *The Journal of Biological Chemistry*, *283*(25), 17227–17237. doi:10.1074/jbc.M802180200 PMID:18413310

Schoch, G. A., Yano, J. K., Wester, M. R., Griffin, K. J., Stout, C. D., & Johnson, E. F. (2004). The Structure of Human Microsomal Cytochrome P450 3A4 Determined by X-ray Crystallography to 2.05-Å Resolution. *The Journal of Biological Chemistry*, *279*(37), 38091–38094. doi:10.1074/jbc.C400293200 PMID:15258162

Schüttelkopf, A. W., & van Aalten, D. M. F. (2004). PRODRG: A tool for high-throughput crystallography of protein-ligand complexes. *Acta Crystallographica. Section D, Biological Crystallography*, *60*(8), 1355–1363. doi:10.1107/S0907444904011679 PMID:15272157

Shoichet, B. K., Kuntz, I. D., & Bodian, D. L. (2004). Molecular docking using shape descriptors. *Journal of Computational Chemistry*, *13*(3), 380–397. doi:10.1002/jcc.540130311

Söding, J. (2005). Protein homology detection by HMM-HMM comparison. *Bioinformatics (Oxford, England)*, *21*(7), 951–960. doi:10.1093/bioinformatics/bti125 PMID:15531603

Sun, H. (2008). Pharmacophore-based virtual screening. *Current Medicinal Chemistry*, *15*(10), 1018–1024. doi:10.2174/092986708784049630 PMID:18393859

Tan, K. P., Nguyen, B. T., Patel, S., Varadarajan, R., & Madhusudhan, M. S. (2013). Depth: A web server to compute depth, cavity sizes, detect potential small-molecule ligand-binding cavities and predict the pKa of ionizable residues in proteins. *Nucleic Acids Research*, *41*(W1), W314–W321. doi:10.1093/nar/gkt503 PMID:23766289

The UniProt Consortium. (2008). The Universal Protein Resource (UniProt). *Nucleic Acids Research*, *36*, D190–D195. PMID:18045787

Wallner, B., & Elofsson, A. (2005). All are not equal: A benchmark of different homology modeling programs. *Protein Science*, *14*(5), 1315–1327. doi:10.1110/ps.041253405 PMID:15840834

Wang, Y., Suzek, T., Zhang, J., Wang, J., He, S., Cheng, T., & Bryant, S. H. et al. (2014). PubChem BioAssay: Update. *Nucleic Acids Research*, *42*(1), D1075–D1082. doi:10.1093/nar/gkt978 PMID:24198245

Weininger, D. (1988). SMILES, a chemical language and information system. 1. Introduction to methodology and encoding rules. *Journal of Chemical Information and Modeling*, *28*(1), 31–63. doi:10.1021/ci00057a005

Willet, P., & Barnard, J. M. (1998). Chemical similarity searching. *Journal of Chemical Information and Computer Sciences*, *38*(6), 983–996. doi:10.1021/ci9800211

Wishart, D. S., Knox, C., Guo, A. C., Shrivastava, S., Hassanali, M., Stothard, P., & Woolsey, J. et al. (2006). DrugBank: A comprehensive resource for in silico drug discovery and exploration. *Nucleic Acids Research*, *34*(90001), D668–D672. doi:10.1093/nar/gkj067 PMID:16381955

Xu, D., Xu, Y., & Uberbacher, E. C. (2000). Computational tools for protein modeling. *Current Protein & Peptide Science*, *1*(1), 1–21. doi:10.2174/1389203003381469 PMID:12369918

Yu, J., Zhou, Y., Tanaka, I., & Yao, M. (2010). Roll: A new algorithm for the detection of protein pockets and cavities with a rolling probe sphere. *Bioinformatics (Oxford, England)*, *26*(1), 46–52. doi:10.1093/bioinformatics/btp599 PMID:19846440

Compilation of References

Abagyan, R. A., & Totrov, M. M. (1997). Contact area difference (CAD): A robust measure to evaluate accuracy of protein models. *Journal of Molecular Biology*, *268*(3), 678–685. PMID:9171291

Abagyan, R., & Totrov, M. (1994). Biased Probability Monte Carlo Conformational Searches and Electrostatic Calculations for Peptides and Proteins. *Journal of Molecular Biology*, *235*(3), 983–1002. doi:10.1006/jmbi.1994.1052 PMID:8289329

Abagyan, R., Totrov, M., & Kuznetsov, D. (2004). ICM—A new method for protein modeling and design: Applications to docking and structure prediction from the distorted native conformation. *Journal of Computational Chemistry*, *15*(5), 488–506. doi:10.1002/jcc.540150503

Abass, K., Turpeinen, M., Rautio, A., Hakkola, J., & Pelkonen, O. (2011). Metabolism of Pesticides by Human Cytochrome P450 Enzymes In Vitro – A Survey. In Insecticides - Advances in Integrated Pest Management (pp 165-194). Rijeka: InTech.

Abid, O. U. R., Babar, T. M., Ali, F. I., Ahmed, S., Wadood, A., Rama, N. M., & Choudhary, M. I. et al. (2010). Identification of novel urease inhibitors by high-throughput virtual and *in-vitro* screening. *ACS Medicinal Chemistry Letters*, *1*(4), 145–149. doi:10.1021/ml100068u PMID:24900188

Adock, S. A., & McCammon, J. A. (2006). Molecular dynamics: Survey of methods for simulating the activity of proteins. *Chemical Reviews*, *106*(5), 1589–21615. doi:10.1021/cr040426m

Agarwal, S., Mishra, N. K., Bhatnagar, S., & Bhatnagar, R. (2010). PemK toxin of Bacillus anthracis is a ribonuclease: An insight into its active site, structure, and function. *The Journal of Biological Chemistry*, *285*(10), 7254–7270. doi:10.1074/jbc.M109.073387 PMID:20022964

Ahmed, M., & Barakat, K. (2015). Baby steps toward modelling the full human programmed Death-1 (PD-1) pathway. *Receptors & Clinical Investigation*, *1*, 1–5.

Alexandrov, N., & Shindyalov, I. (2003). PDP: Protein domain parser. *Bioinformatics (Oxford, England)*, *19*(3), 429–430. doi:10.1093/bioinformatics/btg006 PMID:12584135

Allenby, G., Bocquel, M.-T., Saunders, M., Kazmer, S., Speck, J., Rosenberger, M., & Chambon, P. et al. (1993). Retinoic acid receptors and retinoid X receptors: Interactions with endogenous retinoic acids. *Proceedings of the National Academy of Sciences of the United States of America*, *90*(1), 30–34. doi:10.1073/pnas.90.1.30 PMID:8380496

Allen, W. J., Balius, T. E., Mukherjee, S., Brozell, S. R., Moustakas, D. T., Lang, P. T., & Rizzo, R. C. et al. (2015). DOCK 6: Impact of new features and current docking performance. *Journal of Computational Chemistry*, *36*(15), 1132–1156. doi:10.1002/jcc.23905 PMID:25914306

Amadei, A., Linssen, A. B., & Berendsen, H. J. (1993). Essential dynamics of proteins. *Proteins*, *17*(4), 412–425. doi:10.1002/prot.340170408 PMID:8108382

Amaro, R., Baron, R., & McCammon, J. A. (2008). An improved relaxed complex scheme for receptor flexibility in computer-aided drug design. *Journal of Computer-Aided Molecular Design*, *22*(9), 693–705. doi:10.1007/s10822-007-9159-2 PMID:18196463

Amzel, L. M. (1997). Loss of translational entropy in binding, folding, and catalysis. *Proteins:* Structure Function and Genetics, *28*(2), 144–149. PMID:9188731

Anand, P., Nagarajan, D., Mukherjee, S., & Chandra, N. (2014). PLIC: Protein-ligand interaction clusters. *Database (Oxford)*, *2014*(0), bau029. doi:10.1093/database/bau029 PMID:24763918

Anderluh, M., Jug, G., Švajger, U., & Obermajer, N. (2012). DC-SIGN antagonists, a potential new class of anti-infectives. *Current Medicinal Chemistry*, *19*(7), 992–1007. doi:10.2174/092986712799320664

Anderson, A. C. (2012). Structure-Based Functional Design of Drugs: From Target to Lead Compound. In V. Espina & L. A. Liotta (Eds.), *Molecular Profiling* (Vol. 823, pp. 359–366). Totowa, NJ: Humana Press. doi:10.1007/978-1-60327-216-2_23

Andricioaei, I., & Karplus, M. (2001). On the calculation of entropy from covariance matrices of the atomic fluctuations. *The Journal of Chemical Physics*, *115*(14), 6289–6292.

Andrusier, N., Nussinov, R., & Wolfson, H. J. (2007). FireDock: Fast interaction refinement in molecular docking. *Proteins*, *69*(1), 139–159. doi:10.1002/prot.21495 PMID:17598144

Antes, I. (2010). DynaDock: A new molecular dynamics-based algorithm for protein–peptide docking including receptor flexibility. *Proteins: Structure, Function, and Bioinformatics*, *78*(5), 1084–1104. doi:10.1002/prot.22629 PMID:20017216

Apostolakis, J., Plückthun, A., & Caflisch, A. (1998). Docking small ligands in flexible binding sites. *Journal of Computational Chemistry*, *19*(1), 21–37. doi:10.1002/(SICI)1096-987X(19980115)19:1<21::AID-JCC2>3.0.CO;2-0

Aqvist, J., Luzhkov, V. B., & Brandsdal, B. O. (2002). Ligand binding affinities from MD simulations. *Accounts of Chemical Research*, *35*(6), 358–365. doi:10.1021/ar010014p PMID:12069620

Aqvist, J., Medina, C., & Samuelsson, J.-E. (1994). A new method for predicting binding affinity in computer-aided drug design. *Protein Engineering*, *7*(3), 385–391. PMID:8177887

Arnold, K., Bordoli, L., Kopp, J., & Schwede, T. (2006). The SWISS-MODEL workspace: a web-based environment for protein structure homology modelling. *Bioinformatics, 22*(2), 195-201. doi: 10.1093/bioinformatics/bti770

Arnold, G. E., & Ornstein, R. L. (1997). Molecular dynamics study of time-correlated protein domain motions and molecular flexibility: Cytochrome P450BM-3. *Biophysical Journal*, *73*(3), 1147–1159. doi:10.1016/S0006-3495(97)78147-5 PMID:9284282

Ashtawy, H. M., & Mahapatra, N. R. (2015). Machine-learning scoring functions for identifying native poses of ligands docked to known and novel proteins. *BMC Bioinformatics*, *16*(Suppl 6), S3. doi:10.1186/1471-2105-16-S6-S3 PMID:25916860

Atanasova, M., Patronov, A., Dimitrov, I., Flower, D. R., & Doytchinova, I. (2013). EpiDOCK: A molecular docking-based tool for MHC class II binding prediction. *Protein Engineering, Design & Selection*, *26*(10), 631–634. doi:10.1093/protein/gzt018 PMID:23661105

Avram, S., Pacureanu, L. M., Seclaman, E., Bora, A., & Kurunczi, L. (2011). PLS-DA - Docking Optimized Combined Energetic Terms (PLSDA-DOCET) protocol: A brief evaluation. *Journal of Chemical Information and Modeling*, *51*(12), 3169–3179. doi:10.1021/ci2002268 PMID:22066983

Baber, J. C., Shirley, W. A., Gao, Y., & Feher, M. (2005). The use of consensus scoring in ligand-based virtual screening. *Journal of Chemical Information and Modeling*, *46*(1), 277–288. doi:10.1021/ci050296y

Baber, J. C., Thompson, D. C., Cross, J. B., & Humblet, C. (2009). GARD: A generally applicable replacement for rmsd. *Journal of Chemical Information and Modeling*, *49*(8), 1889–1900. PMID:19618919

Bachwani, M., & Kumar, R. (2011). Molecular docking: A Review. *International Journal of Research in Ayurveda and Pharmacy*, *2*(6), 1746–1751.

Bailey, D., & Brown, D. (2001). High-throughput chemistry and structure-based design: Survival of the smartest. *Drug Discovery Today*, *6*(2), 57–59. doi:10.1016/S1359-6446(00)01596-8 PMID:11166243

Bajaj, C., Chowdhury, R., & Siddavanahalli, V. (2011). F2Dock: Fast Fourier protein-protein docking. *IEEE/ACM Transactions on Computational Biology and Bioinformatics*, *8*(1), 45–58. doi:10.1109/TCBB.2009.57 PMID:21071796

Bajorath, J. (2002). Integration of virtual and high-throughput screening. *Nature Reviews. Drug Discovery*, *1*(11), 882–894. doi:10.1038/nrd941 PMID:12415248

Bajorath, J. (2008). Computational analysis of ligand relationships within target families. *Current Opinion in Chemical Biology*, *12*(3), 352–358. doi:10.1016/j.cbpa.2008.01.044 PMID:18312862

Balius, T. E., Mukherjee, S., & Rizzo, R. C. (2011). Implementation and evaluation of a docking-rescoring method using molecular footprint comparisons. *Journal of Computational Chemistry*, *32*(10), 2273–2289. doi:10.1002/jcc.21814 PMID:21541962

Ballester, P. J., & Mitchell, J. B. (2010). A machine learning approach to predicting protein-ligand binding affinity with applications to molecular docking. *Bioinformatics (Oxford, England)*, *26*(9), 1169–1175. doi:10.1093/bioinformatics/btq112 PMID:20236947

Ballester, P. J., & Mitchell, J. B. (2010). A machine learning approach to predicting protein–ligand binding affinity with applications to molecular docking. *Bioinformatics (Oxford, England)*, *26*(9), 1169–1175. PMID:20236947

Ballester, P. J., & Mitchell, J. B. (2011). Comments on leave-cluster-out cross-validation is appropriate for scoring functions derived from diverse protein data sets: Significance for the validation of scoring functions. *Journal of Chemical Information and Modeling*, *51*(8), 1739–1741. PMID:21591735

Ballester, P. J., Schreyer, A., & Blundell, T. L. (2014). Does a more precise chemical description of protein–ligand complexes lead to more accurate prediction of binding affinity? *Journal of Chemical Information and Modeling*, *54*(3), 944–955. PMID:24528282

Banitt, I., & Wolfson, H. J. (2011). ParaDock: A flexible non-specific DNA--rigid protein docking algorithm. *Nucleic Acids Research*, *39*(20), e135. doi:10.1093/nar/gkr620 PMID:21835777

Barage, S. H., Jalkute, C. B., Dhanavade, M. J., & Sonawane, K. D. (2014). Simulated interactions between endothelin converting enzyme and Aβ peptide: Insights into subsite recognition and cleavage mechanism. *International Journal of Peptide Research and Therapeutics*, *20*(4), 409–420. doi:10.1007/s10989-014-9403-2

Barakat, K., & Tuszynski, J. (2011). Virtual Screening for DNA Repair Inhibitors. *DN A Repair-On the Pathways to Fixing DNA Damage and Errors, 1*, 287–312.

Barakat, K. H., Houghton, M., Tyrrell, D. L., & Tuszynski, J. (2014). Rational Drug Design : One Target, Many Paths to It. *International Journal of Computational Models and Algorithms in Medicine*, *4*(1), 59–85. doi:10.4018/ijcmam.2014010104

Barakat, K. H., Mane, J. Y., & Tuszynski, J. A. (2011). Virtual Screening : An Overview on Methods and Applications. In L. Liu, D. Wei, Y. Li, & H. Lei (Eds.), *Handbook of Research on Computational and Systems Biology: Interdisciplinary Applications* (pp. 28–60). Hershey, PA: Medical Information Science Reference. doi:10.4018/978-1-60960-491-2.ch002

Barakat, K., Gajewski, M., & Tuszynski, J. (2012). DNA Repair Inhibitors: Our Last Disposal to Improve Cancer Therapy. *Current Topics in Medicinal Chemistry*. doi:10.2174/156802612801319070 PMID:22690684

Barakat, K., Issack, B. B., Stepanova, M., & Tuszynski, J. (2011). Effects of temperature on the p53-DNA binding interactions and their dynamical behavior: Comparing the wild type to the R248Q mutant. *PLoS ONE*, *6*(11), e27651. doi:10.1371/journal.pone.0027651 PMID:22110706

Barakat, K., Mane, J., Friesen, D., & Tuszynski, J. (2010). Ensemble-based virtual screening reveals dual-inhibitors for the p53-MDM2/MDMX interactions. *Journal of Molecular Graphics & Modelling*, *28*(6), 555–568. doi:10.1016/j.jmgm.2009.12.003 PMID:20056466

Bar-Haim, S., Aharon, A., Ben-Moshe, T., Marantz, Y., & Senderowitz, H. (2009). Selex-CS: A new consensus scoring algorithm for hit discovery and lead optimization. *Journal of Chemical Information and Modeling*, *49*(3), 623–633. PMID:19231809

Baroni, M., Cruciani, G., Sciabola, S., Perruccio, F., & Mason, J. S. (2007). A common reference framework for analyzing/comparing proteins and ligands. Fingerprints for Ligands and Proteins (FLAP): Theory and application. *Journal of Chemical Information and Modeling*, *47*(2), 279–294. doi:10.1021/ci600253e PMID:17381166

Baron, R., Hunenberger, P. H., & McCammon, J. A. (2009). Absolute single-molecule entropies from quasi-harmonic analysis of microsecond molecular dynamics: Correction terms and convergence properties. *Journal of Chemical Theory and Computation*, *5*(12), 3150–3160. PMID:20011626

Baron, R., & McCammon, J. A. (2008). (Thermo) dynamic role of receptor flexibility, entropy, and motional correlation in protein–ligand binding. *ChemPhysChem*, *9*(7), 983–988. PMID:18418822

Baron, R., van Gunsteren, W. F., & Hünenberger, P. H. (2006). Estimating the configurational entropy from molecular dynamics simulations: Anharmonicity and correlation corrections to the quasi-harmonic approximation. *Trends in Physical Chemistry*, *11*, 87–122.

Barril, X., & Fradera, X. (2006). Incorporating protein flexibility into docking and structure-based drug design. *Expert Opinion Drug Discovery*, *1*(4), 1–15. doi:10.1517/17460441.1.4.335 PMID:23495904

Bates, P. A., Kelley, L. A., MacCallum, R. M., & Sternberg, M. J. (2001). Enhancement of protein modeling by human intervention in applying the automatic programs 3D-JIGSAW and 3D-PSSM. *Proteins*, *45*(S5), 39–46. doi:10.1002/prot.1168 PMID:11835480

Baxter, C. A., Murray, C. W., Clark, D. E., Westhead, D. R., & Eldridge, M. D. (1998). Flexible docking using Tabu search and an empirical estimate of binding affinity. *Proteins*, *33*(3), 367–382. doi:10.1002/(SICI)1097-0134(19981115)33:3<367::AID-PROT6>3.0.CO;2-W PMID:9829696

Bayden, A. S., Fornabaio, M., Scarsdale, J. N., & Kellogg, G. E. (2009). Web application for studying the free energy of binding and protonation states of protein-ligand complexes based on HINT. *Journal of Computer-Aided Molecular Design*, *23*(9), 621–632. doi:10.1007/s10822-009-9270-7 PMID:19554265

Bello, M., Martinez-Archundia, M., & Correa-Basurto, J. (2013). Automated docking for novel drug discovery. *Expert Opinion on Drug Discovery*, *8*(7), 821–834. doi:10.1517/17460441.2013.794780

Berendsen, H. J. C., van der Spoel, D., & van Drunen, R. (1995). GROMACS: A message-passing parallel molecular dynamics implementation. *Computer Physics Communications*, *91*(1-3), 43–56. doi:10.1016/0010-4655(95)00042-E

Berman, H. M., Kleywegt, G. J., Nakamura, H., & Markley, J. L. (2014). The protein data bank archive as an open data resource. *Journal of Computer-Aided Molecular Design*, *28*(10), 1009–1014. doi:10.1007/s10822-014-9770-y

Berman, H. M., Westbrook, J., Feng, Z., Gilliland, G., Bhat, T. N., Weissig, H., & Bourne, P. E. et al. (2000). The protein data bank. *Nucleic Acids Research*, *28*(1), 235–242. doi:10.1093/nar/28.1.235

Bernauer, J., Aze, J., Janin, J., & Poupon, A. (2007). A new protein-protein docking scoring function based on interface residue properties. *Bioinformatics (Oxford, England)*, *23*(5), 555–562. doi:10.1093/bioinformatics/btl654 PMID:17237048

Berne, B. J., Weeks, J. D., & Zhou, R. (2009). Dewetting and hydrophobic interaction in physical and biological systems. *Annual Review of Physical Chemistry*, *60*, 85. PMID:18928403

Bernstein, F. C., Koetzle, T. F., Williams, G. J., Meyer, E. E. Jr, Brice, M. D., Rodgers, J. R., & Tasumi, M. et al. (1977). The Protein Data Bank: A Computer-based Archival File For Macromolecular Structures. *Journal of Molecular Biology*, *112*(3), 535–542. doi:10.1016/S0022-2836(77)80200-3 PMID:875032

Betts, M. J., & Sternberg, M. J. (1999). An analysis of conformational changes on protein-protein association: Implications for predictive docking. *Protein Engineering*, *12*(4), 271–283. doi:10.1093/protein/12.4.271 PMID:10325397

Betzi, S., Suhre, K., Chetrit, B., Guerlesquin, F., & Morelli, X. (2006). GFscore: A general nonlinear consensus scoring function for high-throughput docking. *Journal of Chemical Information and Modeling*, *46*(4), 1704–1712. doi:10.1021/ci0600758 PMID:16859302

Betzi, S., Suhre, K., Chétrit, B., Guerlesquin, F., & Morelli, X. (2006). GFScore: A general nonlinear consensus scoring function for high-throughput docking. *Journal of Chemical Information and Modeling*, *46*(4), 1704–1712. PMID:16859302

Bhattacharya, S., Das, A., Dasgupta, R., & Bagchi, A. (2014). Analyses of the presence of mutations in Dystrophin protein to predict their relative influences in the onset of Duchenne Muscular Dystrophy. *Cellular Signalling*, *26*(12), 2857–2864. doi:10.1016/j.cellsig.2014.09.006 PMID:25220406

Bhattacharya, S., Das, A., Ghosh, S., Dasgupta, R., & Bagchi, A. (2014). Hypoglycosylation of dystroglycan due to T192M mutation: A molecular insight behind the fact. *Gene*, *537*(1), 108–114. doi:10.1016/j.gene.2013.11.071 PMID:24361964

Biasini, M., Bienert, T., Waterhouse, A., Arnold, K., Studer, G., Schmidt, T., & Schwede, T. et al. (2014). SWISS-MODEL: Modelling protein tertiary and quaternary structure using evolutionary information. *Nucleic Acids Research*, *42*(1), W252–W258. doi:10.1093/nar/gku340 PMID:24782522

Bikadi, Z., & Hazai, E. (2009). Application of the PM6 semi-empirical method to modeling proteins enhances docking accuracy of AutoDock. *Journal of Chemical Information*, *1*, 15. PMID:20150996

Bikadi, Z., & Hazai, E. (2009). Application of the PM6 semi-empirical method to modeling proteins enhances docking accuracy of AutoDock. *Journal of Cheminformatics*, *1*(1), 15. doi:10.1186/1758-2946-1-15 PMID:20150996

Bissantz, C., Folkers, G., & Rognan, D. (2000). Protein-based virtual screening of chemical databases. 1. Evaluation of different docking/scoring combinations. *Journal of Medicinal Chemistry*, *43*(25), 4759–4767. doi:10.1021/jm0010441 PMID:11123984

Bissantz, C., Folkers, G., & Rognan, D. (2000). Protein-based virtual screening of chemical databases. 1. evaluation of different docking/scoring combinations. *Journal of Medicinal Chemistry*, *43*(25), 4759–4767. PMID:11123984

Bland, N. D., Wang, C., Tallman, C., Gustafson, A. E., Wang, Z., Ashton, T. D., & Pollastri, M. P. et al. (2011). Pharmacological validation of Trypanosoma brucei phosphodiesterases B1 and B2 as druggable targets for African sleeping sickness. *Journal of Medicinal Chemistry*, *54*(23), 8188–8194. doi:10.1021/jm201148s PMID:22023548

Blaney, J. M., & Dixon, J. S. (1993). A Good Ligand is Hard to Find: Automatic Docking Methods. *Perspectives in Drug Discovery and Design*, *15*(2), 301–319. doi:10.1007/BF02174531

Bleicher, K. H., Bohm, H. J., Muller, K., & Alanine, A. I. (2003). Hit and lead generation: Beyond high-throughput screening. *Nature Reviews. Drug Discovery*, *2*(5), 369–378. doi:10.1038/nrd1086 PMID:12750740

Block, P., Sotriffer, C. A., Dramburg, I., & Klebe, G. (2006). AffinDB: A freely accessible database of affinities for protein–ligand complexes from the pdb. *Nucleic Acids Research*, *34*(suppl 1), D522–D526. PMID:16381925

Blundell, T. L., Jhoti, H., & Abell, C. (2002). High-throughput crystallography for lead discovery in drug design. *Nat Rev Drug Discov*, *1*(1), 45-54.

Boehr, D. D., Nussinov, R., & Wright, P. E. (2009). The role of dynamic conformational ensembles in biomolecular recognition. *Nature Chemical Biology*, *5*(11), 789–796. doi:10.1038/nchembio.232 PMID:19841628

Bohm. (1998). Prediction of binding constants of protein ligands: a fast method for the prioritization of hits obtained from de novo design or 3D database search programs. *Journal of Computer-Aided Molecular Design*, *12*, 309-309.

Bohm, H. J. (1992a). The computer program LUDI: A new method for the de novo design of enzyme inhibitors. *Journal of Computer-Aided Molecular Design*, *6*(1), 61–78. doi:10.1007/BF00124387 PMID:1583540

Bohm, H. J. (1992b). LUDI: Rule-based automatic design of new substituents for enzyme inhibitor leads. *Journal of Computer-Aided Molecular Design*, *6*(6), 593–606. doi:10.1007/BF00126217 PMID:1291628

Bohm, H. J. (1994). The development of a simple empirical scoring function to estimate the binding constant for a protein-ligand complex of known three-dimensional structure. *Journal of Computer-Aided Molecular Design*, *8*(3), 243–256. doi:10.1007/BF00126743 PMID:7964925

Bolon, D. N., Grant, R. A., Baker, T. A., & Sauer, R. T. (2005). Specificity versus stability in computational protein design. *Proceedings of the National Academy of Sciences of the United States of America*, *102*(36), 12724–12729. PMID:16129838

Bolton, E., Wang, Y., Thiessen, P. A., & Bryant, S. H. (2008). PubChem: Integrated Platform of Small Molecules and Biological Activities. In R. A. Wheeler & D. C. Spellmeyer (Eds.), *Annual Reports in Computational Chemistry* (pp. 217–241). Oxford, UK: Elsevier. doi:10.1016/S1574-1400(08)00012-1

Bonneau, R., & Baker, D. (2001). Ab initio protein structure prediction: Progress and prospects. *Annual Review of Biophysics and Biomolecular Structure*, *30*(1), 173–189. doi:10.1146/annurev.biophys.30.1.173 PMID:11340057

Bonvin, A. (2013). Coming to peace with protein complexes? 5th CAPRI evaluation meeting, April 17–19th 2013 – Utrecht. *Proteins: Structure, Function, and Bioinformatics*, *81*(12), 2073–2074.

Borišek, J., Drgan, V., Minovski, N., & Novič, M. (2014). Mechanistic interpretation of artificial neural network-based QSAR model for prediction of cathepsin K inhibition potency. *Journal of Chemometrics*, *28*(4), 272–281. doi:10.1002/cem.2617

Borišek, J., Vizovišek, M., Sosnowski, P., Turk, B., Turk, D., Mohar, B., & Novič, M. (2015). Development of N-(Functionalized Benzoyl)-homocycloleucyl-glycinonitriles as Potent Cathepsin K Inhibitors. *Journal of Medicinal Chemistry*, *58*(17), 6928–6937. doi:10.1021/acs.jmedchem.5b00746

Borkotoky, S. (2012). Docking studies on HIV integrase inhibitors based on potential ligand binding sites. *International Journal of Bioscience. Biochemistry and Bioinformatics*, *2*(3), 21–29.

Bortolato, A., Fanton, M., Mason, J. S., & Moro, S. (2013). Molecular Docking Methodologies. In L. Monticelli & E. Salonen (Eds.), *Biomolecular Simulations* (Vol. 924, pp. 339–360). Totowa, NJ: Humana Press. doi:10.1007/978-1-62703-017-5_13

Bottegoni, G., Kufareva, I., Totrov, M., & Abagyan, R. (2009). Four-Dimensional Docking: A Fast and Accurate Account of Discrete Receptor Flexibility in Ligand Docking. *Journal of Medicinal Chemistry*, *52*(2), 397–406. doi:10.1021/jm8009958 PMID:19090659

Bouvier, G., Evrard-Todeschi, N., Girault, J. P., & Bertho, G. (2010). Automatic clustering of docking poses in virtual screening process using self-organizing map. *Bioinformatics (Oxford, England)*, *26*(1), 53–60. doi:10.1093/bioinformatics/btp623 PMID:19910307

Boyd, D. B., & Lipkowitz, K. B. (2001). *Reviews in Computational Chemistry* (Vol. 17). New York: Wiley-VCH.

Bradford, J. R., & Westhead, D. R. (2005). Improved prediction of protein-protein binding sites using a support vector machines approach. *Bioinformatics (Oxford, England)*, *21*(8), 1487–1494. doi:10.1093/bioinformatics/bti242 PMID:15613384

Brady, G. P. Jr, & Stouten, P. F. (2000). Fast prediction and visualization of protein binding pockets with PASS. *Journal of Computer-Aided Molecular Design*, *14*(4), 383–401. doi:10.1023/A:1008124202956 PMID:10815774

B-Rao, C., Subramanian, J., & Sharma, S. D. (2009). Managing protein flexibility in docking and its applications. *Drug Discovery Today*, *14*(7–8), 394–400. PMID:19185058

Brás, N., Cerqueira, N., Sousa, S., Fernandes, P., & Ramos, M. (2014). Protein ligand docking in drug discovery. *Protein Modelling*, 249.

Bras, N. F., Cerqueira, N. M. F. S. A., Sousa, S. F., Fernandes, P. A., & Ramos, M. J. (2014). Protein ligand docking in drug discovery. In N. S. Gabor (Ed.), *Protein modeling* (pp. 249–286). Switzerland: Springer International Publishing.

Breiten, B., Lockett, M. R., Sherman, W., Fujita, S., Al-Sayah, M., Lange, H., & Whitesides, G. M. et al. (2013). Water networks contribute to enthalpy/entropy compensation in protein–ligand binding. *Journal of the American Chemical Society*, *135*(41), 15579–15584.

Brenke, R., Hall, D. R., Chuang, G. Y., Comeau, S. R., Bohnuud, T., Beglov, D., & Kozakov, D. et al. (2012). Application of asymmetric statistical potentials to antibody-protein docking. *Bioinformatics (Oxford, England)*, *28*(20), 2608–2614. doi:10.1093/bioinformatics/bts493 PMID:23053206

Brint, A. T., & Willett, P. (1987). Algorithms for the Identification of Three-Dimensional Maximal Common Substructures. *Journal of Chemical Information and Computer Sciences*, *27*(4), 152–158. doi:10.1021/ci00056a002

Bron, C., & Kerbosch, J. (1973). Algorithm 457: Finding All Cliques of an Undirected Graph. *Communications of the ACM*, *16*(9), 575–576. doi:10.1145/362342.362367

Brooijmans, N., & Kuntz, I. D. (2003). Molecular recognition and docking algorithms. *Annual Review of Biophysics and Biomolecular Structure*, *32*(1), 335–373. doi:10.1146/annurev.biophys.32.110601.142532 PMID:12574069

Brooks, B. R., Brooks, C. L. III, MacKerell, A. D. Jr, Nilsson, L., Petrella, R. J., Roux, B., & Karplus, M. et al. (2010). CHARMM: The Biomolecular Simulation Program. *Journal of Computational Chemistry*, *30*(10), 1545–1614. doi:10.1002/jcc.21287 PMID:19444816

Brooks, B. R., Bruccoleri, R. E., Olafson, B. D., States, D. J., Swaminathan, S., & Karplus, M. (1983). CHARMM: A program for macromolecular energy, minimization, and dynamics calculations. *Journal of Computational Chemistry*, *4*(2), 187–217.

Brooks, B. R., Bruccoleri, R. E., Olafson, B. D., States, D. J., Swaminathan, S., & Karplus, M. (1983). CHARMM: A Program for Macromolecular Energy, Minimization, and Dynamics Calculations. *Journal of Computational Chemistry*, *4*(2), 187–217. doi:10.1002/jcc.540040211

Brylinski, M. (2013). Nonlinear scoring functions for similarity-based ligand docking and binding affinity prediction. *Journal of Chemical Information and Modeling*, *53*(11), 3097–3112. doi:10.1021/ci400510e PMID:24171431

Brylinski, M., Lee, S. Y., Zhou, H., & Skolnick, J. (2011). The utility of geometrical and chemical restraint information extracted from predicted ligand-binding sites in protein structure refinement. *Journal of Structural Biology*, *173*(3), 558–569. doi:10.1016/j.jsb.2010.09.009 PMID:20850544

Brylinski, M., & Skolnick, J. (2010). Comprehensive structural and functional characterization of the human kinome by protein structure modeling and ligand virtual screening. *Journal of Chemical Information and Modeling*, *50*(10), 1839–1854. doi:10.1021/ci100235n PMID:20853887

Bryngelson, J. D., Onuchic, J. N., Socci, N. D., & Wolynes, P. G. (1995). Funnels, pathways, and the energy landscape of protein folding: A synthesis. *Proteins: Structure, Function, and Bioinformatics*, *21*(3), 167–195. PMID:7784423

Burkhard, P., Taylor, P., & Walkinshaw, M. D. (1998). An example of a protein ligand found by database mining: Description of the docking method and its verification by a 2.3 A X-ray structure of a thrombin-ligand complex. *Journal of Molecular Biology*, *277*(2), 449–466. doi:10.1006/jmbi.1997.1608 PMID:9514757

Bustanji, Y., Al-Masri, I. M., Qasem, A., Al-Bakri, A. G., & Taha, M. O. (2009). *In silico* screening for non-nucleoside HIV-1 reverse transcriptase inhibitors using physicochemical filters and high-throughput docking followed by *in vitro* evaluation. *Chemical Biology & Drug Design*, *74*(3), 258–265. doi:10.1111/j.1747-0285.2009.00852.x PMID:19703027

Buzko, O. V., Bishop, A. C., & Shokat, K. M. (2002). Modified AutoDock for accurate docking of protein kinase inhibitors. *Journal of Computer-Aided Molecular Design*, *16*(2), 113–127. doi:10.1023/A:1016366013656 PMID:12188021

Bystroff, C., & Shao, Y. (2002). Fully automated ab initio protein structure prediction using I-SITES, HMMSTR and ROSETTA. *Bioinformatics (Oxford, England)*, *18*(Suppl 1), S54–S61. doi:10.1093/bioinformatics/18.suppl_1.S54 PMID:12169531

Böhm, H. J. (1992). The computer program LUDI: A new method for the de novo design of enzyme inhibitors. *Journal of Computer-Aided Molecular Design, 6*, 61-78.

Böhm, H. J. (1994). The development of a simple empirical scoring function to estimate the binding constant for a protein-ligand complex of known three-dimensional structure. *Journal of Computer-Aided Molecular Design, 8*, 243-256.

Caflisch, A., Niederer, P., & Anliker, M. (1992). Monte Carlo docking of oligopeptides to proteins. *Proteins*, *13*(3), 223–230. doi:10.1002/prot.340130305

Camacho, C. J., & Gatchell, D. W. (2003). Successful discrimination of protein interactions. *Proteins*, *52*(1), 92–97. doi:10.1002/prot.10394 PMID:12784373

Camacho, C. J., Ma, H., & Champ, P. C. (2006). Scoring a diverse set of high-quality docked conformations: A metascore based on electrostatic and desolvation interactions. *Proteins*, *63*(4), 868–877. doi:10.1002/prot.20932 PMID:16506242

Camacho, C. J., & Vajda, S. (2002). Protein-protein association kinetics and protein docking. *Current Opinion in Structural Biology*, *12*(1), 36–40. doi:10.1016/S0959-440X(02)00286-5 PMID:11839487

Campbell, A. J., Lamb, M. L., & Joseph-McCarthy, D. (2014). Ensemble-based docking using biased molecular dynamics. *Journal of Chemical Information and Modeling*, *54*(7), 2127–2138. doi:10.1021/ci400729j PMID:24881672

Capdeville, R., Buchdunger, E., Zimmermann, J., & Matter, A. (2002). Glivec (STI571, imatinib), a rationally developed, targeted anticancer drug. *Nature Reviews. Drug Discovery*, *1*(7), 493–502. doi:10.1038/nrd839 PMID:12120256

Carlson, H. A. (2002). Protein flexibility and drug design: How to hit a moving target. *Current Opinion in Chemical Biology*, *6*(4), 447–452. doi:10.1016/S1367-5931(02)00341-1 PMID:12133719

Case, D. A., Cheatham, T. E. III, Darden, T., Gohlke, H., Luo, R., Merz, K. M. Jr, & Woods, R. et al. (2005). The Amber biomolecular simulation programs. *Journal of Computational Chemistry*, *26*(16), 1668–1688. doi:10.1002/jcc.20290

Cavasatto, C. N., Kovacs, J. A., & Abagyan, R. A. (2005). Representing receptor flexibility in ligand docking through relevant normal modes. *Journal of the American Chemical Society*, *127*(26), 9632–9640. doi:10.1021/ja042260c PMID:15984891

Cavasotto, C. N., & Abagyan, R. A. (2004). Protein flexibility in ligand docking and virtual screening to protein kinases. *Journal of Molecular Biology*, *337*(1), 209–225. doi:10.1016/j.jmb.2004.01.003 PMID:15001363

Cavasotto, C. N., Orry, A. J. W., & Abagyan, R. A. (2005). The challenge of considering receptor flexibility in ligand docking and virtual screening. *Current Computer-aided Drug Design*, *1*(4), 423–440. doi:10.2174/157340905774330291

Cavasotto, C. N., & Phatak, S. S. (2011). Docking methods for structure-based library design. *Methods in Molecular Biology (Clifton, N.J.)*, *685*, 155–174. doi:10.1007/978-1-60761-931-4_8 PMID:20981523

Cerqueira, N. M. F. S. A., Bras, N. F., Fernandes, P. A., & Ramos, M. J. (2009). MADAMM: A multistaged docking with an automated molecular modeling protocol. *Proteins*, *74*(1), 192–206. doi:10.1002/prot.22146 PMID:18618708

Ceulemans, H., & Russell, R. B. (2004). Fast fitting of atomic structures to low-resolution electron density maps by surface overlap maximization. *Journal of Molecular Biology*, *338*(4), 783–793. doi:10.1016/j.jmb.2004.02.066 PMID:15099745

Chakravarty, S., Wang, L., & Sanchez, R. (2005). Accuracy of structure-derived properties in simple comparative models of protein structures. *Nucleic Acids Res*, *33*(1), 244-259. doi: 10.1093/nar/gki162

Chang, D. T., Oyang, Y. J., & Lin, J. H. (2005). MEDock: a web server for efficient prediction of ligand binding sites based on a novel optimization algorithm. *Nucleic Acids Res*, *33*, W233-238.

Chang, M. W., Belew, R. K., Carroll, K. S., Olson, A. J., & Goodsell, D. S. (2008). Empirical entropic contributions in computational docking: Evaluation in APS reductase complexes. *Journal of Computational Chemistry*, *29*(11), 1753–1761. doi:10.1002/jcc.20936 PMID:18351616

Charifson, P. S., Corkery, J. J., Murcko, M. A., & Walters, W. P. (1999). Consensus scoring: A method for obtaining improved hit rates from docking databases of three-dimensional structures into proteins. *Journal of Medicinal Chemistry*, *42*(25), 5100–5109. doi:10.1021/jm990352k

Chaves, R. C., & Pellequer, J. L. (2013). DockAFM: Benchmarking protein structures by docking under AFM topographs. *Bioinformatics (Oxford, England)*, *29*(24), 3230–3231. doi:10.1093/bioinformatics/btt561 PMID:24078683

Chen, R., & Weng, Z. (2002). Docking unbound proteins using shape complementarity, desolvation, and electrostatics. *Proteins: Structure, Function and Genetics*, *47*, 281–294.

Chen, R., Li, L., & Weng, Z. (2003). ZDOCK: An Initial-Stage Protein-Docking Algorithm. *Proteins*, *52*, 80–87.

Chen, Y. & Pohlhaus, D. T. (2010). In silico docking and scoring of fragments. *Drug Discovery Today: Technologies | Fragment based drug discovery*, *7*(3), 149-156.

Chen, C. Y.-C., & Tou, W. (2013). How to design a drug for the disordered proteins? *Drug Discovery Today*, *18*(19-20), 910–915. doi:10.1016/j.drudis.2013.04.008 PMID:23643489

Chène, P. (2003). Inhibiting the p53–MDM2 interaction: An important target for cancer therapy. *Nature Reviews. Cancer*, *3*(2), 102–109. doi:10.1038/nrc991 PMID:12563309

Cheng, T. M., Blundell, T. L., & Fernandez-Recio, J. (2007). pyDock: Electrostatics and desolvation for effective scoring of rigid-body protein-protein docking. *Proteins*, *68*(2), 503–515. doi:10.1002/prot.21419 PMID:17444519

Cheng, T., Li, Q., Zhou, Z., Wang, Y., & Bryant, S. H. (2012). Structure-based virtual screening for drug discovery: A problem-centric review. *The AAPS Journal*, *14*(1), 133–141. PMID:22281989

Cheng, T., Li, X., Li, Y., Liu, Z., & Wang, R. (2009). Comparative assessment of scoring functions on a diverse test set. *Journal of Chemical Information and Modeling*, *49*(4), 1079–1093. doi:10.1021/ci9000053

Chen, H.-M., Liu, B.-F., Huang, H.-L., Hwang, S.-F., & Ho, S.-Y. (2007). SODOCK: Swarm optimization for highly flexible protein–ligand docking. *Journal of Computational Chemistry*, *28*(2), 612–623. doi:10.1002/jcc.20542 PMID:17186483

Chen, J. H., Linstead, E., Swamidass, S. J., Wang, D., & Baldi, P. (2007). ChemDB update-full-text search and virtual chemical space. *Bioinformatics (Oxford, England)*, *23*(17), 2348–2351. doi:10.1093/bioinformatics/btm341 PMID:17599932

Chen, R., Li, L., & Weng, Z. (2003). ZDOCK: An initial-stage protein-docking algorithm. *Proteins*, *52*(1), 80–87. doi:10.1002/prot.10389 PMID:12784371

Chen, R., Tong, W., Mintseris, J., Li, L., & Weng, Z. (2003). ZDOCK predictions for the CAPRI challenge. *Proteins*, *52*(1), 68–73. doi:10.1002/prot.10388 PMID:12784369

Chen, R., & Weng, Z. (2002). Docking unbound proteins using shape complementarity, desolvation, and electrostatics. *Proteins*, *47*(3), 281–294. doi:10.1002/prot.10092

Chen, R., & Weng, Z. (2003). A novel shape complementarity scoring function for protein-protein docking. *Proteins*, *51*(3), 397–408. doi:10.1002/prot.10334 PMID:12696051

Chen, Y. (2009). Weighted Equation and Rules-A Novel Concept for Evaluating Protein-Ligand Interaction. *Journal of Biomolecular Structure & Dynamics*, *27*(3), 271–282. doi:10.1080/07391102.2009.10507315 PMID:19795911

Chen, Y.-C. (2015). Beware of docking! *Trends in Pharmacological Sciences*, *36*(2), 78–95. doi:10.1016/j.tips.2014.12.001

Chermak, E., Petta, A., Serra, L., Vangone, A., Scarano, V., Cavallo, L., & Oliva, R. (2015). CONSRANK: A server for the analysis, comparison and ranking of docking models based on inter-residue contacts. *Bioinformatics (Oxford, England)*, *31*(9), 1481–1483. doi:10.1093/bioinformatics/btu837 PMID:25535242

Chiba, S., Harano, Y., Roth, R., Kinoshita, M., & Sakurai, M. (2012). Evaluation of protein-ligand binding free energy focused on its entropic components. *Journal of Computational Chemistry*, *33*(5), 550–560.

Choi, Y. S., Han, S. K., Kim, J., Yang, J. S., Jeon, J., Ryu, S. H., & Kim, S. (2010). ConPlex: a server for the evolutionary conservation analysis of protein complex structures. *Nucleic Acids Res, 38*, W450-456. doi: 10.1093/nar/gkq328

Chothia, C. L. A. (1986). The relation between the divergence of sequence and structure in proteins. *The EMBO Journal*, *5*, 823–836. PMID:3709526

Chourasia, M., Sastry, G. M., & Sastry, G. N. (2011). Aromatic–Aromatic Interactions Database, A 2 ID: An analysis of aromatic π-networks in proteins. *International Journal of Biological Macromolecules*, *48*(4), 540–552. doi:10.1016/j.ijbiomac.2011.01.008 PMID:21255607

Chowdhury, R., Rasheed, M., Keidel, D., Moussalem, M., Olson, A., Sanner, M., & Bajaj, C. (2013). Protein-protein docking with F(2)Dock 2.0 and GB-rerank. *PLoS ONE, 8*(3), e51307. doi:10.1371/journal.pone.0051307 PMID:23483883

Chuang, C. H., Cheng, T. C., Leu, Y. L., Chuang, K. H., Tzou, S. C., & Chen, C. S. (2015). Discovery of Akt kinase inhibitors through structure-based virtual screening and their evaluation as potential anticancer agents. *International Journal of Molecular Sciences, 16*(2), 3202–3212. doi:10.3390/ijms16023202 PMID:25648320

Chuang, G.-Y., Kozakov, D., Brenke, R., Comeau, S. R., & Vajda, S. (2008). DARS (Decoys As the Reference State) potentials for protein-protein docking. *Biophysical Journal, 95*(9), 4217–4227. doi:10.1529/biophysj.108.135814 PMID:18676649

Chuaqui, C., Deng, Z., & Singh, J. (2005). Interaction profiles of protein kinase-inhibitor complexes and their application to virtual screening. *Journal of Medicinal Chemistry, 48*(1), 121–133. doi:10.1021/jm049312t PMID:15634006

Cimerman, N., Prebanda, M. T., Turk, B., Popovic, T., Dolenc, I., & Turk, V. (1999). Interaction of Cystatin C variants with papain and human cathepsins B, H and L. *Journal of Enzyme Inhibition, 14*(2), 167–174. doi:10.3109/14756369909036552 PMID:10445041

Clark, A. M., & Labute, P. (2007). 2D depiction of protein-ligand complexes. *Journal of Chemical Information and Modeling, 47*(5), 1933–1944. doi:10.1021/ci7001473 PMID:17715911

Clark, K. P., & Ajay, . (1995). Flexible ligand docking without parameter adjustment across four ligandreceptor complexes. *Journal of Computational Chemistry, 16*(10), 1210–1226. doi:10.1002/jcc.540161004

Clark, R. D., Strizhev, A., Leonard, J. M., Blake, J. F., & Matthew, J. B. (2002). Consensus scoring for ligand/protein interactions. *Journal of Molecular Graphics & Modelling, 20*(4), 281–295. doi:10.1016/S1093-3263(01)00125-5 PMID:11858637

Claussen, H., Buning, C., Rarey, M., & Lengauer, T. (2001). FlexE: Efficient molecular docking considering protein structure variations. *Journal of Molecular Biology, 308*(2), 377–395. doi:10.1006/jmbi.2001.4551 PMID:11327774

Cohen, P., & Alessi, D. R. (2013). Kinase drug discovery– What's next in the field? *ACS Chemical Biology, 18*(8), 96–104. doi:10.1021/cb300610s PMID:23276252

Cole, J. C., Murray, C. W., Nissink, J. W. M., Taylor, R. D., & Taylor, R. (2005). Comparing protein-ligand docking programs is difficult. *Proteins: Structure, Function, and Bioinformatics, 60*(3), 325–332. doi:10.1002/prot.20497 PMID:15937897

Cole, J. C., Nissink, J. W. M., & Taylor, R. (2005). Protein–ligand docking and virtual screening with GOLD. In J. Alvarez & B. Shoichet (Eds.), *Virtual screening in drug discovery* (pp. 379–416). Boca Raton, FL: Taylor & Francis Group. doi:10.1201/9781420028775.ch15

Combet, C., Jambon, M., Deléage, G., & Geourjon, C. (2002). Geno3D: Automatic comparative molecular modelling of protein. *Bioinformatics (Oxford, England), 18*(1), 213–214. doi:10.1093/bioinformatics/18.1.213 PMID:11836238

Combs, S. A., Deluca, S. L., DeLuca, S. H., Lemmon, G. H., Nannemann, D. P., Nguyen, E. D., & Meiler, J. et al. (2013). Small-molecule ligand docking into comparative models with Rosetta. *Nature Protocols, 8*(7), 1277–1298. doi:10.1038/nprot.2013.074 PMID:23744289

Comeau, S. R., Gatchell, D. W., Vajda, S., & Camacho, C. J. (2004a). ClusPro: a fully automated algorithm for protein-protein docking. *Nucleic Acids Res, 32*, W96-99. doi: 10.1093/nar/gkh354

Comeau, S. R., Gatchell, D. W., Vajda, S., & Camacho, C. J. (2004b). ClusPro: An automated docking and discrimination method for the prediction of protein complexes. *Bioinformatics (Oxford, England), 20*(1), 45–50. doi:10.1093/bioinformatics/btg371 PMID:14693807

Cornell, W. D., Cieplak, P., Bayly, C. I., Gould, I. R., Merz, K. M., Ferguson, D. M., & Kollman, P. A. et al. (1995). A second generation force field for the simulation of proteins, nucleic acids, and organic molecules. *Journal of the American Chemical Society*, *117*(19), 5179–5197. doi:10.1021/ja00124a002

Cortes Cabrera, A., Klett, J., Dos Santos, H. G., Perona, A., Gil-Redondo, R., Francis, S. M., & Morreale, A. et al. (2012). CRDOCK: An ultrafast multipurpose protein-ligand docking tool. *Journal of Chemical Information and Modeling*, *52*(8), 2300–2309. doi:10.1021/ci300194a PMID:22764680

Cosconati, S., Forli, S., Perryman, A. L., Harris, R., Goodsell, D. S., & Olson, A. J. (2010). Virtual screening with AutoDock: Theory and practice. *Expert Opinion on Drug Discovery*, *5*(6), 597–607. doi:10.1517/17460441.2010.484460

Coupez, B., & Lewis, R. A. (2006). Docking and scoring--theoretically easy, practically impossible? *Current Medicinal Chemistry*, *13*(25), 2995–3003. doi:10.2174/092986706778521797 PMID:17073642

Craig, I. R., Essex, J. W., & Spiegel, K. (2010). Ensemble Docking into Multiple Crystallographically Derived Protein Structures: An Evaluation Based on the Statistical Analysis of Enrichments. *Journal of Chemical Information and Modeling*, *50*(4), 511–524. doi:10.1021/ci900407c PMID:20222690

Crescenzi, O., Tomaselli, S., Guerrini, R., Salvadori, S., D'Ursi, A. M., Temussi, P. A., & Picone, D. (2002). Solution structure of the Alzheimer amyloid beta-peptide (1–42) in an apolar microenvironment similarity with a virus fusion domain. *European Journal of Biochemistry*, *269*(22), 5642–5648. doi:10.1046/j.1432-1033.2002.03271.x PMID:12423364

Crisman, T. J., Sisay, M. T., & Bajorath, J. (2008). Ligand-target interaction-based weighting of substructures for virtual screening. *Journal of Chemical Information and Modeling*, *48*(10), 1955–1964. doi:10.1021/ci800229q PMID:18821751

Cross, J. B., Thompson, D. C., Rai, B. K., Baber, J. C., Fan, K. Y., Hu, Y., & Humblet, C. (2009). Comparison of several molecular docking programs: Pose prediction and virtual screening accuracy. *Journal of Chemical Information and Modeling*, *49*(6), 1455–1474. doi:10.1021/ci900056c PMID:19476350

Crowley, P. B., & Golovin, A. (2005). Cation–π interactions in protein–protein interfaces. *Proteins: Structure, Function, and Bioinformatics*, *59*(2), 231–239. doi:10.1002/prot.20417 PMID:15726638

Da, C., & Kireev, D. (2014). Structural Protein–Ligand Interaction Fingerprints (SPLIF) for Structure-Based Virtual Screening: Method and Benchmark Study. *Journal of Chemical Information and Modeling*, *54*(9), 2555–2561. doi:10.1021/ci500319f PMID:25116840

Dalby, A., Nourse, J. G., Hounshell, W. D., Gushurst, A. K. I., Grier, D. L., Leland, B. A., & Laufer, J. (1992). Description of several chemical structure file formats used by computer programs developed at Molecular Design Limited. *Journal of Chemical Information and Modeling*, *32*(3), 244–255. doi:10.1021/ci00007a012

Dalkas, G. A., Teheux, F., Kwasigroch, J. M., & Rooman, M. (2014). Cation–π, amino–π, π–π, and H-bond interactions stabilize antigen–antibody interfaces. *Proteins: Structure, Function, and Bioinformatics*, *82*(9), 1734–1746. doi:10.1002/prot.24527 PMID:24488795

Damm, K. L., & Carlson, H. A. (2007). Exploring experimental sources of multiple protein conformations in structure-based drug design. *Journal of the American Chemical Society*, *129*(26), 8225–8235. doi:10.1021/ja0709728 PMID:17555316

Danishuddin, M., & Khan, A. U. (2015). Structure based virtual screening to discover putative drug candidates: Necessary considerations and successful case studies. *Methods (San Diego, Calif.)*, *71*, 135–145. PMID:25448480

Das, A., Bhattacharya, S., Bagchi, A., & Dasgupta, R. (2013). Structural insight into the formin binding property of human formin binding protein 4. *Journal of Proeins and Proteomics*, *4*(December), 93–100.

Davis, I. W., & Baker, D. (2009). RosettaLigand Docking with Full Ligand and Receptor Flexibility. *Journal of Molecular Biology*, *385*(2), 381–392. doi:10.1016/j.jmb.2008.11.010 PMID:19041878

de Beer, T. A. P., Berka, K., Thornton, J. M., & Laskowski, R. A. (2014). PDBsum additions. *Nucleic Acids Research*, *42*(D1), D292–D296. doi:10.1093/nar/gkt940 PMID:24153109

de Groot, M. J., Ackland, M. J., Horne, V. A., Alex, A. A., & Jones, B. C. (1999). Novel Approach To Predicting P450-Mediated Drug Metabolism: Development of a Combined Protein and Pharmacophore Model for CYP2D6. *Journal of Medicinal Chemistry*, *42*(9), 1515–1524. doi:10.1021/jm981118h PMID:10229622

de Vries, S. J., & Bonvin, A. M. (2011). CPORT: A consensus interface predictor and its performance in prediction-driven docking with HADDOCK. *PLoS ONE*, *6*(3), e17695. doi:10.1371/journal.pone.0017695 PMID:21464987

de Vries, S. J., Schindler, C. E., Chauvot de Beauchene, I., & Zacharias, M. (2015). A web interface for easy flexible protein-protein docking with ATTRACT. *Biophysical Journal*, *108*(3), 462–465. doi:10.1016/j.bpj.2014.12.015 PMID:25650913

De Vries, S. J., van Dijk, A. D. J., Krzeminski, M., van Dijk, M., Thureau, A., Hsu, V., & Bonvin, A. M. J. J. et al. (2007). HADDOCK versus HADDOCK: New features and performance of HADDOCK2.0 on the CAPRI targets. *Proteins*, *69*(4), 726–733. doi:10.1002/prot.21723 PMID:17803234

de Vries, S. J., van Dijk, M., & Bonvin, A. M. (2010). The HADDOCK web server for data-driven biomolecular docking. *Nature Protocols*, *5*(5), 883–897. doi:10.1038/nprot.2010.32 PMID:20431534

De Vries, S., & Zacharias, M. (2013). Flexible docking and refinement with a coarse-grained protein model using ATTRACT. *Proteins: Structure. Function and Bioinformatics*, *81*(12), 2167–2174. doi:10.1002/prot.24400

Delano, W. L. (2002). *The PyMOL Molecular Graphics System*. Retrieved from http://www.pymol.org

Deng, W., Breneman, C., & Embrechts, M. J. (2004). Predicting protein-ligand binding affinities using novel geometrical descriptors and machine-learning methods. *Journal of Chemical Information and Computer Sciences*, *44*(2), 699–703. PMID:15032552

Deng, Z., Chuaqui, C., & Singh, J. (2004). Structural interaction fingerprint (SIFt): A novel method for analyzing three-dimensional protein-ligand binding interactions. *Journal of Medicinal Chemistry*, *47*(2), 337–344. doi:10.1021/jm030331x PMID:14711306

Desaphy, J., Raimbaud, E., Ducrot, P., & Rognan, D. (2013). Encoding protein–ligand interaction patterns in fingerprints and graphs. *Journal of Chemical Information and Modeling*, *53*(3), 623–637. doi:10.1021/ci300566n PMID:23432543

DesJarlais, R. L., Seibel, G. L., Kuntz, I. D., Furth, P. S., Alvarez, J. C., Babe, L. M., & Craik, Y., C. S. (1990). Structure-based design of nonpeptide inhibitors specific for the Human Immunodeficiency Virus 1 protease.*Proceedings of the National Academy of Sciences*, *87*,6644-6448. doi:10.1073/pnas.87.17.6644

DesJarlais, R. L., Sheridan, R. P., Dixon, J. S., Kuntz, I. D., & Venkataraghavan, R. (1986). Docking flexible ligands to macromolecular receptors by molecular shape. *Journal of Medicinal Chemistry*, *29*(11), 2149–2153. doi:10.1021/jm00161a004

Desmet, J., Maeyer, M. D., Hazes, B., & Lasters, I. (1992). The dead-end elimination theorem and its use in protein sidechain positioning. *Nature*, *356*(6369), 539–542. doi:10.1038/356539a0 PMID:21488406

Desmet, J., Wilson, I. A., Joniau, M., De Maeyer, M., & Lasters, I. (1997). Computation of the binding of fully flexible peptides to proteins with flexible side chains. *The FASEB Journal*, *11*(2), 164–172. PMID:9039959

DeWitte, R. S., & Shakhnovich, E. I. (1996). Smog: De novo design method based on simple, fast, and accurate free energy estimates. 1. Methodology and supporting evidence. *Journal of the American Chemical Society*, *118*(47), 11733–11744.

DeWitte, R. S., & Shakhnovich, E. I. (1996). SMoG: De Novo Design Method Based on Simple, Fast, and Accurate Free Energy Estimates. 1. Methodology and Supporting Evidence. *Journal of the American Chemical Society*, *118*(47), 11733–11744. doi:10.1021/ja960751u

Dhanavade, M. J., Jalkute, C. B., Barage, S. H., & Sonawane, K. D. (2013). Homology modeling, molecular docking and MD simulation studies to investigate role of cysteine protease from *Xanthomonas campestris* in degradation of Aβ peptide. *Computers in Biology and Medicine*, *43*(12), 2063–2070. doi:10.1016/j.compbiomed.2013.09.021 PMID:24290922

Dhanavade, M. J., & Sonawane, K. D. (2014). Insights into the molecular interactions between aminopeptidase and amyloid beta peptide using molecular modeling techniques. *Amino Acids*, *46*(8), 1853–1866. doi:10.1007/s00726-014-1740-0 PMID:24729013

Dhanik, A., McMurray, J. S., & Kavraki, L. E. (2013). DINC: A new AutoDock-based protocol for docking large ligands. *BMC Structural Biology*, *13*(Suppl 1), S11. doi:10.1186/1472-6807-13-S1-S11 PMID:24564952

Di Nola, A., Roccatano, D., & Berendsen, H. J. (1994). Molecular dynamics simulation of the docking of substrates to proteins. *Proteins*, *19*(3), 174–182. doi:10.1002/prot.340190303 PMID:7937732

Dias, R., de Azevedo, J., & Walter, F. (2008). Molecular docking algorithms. *Current Drug Targets*, *9*(12), 1040–1047. PMID:19128213

Dias, R., & de Azevedo, W. F. Jr. (2008). Molecular Docking Algorithms. *Current Drug Targets*, *9*(12), 1040–1047. doi:10.2174/138945008786949432 PMID:19128213

Dickson, M., & Gagnon, J. P. (2004). Key factors in the rising cost of new drug discovery and development. *Nature Reviews. Drug Discovery*, *3*(5), 417–429. PMID:15136789

Dijk, M. V., & Bonvin, A. M. J. J. (2010). Pushing the limits of what is achievable in protein–DNA docking: Benchmarking HADDOCK's performance. *Nucleic Acids Research*, *38*(17), 5634–5647. doi:10.1093/nar/gkq222 PMID:20466807

Diller, D. J., & Merz, K. M. Jr. (2001). High throughput docking for library design and library prioritization. *Proteins*, *43*(2), 113–124. doi:10.1002/1097-0134(20010501)43:2<113::AID-PROT1023>3.0.CO;2-T PMID:11276081

Dill, K. A., Truskett, T. M., Vlachy, V., & Hribar-Lee, B. (2005). Modeling water, the hydrophobic effect, and ion solvation. *Annual Review of Biophysics and Biomolecular Structure*, *34*, 173–199. PMID:15869376

DiscoveryStudio. (2007). *Version 2.5, Software Package AI.* Retrieved from http://www.accelrys.com

Dobes, P., Fanfrlik, J., Rezac, J., Otyepka, M., & Hobza, P. (2011). Transferable scoring function based on semiempirical quantum mechanical PM6-DH2 method: CDK2 with 15 structurally diverse inhibitors. *Journal of Computer-Aided Molecular Design*, *25*(3), 223–235. doi:10.1007/s10822-011-9413-5

Doman, T. N., McGovern, S. L., Witherbee, B. J., Kasten, T. P., Kurumbail, R., Stallings, W. C., & Shoichet, B. K. et al. (2002). Molecular docking and high-throughput screening for novel inhibitors of protein tyrosine phosphatase-1B. *Journal of Medicinal Chemistry*, *45*(11), 2213–2221. doi:10.1021/jm010548w PMID:12014959

Dominguez, C., Boelens, R., & Bonvin, A. M. J. J. (2003). HADDOCK: A protein-protein docking approach based on biochemical or biophysical information. *Journal of the American Chemical Society*, *125*(7), 1731–1737. doi:10.1021/ja026939x PMID:12580598

Dominy, B. N., & Shakhnovich, E. I. (2004). Native atom types for knowledge-based potentials: Application to binding energy prediction. *Journal of Medicinal Chemistry*, *47*(18), 4538–4558. PMID:15317465

Dossetter, A. G., Beeley, H., Bowyer, J., Cook, C. R., Crawford, J. J., Finlayson, J. E., & Vickers, M. et al. (2012). (1R,2R)-N-(1-cyanocyclopropyl)-2-(6-methoxy-1,3,4,5-tetrahydropyrido[4,3-b]indole -2-carbonyl)cyclohexanecarboxamide (AZD4996): A potent and highly selective cathepsin K inhibitor for the treatment of osteoarthritis. *Journal of Medicinal Chemistry*, *55*(14), 6363–6374. doi:10.1021/jm3007257

Dror, O., Shulman-Peleg, A., Nussinov, R., & Wolfson, H. J. (2004). Predicting molecular interactions in silico: I. A guide to pharmacophore identification and its applications to drug design. *Current Medicinal Chemistry*, *11*(1), 71–90. doi:10.2174/0929867043456287 PMID:14754427

Duhovny, D., Nussinov, R., & Wolfson, H. (2002). Efficient Unbound Docking of Rigid Molecules. *Algorithms in Bioinformatics*, 185–200.

Duhovny, D., Nussinov, R., & Wolfson, H. (2002). Efficient Unbound Docking of Rigid Molecules. In *Algorithms in Bioinformatics* (pp. 185–200). Springer Berlin Heidelberg; doi:10.1007/3-540-45784-4_14

Dunbar, J. B. Jr, Smith, R. D., Yang, C.-Y., Ung, P. M.-U., Lexa, K. W., Khazanov, N. A., & Carlson, H. A. et al. (2011). CSAR benchmark exercise of 2010: Selection of the protein–ligand complexes. *Journal of Chemical Information and Modeling*, *51*(9), 2036–2046. PMID:21728306

Dundas, J., Ouyang, Z., Tseng, J., Binkowski, A., Turpaz, Y., & Liang, J. (2006). CASTp: Computed atlas of surface topography of proteins with structural and topographical mapping of functionally annotated residues. *Nucleic Acids Research*, *34*(Web Server), W116–W118. doi:10.1093/nar/gkl282 PMID:16844972

Durham, E., Dorr, B., Woetzel, N., Staritzbichler, R., & Meiler, J. (2009). Solvent accessible surface area approximations for rapid and accurate protein structure prediction. *Journal of Molecular Modeling*, *15*(9), 1093–1108. doi:10.1007/s00894-009-0454-9 PMID:19234730

Durrant, J. D., & McCammon, J. A. (2010). Computer-aided drug-discovery techniques that account for receptor flexibility. *Current Opinion in Pharmacology*, *10*(6), 77–0774. doi:10.1016/j.coph.2010.09.001 PMID:20888294

Durrant, J. D., & McCammon, J. A. (2010). NNScore: A neural-network-based scoring function for the characterization of protein- ligand complexes. *Journal of Chemical Information and Modeling*, *50*(10), 1865–1871. PMID:20845954

Durrant, J. D., & McCammon, J. A. (2011). Molecular dynamics simulations and drug discovery. *BMC Biology*, *9*(1), 71. doi:10.1186/1741-7007-9-71 PMID:22035460

Eberhart, R., & Kennedy, J. (1995). A new optimizer using particle swarm theory. In *Proceedings of the Sixth International Symposium on Micro Machine and Human Science (MHS '95)*. Nagoya, Japan. IEEE. doi:10.1109/MHS.1995.494215

Ehrlich, L. P., & Wade, R. C. (2001). Protein–Protein Docking. In *Reviews in Computational Chemistry* (17th ed.; pp. 61–92). New York: Wiley-VCH. doi:10.1002/0471224413.ch2

Eisenberg, D., Luthy, R., & Bowie, J. U. (1997). VERIFY3D: assessment of protein models with three-dimensional profiles. *Methods Enzymol*, *277*, 396-404. doi: S0076-6879(97)77022-8

Eisenberg, D., & Mclachlan, A. (1986). Solvation energy in protein folding and binding. *Nature*, *319*(6050), 199–203. PMID:3945310

Eisenhaber, F., Lijnzaad, P., Argos, P., Sander, C., & Scharf, M. (1995). The double cubic lattice method: Efficient approaches to numerical integration of surface area and volume and to dot surface contouring of molecular assemblies. *Journal of Computational Chemistry*, *16*(3), 273–284. doi:10.1002/jcc.540160303

Elcock, A. A. H., Sept, D., & McCammon, J. A. (2001). Computer simulation of protein-protein interactions. *The Journal of Physical Chemistry B*, *105*(8), 1504–1518. doi:10.1021/jp003602d

Eldridge, M. D., Murray, C. W., Auton, T. R., Paolini, G. V., & Mee, R. P. (1997). Empirical scoring functions: I. The development of a fast empirical scoring function to estimate the binding affinity of ligands in receptor complexes. *Journal of Computer-Aided Molecular Design*, *11*(5), 425–445. doi:10.1023/A:1007996124545 PMID:9385547

Eldridge, M. D., Murray, C. W., Auton, T. R., Paolini, G. V., & Mee, R. P. (1997). Empirical scoring functions: I. the development of a fast empirical scoring function to estimate the binding affinity of ligands in receptor complexes. *Journal of Computer-Aided Molecular Design*, *11*(5), 425–445. PMID:9385547

Elokely, K. M., & Doerksen, R. J. (2013). Docking challenge: Protein sampling and molecular docking performance. *Journal of Chemical Information and Modeling*, *53*(8), 1934–1945. doi:10.1021/ci400040d PMID:23530568

Emekli, U., Schneidman-Duhovny, D., Wolfson, H. J., Nussinov, R., & Haliloglu, T. (2008). HingeProt: Automated prediction of hinges in protein structures. *Proteins*, *70*(4), 1219–1227. doi:10.1002/prot.21613 PMID:17847101

Erlanson, D. A., & Jahnke, W. (2006). The Concept of Fragment-based Drug Discovery. *Fragment-based approaches in drug discovery*, 1-10.

Esquieu, D., Péloponèse, J.-M., Opi, S., Gregoire, C., de Mareuil, J., Watkins, J., & Loret, E. P. et al. (2003). Discovery of a Tat HIV-1 inhibitor through computer-aided drug design. *Spectroscopy*, *17*(4), 639–645. doi:10.1155/2003/283731

Eswar, N., Marti-Renom, M.A., Webb, B., Madhusudhan, M.S., Eramian, D., Shen, M., Pieper, U., & Sali. A. (2006). Comparative Protein Structure Modeling With MODELLER. *Current Protocols in Bioinformatics*, *15*, 5.6.1-5.6.30.

Evers, A., Hessler, G., Matter, H., & Klabunde, T. (2005). Virtual screening of biogenic amine-binding G-protein coupled receptors: Comparative evaluation of protein- and ligand-based virtual screening protocols. *Journal of Medicinal Chemistry*, *48*(17), 5448–5465. doi:10.1021/jm050090o PMID:16107144

Eweas, A. F., Maghrabi, I. A., & Namarneh, A. I. (2014). Advances in molecular modeling and docking as a tool for modern drug discovery. *Der Pharma Chemica*, *6*(6), 211–228.

Ewing, T. J. A., Makino, S., Skillman, A. G., & Kuntz, I. D. (2001). DOCK 4.0: Search strategies for automated molecular docking of flexible molecule databases. *Journal of Computer-Aided Molecular Design*, *15*(5), 411–428. doi:10.1023/A:1011115820450

Fabbro, D. (2015). 25 years of small molecular weight kinase inhibitors: Potentials and limitations. *Molecular Pharmacology*, *87*(5), 766–775. doi:10.1124/mol.114.095489 PMID:25549667

Fan, H., Schneidman-Duhovny, D., Irwin, J. J., Dong, G., Shoichet, B. K., & Sali, A. (2011). Statistical potential for modeling and ranking of protein-ligand interactions. *Journal of Chemical Information and Modeling*, *51*(12), 3078–3092. doi:10.1021/ci200377u PMID:22014038

Favia, A. D., Bottegoni, G., Nobeli, I., Bisignano, P., & Cavalli, A. (2011). SERAPhiC: A Benchmark for in Silico Fragment-Based Drug Design. *Journal of Chemical Information and Modeling*, *51*(11), 2882–2896. doi:10.1021/ci2003363 PMID:21936510

Feher, M. (2006). Consensus scoring for protein–ligand interactions. *Drug Discovery Today*, *11*(9), 421–428. doi:10.1016/j.drudis.2006.03.009

Feig, M., & Brooks,, C. L. III (2004). Recent advances in the development and application of implicit solvent models in biomolecule simulations. *Current Opinion in Structural Biology*, *14*(2), 217–224. PMID:15093837

Feig, M., Onufriev, A., Lee, M. S., Im, W., Case, D. A., & Brooks, C. L. (2004). Performance comparison of generalized born and Poisson methods in the calculation of electrostatic solvation energies for protein structures. *Journal of Computational Chemistry*, *25*(2), 265–284. doi:10.1002/jcc.10378 PMID:14648625

Feixas, F., Lindert, S., Sinko, W., & McCammon, J. A. (2014). Exploring the role of receptor flexibility in structure-based drug discovery. *Biophysical Chemistry*, *186*, 31–45. doi:10.1016/j.bpc.2013.10.007 PMID:24332165

Fenollosa, C., Oton, M., Andrio, P., Cortes, J., Orozco, M., & Goni, J. R. (2015). SEABED: Small molEcule activity scanner weB servicE baseD. *Bioinformatics (Oxford, England)*, *31*(5), 773–775. doi:10.1093/bioinformatics/btu709 PMID:25348211

Fernandez-Recio, J., Totrov, M., & Abagyan, R. (2002). Soft protein – protein docking in internal coordinates. *Protein Science*, *11*(2), 280–291. doi:10.1110/ps.19202 PMID:11790838

Fernandez-Recio, J., Totrov, M., & Abagyan, R. (2003). ICM-DISCO docking by global energy optimization with fully flexible side-chains. *Proteins*, *52*(1), 113–117. doi:10.1002/prot.10383 PMID:12784376

Fernandez-Recio, J., Totrov, M., Skorodumov, C., & Abagyan, R. (2005). Optimal docking area: A new method for predicting protein-protein interaction sites. *Proteins*, *58*(1), 134–143. doi:10.1002/prot.20285 PMID:15495260

Ferrari, A. M., Wei, B. Q., Costantino, L., & Shoichet, B. K. (2004). Soft Docking and Multiple Receptor Conformations in Virtual Screening. *Journal of Medicinal Chemistry*, *47*(21), 5076–5084. doi:10.1021/jm049756p PMID:15456251

Ferreira, L. G., Dos Santos, R. N., Oliva, G., & Andricopulo, A. D. (2015). Molecular docking and structure-based drug design strategies. *Molecules (Basel, Switzerland)*, *20*(7), 13384–13421. doi:10.3390/molecules200713384 PMID:26205061

Ferreira, R. S., Simeonov, A., Jadhav, A., Eidam, O., Mott, B. T., Keiser, M. J., & Shoichet, B. K. et al. (2010). Complementarity between a docking and a high-throughput screen in discovering new cruzain inhibitors. *Journal of Medicinal Chemistry*, *53*(13), 4891–4905. doi:10.1021/jm100488w PMID:20540517

Finkelstein, A. V., & Janin, J. l. (1989). The price of lost freedom: Entropy of bimolecular complex formation. *Protein Engineering*, *3*(1), 1–3. PMID:2813338

Fischer, E. (1894). Einfluss der Configuration auf die Wirkung der Enzyme. *Berichte der deutschen chemischen Gesellschaft, 27*(3), 2985-2993.

Fischer, D., Norel, R., Wolfson, H., & Nussinov, R. (1993). Surface motifs by a computer vision technique: Searches, detection, and implications for protein-ligand recognition. *Proteins*, *16*(3), 278–292. doi:10.1002/prot.340160306 PMID:8394000

Fischer, E. (1894). Einfluss der configuration auf die wirkungderenzyme. *Berichte der Deutschen Chemischen Gesellschaft*, *27*(3), 2985–2993. doi:10.1002/cber.18940270364

Fraczek, T., Siwek, A., & Paneth, P. (2013). Assessing molecular docking tools for relative biological activity prediction: A case study of triazole HIV-1 NNRTIs. *Journal of Chemical Information and Modeling*, *53*(12), 3326–3342. doi:10.1021/ci400427a PMID:24266618

Francoijs, C. J., Klomp, J. P., & Knegtel, R. M. (2000). Sequence annotation of nuclear receptor ligand-binding domains by automated homology modeling. *Protein Engineering*, *13*(6), 391–394. doi:10.1093/protein/13.6.391 PMID:10877848

Frauenfelder, H., Sligar, S. G., & Wolynes, P. G. (1991). The energy landscapes and motions of proteins. *Science*, *254*(5038), 1598–1603. doi:10.1126/science.1749933 PMID:1749933

Friesner, R. A., Banks, J. L., Murphy, R. B., Halgren, T. A., Klicic, J. J., & Mainz, D. T. et al. others. (2004). Glide: A new approach for rapid, accurate docking and scoring. 1. method and assessment of docking accuracy. *Journal of Medicinal Chemistry*, *47*(7), 1739–1749. PMID:15027865

Friesner, R. A., Banks, J. L., Murphy, R. B., Halgren, T. A., Klicic, J. J., Mainz, D. T., & Perry, J. K. et al. (2004). Glide: A new approach for rapid, accurate docking and scoring. 1. Method and assessment of docking accuracy. *Journal of Medicinal Chemistry*, *47*(7), 1739–1749. doi:10.1021/jm0306430

Friesner, R. A., Murphy, R. B., Repasky, M. P., Frye, L. L., Greenwood, J. R., Halgren, T. A., & Mainz, D. T. et al. (2006). Extra precision glide: Docking and scoring incorporating a model of hydrophobic enclosure for protein-ligand complexes. *Journal of Medicinal Chemistry*, *49*(21), 6177–6196. doi:10.1021/jm051256o

Frishman, D., & Argos, P. (1995). Knowledge-based protein secondary structure assignment. *Proteins*, *23*(4), 566–579. doi:10.1002/prot.340230412 PMID:8749853

Fuhrmann, J., Rurainski, A., Lenhof, H.-P., & Neumann, D. (2010). A new Lamarckian genetic algorithm for flexible ligand-receptor docking. *Journal of Computational Chemistry*, *31*(9), 1911–1918. PMID:20082382

Fukunishi, Y. (2010). Post Processing of Protein-Compound Docking for Fragment-Based Drug Discovery (FBDD): In-Silico Structure-Based Drug Screening and Ligand-Binding Pose Prediction. *Current Topics in Medicinal Chemistry*, *10*(6), 680–694. doi:10.2174/156802610791111452 PMID:20337587

Gabb, A. H. A., Jackson, R. M., & Sternberg, M. J. E. (1997). Modeling protein docking using shape complementarity, electrostatics and biochemical information. *Journal of Molecular Biology*, *272*(1), 106–120. doi:10.1006/jmbi.1997.1203

Gabel, J., Desaphy, J., & Rognan, D. (2014). Beware of machine learning-based scoring functionsr on the danger of developing black boxes. *Journal of Chemical Information and Modeling*, *54*(10), 2807–2815. PMID:25207678

Gajda, M. J., Tuszynska, I., Kaczor, M., Bakulina, A. Y., & Bujnicki, J. M. (2010). FILTREST3D: Discrimination of structural models using restraints from experimental data. *Bioinformatics (Oxford, England)*, *26*(23), 2986–2987. doi:10.1093/bioinformatics/btq582 PMID:20956242

Gane, P. J., & Dean, P. M. (2000). Recent advances in structure-based rational drug design. *Current Opinion in Structural Biology*, *10*(4), 401–404. doi:10.1016/S0959-440X(00)00105-6 PMID:10981625

Gani, O. A. (2007). Signposts of docking and scoring in drug design. *Chemical Biology & Drug Design*, *70*(4), 360–365. doi:10.1111/j.1747-0285.2007.00571.x PMID:17937781

Gao, M., & Skolnick, J. (2009). From nonspecific DNA-protein encounter complexes to the prediction of DNA-protein interactions. *PLoS Computational Biology*, *5*(3), e1000341. doi:10.1371/journal.pcbi.1000341 PMID:19343221

Gao, Z., Li, H., Zhang, H., Liu, X., Kang, L., Luo, X., & Jiang, H. et al. (2008). PDTD: A web-accessible protein database for drug target identification. *BMC Bioinformatics*, *9*(1), 104. doi:10.1186/1471-2105-9-104 PMID:18282303

Garcia-Godoy, M. J., Lopez-Camacho, E., Garcia-Nieto, J., & Aldana-Montes, A. J. (2015). Solving molecular docking problems with multi-objective metaheuristics. *Molecules (Basel, Switzerland)*, *20*(6), 10154–10183. doi:10.3390/molecules200610154 PMID:26042856

Gardiner, E. J., Willett, P., & Artymiuk, P. J. (2003). GAPDOCK: A Genetic Algorithm Approach to Protein Docking in CAPRI round 1. *Proteins*, *52*(1), 10–14. doi:10.1002/prot.10386 PMID:12784360

Garzon, J. I., Lopez-Blanco, J. R., Pons, C., Kovacs, J., Abagyan, R., Fernandez-Recio, J., & Chacon, P. (2009). FRODOCK: A new approach for fast rotational protein-protein docking. *Bioinformatics (Oxford, England)*, *25*(19), 2544–2551. doi:10.1093/bioinformatics/btp447 PMID:19620099

Ghanemi, A. (2015). Targeting G protein coupled receptor-related pathways as emerging molecular therapies. *Saudi Pharmaceutical Journal*, *23*(2), 115–129. doi:10.1016/j.jsps.2013.07.007 PMID:25972730

Ghoorah, A. W., Devignes, M. D., Smail-Tabbone, M., & Ritchie, D. W. (2011). Spatial clustering of protein binding sites for template based protein docking. *Bioinformatics (Oxford, England)*, *27*(20), 2820–2827. doi:10.1093/bioinformatics/btr493 PMID:21873637

Gilad, Y., & Senderowitz, H. (2014). Docking Studies on DNA Intercalators. *Journal of Chemical Information and Modeling*, *54*(1), 96–107. doi:10.1021/ci400352t PMID:24303988

Ginalski, K., Rychlewski, L., Baker, D., & Grishin, N. V. (2004). Protein structure prediction for the male-specific region of the human Y chromosome. *Proc Natl Acad Sci U S A, 101*(8), 2305-2310. doi: 101/8/2305

gkq535 [pii]

Glaser, F., Morris, R. J., Najmanovich, R. J., Laskowski, R. A., & Thornton, J. M. (2006). A method for localizing ligand binding pockets in protein structures. *Proteins*, *62*(2), 479–488. doi:10.1002/prot.20769 PMID:16304646

Glenner, G. G., & Wong, C. W. (1984). Alzheimer's disease: Initial report of the purification and Characterization of a novel cerebrovascular Amyloid protein. *Biochemical and Biophysical Research Communications*, *120*(3), 885–890. doi:10.1016/S0006-291X(84)80190-4 PMID:6375662

Gógl, G., Schneider, K. D., Yeh, B. J., Alam, N., Nguyen Ba, A. N., Moses, A. M., & Weiss, E. L. et al. (2015). The Structure of an NDR/LATS Kinase–Mob Complex Reveals a Novel Kinase–Coactivator System and Substrate Docking Mechanism. *PLoS Biology*, *13*(5), e1002146. doi:10.1371/journal.pbio.1002146 PMID:25966461

Gogos, A., Mu, H., Bahna, F., Gomez, C. A., & Shapiro, L. (2003). Crystal structure of YdcE protein from Bacillus subtilis. *Proteins*, *53*(2), 320–322. doi:10.1002/prot.10457 PMID:14517982

Gohlke, H., Hendlich, M., & Klebe, G. (2000). Knowledge-based scoring function to predict protein-ligand interactions. *Journal of Molecular Biology*, *295*(2), 337–356. doi:10.1006/jmbi.1999.3371 PMID:10623530

Gohlke, H., & Klebe, G. (2001). Statistical potentials and scoring functions applied to protein–ligand binding. *Current Opinion in Structural Biology*, *11*(2), 231–235. doi:10.1016/S0959-440X(00)00195-0

Gohlke, H., & Klebe, G. (2002). Approaches to the description and prediction of the binding affinity of smallmolecule ligands to macromolecular receptors. *Angewandte Chemie International Edition in English*, *41*(15), 2644–2676. doi:10.1002/1521-3773(20020802)41:15<2644::AID-ANIE2644>3.0.CO;2-O PMID:12203463

Gohlke, H., & Klebe, G. (2002). DrugScore meets CoMFA: Adaptation of fields for molecular comparison (AFMoC) or how to tailor knowledge-based pair-potentials to a particular protein. *Journal of Medicinal Chemistry*, *45*(19), 4153–4170. doi:10.1021/jm020808p PMID:12213058

Gonzalez-Ruiz, D., & Gohlke, H. (2009). Steering protein–ligand dockingwith quantitative NMR chemical shift perturbations. *Journal of Chemical Information and Modeling*, *49*(10), 2260–2271. doi:10.1021/ci900188r PMID:19795907

Goodford, P. J. (1985). A computational procedure for determining energetically favorable binding sites on biologically important macromolecules. *Journal of Medicinal Chemistry*, *28*(7), 849–857. doi:10.1021/jm00145a002 PMID:3892003

Goodsell, D. S., Lauble, H., Stout, C. D., & Olson, A. J. (1993). Automated docking in crystallography: Analysis of the substrates of aconitase. *Proteins*, *17*(1), 1–10. doi:10.1002/prot.340170104 PMID:8234239

Goodsell, D. S., Morris, G. M., & Olson, A. J. (1996). Automated docking of flexible ligands: Applications of autodock. *Journal of Molecular Recognition*, *9*(1), 1–5. PMID:8723313

Goodsell, D. S., Morris, G. M., & Olson, A. J. (1996). Automated docking of flexible ligands: Applications of AutoDock. *Journal of Molecular Recognition: JMR*, *9*(1), 1–5. doi:10.1002/(SICI)1099-1352(199601)9:1<1::AID-JMR241>3.0.CO;2-6 PMID:8723313

Goodsell, S., & Olson, A. J. (1990). Automated docking of substrates to proteins by simulated annealing. *Proteins*, *8*(3), 195–202. doi:10.1002/prot.340080302

Gouras, G. K., Xu, H., Jovanovic, J. N., Buxbaum, J. D., Wang, R., Greengard, P., & Gandy, S. et al. (1998). Generation and regulation of beta-amyloid peptide variants by neurons. *Journal of Neurochemistry*, *71*(5), 1920–1925. doi:10.1046/j.1471-4159.1998.71051920.x PMID:9798916

Gowthaman, R., Deeds, E. J., & Karanicolas, J. (2013). Structural properties of non-traditional drug targets present new challenges for virtual screening. *Journal of Chemical Information and Modeling*, *53*(8), 2073–2081. doi:10.1021/ci4002316

Grant, J. A., Pickup, B. T., & Nicholls, A. (2001). A smooth permittivity function for Poisson–Boltzmann solvation methods. *Journal of Computational Chemistry*, *22*(6), 608–640. doi:10.1002/jcc.1032

Gray, J. J., Moughon, S., Wang, C., Schueler-Furman, O., Kuhlman, B., Rohl, C., & Baker, D. (2003). Protein–Protein Docking with Simultaneous Optimization of Rigid-body Displacement and Side-chain Conformations. *Journal of Molecular Biology*, *331*(1), 281–299. doi:10.1016/S0022-2836(03)00670-3 PMID:12875852

Grinter, S. Z., Liang, Y., Huang, S.-Y., Hyder, S. M., & Zou, X. (2011). An inverse docking approach for identifying new potential anti-cancer targets. *Journal of Molecular Graphics & Modelling*, *29*(6), 795–799. doi:10.1016/j.jmgm.2011.01.002 PMID:21315634

Grinter, S. Z., & Zou, X. (2014). Challenges, Applications, and Recent Advances of Protein-Ligand Docking in Structure-Based Drug Design. *Molecules (Basel, Switzerland)*, *19*(7), 10150–10176. doi:10.3390/molecules190710150

Grosdidier, A., Zoete, V., & Michielin, O. (2011b). SwissDock, a protein-small molecule docking web service based on EADock DSS. *Nucleic Acids Research, 39*, W270–7. http://doi.org/10.1093/nar/gkr366

Grosdidier, A., Zoete, V., & Michielin, O. (2007). EADock: A multiobjective evolutionary framework for the docking of small molecules into protein active site. *Proteins: Structure, Function, and Bioinformatics*, *67*(4), 1010–1025. doi:10.1002/prot.21367

Grosdidier, A., Zoete, V., & Michielin, O. (2009). Blind docking of 260 protein-ligand complexes with EADock 2.0. *Journal of Computational Chemistry*, *00*, 1–10. PMID:19130502

Grosdidier, A., Zoete, V., & Michielin, O. (2011). SwissParam: A fast force field generation tool for small organic molecule. *Journal of Computational Chemistry*, *32*(11), 2359–2368. doi:10.1002/jcc.21816 PMID:21541964

Grosdidier, A., Zoete, V., & Michielin, O. (2011a). Fast docking using the CHARMM force field with EADock DSS. *Journal of Computational Chemistry*, *32*(10), 2149–2159. doi:10.1002/jcc.21797 PMID:21541955

Grosdidier, S., & Fernández-Recio, J. (2012). Protein-protein docking and hot-spot prediction for drug discovery. *Current Pharmaceutical Design*, *18*(30), 4607–4618. doi:10.2174/138161212802651599 PMID:22650255

Gschwend, D. A., Good, A. C., & Kuntz, I. D. Molecular docking towards drug discovery. *Journal of Molecular Recognition : JMR*, *9*(2), 175–86. http://doi.org/10.1002/(SICI)1099-1352(199603)9:2<175::AID-JMR260>3.0.CO;2-D

Gschwend, D. A., Good, A. C., & Kuntz, I. D. (1996). Molecular docking towards drug discovery. *Journal of Molecular Recognition*, *9*(2), 175–186. doi:10.1002/(SICI)1099-1352(199603)9:2<175::AID-JMR260>3.0.CO;2-D

Guex, N., & Peitsch, M. C. (1997). SWISS-MODEL and the Swiss-PdbViewer: An environment for comparative protein modeling. *Electrophoresis*, *18*(15), 2714–2723. doi:10.1002/elps.1150181505 PMID:9504803

Gu, J., Yang, X., Kang, L., Wu, J., & Wang, X. (2015). MoDock: A multi-objective strategy improves the accuracy for molecular docking. *Algorithms for Molecular Biology; AMB*, *10*(1), 8. doi:10.1186/s13015-015-0034-8 PMID:25705248

Guo, F., Li, S. C., Ma, W., & Wang, L. (2013). Detecting protein conformational changes in interactions via scaling known structures. Lecture Notes in Computer Science, 7821(10), 58–74.

Guo, L., Yan, Z., Zheng, X., Hu, L., Yang, Y., & Wang, J. (2014). A comparison of various optimization algorithms of protein–ligand docking programs by fitness accuracy. *Journal of Molecular Modeling*, *20*(7), 1–10. PMID:24935106

Guo, W., Wisniewski, J. A., & Ji, H. (2014). Hot spot-based design of small-molecule inhibitors for protein–protein interactions. *Bioorganic & Medicinal Chemistry Letters*, *24*(11), 2546–2554. doi:10.1016/j.bmcl.2014.03.095 PMID:24751445

Gupta, A., Gandhimathi, A., Sharma, P., & Jayaram, B. (2007). ParDOCK: An all atom energy based Monte Carlo docking protocol for protein-ligand complexes. *Protein and Peptide Letters*, *14*(7), 632–646. doi:10.2174/092986607781483831 PMID:17897088

Gutiérrez-de Terán, H., & Aqvist, J. (2012). Linear interaction energy: method and applications in drug design. In *Computational drug discovery and design* (pp. 305–323). Springer.

Guvench, O., & Brooks, C. L. (2004). Efficient approximate all-atom solvent accessible surface area method parameterized for folded and denatured protein conformations. *Journal of Computational Chemistry*, *25*(8), 1005–1014. PMID:15067676

Gu, W. G., Zhang, X., & Yuan, J. F. (2014). Anti-HIV drug development through computational methods. *The AAPS Journal*, *16*(4), 674–680. doi:10.1208/s12248-014-9604-9 PMID:24760437

Halgren, T. (2007). New Method for Fast and Accurate Binding-site Identification and Analysis. *Chemical Biology & Drug Design*, *69*(2), 146–148. doi:10.1111/j.1747-0285.2007.00483.x PMID:17381729

Halgren, T. A. (1996). Merck molecular force field. I. Basis, form, scope, parameterization, and performance of MMFF94. *Journal of Computational Chemistry*, *17*(5-6), 490–519. doi:10.1002/(SICI)1096-987X(199604)17:5/6<490::AID-JCC1>3.0.CO;2-P

Halgren, T. A., Murphy, R. B., Friesner, R. A., Beard, H. S., Frye, L. L., Pollard, W. T., & Banks, J. L. (2004). Glide: A new approach for rapid, accurate docking and scoring. 2. Enrichment factors in database screening. *Journal of Medicinal Chemistry*, *47*(7), 1750–1759. doi:10.1021/jm030644s PMID:15027866

Hall, T. A. (1999). BioEdit: A user-friendly biological sequence alignment editor and analysis program for Windows 95/98/NT. *Nucleic Acids Symposium Series*, *41*, 95–98.

Halperin, I., Ma, B., Wolfson, H., & Nussinov, R. (2002). Principles of docking: An overview of search algorithms and a guide to scoring functions. *Proteins*, *47*, 09–443.

Halperin, I., Ma, B., Wolfson, H., & Nussinov, R. (2002). Principles of docking: An overview of search algorithms and a guide to scoring functions. *Proteins*, *47*(4), 409–443. doi:10.1002/prot.10115 PMID:12001221

Hamzeh-Mivehroud, M., Mahmoudpour, A., & Dastmalchi, S. (2012). Identification of new peptide ligands for epidermal growth factor receptor using phage display and computationally modeling their mode of binding. *Chemical Biology & Drug Design*, *79*(3), 246–259. doi:10.1111/j.1747-0285.2011.01282.x PMID:22136656

Hamzeh-Mivehroud, M., Rahmani, S., Feizi, M. A., Dastmalchi, S., & Rashidi, M. R. (2014). In vitro and in silico studies to explore structural features of flavonoids for aldehyde oxidase inhibition. *Archiv der Pharmazie*, *347*(10), 738–747. doi:10.1002/ardp.201400076 PMID:24956178

Hamzeh-Mivehroud, M., Rahmani, S., Rashidi, M. R., Hosseinpour Feizi, M. A., & Dastmalchi, S. (2013). Structure-based investigation of rat aldehyde oxidase inhibition by flavonoids. *Xenobiotica*, *43*(8), 661–670. doi:10.3109/00498254.2012.755228 PMID:23282065

Hardy, J., & Selkoe, D. J. (2002). The Amyloid hypothesis of Alzheimer's disease: Progress and problems on the road to therapeutics. *Science*, *297*(5580), 353–356. doi:10.1126/science.1072994 PMID:12130773

Harpsoe, K., Liljefors, T., & Balle, T. (2008). Prediction of the binding mode of biarylpropylsulfonamide allosteric AMPA receptor modulators based on docking, GRID molecular interaction fields and 3D-QSAR analysis. *Journal of Molecular Graphics & Modelling*, *26*(5), 874–883. doi:10.1016/j.jmgm.2007.06.002 PMID:17644444

Harris, R. A. (2013). Functional Validation of Virtual Screening for Novel Agents with General Anesthetic Action at Ligand-Gated Ion Channels. *Molecular Pharmacology*, *84*(5), 670–678. doi:10.1124/mol.113.087692 PMID:23950219

Hartshorn, M. J., Verdonk, M. L., Chessari, G., Brewerton, S. C., Mooij, W. T. M., Mortenson, P. N., & Murray, C. W. (2007). Diverse, high-quality test set for the validation of protein-ligand docking performance. *Journal of Medicinal Chemistry*, *50*(4), 726–741. doi:10.1021/jm061277y

Hart, T. N., & Read, R. J. (1992). A multiple-start Monte Carlo docking method. *Proteins*, *13*(3), 206–222. doi:10.1002/prot.340130304 PMID:1603810

Havranek, J. J., & Harbury, P. B. (2003). Automated design of specificity in molecular recognition. *Nature Structural Biology*, *10*(1), 45–52. PMID:12459719

Hawkins, G. D., Cramer, C. J., & Truhlar, D. G. (1995). Pairwise solute descreening of solute charges from a dielectric medium. *Chemical Physics Letters*, *246*(1-2), 122–129. doi:10.1016/0009-2614(95)01082-K

Hawkins, P. C. D., & Nicholls, A. (2012). Conformer generation with OMEGA: Learning from the data set and the analysis of failures. *Journal of Chemical Information and Modeling*, *52*(11), 2919–2936. doi:10.1021/ci300314k

Heidenreich, B., Rachakonda, P. S., Hemminki, K., & Kumar, R. (2014). TERT promoter mutations in cancer development. *Current Opinion in Genetics & Development*, *24*, 30–37. doi:10.1016/j.gde.2013.11.005 PMID:24657534

Hensen, C., Hermann, J. C., Nam, K., Ma, S., Gao, J., & Holtje, H.-D. (2004). A combined QM/MM approach to protein-ligand interactions: Polarization effects of the HIV-1 protease on selected high affinity inhibitors. *Journal of Medicinal Chemistry*, *47*(27), 6673–6680. doi:10.1021/jm0497343

Heo, L., Shin, W. H., Lee, M. S., & Seok, C. (2014). GalaxySite: ligand-binding-site prediction by using molecular docking. *Nucleic Acids Res, 42*, W210-214.

Hermans, J., & Wang, L. (1997). Inclusion of loss of translational and rotational freedom in theoretical estimates of free energies of binding. application to a complex of benzene and mutant t4 lysozyme. *Journal of the American Chemical Society*, *119*(11), 2707–2714.

Hernández-santoyo, A., Tenorio-barajas, A. Y., Altuzar, V., Vivanco-cid, H., & Mendoza-barrera, C. (2013). Protein-Protein and Protein-Ligand Docking, Protein Engineering - Technology and Application (D. T. Ogawa, Ed.). Academic Press.

Hernández-Santoyo, A., Tenorio-Barajas, A. Y., Altuzar, V., Vivanco-Cid, H., & Mendoza-Barrera, C. (2013). *Protein-Protein and Protein-Ligand Docking*. Academic Press.

Hernández-Santoyo, A., Tenorio-Barajas, A. Y., Altuzar, V., Vivano-Cid, H. & Mendoza-Barrera, C. (2013). *Protein-Protein and Protein-Ligand Docking,* Croatia, Europe: INTECH

He, S., & Lai, L. (2011). Molecular docking and competitive binding studydiscovered different binding modes of microsomal prostaglandin E synthase-1 inhibitors. *Journal of Chemical Information and Modeling, 51*(12), 3254–3261. doi:10.1021/ci200427k PMID:22077876

Hess, B., Kutzner, C., van der Spoel, D., & Lindahl, E. (2008). GROMACS 4: Algorithms for Highly Efficient, Load-Balanced, and Scalable Molecular Simulation. *Journal of Chemical Theory and Computation, 4*(3), 435–447. doi:10.1021/ct700301q PMID:26620784

Heusser, S. A., Howard, R. J., Borghese, C. M., Cullins, M. A., Carlsson, J., Hou, T., & Xu, X. (2004). Recent Development and Application of Virtual Screening in Drug Discovery: An Overview. *Current Pharmaceutical Design, 10*(9), 1011–1033. doi:10.2174/1381612043452721 PMID:15078130

Hoi, P. M., Li, S., Vong, C. T., Tseng, H. H. L., Kwan, Y. W., & Lee, S. M.-Y. (2015). Recent advances in structure-based drug design and virtual screening of VEGFR tyrosine kinase inhibitors. *Methods (San Diego, Calif.), 71*, 85–91. doi:10.1016/j.ymeth.2014.09.004 PMID:25239735

Holt, P. A., Chaires, J. B., & Trent, J. O. (2008). Molecular docking of intercalators and groove-binders to nucleic acids using Autodock and Surflex. *Journal of Chemical Information and Modeling, 48*(8), 1602–1615. doi:10.1021/ci800063v

Homeyer, N., & Gohlke, H. (2012). Free Energy Calculations by the Molecular Mechanics Poisson−Boltzmann Surface Area Method. *Molecular Informatics, 31*(2), 114–122. doi:10.1002/minf.201100135

Hong, Y., Chintapalli, S. V., Ko, K. D., Bhardwaj, G., Zhang, Z., van Rossum, D., & Patterson, R. L. (2011). Predicting protein folds with fold-specific PSSM libraries. *PLoS ONE, 6*(6), e20557. doi:10.1371/journal.pone.0020557 PMID:21698189

Höppner, F. (1999). *Fuzzy cluster analysis: methods for classification, data analysis and image recognition.* John Wiley & Sons.

Houston, D. R., & Walkinshaw, M. D. (2013). Consensus docking: Improving the reliability of docking in a virtual screening context. *Journal of Chemical Information and Modeling, 53*(2), 384–390. doi:10.1021/ci300399w PMID:23351099

Hou, T., Qiao, X., Zhang, W., & Xu, X. (2002). Empirical aqueous solvation models based on accessible surface areas with implicit electrostatics. *The Journal of Physical Chemistry B, 106*(43), 11295–11304.

Hou, T., Wang, J., Chen, L., & Xu, X. (1999). Automated docking of peptides and proteins by using a genetic algorithm combined with a tabu search. *Protein Engineering, 12*(8), 639–648. doi:10.1093/protein/12.8.639 PMID:10469824

Hou, T., Wang, J., Li, Y., & Wang, W. (2011). Assessing the Performance of the MM/PBSA and MM/GBSA Methods. 1. The Accuracy of Binding Free Energy Calculations Based on Molecular Dynamics Simulations. *Journal of Chemical Information and Modeling, 51*(1), 69–82. doi:10.1021/ci100275a PMID:21117705

Hou, T., Zhang, W., Huang, Q., & Xu, X. (2005). An extended aqueous solvation model based on atom-weighted solvent accessible surface areas: SAWSA v2. 0 model. *Journal of Molecular Modeling, 11*(1), 26–40. PMID:15565273

Hruby, V. J., al-Obeidi, F., & Kazmierski, W. (1990). Emerging approaches in the molecular design of receptor-selective peptide ligands: Conformational, topographical and dynamic considerations. *The Biochemical Journal, 268*(2), 249–262. doi:10.1042/bj2680249 PMID:2163604

Hsin, K. Y., Ghosh, S., & Kitano, H. (2013). Combining machine learning systems and multiple docking simulation packages to improve docking prediction reliability for network pharmacology. *PLoS ONE*, *8*(12), e83922. doi:10.1371/journal.pone.0083922 PMID:24391846

Hsu, R. L., Lee, K. T., Wang, J. H., Lily, Y., Lee, L., Rita, P., & Chen, Y. (2009). Amyloid-degrading ability of nattokinase from *Bacillus subtilis* Natto. *Journal of Agricultural and Food Chemistry*, *57*(2), 503–508. doi:10.1021/jf803072r PMID:19117402

Hu, X., Eberhart, R., & Shi, Y. (2004). *Recent advances in particle swarm.* IEEE Congress on Evolutionary Computation 2004, Portland, OR.

Huang, T.-T., Hwang, J.-K., Chen, C.-H., Chu, C.-S., Lee, C.-W., & Chen, C.-C. (2015). (PS)2: protein structure prediction server version 3.0. *Nucleic Acids Research, 454*(Ext 56921), 1–5.

Huang, N., Kalyanaraman, C., Bernacki, K., & Jacobson, M. P. (2006). Molecular mechanics methods for predicting protein-ligand binding. *Physical Chemistry Chemical Physics*, *8*(44), 5166–5177. doi:10.1039/B608269F

Huang, N., Shoichet, B. K., & Irwin, J. J. (2006). Benchmarking sets for molecular docking. *Journal of Medicinal Chemistry*, *49*(23), 6789–6801. PMID:17154509

Huang, S. (2015). Exploring the potential of global protein-protein docking: An overview and critical assessment of current programs for automatic ab initio docking. *Drug Discovery Today*, *20*(8), 969–977. doi:10.1016/j.drudis.2015.03.007 PMID:25801181

Huang, S. Y. (2014). Search strategies and evaluation in protein-protein docking: Principles, advances and challenges. *Drug Discovery Today*, *19*(8), 1081–1096. doi:10.1016/j.drudis.2014.02.005 PMID:24594385

Huang, S. Y., Grinter, S. Z., & Zou, X. (2010). Scoring functions and their evaluation methods for protein-ligand docking: Recent advances and future directions. *Physical Chemistry Chemical Physics*, *12*(40), 12899–12908. doi:10.1039/c0cp00151a PMID:20730182

Huang, S. Y., & Zou, X. (2010a). Advances and challenges in protein-ligand docking. *International Journal of Molecular Sciences*, *11*(8), 3016–3034. doi:10.3390/ijms11083016 PMID:21152288

Huang, S. Y., & Zou, X. (2010b). Inclusion of solvation and entropy in the knowledge-based scoring function for protein-ligand interactions. *Journal of Chemical Information and Modeling*, *50*(2), 262–273. doi:10.1021/ci9002987 PMID:20088605

Huang, S. Y., & Zou, X. (2011). Construction and test of ligand decoy sets using MDock: Community structure-activity resource benchmarks for binding mode prediction. *Journal of Chemical Information and Modeling*, *51*(9), 2107–2114. doi:10.1021/ci200080g PMID:21755952

Huang, S.-Y., Grinter, S. Z., & Zou, X. (2010). Scoring functions and their evaluation methods for protein–ligand docking: Recent advances and future directions. *Physical Chemistry Chemical Physics*, *12*(40), 12899–12908. PMID:20730182

Huang, S.-Y., & Zou, X. (2006). An iterative knowledge-based scoring function to predict protein–ligand interactions: I. Derivation of interaction potentials. *Journal of Computational Chemistry*, *27*(15), 1866–1875. doi:10.1002/jcc.20504

Huang, S.-Y., & Zou, X. (2006). An iterative knowledge-based scoring function to predict protein–ligand interactions: I. derivation of interaction potentials. *Journal of Computational Chemistry*, *27*(15), 1866–1875. PMID:16983673

Huang, S.-Y., & Zou, X. (2007). Efficient molecular docking of NMR structures: Application to HIV-1 protease. *Protein Science*, *16*(1), 43–51. doi:10.1110/ps.062501507 PMID:17123961

Huang, S.-Y., & Zou, X. (2007). Ensemble Docking of Multiple Protein Structures: Considering Protein Structural Variations in Molecular Docking. *PROTEINS: Structure, Function, and Bioinformatics*, *66*(2), 399–421. doi:10.1002/prot.21214 PMID:17096427

Huang, S.-Y., & Zou, X. (2014). A knowledge-based scoring function for protein-RNA interactions derived from a statistical mechanics-based iterative method. *Nucleic Acids Research*, *42*(7), e55–e55. doi:10.1093/nar/gku077

Huey, R., Morris, G. M., Olson, A. J., & Goodsell, D. S. (2007). A semiempirical free energy force field with charge-based desolvation. *Journal of Computational Chemistry*, *28*(6), 1145–1152. doi:10.1002/jcc.20634

Humphrey, W., Dalke, A., & Schulten, K. (1996). VMD - Visual Molecular Dynamics. *Journal of Molecular Graphics*, *14*(1), 33–38. doi:10.1016/0263-7855(96)00018-5 PMID:8744570

Hung, C. L., & Chen, C. C. (2014). Computational approaches for drug discovery. *Drug Development Research*, *75*(6), 412–418. doi:10.1002/ddr.21222 PMID:25195585

Hwang, H., Vreven, T., Pierce, B. G., Hung, J. H., & Weng, Z. (2010). Performance of ZDOCK and ZRANK in CAPRI rounds 13-19. *Proteins*, *78*(15), 3104–3110. doi:10.1002/prot.22764 PMID:20936681

Ian, I.-F., Pablo, C. V. J., Normande, C.-I., & Absalom, Z.-C. et al. (2013). Identification of pharmacological targets combining docking and molecular dynamics simulations. *American Journal of Agricultural and Biological Sciences*, *8*(1), 89–106. doi:10.3844/ajabssp.2013.89.106

Irwin, J. J. (2008). Community benchmarks for virtual screening. *Journal of Computer-Aided Molecular Design*, *22*(3-4), 193–199. PMID:18273555

Irwin, J. J., & Shoichet, B. K. (2005). ZINC--a free database of commercially available compounds for virtual screening. *Journal of Chemical Information and Modeling*, *45*(1), 177–182. doi:10.1021/ci049714+ PMID:15667143

Irwin, J. J., Shoichet, B. K., Mysinger, M. M., Huang, N., Colizzi, F., Wassam, P., & Cao, Y. (2009). Automated docking screens: A feasibility study. *Journal of Medicinal Chemistry*, *52*(18), 5712–5720. doi:10.1021/jm9006966 PMID:19719084

Irwin, J. J., Sterling, T., Mysinger, M. M., Bolstad, E. S., & Coleman, R. G. (2012). ZINC: A Free Tool to Discover Chemistry for Biology. *Journal of Chemical Information and Modeling*, *52*(7), 1757–1768. doi:10.1021/ci3001277 PMID:22587354

Ishchenko, A. V., & Shakhnovich, E. I. (2002). Small molecule growth 2001 (SMoG2001): An improved knowledge-based scoring function for protein-ligand interactions. *Journal of Medicinal Chemistry*, *45*(13), 2770–2780. doi:10.1021/jm0105833

Istyastono, E. P., Kooistra, A. J., Vischer, H. F., Kuijer, M., Roumen, L., Nijmeijer, S., & de Graaf, C. et al. (2015). Structure-based virtual screening for fragment-like ligands of the G protein-coupled histamine H4 receptor. *MedChemComm*, *6*(6), 1003–1017. doi:10.1039/C5MD00022J

Iwata, N., Tsubuki, S., Takaki, Y., Shirotani, K., Lu, B., Gerard, N. P., & Saido, T. C. et al. (2001). Metabolic regulation of brain Aβ by neprilysin. *Science*, *292*(5521), 1550–1552. doi:10.1126/science.1059946 PMID:11375493

Jackson, R. M., Gabb, H. A., & Sternberg, M. J. (1998). Rapid refinement of protein interfaces incorporating solvation: Application to the docking problem. *Journal of Molecular Biology*, *276*(1), 265–285. doi:10.1006/jmbi.1997.1519 PMID:9514726

Jahn, A., Hinselmann, G., Fechner, N., & Zell, A. (2009). Optimal assignment methods for ligand-based virtual screening. *Journal of Cheminformatics*, *1*(1), 14. doi:10.1186/1758-2946-1-14 PMID:20150995

Jain, A. (2008). Bias, reporting, and sharing: Computational evaluations of docking methods. *Journal of Computer-Aided Molecular Design*, *22*(3-4), 201–212. doi:10.1007/s10822-007-9151-x

Jain, A. N. (2003). Surflex: Fully automatic flexible molecular docking using a molecular similarity-based search engine. *Journal of Medicinal Chemistry*, *46*(4), 499–511. doi:10.1021/jm020406h PMID:12570372

Jain, A. N. (2006). Scoring functions for protein-ligand docking. *Current Protein & Peptide Science*, *7*(5), 407–420. doi:10.2174/138920306778559395 PMID:17073693

Jain, A. N. (2007). Surflex-Dock 2.1: Robust performance from ligand energetic modeling, ring flexibility, and knowledge-based search. *Journal of Computer-Aided Molecular Design*, *21*(5), 281–306. doi:10.1007/s10822-007-9114-2 PMID:17387436

Jain, E., Bairoch, A., Duvaud, S., Phan, I., Redaschi, N., Suzek, B. E., & Gasteiger, E. et al. (2009). Infrastructure for the life sciences: Design and implementation of the UniProt website. *BMC Bioinformatics*, *10*(1), 136. doi:10.1186/1471-2105-10-136 PMID:19426475

Jalkute, C. B., Barage, S. H., Dhanavade, M. J., & Sonawane, K. D. (2014). Identification of Angiotensin Converting Enzyme Inhibitor: An InSilico Perspective. *International Journal of Peptide Research and Therapeutics*, *21*(1), 107–115. doi:10.1007/s10989-014-9434-8

Jalkute, C. B., Barage, S. H., & Sonawane, K. D. (2015). Insight into molecular interactions of Aβ peptide and gelatinase from *Enterococcus faecalis*: A molecular modeling approach. *RSC Advances*, *5*(14), 10488–10496. doi:10.1039/C4RA09354B

Jalkute, C. B., & Sonawane, K. D. (2015). Evaluation of a possible role of *Stigmatella aurantiaca* ACE in Aβ peptide degradation: A molecular modeling approach. *Journal of Molecular Microbiology and Biotechnology*, *25*(1), 26–36. doi:10.1159/000370114 PMID:25677850

Janin, J. (1995). Principles of protein-protein recognition from structure to thermodynamics. *Biochimie*, *77*(7), 497–505. PMID:8589061

Janin, J. (1996). Quantifying biological specificity: The statistical mechanics of molecular recognition. *Proteins: Structure, Function, and Bioinformatics*, *25*(4), 438–445. PMID:8865339

Janin, J. (2013). The targets of CAPRI rounds 20-27. *Proteins: Structure, Function, and Bioinformatics*, *81*(12), 2075–2081. doi:10.1002/prot.24375

Janin, J., Henrick, K., Moult, J., Eyck, L., Ten, , Sternberg, M. J. E., & Vajda, S. (2003). CAPRI: A critical assessment of PRedicted interactions. *Proteins*, *52*(1), 2–9. doi:10.1002/prot.10381 PMID:12784359

Janin, J., Wodak, S. J., Lensink, M. F., & Velankar, S. (2015). Assessing Structural Predictions of Protein–Protein Recognition: The CAPRI Experiment. *Reviews in Computational Chemistry*, *28*, 137–173.

Jardin, C., Stefani, A. G., Eberhardt, M., Huber, J. B., & Sticht, H. (2013). An information-theoretic classification of amino acids for the assessment of interfaces in protein-protein docking. *Journal of Molecular Modeling*, *19*(9), 3901–3910. doi:10.1007/s00894-013-1916-7 PMID:23828247

Jayaram, B., Singh, T., Mukherjee, G., Mathur, A., Shekhar, S., & Shekhar, V. (2012). Sanjeevini: A freely accessible web-server for target directed lead molecule discovery. *BMC Bioinformatics*, *13*(Suppl 17), S7. doi:10.1186/1471-2105-13-S17-S7 PMID:23282245

Jenkins, J. L., Kao, R. Y. T., & Shapiro, R. (2003). Virtual screening to enrich hit lists from high-throughput screening: A case study on small-molecule inhibitors of angiogenin. *Proteins*, *50*(1), 81–93. doi:10.1002/prot.10270 PMID:12471601

Jerabek, P., Florian, J., Stiborova, M., & Martinek, V. (2014). Flexible docking-based molecular dynamics/steered molecular dynamics calculations of protein-protein contacts in a complex of cytochrome P450 1A2 with cytochrome b5. *Biochemistry*, *53*(42), 6695–6705. doi:10.1021/bi500814t PMID:25313797

Jiang, F., & Kim, S. H. (1991). "Soft docking": Matching of molecular surface cubes. *Journal of Molecular Biology*, *219*(1), 79–102. doi:10.1016/0022-2836(91)90859-5 PMID:2023263

Jimenez-Garcia, B., Pons, C., & Fernandez-Recio, J. (2013). pyDockWEB: A web server for rigid-body protein-protein docking using electrostatics and desolvation scoring. *Bioinformatics (Oxford, England)*, *29*(13), 1698–1699. doi:10.1093/bioinformatics/btt262 PMID:23661696

Jimenez-Garcia, B., Pons, C., Svergun, D. I., Bernado, P., & Fernandez-Recio, J. (2015). pyDockSAXS: Protein-protein complex structure by SAXS and computational docking. *Nucleic Acids Research*, *43*(W1), W356-361. doi:10.1093/nar/gkv368 PMID:25897115

jmbi.2000.3741 [pii]

Jones, D. T. (1999). GenTHREADER: an efficient and reliable protein fold recognition method for genomic sequences. *J Mol Biol, 287*(4), 797-815. doi: S0022-2836(99)92583-4

Jones, G., & Willett, P. (1995). Docking small-molecule ligands into active sites. *Current Opinion in Biotechnology*, *6*(6), 652–656. doi:10.1016/0958-1669(95)80107-3

Jones, G., Willett, P., & Glen, R. C. (1995). Molecular recognition of receptor sites using a genetic algorithm with a description of desolvation. *Journal of Molecular Biology*, *245*(1), 43–53. doi:10.1016/S0022-2836(95)80037-9 PMID:7823319

Jones, G., Willett, P., Glen, R. C., Leach, A. R., & Taylor, R. (1997). Development and validation of a genetic algorithm for flexible docking. *Journal of Molecular Biology*, *267*(3), 727–748. doi:10.1006/jmbi.1996.0897 PMID:9126849

Jones, J. E. (1924). On the Determination of Molecular Fields. II. From the Equation of State of a Gas. *Proceedings - Royal Society. Mathematical, Physical and Engineering Sciences*, *106*(738), 463–477. doi:10.1098/rspa.1924.0082

Jorgensen, W. L. (2004). The many roles of computation in drug discovery. *Science*, *303*(5665), 1813–1818. doi:10.1126/science.1096361 PMID:15031495

Jorgensen, W. L., & Tirado-Rives, J. (1988). The OPLS [optimized potentials for liquid simulations] potential functions for proteins, energy minimizations for crystals of cyclic peptides and crambin. *Journal of the American Chemical Society*, *110*(6), 1657–1666. doi:10.1021/ja00214a001

Jug, G., Anderluh, M., & Tomašič, T. (2015). Comparative evaluation of several docking tools for docking small molecule ligands to DC-SIGN. *Journal of Molecular Modeling*, *21*(6), 164–176. doi:10.1007/s00894-015-2713-2

Kalé, L., Skeel, R., Bhandarkar, M., Brunner, R., Gursoy, A., Krawetz, N., & Schulten, K. (1999). NAMD2: Greater Scalability for Parallel Molecular Dynamics. *Journal of Computational Physics*, *151*(1), 283–312. doi:10.1006/jcph.1999.6201

Kang, L., Li, H., Jiang, H., & Wang, X. (2009). An improved adaptive genetic algorithm for protein–ligand docking. *Journal of Computer-Aided Molecular Design*, *23*(1), 1–12. doi:10.1007/s10822-008-9232-5 PMID:18777161

Kann, M. G. (2007). Protein interactions and disease: Computational approaches to uncover the etiology of diseases. *Briefings in Bioinformatics*, *8*(5), 333–346. doi:10.1093/bib/bbm031 PMID:17638813

Kantardjiev, A. A. (2011). GPU.proton.DOCK: Genuine Protein Ultrafast proton equilibria consistent DOCKing. *Nucleic Acids Res, 39*, W223-228. doi: 10.1093/nar/gkr412

Kantardjiev, A. A. (2012). Quantum.Ligand.Dock: protein-ligand docking with quantum entanglement refinement on a GPU system. *Nucleic Acids Res, 40*, W415-422. doi: 10.1093/nar/gks515

Kapetanovic, I. M. (2008). Computer-aided drug discovery and development (CADDD): In silico-chemico-biological approach. *Chemico-Biological Interactions, 171*(2), 165–176. doi:10.1016/j.cbi.2006.12.006 PMID:17229415

Karpen, M. E., Tobias, D. J., & Brooks, C. L. (1993). Statistical clustering techniques for the analysis of long molecular dynamics trajectories: Analysis of 2.2-ns trajectories of YPGDV. *Biochemistry, 32*(2), 412–420. doi:10.1021/bi00053a005 PMID:8422350

Karplus, M., & Weaver, D. L. (1976). Protein-folding dynamics. *Nature, 260*, 404–406. PMID:1256583

Kastenholz, M. A., Pastor, M., Cruciani, G., Haaksma, E. E., & Fox, T. (2000). GRID/CPCA: A new computational tool to design selective ligands. *Journal of Medicinal Chemistry, 43*(16), 3033–3044. doi:10.1021/jm000934y PMID:10956211

Katchalski-Katzir, E., Shariv, I., Eisenstein, M., Friesem, A. A., Aflalo, C., & Vakser, I. A. (1992). Molecular surface recognition: Determination of geometric fit between proteins and their ligands by correlation techniques. *Proceedings of the National Academy of Sciences of the United States of America, 89*(6), 2195–2199. doi:10.1073/pnas.89.6.2195 PMID:1549581

Katritch, V., Jaakola, V.-P., Lane, J. R., Lin, J., IJzerman, A. P., Yeager, M., & Abagyan, R. et al. (2010). Structure-Based Discovery of Novel Chemotypes for Adenosine A_{2A}Receptor Antagonists. *Journal of Medicinal Chemistry, 53*(4), 1799–1809. doi:10.1021/jm901647p PMID:20095623

Kearsley, S. K., Underwood, D. J., Sheridan, R. P., & Miller, M. D. (1994). Flexibases: A way to enhance the use of molecular docking methods. *Journal of Computer-Aided Molecular Design, 8*(5), 565–582. doi:10.1007/BF00123666

Kelley, L. A., MacCallum, R. M., & Sternberg, M. J. (2000). Enhanced genome annotation using structural profiles in the program 3D-PSSM. *Journal of Molecular Biology, 299*(2), 499–520. doi:10.1006/jmbi.2000.3741 PMID:10860755

Kelley, L. A., Mezulis, S., Yates, C. M., Wass, M. N., & Sternberg, M. J. E. (2015). The Phyre2 web portal for protein modeling, prediction and analysis. *Nature Protocols, 10*(6), 845–858. doi:10.1038/nprot.2015.053 PMID:25950237

Kelly, M. D., & Mancera, R. L. (2004). Expanded interaction fingerprint method for analyzing ligand binding modes in docking and structure-based drug design. *Journal of Chemical Information and Computer Sciences, 44*(6), 1942–1951. doi:10.1021/ci049870g PMID:15554663

Kenneth Morrow, J., & Zhang, S. (2012). Computational Prediction of Protein Hot Spot Residues. *Current Pharmaceutical Design, 18*(9), 1255–1265. doi:10.2174/138161212799436412 PMID:22316154

Keskin, O., Ma, B., & Nussinov, R. (2005). Hot regions in protein-protein interactions: The organization and contribution of structurally conserved hot spot residues. *Journal of Molecular Biology, 345*(5), 1281–1294. doi:10.1016/j.jmb.2004.10.077 PMID:15644221

Khamis, M. A., Gomaa, W., & Ahmed, W. F. (2015). Machine learning in computational docking. *Artificial Intelligence in Medicine, 63*(3), 135–152. doi:10.1016/j.artmed.2015.02.002 PMID:25724101

Kharb, R., Meenakshi, R., Sharma, P. C., & Yar, M. S. (2011). Therapeutic importance of peptidomimetics in medici nal chemistry. *Journal of Chemical and Pharmaceutical Research, 3*(6), 173–186.

Kim, D., Xu, D., Guo, J. T., Ellrott, K., & Xu, Y. (2003). PROSPECT II: Protein structure prediction program for genome-scale applications. *Protein Engineering, 16*(9), 641–650. doi:10.1093/protein/gzg081 PMID:14560049

Kim, J. T., Hamilton, A. D., Bailey, C. M., Domaoal, R. A., Domoal, R. A., Wang, L., & Jorgensen, W. L. (2006). FEP-guided selection of bicyclic heterocycles in lead optimization for non-nucleoside inhibitors of HIV-1 reverse transcriptase. *Journal of the American Chemical Society*, *128*(48), 15372–15373. doi:10.1021/ja066472g PMID:17131993

Kim, S.-K., Gao, Z.-G., Jeong, L. S., & Jacobson, K. A. (2006). Docking studies of agonists and antagonists suggest an activation pathway of the A3 adenosine receptor. *Journal of Molecular Graphics & Modelling*, *25*(4), 562–577. doi:10.1016/j.jmgm.2006.05.004 PMID:16793299

Kinnings, S. L., Liu, N., Tonge, P. J., Jackson, R. M., Xie, L., & Bourne, P. E. (2011). A machine learning-based method to improve docking scoring functions and its application to drug repurposing. *Journal of Chemical Information and Modeling*, *51*(2), 408–419. doi:10.1021/ci100369f PMID:21291174

Kirchmair, J., Markt, P., Distinto, S., Wolber, G., & Langer, T. (2008). Evaluation of the performance of 3D virtual screening protocols: RMSD comparisons, enrichment assessments, and decoy selection - What can we learn from earlier mistakes? *Journal of Computer-Aided Molecular Design*, *22*(3-4), 213–228. doi:10.1007/s10822-007-9163-6

Kirkpatrick, S., Gelatt, C. D., & Vecchi, M. P. (1983). Optimization by simulation annealing. *Science*, *220*(4598), 671–680. doi:10.1126/science.220.4598.671

Kitao, A., Hayward, S., & Go, N. (1998). Energy landscape of a native protein: Jumping-among-minima model. *Proteins*, *33*(4), 496–517. doi:10.1002/(SICI)1097-0134(19981201)33:4<496::AID-PROT4>3.0.CO;2-1 PMID:9849935

Kitchen, D. B., Decornez, H., Furr, J. R., & Bajorath, J. (2004). Docking and scoring in virtual screening for drug discovery: methods and applications. *Nat Rev Drug Discov, 3*(11), 935-949.

Kitchen, D. B., Decornez, H., Furr, J. R., & Bajorath, J. (2004). Docking and scoring in virtual screening for drug discovery: Methods and applications. *National Review*, *3*, 935–949.

Kitchen, D. B., Decornez, H., Furr, J. R., & Bajorath, J. (2004). Docking and scoring in virtual screening for drug discovery: Methods and applications. *Nature Reviews. Drug Discovery*, *3*(11), 935–949. doi:10.1038/nrd1549 PMID:15520816

Klebe, G. (2006). Virtual ligand screening: Strategies, perspectives and limitations. *Drug Discovery Today*, *11*(13-14), 580–594. doi:10.1016/j.drudis.2006.05.012 PMID:16793526

Knegtel, R. M., Kuntz, I. D., & Oshiro, C. M. (1997). Molecular docking to ensembles of protein structures. *Journal of Molecular Biology*, *266*(2), 424–440. doi:10.1006/jmbi.1996.0776 PMID:9047373

Kohonen, T. (1998). The self-organizing map. *Neurocomputing*, *21*(1), 1–6. doi:10.1016/S0925-2312(98)00030-7

Kojima, K., Konopleva, M., McQueen, T., O'Brien, S., Plunkett, W., & Andreeff, M. (2006). Mdm2 inhibitor Nutlin-3a induces p53-mediated apoptosis by transcription-dependent and transcription-independent mechanisms and may overcome Atm-mediated resistance to fludarabine in chronic lymphocytic leukemia. *Blood*, *108*(3), 993–1000. doi:10.1182/blood-2005-12-5148 PMID:16543464

Kokh, D. B., Wade, R. C., & Wenzel, W. (2011). Receptor flexibility in small-molecule docking calculations. *WIREs Computational Molecular Science*, *1*(2), 298–314. doi:10.1002/wcms.29

Kolinski, A., Rotkiewicz, P., Ilkowski, B., & Skolnick, J. (1999). A method for the improvement of threading-based protein models. *Proteins*, *37*(4), 592–610. doi:10.1002/(SICI)1097-0134(19991201)37:4<592::AID-PROT10>3.0.CO;2-2 PMID:10651275

Kollman, P. A. (1996). AMBER 5.0. San Francisco, CA: University of California.

Kollman, P. (1993). Free energy calculations: Applications to chemical and biochemical phenomena. *Chemical Reviews*, *93*(7), 2395–2417.

Kollman, P. A., Massova, I., Reyes, C., Kuhn, B., Huo, S., & Chong, L. et al. others. (2000). Calculating structures and free energies of complex molecules: Combining molecular mechanics and continuum models. *Accounts of Chemical Research*, *33*(12), 889–897. PMID:11123888

Kollman, P. A., Massova, I., Reyes, C., Kuhn, B., Huo, S., Chong, L., & Cheatham, T. E. (2000). Calculating Structures and Free Energies of Complex Molecules: Combining Molecular Mechanics and Continuum Models. *Accounts of Chemical Research*, *33*(12), 889–897. doi:10.1021/ar000033j PMID:11123888

Konc, J., Konc, J. T., Penca, M., & Janezic, D. (2011). Binding-sites Prediction Assisting Protein-protein Docking. *Acta Chimica Slovenica*, *58*(3), 396–401. PMID:24062097

Kongsted, J., & Ryde, U. (2009). An improved method to predict the entropy term with the MM/PBSA approach. *Journal of Computer-Aided Molecular Design*, *23*(2), 63–71. doi:10.1007/s10822-008-9238-z

Kontoyianni, M., Madhav, P., Suchanek, E., & Seibel, W. (2008). Theoretical and practical considerations in virtual screening: A beaten field? *Current Medicinal Chemistry*, *15*(2), 107–116. doi:10.2174/092986708783330566 PMID:18220766

Korb, O., Möller, H. M., & Exner, T. E. (2010). NMR-guided molecular docking of a protein-peptide complex based on ant colony optimization. *ChemMedChem*, *5*(7), 1001–1006. doi:10.1002/cmdc.201000090 PMID:20486157

Korb, O., Stutzle, T., & Exner, T. E. (2006). PLANTS: Application of Ant Colony optimization to Structure-Based Drug Design. *ANTS 2006. LNCS*, *4150*, 247–258.

Korb, O., Stützle, T., & Exner, T. E. (2007). An ant colony optimization approach to flexible protein–ligand docking. *Swarm Intelligence*, *1*(2), 115–134. doi:10.1007/s11721-007-0006-9

Korb, O., Stützle, T., & Exner, T. E. (2009). Empirical scoring functions for advanced protein-ligand docking with PLANTS. *Journal of Chemical Information and Modeling*, *49*(1), 84–96. doi:10.1021/ci800298z PMID:19125657

Kortagere, S., Krasowski, M. D., Reschly, E. J., Venkatesh, M., Mani, S., & Ekins, S. (2010). Evaluation of Computational Docking to Identify Pregnane X Receptor Agonists in the ToxCast Database. *Environmental Health Perspectives*, *118*(10), 1412–1417. doi:10.1289/ehp.1001930 PMID:20558333

Kortemme, T., & Baker, D. (2004). Computational design of protein-protein interactions. *Current Opinion in Chemical Biology*, *8*(1), 91–97. doi:10.1016/j.cbpa.2003.12.008 PMID:15036162

Kortemme, T., Joachimiak, L. A., Bullock, A. N., Schuler, A. D., Stoddard, B. L., & Baker, D. (2004). Computational redesign of protein-protein interaction specificity. *Nature Structural & Molecular Biology*, *11*(4), 371–379. PMID:15034550

Koshland, D. E. (1958). Application of a Theory of Enzyme Specificity to Protein Synthesis. *Proceedings of the National Academy of Sciences of the United States of America*, *44*(2), 98–104. doi:10.1073/pnas.44.2.98 PMID:16590179

Koshland, D. E. Jr. (1963). Correlation of Structure and Function in Enzyme Action. *Science*, *142*(3599), 1533–1541. doi:10.1126/science.142.3599.1533 PMID:14075684

Kozakov, D., Beglov, D., Bohnuud, T., Mottarella, S. E., Xia, B., Hall, D. R., & Vajda, S. (2013). How good is automated protein docking? *Proteins: Structure. Function and Bioinformatics*, *81*(12), 2159–2166. doi:10.1002/prot.24403

Kozakov, D., Brenke, R., Comeau, S. R., & Vajda, S. (2006). PIPER: An FFT-based protein docking program with pairwise potentials. *Proteins*, *65*(2), 392–406. doi:10.1002/prot.21117 PMID:16933295

Kramer, B., Rarey, M., & Lengauer, T. (1999). Evaluation of the FLEXX incremental construction algorithm for protein-ligand docking. *Proteins*, *37*(2), 228–241. doi:10.1002/(SICI)1097-0134(19991101)37:2<228::AID-PROT8>3.0.CO;2-8 PMID:10584068

Kramer, B., Rarey, M., & Lengauer, T. (1999). Evaluation of the flexx incremental construction algorithm for protein–ligand docking. *Proteins: Structure, Function, and Bioinformatics*, *37*(2), 228–241. PMID:10584068

Kramer, C., & Gedeck, P. (2010). Leave-cluster-out cross-validation is appropriate for scoring functions derived from diverse protein data sets. *Journal of Chemical Information and Modeling*, *50*(11), 1961–1969. PMID:20936880

Kramer, C., & Gedeck, P. (2011). Global free energy scoring functions based on distance-dependent atom-type pair descriptors. *Journal of Chemical Information and Modeling*, *51*(3), 707–720. doi:10.1021/ci100473d

Krammer, A., Kirchhoff, P. D., Jiang, X., Venkatachalam, C. M., & Waldman, M. (2005). LigScore: A novel scoring function for predicting binding affinities. *Journal of Molecular Graphics & Modelling*, *23*(5), 395–407. doi:10.1016/j.jmgm.2004.11.007 PMID:15781182

Krieger, E., Nabuurs, S. B., & Vriend, G. (2003). Homology modeling. *Methods of Biochemical Analysis*, *44*, 509–524. PMID:12647402

Kroeger, M. B., Rizzo, R. C., Hose, B. M., Tirado-Rives, J., & Smith, R. H. (2003). Molecular modeling calculations of HIV-1 reverse transcriptase nonnucleoside inhibitors: Correlation of binding energy with biological activity for novel 2-aryl-substituted benzimidazole analogues. *Journal of Medicinal Chemistry*, *46*(10), 1940–1947. doi:10.1021/jm020271f PMID:12723956

Kroemer, R. T. (2003). Molecular modelling probes: Docking and scoring. *Biochemical Society Transactions*, *31*(Pt 5), 980–984. doi:10.1042/bst0310980 PMID:14505464

Kroemer, R. T. (2007). Structure-based drug design: Docking and scoring. *Current Protein & Peptide Science*, *8*, 312–328. doi:10.2174/138920307781369382

Kroemer, R. T., Vulpetti, A., McDonald, J. J., Rohrer, D. C., Trosset, J.-Y., Giordanetto, F., & Stouten, P. F. et al. (2004). Assessment of docking poses: Interactions-based accuracy classification (IBAC) versus crystal structure deviations. *Journal of Chemical Information and Computer Sciences*, *44*(3), 871–881. PMID:15154752

Krovat, E. M., Steindl, T., & Langer, T. (2005). Recent advances in docking and scoring. *Current Computer-aided Drug Design*, *1*(1), 93–102. doi:10.2174/1573409052952314

Kubinyi, H. (2006). *Computer Applications in Pharmaceutical Research and Development*. New York: John Wiley.

Kulikova, T., Aldebert, P., Althorpe, N., Baker, W., Bates, K., Browne, P., & Apweiler, R. et al. (2004). The EMBL Nucleotide Sequence Database. *Nucleic Acids Research*, *32*, D27–D30. doi:10.1093/nar/gkh120 PMID:14681351

Kumalo, H. M., Bhakat, S., & Soliman, M. E. (2015). Theory and applications of covalent docking in drug discovery: Merits and pitfalls. *Molecules (Basel, Switzerland)*, *20*(2), 1984–2000. doi:10.3390/molecules20021984

Kumar, A., Goyal, R., Kumar, S., Jain, S., Jain, N., & Kumar, P. (2015). Estrogenic and Anti-Alzheimer's studies of *Zingiber officinalis* as well as *Amomum subulatumRoxb.*: The success story of dry techniques. *Medicinal Chemistry Research*, *24*(3), 1089–1097. doi:10.1007/s00044-014-1166-y

Kumar, A., Kumar, S., Jain, S., Kumar, P., & Goyal, R. (2013). Study of binding of pyridoacridine alkaloids on Topoisomerase II using *In-Silico* tools. *Medicinal Chemistry Research*, *22*(11), 5431–5441. doi:10.1007/s00044-013-0496-5

Kuntz, I. D., Blaney, J. M., Oatley, S. J., Langridge, R., & Ferrin, T. E. (1982). A geometric approach to macromolecule-ligand interactions. *Journal of Molecular Biology*, *161*(2), 269–288. doi:10.1016/0022-2836(82)90153-X PMID:7154081

Kurcinski, M., Jamroz, M., Blaszczyk, M., Kolinski, A., & Kmiecik, S. (2015). CABS-dock web server for the flexible docking of peptides to proteins without prior knowledge of the binding site. *Nucleic Acids Research*, *43*(W1), W419-424. doi:10.1093/nar/gkv456 PMID:25943545

Labbe, C. M., Rey, J., Lagorce, D., Vavrusa, M., Becot, J., Sperandio, O., & Miteva, M. A. et al. (2015). MTiOpen-Screen: A web server for structure-based virtual screening. *Nucleic Acids Research*, *43*(W1), W448-454. doi:10.1093/nar/gkv306 PMID:25855812

La, D., Esquivel-Rodriguez, J., Venkatraman, V., Li, B., Sael, L., Ueng, S., & Kihara, D. et al. (2009). 3D-SURFER: Software for high throughput protein surface comparison and analysis. *Bioinformatics (Oxford, England)*, *25*(21), 2843–2844. doi:10.1093/bioinformatics/btp542 PMID:19759195

Lafleur, K., Huang, D., Zhou, T., Caflisch, A., & Nevado, C. (2009). Structure based optimization of potent and selective inhibitors of the tyrosine kinase erythropoietin producing human hepatocellular carcinoma receptor B4 (EphB4). *Journal of Medicinal Chemistry*, *52*(20), 6433–6446. doi:10.1021/jm9009444 PMID:19788238

Lahana, R. (1999). How many leads from HTS. *Drug Discovery Today*, *4*(10), 447–448. doi:10.1016/S1359-6446(99)01393-8 PMID:10481138

Lambert, C., Leonard, N., De Bolle, X., & Depiereux, E. (2002). ESyPred3D: Prediction of proteins 3D structures. *Bioinformatics (Oxford, England)*, *18*(9), 1250–1256. doi:10.1093/bioinformatics/18.9.1250 PMID:12217917

Lang, C. A., Ray, S. S., Liu, M., Singh, A. K., & Cuny, G. D. (2015). Discovery of LRRK2 inhibitors using sequential in silico joint pharmacophore space (JPS) and ensemble docking. *Bioorganic & Medicinal Chemistry Letters*, *25*(13), 2713–2719. doi:10.1016/j.bmcl.2015.04.027 PMID:25998502

Langer, T., & Hoffmann, R. D. (2001). Virtual screening: An effective tool for lead structure discovery? *Current Pharmaceutical Design*, *7*(7), 509–527. doi:10.2174/1381612013397861 PMID:11375766

Langer, T., & Krovat, E. M. (2003). Chemical feature-based pharmacophores and virtual library screening for discovery of new leads. *Current Opinion in Drug Discovery & Development*, *6*, 370–376. PMID:12833670

Laskowski, R. A. (1995). SURFNET: A program for visualizing molecular surfaces, cavities, and intermolecular interactions. *Journal of Molecular Graphics*, *13*(5), 323–330. doi:10.1016/0263-7855(95)00073-9 PMID:8603061

Laskowski, R. A., MacArthur, M. W., Moss, D. S., & Thornton, J. M. (1993). PROCHECK: A program to check the stereochemical quality of protein structures. *Journal of Applied Crystallography*, *26*(2), 283–291. doi:10.1107/S0021889892009944

Lauck, F., Smith, C. A., Friedland, G. F., Humphris, E. L., & Kortemme, T. (2010). RosettaBackrub--a web server for flexible backbone protein structure modeling and design. *Nucleic Acids Res*, *38*, W569-575. doi: 10.1093/nar/gkq369

Laurie, A. T., & Jackson, R. M. (2005). Q-SiteFinder: An energy-based method for the prediction of protein-ligand binding sites. *Bioinformatics (Oxford, England)*, *21*(9), 1908–1916. doi:10.1093/bioinformatics/bti315 PMID:15701681

Laurini, E., Col, V. D., Mamolo, M. G., Zampieri, D., Posocco, P., Fermeglia, M., & Pricl, S. et al. (2011). Homology Model and Docking-Based Virtual Screening for Ligands of the sigma1 Receptor. *ACS Med Chem Lett*, *2*(11), 834–839. doi:10.1021/ml2001505 PMID:24900272

Lavecchia, A., & Di Giovanni, C. (2013). Virtual screening strategies in drug discovery: A critical review. *Current Medicinal Chemistry*, *20*(23), 2839–2860. doi:10.2174/09298673113209990001 PMID:23651302

Leach, A. R. (1994). Ligand docking to proteins with discrete side-chain flexibility. *Journal of Molecular Biology*, *235*(1), 345–356. doi:10.1016/S0022-2836(05)80038-5 PMID:8289255

Leach, A. R., & Kuntz, I. D. (1992). Conformational analysis of flexible ligands in macromolecular receptor sites. *Journal of Computational Chemistry*, *13*(6), 730–748. doi:10.1002/jcc.540130608

Leach, A. R., Shoichet, B. K., & Peishoff, C. E. (2006). Prediction of protein-ligand interactions. Docking and scoring: Successes and gaps. *Journal of Medicinal Chemistry*, *49*(20), 5851–5855. doi:10.1021/jm060999m PMID:17004700

Learning, C. M. (2010). Pharmacophore-Based Interaction Fingerprint for in Silico Screening Sato, Tomohiro; Honma, Teruki; Yokoyama. *Journal of Chemical Information and Modeling*, *50*(1), 170–185. doi:10.1021/ci900382e PMID:20038188

Lee, P. H., Kuo, K. L., Chu, P. Y., Liu, E. M., & Lin, J. H. (2009). SLITHER: a web server for generating contiguous conformations of substrate molecules entering into deep active sites of proteins or migrating through channels in membrane transporters. *Nucleic Acids Res*, *37*, W559-564. doi: 10.1093/nar/gkp359

Lee, H. S., & Zhang, Y. (2012). BSP-SLIM: A blind low-resolution ligand-protein docking approach using predicted protein structures. *Proteins*, *80*(1), 93–110. doi:10.1002/prot.23165 PMID:21971880

Lee, H., Heo, L., Lee, M. S., & Seok, C. (2015). GalaxyPepDock: A protein-peptide docking tool based on interaction similarity and energy optimization. *Nucleic Acids Research*, *43*(W1), W431-435. doi:10.1093/nar/gkv495 PMID:25969449

Lee, J., & Seok, C. (2008). A statistical rescoring scheme for protein–ligand docking: Consideration of entropic effect. *Proteins: Structure, Function, and Bioinformatics*, *70*(3), 1074–1083. PMID:18076034

Legge, F. S., Budi, A., Treutlein, H., & Yarovsky, I. (2006). Protein flexibility: Multiple molecular dynamics simulations of insulin chain B. *Biophysical Chemistry*, *119*(2), 146–157. doi:10.1016/j.bpc.2005.08.002 PMID:16129550

Leman, J. K., Alford, R. F., & Gray, J. J. (2015). Rosetta-MPDock: A Novel Computational Tool for Protein-Protein Docking within the Membrane Bilayer. *Biophysical Journal*, *108*(2), 250a. doi:10.1016/j.bpj.2014.11.1382

Lengauer, T., & Rarey, M. (1996). Computational methods for biomolecular docking. *Current Opinion in Structural Biology*, *6*(3), 402–406. doi:10.1016/S0959-440X(96)80061-3

Lensink, M. F., & Wodak, S. J. (2013). Docking, scoring, and affinity prediction in CAPRI. *Proteins: Structure, Function, and Bioinformatics*, *81*(12), 2082–2095. doi:10.1002/prot.24428 PMID:24115211

Lesk, A. M. (1997). CASP2: Report on ab initio predictions. *Proteins*, *29*(S1Suppl 1), 151–166. doi:10.1002/(SICI)1097-0134(1997)1+<151::AID-PROT20>3.0.CO;2-M PMID:9485507

Lesk, V. I., & Sternberg, M. J. (2008). 3D-Garden: A system for modelling protein-protein complexes based on conformational refinement of ensembles generated with the marching cubes algorithm. *Bioinformatics (Oxford, England)*, *24*(9), 1137–1144. doi:10.1093/bioinformatics/btn093 PMID:18326508

Levinthal, C., Wodak, S. J., Kahn, P., & Dadivanian, A. K. (1975). Hemoglobin interaction in sickle cell fibers. I: Theoretical approaches to the molecular contacts. *Proceedings of the National Academy of Sciences of the United States of America*, *72*(4), 1330–1334. doi:10.1073/pnas.72.4.1330 PMID:1055409

Levitt, D. G., & Banaszak, L. J. (1992). POCKET: A computer graphics method for identifying and displaying protein cavities and their surrounding amino acids. *Journal of Molecular Graphics*, *10*(4), 229–234. doi:10.1016/0263-7855(92)80074-N PMID:1476996

Levy, R. M., & Gallicchio, E. (1998). Computer simulations with explicit solvent: Recent progress in the thermodynamic decomposition of free energies and in modeling electrostatic effects. *Annual Review of Physical Chemistry, 49*(1), 531–567. PMID:9933909

Levy, R. M., & Karplus, M. (1979). Vibrational Approach to the Dynamics of an alpha-Helix. *Biopolymers, 18*(10), 2465–2495. doi:10.1002/bip.1979.360181008

Levy, Y., & Onuchic, J. N. (2006). Water mediation in protein folding and molecular recognition. *Annual Review of Biophysics and Biomolecular Structure, 35*, 389–415. PMID:16689642

Levy, Y., Wolynes, P. G., & Onuchic, J. N. (2004). Protein topology determines binding mechanism. *Proceedings of the National Academy of Sciences of the United States of America, 101*(2), 511–516. PMID:14694192

Lexa, K. W., & Carlson, H. A. (2012). Protein flexibility in docking and surface mapping. *Quarterly Reviews of Biophysics, 45*(3), 301–343. doi:10.1017/S0033583512000066 PMID:22569329

Li, H., Gao, Z., Kang, L., Zhang, H., Yang, K., Yu, K., . . . Jiang, H. (2006). TarFisDock: a web server for identifying drug targets with docking approach. *Nucleic Acids Res, 34*, W219-224. doi: 10.1093/nar/gkl114

Li, L., Chen, R., & Weng, Z. (2003). RDOCK:Refinement of Rigid-body Protein Docking Predictions. *Proteins: Struc. Funct. Gen, 53*(061/14), 693–707.

Li, X., & Liang, J. (2006). Knowledge-based energy functions for computational studies of proteins. In Y. Xu., D. Xu & J. Liang (Eds.), Computational methods for protein structure prediction and modeling (pp. 71-124). New York, NY: Springer.

Liang, S., Liu, S., Zhang, C., & Zhou, Y. (2007). A simple reference state makes a significant improvement in near-native selections from structurally refined docking decoys. *Proteins, 69*(2), 244–253. doi:10.1002/prot.21498 PMID:17623864

Liao, C., & Nicklaus, M. C. (2010). Computer tools in the discovery of HIV-1 integrase inhibitors. *Future Medicinal Chemistry, 2*(7), 1123–1140. doi:10.4155/fmc.10.193 PMID:21426160

Liebeschuetz, J. W., Cole, J. C., & Korb, P. (2012). Pose prediction and virtual screening performance of GOLD scoring function in a standardized test. *Journal of Computer-Aided Molecular Design, 26*(6), 737–748. doi:10.1007/s10822-012-9551-4

Li, G. B., Yang, L. L., Wang, W. J., Li, L. L., & Yang, S. Y. (2013). ID-Score: A new empirical scoring function based on a comprehensive set of descriptors related to protein-ligand interactions. *Journal of Chemical Information and Modeling, 53*(3), 592–600. doi:10.1021/ci300493w PMID:23394072

Li, G.-B., Yang, L.-L., Wang, W.-J., Li, L.-L., & Yang, S.-Y. (2013). Id-score: A new empirical scoring function based on a comprehensive set of descriptors related to protein–ligand interactions. *Journal of Chemical Information and Modeling, 53*(3), 592–600. PMID:23394072

Li, H., Leung, K. S., Ballester, P. J., & Wong, M. H. (2014). istar: A web platform for large-scale protein-ligand docking. *PLoS ONE, 9*(1), e85678. doi:10.1371/journal.pone.0085678 PMID:24475049

Li, H., Zhang, H., Zheng, M., Luo, J., Kang, L., Liu, X., & Jiang, H. et al. (2009). An effective docking strategy for virtual screening based on multi-objective optimization algorithm. *BMC Bioinformatics, 10*(1), 58. doi:10.1186/1471-2105-10-58 PMID:19210777

Li, L., Chen, R., & Weng, Z. (2003). RDOCK: Refinement of rigid-body protein docking predictions. *Proteins, 53*(3), 693–707. doi:10.1002/prot.10460 PMID:14579360

Li, L., Wang, B., & Meroueh, S. O. (2011). Support Vector Regression Scoring of Receptor–Ligand Complexes for Rank-Ordering and Virtual Screening of Chemical Libraries. *Journal of Chemical Information and Modeling*, *51*(9), 2132–2138. doi:10.1021/ci200078f PMID:21728360

Lill, M. A. (2011). Efficient incorporation of protein flexibility and dynamics into molecular docking simulations. *Biochemistry*, *50*(28), 6157–6169. doi:10.1021/bi2004558 PMID:21678954

Lindahl, E., & Delarue, M. (2005). Refinement of docked protein-ligand and protein-DNA structures using low frequency normal mode amplitude optimization. *Nucleic Acids Research*, *33*(14), 4496–4506. doi:10.1093/nar/gki730 PMID:16087736

Lindstrom, A., Edvinsson, L., Johansson, A., Andersson, C. D., Andersson, I. E., Raubacher, F., & Linusson, A. (2011). Postprocessing of docked protein-ligand complexes using implicit solvation models. *Journal of Chemical Information and Modeling*, *51*(2), 267–282. PMID:21309544

Lin, J.-H., Perryman, A. L., Schames, J. R., & McCammon, J. A. (2002). Computational drug design accommodating receptor flexibility: The relaxed complex scheme. *Journal of the American Chemical Society*, *124*(20), 5632–5633. doi:10.1021/ja0260162 PMID:12010024

Lionta, E., Spyrou, G., Vassilatis, D. K., & Cournia, Z. (2014). Structure-based virtual screening for drug discovery: Principles, applications and recent advances. *Current Topics in Medicinal Chemistry*, *14*(16), 1923–1938. doi:10.2174/1568026614666140929124445 PMID:25262799

Liu, X., Ouyang, S., Yu, B., Liu, Y., Huang, K., Gong, J., . . . Jiang, H. (2010). PharmMapper server: a web server for potential drug target identification using pharmacophore mapping approach. *Nucleic Acids Res*, *38*, W609-614. doi: 10.1093/nar/gkq300

Liu, B., Li, S., & Hu, J. (2004). Technological advances in high-throughput screening. *American Journal of Pharmacogenomics: Genomics-Related Research in Drug Development and Clinical Practice*, *4*(4), 263–276. doi:10.2165/00129785-200404040-00006 PMID:15287820

Liu, J., & Wang, R. (2015). Classification of current scoring functions. *Journal of Chemical Information and Modeling*, *55*(3), 475–482. doi:10.1021/ci500731a PMID:25647463

Liu, J., & Wang, R. (2015). On classification of current scoring functions. *Journal of Chemical Information and Modeling*, *55*(3), 475–482. PMID:25647463

Liu, M., & Wang, S. (1999). MCDOCK: A Monte Carlo simulation approach to the molecular docking problem. *Journal of Computer-Aided Molecular Design*, *13*(5), 435–451. doi:10.1023/A:1008005918983 PMID:10483527

Liu, S., & Vakser, I. (2011). DECK: Distance and environment-dependent, coarse-grained, knowledge-based potentials for protein-protein docking. *BMC Bioinformatics*, *12*(1), 280. doi:10.1186/1471-2105-12-280 PMID:21745398

Liu, Y., Zhao, L., Li, W., Zhao, D., Song, M., & Yang, Y. (2013). FIPSDock: A new molecular docking technique driven by fully informed swarm optimization algorithm. *Journal of Computational Chemistry*, *34*(1), 67–75. doi:10.1002/jcc.23108 PMID:22961860

Liu, Z., Dominy, B. N., & Shakhnovich, E. I. (2004). Structural mining: Self-consistent design on flexible protein-peptide docking and transferable binding affinity potential. *Journal of the American Chemical Society*, *126*(27), 8515–8528. PMID:15238009

Li, X., Li, Y., Cheng, T., Liu, Z., & Wang, R. (2010). Evaluation of the performance of four molecular docking programs on a diverse set of protein-ligand complexes. *Journal of Computational Chemistry*, *31*(11), 2109–2125. doi:10.1002/jcc.21498 PMID:20127741

Li, Y. (2013). Conformational sampling in template-free protein loop structure modeling: An overview. *Comput Struct Biotechnol J*, *5*(6), e201302003. doi:10.5936/csbj.201302003 PMID:24688696

Li, Y., Han, L., Liu, Z., & Wang, R. (2014). Comparative assessment of scoring functions on an updated benchmark: II. Evaluation methods and general results. *Journal of Chemical Information and Modeling*, *54*, 1717–1736. PMID:24708446

Li, Y., Liu, Z., Li, J., Han, L., Liu, J., Zhao, Z.-X., & Wang, R. (2014). Comparative assessment of scoring functions on an updated benchmark: I. Compilation of the test set. *Journal of Chemical Information and Modeling*, *54*, 1700–1716. PMID:24716849

Li, Y., Liu, Z., & Wang, R. (2010). Test MM-PB/SA on True Conformational Ensembles of Protein-Ligand Complexes. *Journal of Chemical Information and Modeling*, *50*(9), 1682–1692. doi:10.1021/ci100036a PMID:20695488

Lobanov, M. Y., Bogatyreva, N. S., & Galzitskaya, O. V. (2008). Radius of gyration as an indicator of protein structure compactness. *Molecular Biology*, *42*(4), 623–628. doi:10.1134/S0026893308040195 PMID:18856071

London, N., Miller, R. M., Krishnan, S., Uchida, K., Irwin, J. J., Eidam, O., & Taunton, J. et al. (2014). Covalent docking of large libraries for the discovery of chemical probes. *Nature Chemical Biology*, *10*(12), 1066–1072. doi:10.1038/nchembio.1666 PMID:25344815

Lorenzen, S., & Zhang, Y. (2007). Monte Carlo refinement of rigid-body protein docking structures with backbone displacement and side-chain optimization. *Protein Science : A Publication of the Protein Society*, *16*(12), 2716–2725.

Lu, L., Lu, H., & Skolnick, J. (2002). MULTIPROSPECTOR: An algorithm for the prediction of protein-protein interactions by multimeric threading. *Proteins*, *49*(3), 350–364. doi:10.1002/prot.10222 PMID:12360525

Luo, J., & Bruice, T. C. (2002). Ten-nanosecond molecular dynamics simulation of the motions of the horse liver alcohol dehydrogenase*PhCH2O- complex. *Proceedings of the National Academy of Sciences of the United States of America*, *99*(26), 16597–16600. doi:10.1073/pnas.262667599 PMID:12481026

Lu, S.-Y., Jiang, Y.-J., Lv, J., Wu, T.-X., Yu, Q.-S., & Zhu, W.-L. (2010). Molecular docking and molecular dynamics simulation studies of GPR40 receptor–agonist interactions. *Journal of Molecular Graphics & Modelling*, *28*(8), 766–774. doi:10.1016/j.jmgm.2010.02.001 PMID:20227312

Lybrand, T. P. (1995). Ligand-protein docking and rational drug design. *Current Opinion in Structural Biology*, *5*(2), 224–228. doi:10.1016/0959-440X(95)80080-8

Lyne, P. D. (2002). Structure-based virtual screening: An overview. *Drug Discovery Today*, *7*(20), 1047–1055. doi:10.1016/S1359-6446(02)02483-2 PMID:12546894

Lyne, P. D., Lamb, M. L., & Saeh, J. C. (2006). Accurate prediction of the relative potencies of members of a series of kinase inhibitors using molecular docking and MM-GBSA scoring. *Journal of Medicinal Chemistry*, *49*(16), 4805–4808. doi:10.1021/jm060522a PMID:16884290

Lyskov, S., & Gray, J. J. (2008). The RosettaDock server for local protein-protein docking. *Nucleic Acids Res*, *36*, W233-238. doi: 10.1093/nar/gkn216

Lyskov, S., Chou, F. C., Conchuir, S. O., Der, B. S., Drew, K., Kuroda, D., & Das, R. et al. (2013). Serverification of molecular modeling applications: The Rosetta Online Server that Includes Everyone (ROSIE). *PLoS ONE*, *8*(5), e63906. PMID:23717507

Ma, B., Kumar, S., Tsai, C. J., & Nussinov, R. (1999). Folding funnels and binding mechanisms. *Protein Engineering*, *12*(9), 713–720. doi:10.1093/protein/12.9.713 PMID:10506280

Ma, B., Shatsky, M., Wolfson, H. J., & Nussinov, R. (2002). Multiple diverse ligands binding at a single protein site: A matter of pre-existing populations. *Protein Science, 11*(2), 184–197. doi:10.1110/ps.21302 PMID:11790828

Macindoe, G., Mavridis, L., Venkatraman, V., Devignes, M. D., & Ritchie, D. W. (2010). HexServer: an FFT-based protein docking server powered by graphics processors. *Nucleic Acids Res, 38*, W445-449. doi: 10.1093/nar/gkq311

Madeira, F., Tinti, M., Murugesan, G., Berrett, E., Stafford, M., Toth, R., & Barton, G. J. et al. (2015). 14-3-3-Pred: Improved methods to predict 14-3-3-binding phosphopeptides. *Bioinformatics (Oxford, England), 31*(14), 2276–2283. doi:10.1093/bioinformatics/btv133 PMID:25735772

Mahajan, A., & Gill, N. S. (2014). A review on molecular docking. *International Journal of Recent Adv Pharm Res., 4*(2), 1–7.

Mahasenan, K. V., & Li, C. (2012). Novel inhibitor discovery through virtual screening against multiple protein conformations generated via ligand-directed modeling: A maternal embryonic leucine zipper kinase example. *Journal of Chemical Information and Modeling, 52*(5), 1345–1355. doi:10.1021/ci300040c PMID:22540736

Maimon, O., & Rokach, L. (2005). *Data mining and knowledge discovery handbook* (Vol. 2). Springer.

Makino, S., & Kuntz, I. D. (1997). Automated flexible ligand docking method and its application for database search. *Journal of Computational Chemistry, 18*(14), 1812–1825.

Malhotra, S., Mathew, O. K., & Sowdhamini, R. (2015). DOCKSCORE: A webserver for ranking protein-protein docked poses. *BMC Bioinformatics, 16*(1), 127. doi:10.1186/s12859-015-0572-6 PMID:25902779

Mallipeddi, P. L., Kumar, G., White, S. W., & Webb, T. R. (2014). Recent advances in computer-aided drug design as applied to anti-influenza drug discovery. *Current Topics in Medicinal Chemistry, 14*(16), 1875–1889. doi:10.2174/1568026614666140929153812 PMID:25262796

Mandal, S., Moudgil, M., & Mandal, S. K. (2009). Rational drug design. *European Journal of Pharmacology, 625*(1-3), 90–100. doi:10.1016/j.ejphar.2009.06.065 PMID:19835861

Mandell, J. G., Roberts, V., Pique, M. E., Kotlovyi, V., Mitchell, J. C., Nelson, E., & Ten Eyck, L. F. et al. (2001). Protein docking using continuum electrostatics and geometric fit. *Protein Engineering, 14*(2), 105–113. doi:10.1093/protein/14.2.105 PMID:11297668

Marcou, G., & Rognan, D. (2007). Optimizing fragment and scaffold docking by use of molecular interaction fingerprints. *Journal of Chemical Information and Modeling, 47*(1), 195–207. doi:10.1021/ci600342e PMID:17238265

Martin, L., Catherinot, V., & Labesse, G. (2006). kinDOCK: a tool for comparative docking of protein kinase ligands. *Nucleic Acids Res, 34*, W325-329. doi: 10.1093/nar/gkl211

Marti-Renom, M. A., Stuart, A. C., Fiser, A., Sanchez, R., Melo, F., & Sali, A. (2000). Comparative protein structure modeling of genes and genomes. *Annu Rev Biophys Biomol Struct, 29*, 291-325.

Mashiach, E., Schneidman-Duhovny, D., Andrusier, N., Nussinov, R., & Wolfson, H. J. (2008). FireDock: a web server for fast interaction refinement in molecular docking. *Nucleic Acids Research, 36*, W229–32.

Mashiach, E., Schneidman-Duhovny, D., Andrusier, N., Nussinov, R., & Wolfson, H. J. (2008). FireDock: a web server for fast interaction refinement in molecular docking. *Nucleic Acids Research, 36*, W229–32. http://doi.org/10.1093/nar/gkn186

Mashiach, E., Nussinov, R., & Wolfson, H. J. (2010). FiberDock: Flexible induced-fit backbone refinement in molecular docking. *Proteins: Structure. Function and Bioinformatics, 78*(6), 1503–1519.

Masters, C. L., Simms, G., Weinman, N. A., Multhaup, G., McDonald, B. L., & Beyreuther, K. (1985). Amyloid plaque core protein in Alzheimer disease and Down syndrome. *Proceedings of the National Academy of Sciences of the United States of America*, *82*(12), 4245–4249. doi:10.1073/pnas.82.12.4245 PMID:3159021

Matsuno, K., Masuda, Y., Uehara, Y., Sato, H., Muroya, A., Takahashi, O., & Asai, A. et al. (2010). Identification of a new series of STAT3 inhibitors by virtual screening. *ACS Med Chem Lett*, *1*(8), 371–375. doi:10.1021/ml1000273 PMID:24900220

Matsuzaki, Y., Uchikoga, N., Ohue, M., Shimoda, T., Sato, T., Ishida, T., & Akiyama, Y. (2013). MEGADOCK 3.0: A high-performance protein-protein interaction prediction software using hybrid parallel computing for petascale supercomputing environments. *Source Code for Biology and Medicine*, *8*(1), 18. doi:10.1186/1751-0473-8-18 PMID:24004986

McCammon, J. A., Gelin, B. R., & Karplus, M. (1977). Dynamics of folded proteins. *Nature*, *267*(5612), 585–590. doi:10.1038/267585a0 PMID:301613

McConkey, B. J., Sobolev, V., & Edelman, M. (2002). The performance of current methods in ligand-protein docking. *Current Science*, *83*, 845–855.

McConkey, B. J., Soboley, V., & Edelmen, M. (2002). The performance of current methods in ligand–protein docking. *Current Science*, *83*(7), 845–956.

McGann, M. (2011). FRED pose prediction and virtual screening accuracy. *Journal of Chemical Information and Modeling*, *51*(3), 578–596. doi:10.1021/ci100436p

McMartin, C., & Bohacek, R. S. (1997). QXP: Powerful, rapid computer algorithms for structure-based drug design. *Journal of Computer-Aided Molecular Design*, *11*(4), 333–344. doi:10.1023/A:1007907728892 PMID:9334900

McQuarrie, D. A. (1976). *StatisticalMechanics*. New York: Harper Collins Publishers.

Mehler, E. L., & Solmajer, T. (1991). Electrostatic effects in proteins: Comparison of dielectric and charge models. *Protein Engineering*, *4*(8), 903–910. doi:10.1093/protein/4.8.903

Meier, R., Pippel, M., Brandt, F., Sippl, W., & Baldauf, C. (2010). ParaDockS: A framework for molecular docking with population-based metaheuristics. *Journal of Chemical Information and Modeling*, *50*(5), 879–889. doi:10.1021/ci900467x PMID:20415499

Meiler, J., & Baker, D. (2006). ROSETTALIGAND: Protein-small molecule docking with full side-chain flexibility. *Proteins*, *65*(3), 538–548. doi:10.1002/prot.21086 PMID:16972285

Mendez, R., Leplae, R., De Maria, L., & Wodak, S. J. (2003). Assessment of blind predictions of protein-protein interactions: Current status of docking methods. *Proteins*, *52*(1), 51–67. doi:10.1002/prot.10393 PMID:12784368

Mendez, R., Leplae, R., Lensink, M. F., & Wodak, S. J. (2005). Assessment of CAPRI predictions in rounds 3-5 shows progress in docking procedures. *Proteins*, *60*(2), 150–169. doi:10.1002/prot.20551 PMID:15981261

Meng, X-Y., Zhang, H-X., Mezei, M, & Cui, M. (2011). Molecular Docking: A powerful approach for structure-based drug discovery. *Current Computational Aided Drug Design*, *7*(2), 146–157.

Meng, E. C., Shoichet, B. K., & Kuntz, I. D. (1992). Automated docking with grid-based energy evaluation. *Journal of Computational Chemistry*, *13*(4), 505–524. doi:10.1002/jcc.540130412

Meng, X. Y., Zhang, H. X., Mezei, M., & Cui, M. (2011). Molecular docking: A powerful approach for structure-based drug discovery. *Current Computer-aided Drug Design*, *7*(2), 146–157. doi:10.2174/157340911795677602 PMID:21534921

Mente, S., Guilmette, E., Salafia, M., & Gray, D. (2015). Dopamine D1 receptor-agonist interactions: A mutagenesis and homology modeling study. *Bioorganic & Medicinal Chemistry Letters*, *25*(10), 2106–2111. doi:10.1016/j.bmcl.2015.03.079 PMID:25881819

Mezei, M. (2003). A new method for mapping macromolecular topography. *Journal of Molecular Graphics & Modelling*, *21*(5), 463–472. doi:10.1016/S1093-3263(02)00203-6 PMID:12543141

Michalewicz, Z. (1996). *GAs: What Are They? Genetic Algorithms + Data Structures = Evolution Programs* (pp. 13–31). Springer Berlin Heidelberg. doi:10.1007/978-3-662-03315-9_2

Mihasan, M. (2012). What in silico molecular docking can do for the 'bench-working biologists'. *Journal of Biosciences*, *37*(6), 1089–1095. doi:10.1007/s12038-012-9273-8

Miller, D. W., & Dill, K. A. (2008). Ligand binding to proteins: The binding landscape model. *Protein Science*, *6*(10), 2166–2179. PMID:9336839

Miller, M. D., Kearsley, S. K., Underwood, D. J., & Sheridan, R. P. (1994). FLOG: A system to select 'quasi-flexible' ligands complementary to a receptor of known three-dimensional structure. *Journal of Computer-Aided Molecular Design*, *8*(2), 153–174. doi:10.1007/BF00119865 PMID:8064332

Minini, L., Álvarez, G., González, M., Cerecettob, H., & Merlino, A. (2015). Molecular docking and molecular dynamics simulation studies of *Trypanosoma cruzi* triosephosphate isomerase inhibitors. Insights into the inhibition mechanism and selectivity. *Journal of Molecular Graphics & Modelling*, *58*, 40–49. doi:10.1016/j.jmgm.2015.02.002 PMID:25829097

Mitchell, J. B., Laskowski, R. A., Alex, A., & Thornton, J. M. (1999). Bleeppotential of mean force describing protein–ligand interactions: I. generating potential. *Journal of Computational Chemistry*, *20*(11), 1165–1176.

Miteva, M. A., Lee, W. H., Montes, M. O., & Villoutreix, B. O. (2005). Fast structure-based virtual ligand screening combining FRED, DOCK, and Surflex. *Journal of Medicinal Chemistry*, *48*(19), 6012–6022. doi:10.1021/jm050262h PMID:16162004

Mitra, P., & Pal, D. (2011). PRUNE and PROBE--two modular web services for protein-protein docking. *Nucleic Acids Res*, *39*, W229-234. doi: 10.1093/nar/gkr317

Miyazawa, S., & Jernigan, R. L. (1985). Estimation of effective interresidue contact energies from protein crystal structures: Quasi-chemical approximation. *Macromolecules*, *18*(3), 534–552.

Mizutani, M. Y., Tomioka, N., & Itai, A. (1994). Rational automatic search method for stable docking models of protein and ligand. *Journal of Molecular Biology*, *243*(2), 310–326. doi:10.1006/jmbi.1994.1656 PMID:7932757

Moal, I. H., Dapkunas, J., & Fernandez-Recio, J. (2015). Inferring the microscopic surface energy of protein-protein interfaces from mutation data. *Proteins*, *83*(4), 640–650. doi:10.1002/prot.24761 PMID:25586563

Moal, I. H., Torchala, M., Bates, P. A., & Fernández-Recio, J. (2013). The scoring of poses in protein-protein docking: Current capabilities and future directions. *BMC Bioinformatics*, *14*(1), 286. doi:10.1186/1471-2105-14-286 PMID:24079540

Mohan, V., Gibbs, A. C., Cummings, M. D., Jaeger, E. P., & DesJarlais, R. L. (2005). Docking: Successes and Challenges. *Current Pharmaceutical Design*, *11*(3), 323–333. doi:10.2174/1381612053382106 PMID:15723628

Moitessier, N., Englebienne, P., Lee, D., Lawandi, J., & Corbeil, C. R. (2008). Towards the development of universal, fast and highly accurate docking/scoring methods: A long way to go. *British Journal of Pharmacology*, *153*(S1), S7–S26. doi:10.1038/sj.bjp.0707515

Moitessier, N., Westhof, E., & Hanessian, S. (2006). Docking of aminoglycosides to hydrated and flexible RNA. *Journal of Medicinal Chemistry, 49*(3), 1023–1033. doi:10.1021/jm0508437

Momany, F. A., & Rone, R. (1992). Validation of the general purpose QUANTA® 3.2/CHARMm® force field. *Journal of Computational Chemistry, 13*(7), 888–900. doi:10.1002/jcc.540130714

Monod, J., Wyman, J., & Changeux, J. P. (1965). On the nature of allosteric transitions: A plausible model. *Journal of Molecular Biology, 12*(1), 88–118. doi:10.1016/S0022-2836(65)80285-6 PMID:14343300

Mooij, W., & Verdonk, M. L. (2005). General and targeted statistical potentials for protein-ligand interactions. *Proteins: Structure, Function, and Bioinformatics, 61*(2), 272–287. doi:10.1002/prot.20588

Moont, G., Gabb, H., & Sternberg, M. J. E. (1999). Use of pair potentials across protein interfaces in screening predicted docked complexes. *Proteins, 35*(3), 364–373. doi:10.1002/(SICI)1097-0134(19990515)35:3<364::AID-PROT11>3.0.CO;2-4 PMID:10328272

Mordalski, S., Kosciolek, T., Kristiansen, K., Sylte, I., & Bojarski, A. J. (2011). Protein binding site analysis by means of structural interaction fingerprint patterns. *Bioorganic & Medicinal Chemistry Letters, 21*(22), 6816–6819. doi:10.1016/j.bmcl.2011.09.027 PMID:21974955

Morris, G. M., Goodsell, D. S., Halliday, R. S., Huey, R., Hart, W. E., Belew, R. K., & Olson, A. J. (1999). Automated docking using a Lamarckian genetic algorithm and an empirical binding free energy function. *Journal of Computational Chemistry, 19*(14), 1639–1662. doi:10.1002/(SICI)1096-987X(19981115)19:14<1639::AID-JCC10>3.0.CO;2-B

Morris, G. M., Goodsell, D. S., Huey, R., & Olson, A. J. (1996). Distributed automated docking of flexible ligands to proteins: Parallel applications of AutoDock 2.4. *Journal of Computer-Aided Molecular Design, 10*(4), 293–304. doi:10.1007/BF00124499 PMID:8877701

Morris, G. M., Green, L. G., Radic, Z., Taylor, P., Sharpless, K. B., Olson, A. J., & Grynszpan, F. (2013). Automated docking with protein flexibility in the design of femtomolar "click chemistry" inhibitors of acetylcholinesterase. *Journal of Chemical Information and Modeling, 53*(4), 898–906. doi:10.1021/ci300545a PMID:23451944

Morris, G. M., Huey, R., Lindstrom, W., Sanner, M. F., Belew, R. K., Goodsell, D. S., & Olson, A. J. (2009). AutoDock4 and AutoDockTools4: Automated docking with selective receptor flexibility. *Journal of Computational Chemistry, 30*(16), 2785–2791. doi:10.1002/jcc.21256 PMID:19399780

Morris, G. M., Huey, R., Lindstrom, W., Sanner, M. F., Belew, R. K., Goodsell, D. S., & Olson, A. J. (2009). AutoDock4 and autodocktools4: Automated docking with selective receptor flexibility. *Journal of Computational Chemistry, 30*(16), 2785–2791. PMID:19399780

Morris, G. M., & Lim-Wilby, M. (2008). Molecular docking. In A. Kukol (Ed.), *Molecular Modeling of Proteins* (Vol. 443, pp. 365–382). Totowa: Humana Press. doi:10.1007/978-1-59745-177-2_19

Morris, G. M., Olson, A. J., & Goodsell, D. S. (2000). Protein-Ligand Docking. In D. E. Clark (Ed.), *Evolutionary Algorithms in Molecular Design* (pp. 31–48). New York, NY: Wiley. doi:10.1002/9783527613168.ch3

Mottarella, S. E., Beglov, D., Beglova, N., Nugent, M. A., Kozakov, D., & Vajda, S. (2014). Docking server for the identification of heparin binding sites on proteins. *Journal of Chemical Information and Modeling, 54*(7), 2068–2078. doi:10.1021/ci500115j PMID:24974889

Motto, I., Bordogna, A., Soshilov, A. A., Denison, M. S., & Bonati, L. (2011). New aryl hydrocarbon receptor homology model targeted to improvedocking reliability. *Journal of Chemical Information and Modeling, 51*(11), 2868–2881. doi:10.1021/ci2001617 PMID:21981577

Moult, J., Fidelis, K., Kryshtafovych, A., Schwede, T., & Tramontano, A. (2014). Critical assessment of methods of protein structure prediction (CASP) - round x. *Proteins: Structure. Function and Bioinformatics*, *82*(SUPPL.2), 1–6. doi:10.1002/prot.24452

Mpamhanga, C. P., Chen, B., McLay, I. M., & Willett, P. (2006). Knowledge-based interaction fingerprint scoring: A simple method for improving the effectiveness of fast scoring functions. *Journal of Chemical Information and Modeling*, *46*(2), 686–698. doi:10.1021/ci050420d PMID:16562999

Muegge, I. (2006). PMF scoring revisited. *Journal of Medicinal Chemistry*, *49*(20), 5895–5902. doi:10.1021/jm050038s

Muegge, I., & Enyedy, I. (2003). Virtual screening. In D. J. Abraham (Ed.), *Burger's Medicinal Chemistry and Drug Discovery, Drug Discovery* (Vol. 1, pp. 243–279). New Jersey: John Wiley & Sons. doi:10.1002/0471266949.bmc006

Muegge, I., & Martin, Y. C. (1999). A general and fast scoring function for protein-ligand interactions: A simplified potential approach. *Journal of Medicinal Chemistry*, *42*(5), 791–804. doi:10.1021/jm980536j PMID:10072678

Muegge, I., & Rarey, M. (2001). Small Molecule Docking and Scoring. In *Reviews in Computational Chemistry* (pp. 1–60). John Wiley & Sons, Inc. doi:10.1002/0471224413.ch1

Mueller-Steiner, S., Zhou, Y., Arai, H., Roberson, E. D., Sun, B., Chen, J., & Gan, L. et al. (2006). Antiamyloidogenic and neuroprotective functions of cathepsin B: Implications for Alzheimer's disease. *Neuron*, *51*(6), 703–714. doi:10.1016/j.neuron.2006.07.027 PMID:16982417

Mukherjee, G., Patra, N., Barua, P., & Jayaram, B. (2011). A fast empirical GAFF compatible partial atomic charge assignment scheme for modeling interactions of small molecules with biomolecular targets. *Journal of Computational Chemistry*, *32*(5), 893–907. doi:10.1002/jcc.21671 PMID:21341292

Muralidharan, A. R., Selvaraj, C., Singh, S. K., Sheu, J. R., Thomas, P. A., & Geraldine, P. (2015). Structure-Based Virtual Screening and Biological Evaluation of a Calpain Inhibitor for Prevention of Selenite-Induced Cataractogenesis in an in Vitro System. *Journal of Chemical Information and Modeling*, *55*(8), 1686–1697. doi:10.1021/acs.jcim.5b00092 PMID:26270943

Murray, C. W., Baxter, C. A., & Frenkel, A. D. (1999). The sensitivity of the results of molecular docking to induced fit effects: Application to thrombin, thermolysin and neuraminidase. *Journal of Computer-Aided Molecular Design*, *13*(6), 547–562. doi:10.1023/A:1008015827877 PMID:10584214

Murray, C. W., & Verdonk, M. L. (2002). The consequences of translational and rotational entropy lost by small molecules on binding to proteins. *Journal of Computer-Aided Molecular Design*, *16*(10), 741–753. PMID:12650591

Musil, D., Zucic, D., Turk, D., Engh, R. A., Mayr, I., Huber, R., & Bode, W. et al. (1991). The refined 2.15 Å X-ray crystal structure of human liver cathepsin B: The structural basis for its specificity. *The EMBO Journal*, *9*, 2321–2330. PMID:1868826

Nabuurs, S. B., Wagener, M., & de Vlieg, J. (2007). A flexible approach to induced fit docking. *Journal of Medicinal Chemistry*, *50*(26), 6507–6518. doi:10.1021/jm070593p PMID:18031000

Namasivayam, V., & Gunther, R. (2007). pso@autodock: A fast flexible molecular docking program based on Swarm intelligence. *Chemical Biology & Drug Design*, *70*(6), 475–484. doi:10.1111/j.1747-0285.2007.00588.x PMID:17986206

Nandigam, R. K., Kim, S., Singh, J., & Chuaqui, C. (2009). Position Specific Interaction Dependent Scoring Technique for Virtual Screening Based on Weighted Protein– Ligand Interaction Fingerprint Profiles. *Journal of Chemical Information and Modeling*, *49*(5), 1185–1192. doi:10.1021/ci800466n PMID:19415918

Nantasenamat, C., Isarankura-Na-Ayudhya, C., Naenna, T., & Prachayasittikul, V. (2009). A practical overview of quantitative structure-activity relationship. *EXCLI Journal*, *8*, 74–88.

Neudert, G., & Klebe, G. (2011). DSX: A knowledge-based scoring function for the assessment of protein-ligand complexes. *Journal of Chemical Information and Modeling*, *51*(10), 2731–2745. doi:10.1021/ci200274q

Ng, H. W., Zhang, W., Shu, M., Luo, H., Ge, W., Perkins, R., & Hong, H. et al. (2014). Competitive molecular docking approach for predicting estrogen receptor subtype α agonists and antagonists. *BMC Bioinformatics*, *15*(Suppl 1), S4. doi:10.1186/1471-2105-15-S11-S4 PMID:25349983

Ng, M. C., Fong, S., & Siu, S. W. (2015). PSOVina: The hybrid particle swarm optimization algorithm for protein-ligand docking. *Journal of Bioinformatics and Computational Biology*, *13*(3), 1541007. doi:10.1142/S0219720015410073 PMID:25800162

Nguyen, E. D., Norn, C., Frimurer, T. M., & Meiler, J. (2013). Assessment and Challenges of Ligand Docking into Comparative Models of G-Protein Coupled Receptors. *PLoS ONE*, *8*(7), e67302. PMID:23844000

Nicholls, A., Mobley, D. L., Guthrie, J. P., Chodera, J. D., Bayly, C. I., Cooper, M. D., & Pande, V. S. (2008). Predicting small-molecule solvation free energies: An informal blind test for computational chemistry. *Journal of Medicinal Chemistry*, *51*(4), 769–779. PMID:18215013

Nicolaou, C. A., & Brown, N. (2013). Multi-objective optimization methods in drug design. *Drug Discovery Today. Technologies*, *10*(3), e427–e435. doi:10.1016/j.ddtec.2013.02.001 PMID:24050140

Nielsen, M., Lundegaard, C., Lund, O., & Petersen, T. N. (2010). CPHmodels-3.0--remote homology modeling using structure-guided sequence profiles. *Nucleic Acids Res, 38*, W576-581. doi: 10.1093/nar/gkq535

Nillegoda, N. B., Kirstein, J., Szlachcic, A., Berynskyy, M., Stank, A., Stengel, F., & Bukau, B. et al. (2015). Crucial HSP70 co-chaperone complex unlocks metazoan protein disaggregation. *Nature*, *524*(7564), 247–251. PMID:26245380

Nisius, B., & Gohlke, H. (2012). Alignment-independent comparison of binding sites based on DrugScore potential fields encoded by 3D Zernike descriptors. *Journal of Chemical Information and Modeling*, *52*(9), 2339–2347. doi:10.1021/ci300244y PMID:22881363

Noguti, T., & Go, N. (1989). Structural basis of hierarchical multiple substates of a protein. I: Introduction. *Proteins*, *5*(2), 97–103. doi:10.1002/prot.340050203 PMID:2748581

Nolan, T. L., Lapinsky, D. J., Talbot, J. N., Indarte, M., Liu, Y., Manepalli, S., & Surratt, C. K. et al. (2011). Identification of a novel selective serotonin reuptake inhibitor by coupling monoamine transporter-based virtual screening and rational molecular hybridization. *ACS Chemical Neuroscience*, *2*(9), 544–552. doi:10.1021/cn200044x PMID:21966587

Norel, R., Fischer, D., Wolfson, H. J., & Nussinov, R. (1994). Molecular surface recognition by a computer vision-based technique. *Protein Engineering*, *7*(1), 39–46. doi:10.1093/protein/7.1.39 PMID:8140093

Norel, R., Petrey, D., Wolfson, H. J., & Nussinov, R. (1999). Examination of shape complementarity in docking of unbound proteins. *Proteins*, *36*(3), 307–317. doi:10.1002/(SICI)1097-0134(19990815)36:3<307::AID-PROT5>3.0.CO;2-R PMID:10409824

Obermajer, N., Sattin, S., Colombo, C., Bruno, M., Švajger, U., Anderluh, M., & Bernardi, A. (2011). Design, synthesis and activity evaluation of mannose-based DC-SIGN antagonists. *Molecular Diversity*, *15*(2), 347–360. doi:10.1007/s11030-010-9285-y

O'Boyle, N. M., Banck, M., James, C. A., Morley, C., Vandermeersch, T., & Hutchison, G. R. (2011). Open Babel: An open chemical toolbox. *Journal of Cheminformatics*, *3*(1), 33. doi:10.1186/1758-2946-3-33 PMID:21982300

Oda, A., Tsuchida, K., Takakura, T., Yamaotsu, N., & Hirono, S. (2006). Comparison of consensus scoring strategies for evaluating computational models of protein-ligand complexes. *Journal of Chemical Information and Modeling*, *46*(1), 380–391. doi:10.1021/ci050283k

Ogungbe, I. V., & Setzer, W. N. (2013). *In-silico* Leishmania target selectivity of antiparasitic terpenoids. *Molecules (Basel, Switzerland)*, *18*(7), 7761–7847. doi:10.3390/molecules18077761 PMID:23823876

Okimoto, N., Futatsugi, N., Fuji, H., Suenaga, A., Morimoto, G., Yanai, R., & Taiji, M. et al. (2009). High-performance drug discovery: Computational screening by combining docking and molecular dynamics simulations. *PLoS Computational Biology*, *5*(10), e1000528. doi:10.1371/journal.pcbi.1000528 PMID:19816553

Olimpieri, P. P., Chailyan, A., Tramontano, A., & Marcatili, P. (2013). Prediction of site-specific interactions in antibody-antigen complexes: The proABC method and server. *Bioinformatics (Oxford, England)*, *29*(18), 2285–2291. doi:10.1093/bioinformatics/btt369 PMID:23803466

Oloff, S., Zhang, S., Sukumar, N., Breneman, C., & Tropsha, A. (2006). Chemometric analysis of ligand receptor complementarity: Identifying complementary ligands based on receptor information (COLIBRI). *Journal of Chemical Information and Modeling*, *46*(2), 844–851. doi:10.1021/ci050065r PMID:16563016

Omori, S., & Kitao, A. (2013). CyClus: A fast, comprehensive cylindrical interface approximation clustering/reranking method for rigid-body protein-protein docking decoys. *Proteins*, *81*(6), 1005–1016. doi:10.1002/prot.24252 PMID:23344972

Oshiro, C. M., Kuntz, I. D., & Dixon, J. S. (1995). Flexible ligand docking using a genetic algorithm. *Journal of Computer-Aided Molecular Design*, *9*(2), 113–130. doi:10.1007/BF00124402 PMID:7608743

Ouyang, X., Zhou, S., Ge, Z., Li, R., & Kwoh, C. K. (2013). CovalentDock Cloud: a web server for automated covalent docking. *Nucleic Acids Res, 41*, W329-332.

Ou-Yang, S. S., Lu, J. Y., Kong, X. Q., Liang, Z. J., Luo, C., & Jiang, H. (2012). Computational drug discovery. *Acta Pharmacologica Sinica*, *33*(9), 1131–1140. doi:10.1038/aps.2012.109 PMID:22922346

Ouyang, X., Zhou, S., Ge, Z., Li, R., & Kwoh, C. K. (2013). CovalentDock Cloud: A web server for automated covalent docking. *Nucleic Acids Research*, *329-332*. doi:10.1093/nar/gkt406

Ouyang, X., Zhou, S., Su, C. T. T., Ge, Z., Li, R., & Kwoh, C. K. (2013). CovalentDock: Automated covalent docking with parameterized covalent linkage energy estimation and molecular geometry constraints. *Journal of Computational Chemistry*, *34*(4), 326–336. doi:10.1002/jcc.23136

Pak, Y., & Wang, S. (2000). Application of a Molecular Dynamics Simulation Method with a Generalized Effective Potential to the Flexible Molecular Docking Problems. *The Journal of Physical Chemistry B*, *104*(2), 354–359. doi:10.1021/jp993073h

Pala, N., Stevaert, A., Dallocchio, R., Dessi, A., Rogolino, D., Carcelli, M., . . . Naesens, L. (2015). Virtual screening and biological validation of novel influenza virus PA endonuclease inhibitors. *ACS Medicinal Chemistry Letters ACS Med. Chem. Lett.*

Palma, P. N., Krippahl, L., Wampler, J. E., & Moura, J. J. (2000). BiGGER: A new (soft) docking algorithm for predicting protein interactions. *Proteins*, *39*(4), 372–384. doi:10.1002/(SICI)1097-0134(20000601)39:4<372::AID-PROT100>3.0.CO;2-Q PMID:10813819

Palmer, J. T., Bryant, C., Wang, D.-X., Davis, D. E., Setti, E. L., Rydzewski, R. M., & Rodan, G. et al. (2005). Design and synthesis of tri-ring P3 benzamide-containing aminonitriles as potent, selective, orally effective inhibitors of cathepsin K. *Journal of Medicinal Chemistry*, *48*(24), 7520–7534. doi:10.1021/jm058198r

Pandya, P., Islam, M. M., Kumar, G. S., Jayaram, B., & Kumar, S. (2010). DNA minor groove binding of small molecules: Experimental and computational evidence. *Journal of Chemical Sciences*, *122*(2), 247–257. doi:10.1007/s12039-010-0029-4

Pang, Y.-P., Perola, E., Xu, K., & Prendergast, F. G. (2001). EUDOC: A computer program for identification of drug interaction sites in macromolecules and drug leads from chemical databases. *Journal of Computational Chemistry*, *22*(15), 1750–1771. doi:10.1002/jcc.1129 PMID:12116409

Papoian, G. A., Ulander, J., & Wolynes, P. G. (2003). Role of water mediated interactions in protein-protein recognition landscapes. *Journal of the American Chemical Society*, *125*(30), 9170–9178. PMID:15369374

Paquet, E., & Viktor, H. L. (2013). Macromolecular structure comparison and docking: An algorithmic review. *Current Pharmaceutical Design*, *19*(12), 2183–2193. doi:10.2174/1381612811319120006 PMID:23016846

Park, B. H., & Levitt, M. (1995). The complexity and accuracy of discrete state models of protein structure. *J Mol Biol*, *249*(2), 493-507. doi: S0022-2836(85)70311-7

Park, J.-Y., & Harris, D. (2003). Construction and Assessment of Models of CYP2E1: Predictions of Metabolism from Docking, Molecular Dynamics, and Density Functional Theoretical Calculations. *Journal of Medicinal Chemistry*, *46*(9), 1645–1660. doi:10.1021/jm020538a PMID:12699383

Parulekar, R. S., Barage, S. H., Jalkute, C. B., Dhanavade, M. J., & Sonawane, K. D. (2013). Homology Modeling, Molecular Docking and DNA Binding Studies of Nucleotide Excision Repair UvrC Protein from *M. tuberculosis*. *The Protein Journal*, *32*(6), 467–476. doi:10.1007/s10930-013-9506-1 PMID:23963890

Patel, P. D., Patel, M. R., Kaushik-Basu, N., & Talele, T. T. (2008). 3D QSAR and molecular docking studies of benzimidazole derivatives as hepatitis C virus NS5B polymerase inhibitors. *Journal of Chemical Information and Modeling*, *48*(1), 42–55. doi:10.1021/ci700266z PMID:18076152

Patil, S. A. (2012). Role of Medicinal Chemist in the Modern Drug Discovery and Development. *Organic Chemistry: Current Research*, *1*, e110.

Paul, N., & Rognan, D. (2002). ConsDock: A new program for the consensus analysis of protein–ligand interactions. *Proteins: Structure, Function, and Bioinformatics*, *47*(4), 521–533. doi:10.1002/prot.10119 PMID:12001231

Pedotti, M., Simonelli, L., Livoti, E., & Varani, L. (2011). Computational Docking of Antibody-Antigen Complexes, Opportunities and Pitfalls Illustrated by Influenza Hemagglutinin. *International Journal of Molecular Sciences*, *12*(1), 226–251. doi:10.3390/ijms12010226 PMID:21339984

Pérez-Nueno, V. I., Rabal, O., Borrell, J. I., & Teixidó, J. (2009). APIF: A new interaction fingerprint based on atom pairs and its application to virtual screening. *Journal of Chemical Information and Modeling*, *49*(5), 1245–1260. doi:10.1021/ci900043r PMID:19364101

Pettersen, E. F., Goddard, T. D., Huang, C. C., Couch, G. S., Greenblatt, D. M., Meng, E. C., & Ferrin, T. E. (2004). UCSF Chimera—a visualization system for exploratory research and analysis. *Journal of Computational Chemistry*, *25*(13), 1605–1612. doi:10.1002/jcc.20084 PMID:15264254

Pevzner, Y., Frugier, E., Schalk, V., Caflisch, A., & Woodcock, H. L. (2014). Fragment-based docking: Development of the CHARMMing Web user interface as a platform for computer-aided drug design. *Journal of Chemical Information and Modeling*, *54*(9), 2612–2620. doi:10.1021/ci500322k PMID:25151852

Pfeffer, P., Neudert, G., & Klebe, G. (2008). DrugScoreFP: Profiling protein-ligand interactions using fingerprint simplicity paired with knowledge-based potential fields. *Chemistry Central Journal*, *2*(Suppl 1), S16. doi:10.1186/1752-153X-2-S1-S16

Pham, T. A., & Jain, A. N. (2008). Customizing scoring functions for docking. *Journal of Computer-Aided Molecular Design*, *22*(5), 269–286. PMID:18273558

Philips, A., Milanowska, K., Lach, G., & Bujnicki, J. M. (2013). LigandRNA: Computational predictor of RNA-ligand interactions. *RNA (New York, N.Y.)*, *19*(12), 1605–1616. doi:10.1261/rna.039834.113 PMID:24145824

Phillips, J. C., Braun, R., Wang, W., Gumbart, J., Tajkhorshid, E., Villa, E., & Schulten, K. et al. (2005). Scalable molecular dynamics with NAMD. *Journal of Computational Chemistry*, *26*(16), 1781–1802. doi:10.1002/jcc.20289 PMID:16222654

Pierce, B. G., Hourai, Y., & Weng, Z. (2011). Accelerating protein docking in ZDOCK using an advanced 3D convolution library. *PLoS ONE*, *6*(9), e24657. doi:10.1371/journal.pone.0024657 PMID:21949741

Pierce, B. G., Wiehe, K., Hwang, H., Kim, B. H., Vreven, T., & Weng, Z. (2014). ZDOCK server: Interactive docking prediction of protein-protein complexes and symmetric multimers. *Bioinformatics (Oxford, England)*, *30*(12), 1771–1773. doi:10.1093/bioinformatics/btu097 PMID:24532726

Pierce, B., Tong, W., & Weng, Z. (2005). M-ZDOCK: A grid-based approach for Cn symmetric multimer docking. *Bioinformatics (Oxford, England)*, *21*(8), 1472–1478. doi:10.1093/bioinformatics/bti229 PMID:15613396

Pierce, B., & Weng, Z. (2007). ZRANK: Reranking Protein Docking Predictions With an Optimized Energy Function. *Proteins*, *67*(4), 1078–1086. doi:10.1002/prot.21373 PMID:17373710

Pirhadi, S., Shiri, F., & Ghasemi, J. B. (2013). Methods and applications of structure based pharmacophores in drug discovery. *Current Topics in Medicinal Chemistry*, *13*(9), 1036–1047. doi:10.2174/1568026611313090006 PMID:23651482

Plewczynski, D., La'zniewski, M., Augustyniak, R., & Ginalski, K. (2011). Can we trust docking results? Evaluation of seven commonly used programs on pdbbind database. *Journal of Computational Chemistry*, *32*(4), 742–755. PMID:20812323

Plewczynski, D., Lazniewski, M., Augustyniak, R., & Ginalski, K. (2010). Can we trust docking results? Evaluation of seven commonly used programs on PDB bind database. *Journal of Computational Chemistry*, *32*(4), 742–755. doi:10.1002/jcc.21643 PMID:20812323

Plewczynski, D., Lazniewski, M., von Grotthuss, M., Rychlewski, L., & Ginalski, K. (2011). VoteDock: Consensus docking method for prediction of protein-ligand interactions. *Journal of Computational Chemistry*, *32*(4), 568–581. doi:10.1002/jcc.21642 PMID:20812324

Poli, G., Giuntini, N., Martinelli, A., & Tuccinardi, T. (2015). Application of a FLAP-Consensus Docking Mixed Strategy for the Identification of New Fatty Acid Amide Hydrolase Inhibitors. *Journal of Chemical Information and Modeling*, *55*(3), 667–675. doi:10.1021/ci5006806 PMID:25746133

Politi, A., Durdagi, S., Moutevelis-Minakakis, P., Kokotos, G., & Mavromoustakos, T. (2010). Development of accurate binding affinity predictions of novel renin inhibitors through molecular docking studies. *Journal of Molecular Graphics & Modelling*, *29*(3), 425–435. doi:10.1016/j.jmgm.2010.08.003 PMID:20855222

Pons, J. L., & Labesse, G. (2009). @TOME-2: a new pipeline for comparative modeling of protein-ligand complexes. *Nucleic Acids Res*, *37*, W485-491. doi: 10.1093/nar/gkp368

Pons, C., Jimenez-Gonzalez, D., Gonzalez-Alvarez, C., Servat, H., Cabrera-Benitez, D., Aguilar, X., & Fernandez-Recio, J. (2012). Cell-Dock: High-performance protein-protein docking. *Bioinformatics (Oxford, England)*, *28*(18), 2394–2396. doi:10.1093/bioinformatics/bts454 PMID:22815362

Pons, C., Solernou, A., Perez-Cano, L., Grosdidier, S., & Fernandez-Recio, J. (2010). Optimization of pyDock for the new CAPRI challenges: Docking of homology-based models, domain-domain assembly and protein-RNA binding. *Proteins*, *78*(15), 3182–3188. doi:10.1002/prot.22773 PMID:20602351

Popov, P., Ritchie, D. W., & Grudinin, S. (2014). DockTrina: Docking triangular protein trimers. *Proteins: Structure. Function and Bioinformatics*, *82*(1), 34–44. doi:10.1002/prot.24344

Porollo, A., & Meller, J. (2010). POLYVIEW-MM: web-based platform for animation and analysis of molecular simulations. *Nucleic Acids Res, 38*, W662-666. doi: 10.1093/nar/gkq445

Pottel, J., Therrien, E., Gleason, J. L., & Moitessier, N. (2014). Docking ligands into flexible and solvated macromolecules. 6. Development and application to the docking of HDACs and other zinc metalloenzymes inhibitors. *Journal of Chemical Information and Modeling*, *54*(1), 254–265. doi:10.1021/ci400550m

Purisima, E. O., & Hogues, H. (2012). Protein-ligand binding free energies from exhaustive docking. *The Journal of Physical Chemistry B*, *116*(23), 6872–6879. doi:10.1021/jp212646s PMID:22432509

Purisima, E., & Sulea, T. (2013). Solvation models: Theory and validation. *Current Pharmaceutical Design*, *20*(20), 3266–3280. PMID:23947651

Qiu, J., Sheffler, W., Baker, D., & Noble, W. S. (2008). Ranking predicted protein structures with support vector regression. *Proteins: Structure, Function, and Bioinformatics*, *71*(3), 1175–1182. doi:10.1002/prot.21809 PMID:18004754

Qiu, W. Q., Walsh, D. M., Ye, Z., Vekrellis, K., Zhang, J., Podlisny, M. B., & Selkoe, D. J. et al. (1998). Insulin-degrading enzyme regulates extra-cellular levels of amyloid β-protein by degradation. *The Journal of Biological Chemistry*, *273*(49), 32730–32738. doi:10.1074/jbc.273.49.32730 PMID:9830016

Rajkhowa, S., & Deka, R. C. (2014). DFT based QSAR/QSPR models in the development of novel anti-tuberculosis drugs targeting Mycobacterium tuberculosis. *Current Pharmaceutical Design*, *20*(27), 4455–4473. doi:10.2174/13816128196661311 18165824 PMID:24245759

Rajkhowa, S., Hussain, I., Hazarika, K. K., Sarmah, P., & Deka, R. C. (2013). Quantitative structure-activity relationships of the antimalarial agent artemisinin and some of its derivatives - a DFT approach. *Combinatorial Chemistry & High Throughput Screening*, *16*(8), 590–602. doi:10.2174/1386207311316080002 PMID:23597248

Rajkhowa, S., Jha, A. N., & Deka, R. C. (2015). Anti-tubercular drug development: Computational strategies to identify potential compounds. *Journal of Molecular Graphics & Modelling*, *62*, 56–68. doi:10.1016/j.jmgm.2015.09.007 PMID:26386453

Ranson, M., Mansoor, W., & Jayson, G. (2002). ZD1839 (IRESSA): A selective EGFR-TK inhibitor. *Expert Review of Anticancer Therapy*, *2*(2), 161–168. doi:10.1586/14737140.2.2.161 PMID:12113238

Rappe, A. K., Casewit, C. J., Colwell, K. S., Goddard III, W. A., & Skiff, W. M. (1992). UFF, a full periodic table force field for molecular mechanics and molecular dynamics simulations. *Journal of the American Chemical Society*, *114*(25), 10024–10035. doi:10.1021/ja00051a040

Rarey, M., Kramer, B., & Lengauer, T. (1999). The particle concept: Placing discrete water molecules during protein-ligand docking predictions. *Proteins*, *34*(1), 17–28. doi:10.1002/(SICI)1097-0134(19990101)34:1<17::AID-PROT3>3.0.CO;2-1 PMID:10336380

Rarey, M., Kramer, B., Lengauer, T., & Klebe, G. (1996). A fast flexible docking method using an incremental construction algorithm. *Journal of Molecular Biology*, *261*(3), 470–489. doi:10.1006/jmbi.1996.0477 PMID:8780787

Rarey, M., Wefing, S., & Lengauer, T. (1996). Placement of medium-sized molecule fragments into active site of proteins. *Journal of Computer-Aided Molecular Design*, *10*(1), 41–54. doi:10.1007/BF00124464

Raub, S., Steffen, A., Kamper, A., & Marian, C. M. (2008). AIScore chemically diverse empirical scoring function employing quantum chemical binding energies of hydrogen-bonded complexes. *Journal of Chemical Information and Modeling*, *48*(7), 1492–1510. doi:10.1021/ci7004669 PMID:18597446

Ravikant, D. V. S., & Elber, R. (2010). PIE-efficient filters and coarse grained potentials for unbound protein-protein docking. *Proteins*, *78*(2), 400–419. doi:10.1002/prot.22550 PMID:19768784

Ravikant, D. V. S., & Elber, R. (2011). Energy design for protein-protein interactions. *The Journal of Chemical Physics*, *135*(6), 065102. doi:10.1063/1.3615722 PMID:21842951

Reddy, A. S., Pati, S. P., Kumar, P. P., Pradeep, H. N., & Sastry, G. N. (2007). Virtual Screening in Drug Discovery–A Computational Perspective. *Current Protein & Peptide Science*, *8*, 329–351. doi:10.2174/138920307781369427 PMID:17696867

Rees, D. C., Congreve, M., Murray, C. W., & Carr, R. (2004). Fragment-based lead discovery. *Nat Rev Drug Discov*, *3*(8), 660-672.

Reich, S., Thomsen, C., & Maultzsch, J. (2004). Carbon nanotubes: basic concepts and physical properties. Wiley-VCH.

Rejto, P. A., & Verkhivker, G. M. (1996). Unraveling principles of lead discovery: From unfrustrated energy landscapes to novel molecular anchors. *Proceedings of the National Academy of Sciences of the United States of America*, *93*(17), 8945–8950. PMID:8799133

Ren, J., Williams, N., Clementi, L., Krishnan, S., & Li, W. W. (2010). Opal web services for biomedical applications. *Nucleic Acids Res, 38*, W724-731. doi: 10.1093/nar/gkq503

Renner, S., Derksen, S., Radestock, S., & Mörchen, F. (2008). Maximum common binding modes (MCBM): Consensus docking scoring using multiple ligand information and interaction fingerprints. *Journal of Chemical Information and Modeling*, *48*(2), 319–332. doi:10.1021/ci7003626 PMID:18211051

Repasky, M. P., Shelley, M., & Friesner, R. A. (2007). Flexible ligand docking with Glide. Current Protocols in Bioinformatics. http://doi.org/ doi:10.1002/0471250953.bi0812s18

Ricci, C. G., & Netz, P. A. (2009). Docking studies on DNA-ligand interactions: Building and application of a protocol to identify the binding mode. *Journal of Chemical Information and Modeling*, *49*(8), 1925–1935. doi:10.1021/ci9001537 PMID:19655805

Rich, D. H., Bursavich, M. G., & Estiarte, M. A. (2002). Discovery of nonpeptide, peptidomimetic peptidase inhibitors that target alternate enzyme active site conformations. *Biopolymers*, *66*(2), 115–125. doi:10.1002/bip.10231 PMID:12325161

Ritchie, D. W., & Kemp, G. J. (2000). Protein docking using spherical polar Fourier correlations. *Proteins*, *39*(2), 178–194. doi:10.1002/(SICI)1097-0134(20000501)39:2<178::AID-PROT8>3.0.CO;2-6 PMID:10737939

Ritchie, D. W., Kozakov, D., & Vajda, S. (2008). Accelerating and focusing protein-protein docking correlations using multi-dimensional rotational FFT generating functions. *Bioinformatics (Oxford, England)*, *24*(17), 1865–1873. doi:10.1093/bioinformatics/btn334 PMID:18591193

Ritchie, D. W., & Venkatraman, V. (2010). Ultra-fast FFT protein docking on graphics processors. *Bioinformatics (Oxford, England)*, *26*(19), 2398–2405. doi:10.1093/bioinformatics/btq444 PMID:20685958

Ritchie, D. W., Venkatraman, V., & Mavridis, L. (2010). Using graphics processors to accelerate protein docking calculations. *Studies in Health Technology and Informatics*, *159*, 146–155. PMID:20543434

Rognan, D., Lauemoller, S. L., Holm, A., Buus, S., & Tschinke, V. (1999). Predicting binding affinities of protein ligands from three-dimensional models: Application to peptide binding to class I major histocompatibility proteins. *Journal of Medicinal Chemistry*, *42*(22), 4650–4658. doi:10.1021/jm9910775

Rognan, D., Lauemøller, S. L., Holm, A., Buus, S., & Tschinke, V. (1999). Predicting binding affinities of protein ligands from three-dimensional models: Application to peptide binding to class i major histocompatibility proteins. *Journal of Medicinal Chemistry*, *42*(22), 4650–4658. PMID:10579827

Rohs, R., Bloch, I., Sklenar, H., & Shakked, Z. (2005). Molecular flexibility in ab initio drug docking to DNA: Binding-site and binding-mode transitions in all-atom Monte Carlo simulations. *Nucleic Acids Research*, *33*(22), 7048–7057. doi:10.1093/nar/gki1008 PMID:16352865

Rosenfeld, R., Vajda, S., & Delisi, C. (1995). Flexible Docking and Design. *Annual Review of Biophysics*, *24*(1), 677–700. doi:10.1146/annurev.bb.24.060195.003333 PMID:7663131

Rost, B. (1999). Twilight zone of protein sequence alignments. *Protein Engineering*, *12*(2), 85–94. doi:10.1093/protein/12.2.85 PMID:10195279

Rost, B., Fariselli, P., & Casadio, R. (1996). Topology prediction for helical transmembrane proteins at 86% accuracy. *Protein Science*, *5*(8), 1704–1718. doi:10.1002/pro.5560050824 PMID:8844859

Roy, K., Kar, S., & Das, R. N. (2015). Other Related Techniques. In K. R. K. N. Das (Ed.), *Understanding the Basics of QSAR for Applications in Pharmaceutical Sciences and Risk Assessment* (pp. 357–425). Boston: Academic Press.

Roy, K., Kar, S., & Das, R. N. (2015). *Understanding the basics of qsar for applications in pharmaceutical sciences and risk assessment*. Elsevier Inc.

Rueda, M., Bottegoni, G., & Abagyan, R. (2009). Consistent improvement of cross-docking results using binding site ensembles generated with elastic network normal modes. *Journal of Chemical Information and Modeling*, *49*(3), 716–725. doi:10.1021/ci8003732 PMID:19434904

Ruiz-Carmona, S., Alvarez-Garcia, D., Foloppe, N., Garmendia-Doval, A. B., Juhos, S., Schmidtke, P., & Morley, S. D. et al. (2014). rDock: A fast, versatile and open source program for docking ligands to proteins and nucleic acids. *PLoS Computational Biology*, *10*, e1003571.

Rush, T. S., Grant, J. A., Mosyak, L., & Nicholls, A. (2005). A Shape-Based 3-D Scaffold Hopping Method and Its Application to a Bacterial Protein–Protein Interaction. *Journal of Medicinal Chemistry*, *48*(5), 1489–1495. doi:10.1021/jm040163o PMID:15743191

Ruvinsky, A. (2007). Calculations of protein-ligand binding entropy of relative and overall molecular motions. *Journal of Computer-Aided Molecular Design*, *21*(7), 361–370. doi:10.1007/s10822-007-9116-0 PMID:17503189

Ruvinsky, A. M. (2007). Role of binding entropy in the refinement of protein–ligand docking predictions: Analysis based on the use of 11 scoring functions. *Journal of Computational Chemistry*, *28*(8), 1364–1372. PMID:17342720

Ruvinsky, A. M., & Kozintsev, A. (2005). New and fast statistical-thermodynamic method for computation of protein-ligand binding entropy substantially improves docking accuracy. *Journal of Computational Chemistry*, *26*(11), 1089–1095. PMID:15929088

Sable, R., & Jois, S. (2015). Surfing the Protein-Protein Interaction Surface Using Docking Methods: Application to the Design of PPI Inhibitors. *Molecules (Basel, Switzerland)*, *20*(6), 11569–11603. doi:10.3390/molecules200611569 PMID:26111183

Saikia, N., Rajkhowa, S., & Deka, R. C. (2013). Density functional and molecular docking studies towards investigating the role of single-wall carbon nanotubes as nanocarrier for loading and delivering of pyrazinamide antitubercular drug on to pncA protein. *Journal of Computer-Aided Molecular Design*, *27*(3), 57–276. doi:10.1007/s10822-013-9638-6

Saladin, A., Rey, J., Thevenet, P., Zacharias, M., Moroy, G., & Tuffery, P. (2014). PEP-SiteFinder: a tool for the blind identification of peptide binding sites on protein surfaces. *Nucleic Acids Res, 42*, W221-226.

Salaniwal, S., Manas, E. S., Alvarez, J. C., & Unwalla, R. J. (2007). Critical evaluation of methods to incorporate entropy loss upon binding in high-throughput docking. *Proteins: Structure, Function, and Bioinformatics*, *66*(2), 422–435. PMID:17068803

Salemme, F. R. (1976). An hypothetical structure for an intermolecular electron transfer complex of cytochromes c and b5. *Journal of Molecular Biology*, *102*(3), 563–568. doi:10.1016/0022-2836(76)90334-X PMID:178879

Sali, A. (1995). Comparative protein modeling by satisfaction of spatial restraints. *Molecular Medicine Today*, *1*(6), 270–277. doi:10.1016/S1357-4310(95)91170-7 PMID:9415161

Sali, A., & Blundell, T. L. (1993). Comparative protein modelling by satisfaction of spatial Restraints. *Journal of Molecular Biology*, *234*(3), 779–815. doi:10.1006/jmbi.1993.1626 PMID:8254673

Salomon-Ferrer, R., Case, D. A., & Walker, R. C. (2012). An overview of the Amber biomolecular simulation package. *WIREs Computational Molecular Science*, *3*(2), 198–210. doi:10.1002/wcms.1121

Sandak, B., Nussinov, R., & Wolfson, H. J. (1995). An automated computer vision and robotics-based technique for 3-D flexible biomolecular docking and matching. *Computer Applications in the Biosciences*, *11*, 87–99. PMID:7796279

Sandal, M., Duy, T. P., Cona, M., Zung, H., Carloni, P., Musiani, F., & Giorgetti, A. (2013). GOMoDo: A GPCRs online modeling and docking webserver. *PLoS ONE*, *8*(9), e74092. doi:10.1371/journal.pone.0074092 PMID:24058518

Sander, C., & Schneider, R. (1991). Database of homology-derived protein structures and the structural meaning of sequence alignment. *Proteins*, *9*(1), 56–68. doi:10.1002/prot.340090107 PMID:2017436

Sauton, N., Lagorce, D., Villoutreix, B. O., & Miteva, M. A. (2008). MS-DOCK: Accurate multiple conformation generator and rigid docking protocol for multi-step virtual ligand screening. *BMC Bioinformatics*, *9*(1), 184. doi:10.1186/1471-2105-9-184

Sayle, R. A., & Milner-White, E. J. (1995). RASMOL: Biomolecular graphics for all. *Trends in Biochemical Sciences*, *20*(9), 374–376. doi:10.1016/S0968-0004(00)89080-5 PMID:7482707

Schapira, M., Raaka, B. M., Samuels, H. H., & Abagyan, R. (2001). *In-silico* discovery of novel Retinoic Acid Receptor agonist structures. *BMC Structural Biology*, *1*(1), 1. doi:10.1186/1472-6807-1-1 PMID:11405897

Scharfe, M., Pippel, M., & Sippl, W. (2012). Development of target-biased scoring functions for protein-ligand docking. *Journal of Cheminformatics*, 4(Suppl 1).

Schellhammer, I., & Rarey, M. (2004). FlexX-Scan: Fast, structure-based virtual screening. *Proteins*, *57*(3), 504–517. doi:10.1002/prot.20217 PMID:15382244

Schindler, C. E. M., de Vries, S. J., & Zacharias, M. (2015). iATTRACT: Simultaneous global and local interface optimization for protein–protein docking refinement. *Proteins: Structure, Function, and Bioinformatics*, *83*(2), 248–258. doi:10.1002/prot.24728 PMID:25402278

Schmidtke, P., Le Guilloux, V., Maupetit, J., & Tufféry, P. (2010). Fpocket: Online tools for protein ensemble pocket detection and tracking. *Nucleic Acids Research*, *38*(Web Server), W582–W589. doi:10.1093/nar/gkq383 PMID:20478829

Schmidt, T., Bergner, A., & Schwede, T. (2014). Modelling three-dimensional protein structures for applications in drug design. *Drug Discovery Today*, *19*(7), 890–897. doi:10.1016/j.drudis.2013.10.027 PMID:24216321

Schnecke, V., & Kuhn, L. A. (2000). Virtual Screening with Solvation and Ligand-Induced Complementarity. *Perspectives in Drug Discovery and Design*, *20*(1), 171–190. doi:10.1023/A:1008737207775

Schneider, G., & Böhm, H.-J. (2002). Virtual screening and fast automated docking methods. *Drug Discovery Today*, *7*(1), 64–70. doi:10.1016/S1359-6446(01)02091-8 PMID:11790605

Schneider, G., & Fechner, U. (2005). Computer-based de novo design of drug-like molecules. *Nature Reviews. Drug Discovery*, *4*(8), 649–663. doi:10.1038/nrd1799 PMID:16056391

Schneider, N., Hindle, S., Lange, G., Klein, R., Albrecht, J., Briem, H., & Rarey, M. et al. (2012). Substantial improvements in large-scale redocking and screening using the novel HYDE scoring function. *Journal of Computer-Aided Design*, *26*(6), 701–723. doi:10.1007/s10822-011-9531-0

Schneider, T., Kruse, T., Wimmer, R., Wiedemann, I., Sass, V., Pag, U., & Kristensen, H.-H. et al. (2010). Plectasin, a Fungal Defensin, Targets the Bacterial Cell Wall Precursor Lipid II. *Science*, *328*(5982), 1168–1172. doi:10.1126/science.1185723 PMID:20508130

Schneidman-Duhovny, D., Inbar, Y., Nussinov, R., & Wolfson, H. J. (2005). PatchDock and SymmDock: Servers for rigid and symmetric docking. *Nucleic Acids Research*, *33*(Web ServerSUPPL. 2), 363–367. doi:10.1093/nar/gki481 PMID:15980490

Schneidman-Duhovny, D., Ruth, N., & Wolfson, H. J. (2007). Automatic prediction of protein interactions with large scale motion. *Proteins*, *69*(4), 764–773. doi:10.1002/prot.21759 PMID:17886339

Schoch, G. A., Yano, J. K., Sansen, S., Dansette, P. M., Stout, C. D., & Johnson, E. F. (2008). Determinants of cytochrome p4502c8 substrate binding - structures of complexes with montelukast, troglitazone, felodipine, and 9-cis-retinoic acid. *The Journal of Biological Chemistry*, *283*(25), 17227–17237. doi:10.1074/jbc.M802180200 PMID:18413310

Schoch, G. A., Yano, J. K., Wester, M. R., Griffin, K. J., Stout, C. D., & Johnson, E. F. (2004). The Structure of Human Microsomal Cytochrome P450 3A4 Determined by X-ray Crystallography to 2.05-Å Resolution. *The Journal of Biological Chemistry*, *279*(37), 38091–38094. doi:10.1074/jbc.C400293200 PMID:15258162

Schroder, J., Klinger, A., Oellien, F., Marhofer, R. J., Duszenko, M., & Selzer, P. M. (2013). Docking-based virtual screening of covalently binding ligands: An orthogonal lead discovery approach. *Journal of Medicinal Chemistry*, *56*(4), 1478–1490. doi:10.1021/jm3013932

Schulz-Gasch, T. & Stahl, M. (2004). Scoring functions for protein–ligand interactions: a critical perspective. *Drug discovery today: technologies | lead optimization, 1*(3), 231-239.

Schüttelkopf, A. W., & van Aalten, D. M. F. (2004). PRODRG: A tool for high-throughput crystallography of protein-ligand complexes. *Acta Crystallographica. Section D, Biological Crystallography*, *60*(8), 1355–1363. doi:10.1107/S0907444904011679 PMID:15272157

Scior, T., Bender, A., Tresadern, G., Medina-Franco, J. L., Martínez-Mayorga, K., Langer, T., & Agrafiotis, D. K. et al. (2012). Recognizing pitfalls in virtual screening: A critical review. *Journal of Chemical Information and Modeling*, *52*(4), 867–881. doi:10.1021/ci200528d PMID:22435959

Scott, W. R. P., Hünenberger, P. H., Tironi, I. G., Mark, A. E., Billeter, S. R., Fennen, J., & van Gunsteren, W. F. et al. (1999). The GROMOS Biomolecular Simulation Program Package. *The Journal of Physical Chemistry A*, *103*(19), 3596–3607. doi:10.1021/jp984217f

Seebeck, B., Reulecke, I., Kamper, A., & Rarey, M. (2008). Modeling of metal interaction geometries for protein-ligand docking. *Proteins: Structure, Function, and Bioinformatics*, *71*(3), 1237–1254. doi:10.1002/prot.21818

Seifert, M. H. (2008). Optimizing the signal-to-noise ratio of scoring functions for protein-ligand docking. *Journal of Chemical Information and Modeling*, *48*(3), 602–612. PMID:18293951

Seifert, M. H. (2009). Robust optimization of scoring functions for a target class. *Journal of Computer-Aided Molecular Design*, *23*(9), 633–644. doi:10.1007/s10822-009-9276-1 PMID:19471858

Seifert, M. H. (2009). Targeted scoring functions for virtual screening. *Drug Discovery Today*, *14*(11), 562–569. PMID:19508918

Sheinerman, F. B., Norel, R., & Honig, B. (2000). Electrostatic aspects of protein-protein interactions. *Current Opinion in Structural Biology*, *10*(2), 153–159. doi:10.1016/S0959-440X(00)00065-8 PMID:10753808

Sherman, W., Day, T., Jacobson, M. P., Friesner, R. A., & Farid, R. (2006). Novel procedure for modeling ligand/receptor induced fit effects. *Journal of Medicinal Chemistry*, *49*(2), 534–553. doi:10.1021/jm050540c PMID:16420040

Shimoda, T., Suzuki, S., Ohue, M., Ishida, T., & Akiyama, Y. (2015). Protein-protein docking on hardware accelerators: Comparison of GPU and MIC architectures. *BMC Systems Biology*, *9*(Suppl 1), S6. doi:10.1186/1752-0509-9-S1-S6 PMID:25707855

Shin, W.-H., Heo, L., Lee, J., Ko, J., Seok, C., & Lee, J. (2011). LigDockCSA: Protein-ligand docking using conformational space annealing. *Journal of Computational Chemistry*, *32*(15), 3226–3232. doi:10.1002/jcc.21905 PMID:21837636

Shirvanyants, D., Alexandrova, A. N., & Dokholyan, N. V. (2011). Rigid substructure search. *Bioinformatics (Oxford, England)*, *27*(9), 1327–1329. doi:10.1093/bioinformatics/btr129 PMID:21460026

Shoemaker, B. A., Zhang, D., Tyagi, M., Thangudu, R. R., Fong, J. H., Marchler-Bauer, A., & Panchenko, A. R. et al. (2012). IBIS (Inferred Biomolecular Interaction Server) reports, predicts and integrates multiple types of conserved interactions for proteins. *Nucleic Acids Research*, *40*(Database issue), D834–D840. doi:10.1093/nar/gkr997 PMID:22102591

Shoichet, B. K., McGovern, S. L., Wei, B., & Irwin, J. J. (2002) Hits, leads and artifacts from virtual and high throughput screening. In *Proceedings of the International Beilstein Workshop*. Molecular Informatics: Confronting Complexity.

Shoichet, B. K. (2004). Virtual screening of chemical libraries. *Nature*, *432*(7019), 862–865. PMID:15602552

Shoichet, B. K., Kuntz, I. D., & Bodian, D. L. (2004). Molecular docking using shape descriptors. *Journal of Computational Chemistry*, *13*(3), 380–397. doi:10.1002/jcc.540130311

Shoichet, B. K., McGovern, S. L., Wei, B., & Irwin, J. J. (2002). Lead discovery using molecular docking. *Current Opinion in Chemical Biology*, *6*(4), 439–446. doi:10.1016/S1367-5931(02)00339-3 PMID:12133718

Sindhu, T., & Srinivasan, P. (2015). Identification of potential dual agonists of FXR and TGR5 using e-pharmacophore based virtual screening. *Molecular BioSystems*, *11*(5), 1305–1318. doi:10.1039/C5MB00137D PMID:25787676

Singh, J., Deng, Z., Narale, G., & Chuaqui, C. (2006). Structural interaction fingerprints: A new approach to organizing, mining, analyzing, and designing protein–small molecule complexes. *Chemical Biology & Drug Design*, *67*(1), 5–12. doi:10.1111/j.1747-0285.2005.00323.x PMID:16492144

Singh, J., Petter, R. C., Baillie, T. A., & Whitty, A. (2011). The resurgence of covalent drugs. *Nature Reviews. Drug Discovery*, *10*(4), 307–317. doi:10.1038/nrd3410 PMID:21455239

Singh, R., & Elizabeth Sobhia, M. (2011). Homology modeling of human CCR2 receptor. *Medicinal Chemistry Research*, *20*(9), 1704–1712. doi:10.1007/s00044-010-9497-9

Singh, T., Biswas, D., & Jayaram, B. (2011). AADS--an automated active site identification, docking, and scoring protocol for protein targets based on physicochemical descriptors. *Journal of Chemical Information and Modeling*, *51*(10), 2515–2527. doi:10.1021/ci200193z PMID:21877713

Singh, Y., Dolphin, G. T., Razkin, J., & Dumy, P. (2006). Synthetic Peptide templates for molecular recognition: Recent advances and applications. *ChemBioChem*, *7*(9), 1298–1314. doi:10.1002/cbic.200600078 PMID:16892470

Sinko, W., Lindert, S., & McCammon, J. A. (2013). Accounting for Receptor Flexibility and Enhanced Sampling Methods in Computer-Aided Drug Design: Accounting for Receptor Flexibility. *Chemical Biology & Drug Design*, *81*(1), 41–49. doi:10.1111/cbdd.12051 PMID:23253130

Sippl, M. J. (1990). Calculation of conformational ensembles from potentials of mena force: An approach to the knowledge-based prediction of local structures in globular proteins. *Journal of Molecular Biology*, *213*(4), 859–883. PMID:2359125

Sippl, M. J., Ortner, M., Jaritz, M., Lackner, P., & Flockner, H. (1996). Helmholtz free energies of atom pair interactions in proteins. *Folding & Design*, *1*(4), 289–298. doi:10.1016/S1359-0278(96)00042-9

Sircar, A., Kim, E. T., & Gray, J. J. (2009). RosettaAntibody: antibody variable region homology modeling server. *Nucleic Acids Res*, *37*, W474-479. doi: 10.1093/nar/gkp387

Skjaerven, L., Martinez, A., & Reuter, N. (2011). Principal component and normal mode analysis of proteins; a quantitative comparison using the GroEL subunit. *Proteins*, *79*(1), 232–243. doi:10.1002/prot.22875 PMID:21058295

Skolnick, J., Kihara, D., & Zhang, Y. (2004). Development and large scale benchmark testing of the PROSPECTOR_3 threading algorithm. *Proteins*, *56*(3), 502–518. doi:10.1002/prot.20106 PMID:15229883

Sliwoski, G., Kothiwale, S., Meiler, J., & Lowe, E. W. (2013). Computational Methods in Drug Discovery. *Pharmacological Reviews*, *66*(1), 334–395. doi:10.1124/pr.112.007336 PMID:24381236

Smith, G. R., & Sternberg, M. J. E. (n.d.). *Prediction of protein – protein interactions by docking methods*. Academic Press.

Smith, G. R., & Sternberg, M. J. E. (2002). Prediction of protein–protein interactions by docking methods. *Current Opinion in Structural Biology*, *12*(1), 28–35. doi:10.1016/S0959-440X(02)00285-3 PMID:11839486

Smith, J. (2005). Erlotinib: Small-molecule targeted therapy in the treatment of non-small-cell lung cancer. *Clinical Therapeutics*, *27*(10), 1513–1534. doi:10.1016/j.clinthera.2005.10.014 PMID:16330289

Smith, J., Vanoye, C. G., George, A. L., Meiler, J., & Sanders, C. R. (2007). Structural Models for the KCNQ1 Voltage-Gated Potassium Channel. *Biochemistry*, *46*(49), 14141–14152. doi:10.1021/bi701597s PMID:17999538

Smith, R. D., Dunbar, J. B. Jr, Ung, P. M.-U., Esposito, E. X., Yang, C.-Y., Wang, S., & Carlson, H. A. (2011). CSAR benchmark exercise of 2010: Combined evaluation across all submitted scoring functions. *Journal of Chemical Information and Modeling*, *51*(9), 2115–2131. PMID:21809884

Söding, J. (2005). Protein homology detection by HMM-HMM comparison. *Bioinformatics (Oxford, England)*, *21*(7), 951–960. doi:10.1093/bioinformatics/bti125 PMID:15531603

Sonawane, K. D., & Barage, S. H. (2014). Structural analysis of membrane bound hECE1 dimer using molecular modeling techniques: Insights into conformational changes and Aβ1–42 peptide binding. *Amino Acids*, *47*(3), 543–559. doi:10.1007/s00726-014-1887-8 PMID:25501500

Sondergaard, C. R., Garrett, A. E., Carstensen, T., Pollastri, G., & Nielsen, J. E. (2009). Structural artifacts in protein-ligand X-ray structures: Implications for the development of docking scoring functions. *Journal of Medicinal Chemistry*, *52*(18), 5673–5684. doi:10.1021/jm8016464 PMID:19711919

Songdej, N., & Rao, A. K. (2015). Hematopoietic transcription factor mutations and inherited platelet dysfunction. *F1000prime Reports*, *7*, 66. doi:10.12703/P7-66 PMID:26097739

Sotriffer, C. A., Gohlke, H., & Klebe, G. (2002). Docking into knowledge-based potential fields: A comparative evaluation of DrugScore. *Journal of Medicinal Chemistry*, *45*(10), 1967–1970. doi:10.1021/jm025507u PMID:11985464

Sotriffer, C. A., Sanschagrin, P., Matter, H., & Klebe, G. (2008). SFCscore: Scoring functions for affinity prediction of protein-ligand complexes. *Proteins: Structure, Function, and Bioinformatics*, *73*(2), 395–419. doi:10.1002/prot.22058

Sotriffer, C., Klebe, G., Stahl, M., & Bohm, H.-J. (2003). Docking and scoring functions/Virtual screening. In D. J. Abraham (Ed.), *Burger's Medicinal Chemistry and Drug Discovery, Drug Discovery* (Vol. 1, pp. 281–416). John Wiley & Sons. doi:10.1002/0471266949.bmc007

Sousa, S. F., Fernandes, P. A., & Ramos, M. J. (2006). Protein-ligand docking: Current status and future challenges. *Proteins*, *65*(1), 15–26. doi:10.1002/prot.21082 PMID:16862531

Sousa, S. F., Ribeiro, A. J., Coimbra, J. T., Neves, R. P., Martins, S. A., Moorthy, N. S., & Ramos, M. J. et al. (2013). Protein-ligand docking in the new millennium--a retrospective of 10 years in the field. *Current Medicinal Chemistry*, *20*(18), 2296–2314. doi:10.2174/0929867311320180002 PMID:23531220

Sousa, S., Cerqueira, N., Fernandes, P., & Ramos, M. (2010). Virtual screening in drug design and development. *Combinatorial Chemistry & High Throughput Screening*, *13*(5), 442–453. PMID:20236061

Sousa, S., Ribeiro, A., Coimbra, J., Neves, R., Martins, S., Moorthy, N., & Ramos, M. et al. (2013). Protein-ligand docking in the new millennium–a retrospective of 10 years in the field. *Current Medicinal Chemistry*, *20*(18), 2296–2314. PMID:23531220

Speck-Planche, A., & Cordeiro, M. N. (2011). Current drug design of anti-HIV agents through the inhibition of C-C chemokine receptor type 5. *Current Computer-aided Drug Design*, *7*(4), 238–248. doi:10.2174/157340911798260287 PMID:22050678

Spitzer, R., & Jain, A. N. (2012). Surflex-Dock: Docking benchmarks and real-world application. *Journal of Computer-Aided Molecular Design*, *26*(6), 687–699. doi:10.1007/s10822-011-9533-y PMID:22569590

Srivastav, V. K., & Tiwari, M. (2015). k-nearest neighbor molecular field analysis based 3D-QSAR and in silico ADME/T studies of cinnamoyl derivatives as HIV-1 integrase inhibitors. *Medicinal Chemistry Research*, *24*(2), 684–700. doi:10.1007/s00044-014-1183-x

Stahl, M., & Rarey, M. (2001). Detailed analysis of scoring functions for virtual screening. *Journal of Medicinal Chemistry*, *44*(7), 1035–1042. doi:10.1021/jm0003992

Stark, J. L., & Powers, R. (2012). Application of NMR and Molecular Docking in Structure-Based Drug Discovery. *Topics in Current Chemistry*, *326*, 1–34. doi:10.1007/128_2011_213 PMID:21915777

Stierand, K., Maaß, P. C., & Rarey, M. (2006). Molecular complexes at a glance: Automated generation of two-dimensional complex diagrams. *Bioinformatics (Oxford, England)*, *22*(14), 1710–1716. doi:10.1093/bioinformatics/btl150 PMID:16632493

Still, W. C., Tempczyk, A., Hawley, R. C., & Hendrickson, T. (1990). Semianalytical treatment of solvation for molecular mechanics and dynamics. *Journal of the American Chemical Society*, *112*(16), 6127–6129. doi:10.1021/ja00172a038

Storn, R., & Price, K. (1995). Differential Evolution - A Simple and Efficient Adaptive Scheme for Global Optimization over Continuous Spaces. Technical Report. Berkley, CA: International Computer Science Institute.

Straatsma, T. P., & McCammon, J. A. (1991). Multiconfiguration thermodynamic integration. *The Journal of Chemical Physics*, *95*(2), 1175–1118. doi:10.1063/1.461148

Stuart, B. G., Coxon, J. M., Morton, J. D., Abell, A. D., McDonald, D. Q., Aitken, S. G., & Bickerstaffe, R. et al. (2011). Molecular modeling: A search for a calpain inhibitor as a new treatment for cataractogenesis. *Journal of Medicinal Chemistry*, *54*(21), 7503–7522. doi:10.1021/jm200471r PMID:21955158

Subramanian, J., Sharma, S., & B-Rao, C. (2006). A Novel Computational Analysis of Ligand-Induced Conformational Changes in the ATP Binding Sites of Cyclin Dependent Kinases. *Journal of Medicinal Chemistry*, *49*(18), 5434–5441. doi:10.1021/jm060172s PMID:16942017

Subramanian, J., Sharma, S., & B-Rao, C. (2008). Modeling and Selection of Flexible Proteins for Structure-Based Drug Design: Backbone and Side Chain Movements in p38 MAPK. *ChemMedChem*, *3*(2), 336–344. doi:10.1002/cmdc.200700255 PMID:18081134

Sukhwani, B., & Herbordt, M. C. (2010). FPGA acceleration of rigid-molecule docking codes. *IET Comput Digit Tech*, *4*(3), 184–195. doi:10.1049/iet-cdt.2009.0013 PMID:21857870

Sun, H. (2008). Pharmacophore-based virtual screening. *Current Medicinal Chemistry*, *15*(10), 1018–1024. doi:10.2174/092986708784049630 PMID:18393859

Szilagyi, A., Grimm, V., Arakaki, K., & Skolnick, J. (2005). Prediction of physical protein – protein. *Most*, *2*, 1–16. PMID:16204844

Tahir, R. A., Sehgal, S. A., Khattak, N. A., Khan Khattak, J. Z., & Mir, A. (2013). Tumor necrosis factor receptor superfamily 10B (TNFRSF10B): An insight from structure modeling to virtual screening for designing drug against head and neck cancer. *Theoretical Biology & Medical Modelling*, *10*(1), 38. doi:10.1186/1742-4682-10-38 PMID:23724937

Takeda-Shitaka, M., Takaya, D., Chiba, C., Tanaka, H., & Umeyama, H. (2004). Protein structure prediction in structure based drug design. *Current Medicinal Chemistry*, *11*(5), 551–558. doi:10.2174/0929867043455837 PMID:15032603

Takeda, T., Corona, R. I., & Guo, J.-T. (2013). A knowledge-based orientation potential for transcription factor-DNA docking. *Bioinformatics (Oxford, England)*, *29*(3), 322–330. doi:10.1093/bioinformatics/bts699 PMID:23220572

Talele, T. T., Khedkar, S. A., & Rigby, A. C. (2010). Successful applications of computer aided drug discovery: Moving drugs from concept to the clinic. *Current Topics in Medicinal Chemistry*, *10*(1), 127–141. doi:10.2174/156802610790232251 PMID:19929824

Talukdar, A. S., & Wilson, D. L. (1999). Modeling and optimization of rotational C-arm stereoscopic X-ray angiography. *IEEE Transactions on Medical Imaging*, *18*(7), 604–616. doi:10.1109/42.790460 PMID:10504094

Tan, K. P., Varadarajan, R., & Madhusudhan, M. S. (2011). DEPTH: a web server to compute depth and predict small-molecule binding cavities in proteins. *Nucleic Acids Res, 39*, W242-248. doi: 10.1093/nar/gkr356

Tanaka, S., & Scheraga, H. A. (1976). Medium- and long-range interaction parameters between amino acids for predicting three-dimensional structures of proteins. *Macromolecules*, *9*(6), 945–950. doi:10.1021/ma60054a013

Tanaka, S., & Scheraga, H. A. (1976). Medium-and long-range interaction parameters between amino acids for predicting three-dimensional structures of proteins. *Macromolecules*, *9*(6), 945–950. PMID:1004017

Tang, Y. T., & Marshall, G. R. (2011). PHOENIX: A scoring function for affinity prediction derived using high-resolution crystal structures and calorimetry measurements. *Journal of Chemical Information and Modeling*, *51*(2), 214–228. doi:10.1021/ci100257s

Tan, K. P., Nguyen, B. T., Patel, S., Varadarajan, R., & Madhusudhan, M. S. (2013). Depth: A web server to compute depth, cavity sizes, detect potential small-molecule ligand-binding cavities and predict the pKa of ionizable residues in proteins. *Nucleic Acids Research*, *41*(W1), W314–W321. doi:10.1093/nar/gkt503 PMID:23766289

Tan, L., Batista, J., & Bajorath, J. (2010). Computational methodologies for compound database searching that utilize experimental protein–ligand interaction information. *Chemical Biology & Drug Design*, *76*(3), 191–200. PMID:20636330

Tanzi, R. E., & Bertram, L. (2005). Twenty years of the Alzheimer's disease Amyloid hypothesis: A genetic perspective. *Cell*, *120*(4), 545–555. doi:10.1016/j.cell.2005.02.008 PMID:15734686

Tautermann, C. S. (2014). GPCR structures in drug design, emerging opportunities with new structures. *Bioorganic & Medicinal Chemistry Letters*, *24*(17), 4073–4079. doi:10.1016/j.bmcl.2014.07.009 PMID:25086683

Taylor, D. W., Zhu, Y., Staals, R. H. J., Kornfeld, J. E., Shinkai, A., van der Oost, J., & Doudna, J. A. et al. (2015). Structures of the CRISPR-Cmr complex reveal mode of RNA target positioning. *Science*, *348*(6234), 581–585. doi:10.1126/science.aaa4535 PMID:25837515

Taylor, J. S., & Burnett, R. M. (2000). DARWIN: A program for docking flexible molecules. *Proteins*, *41*(2), 173–191. doi:10.1002/1097-0134(20001101)41:2<173::AID-PROT30>3.0.CO;2-3 PMID:10966571

Taylor, R. D., Jewsbury, P. J., & Essex, J. W. (2002). A review of protein-small molecule docking methods. *Journal of Computer-Aided Molecular Design*, *16*(3), 151–166. doi:10.1023/A:1020155510718 PMID:12363215

Teague, S. J. (2003). Implications of protein flexibility for drug discovery. *Nature Reviews. Drug Discovery*, *2*(7), 527–541. doi:10.1038/nrd1129 PMID:12838268

ten Brink, T., & Exner, T. E. (2009). Influence of protonation, tautomeric, and stereoisomeric states on proteinligand docking results. *Journal of Chemical Information and Modeling*, *49*(6), 1535–1546. doi:10.1021/ci800420z PMID:19453150

Teodorescu, O., Galor, T., Pillardy, J., & Elber, R. (2004). Enriching the sequence substitution matrix by structural information. *Proteins*, *54*(1), 41–48. doi:10.1002/prot.10474 PMID:14705022

Teodoro, M. L., Phillips, G. N., & Kavraki, L. E. (2001). *Molecular docking: A problem with thousands of degrees of freedom.* Paper presented at the Robotics and Automation, 2001. doi:10.1109/ROBOT.2001.932674

Teodoro, M. L., & Kavraki, L. E. (2003). Conformational flexibility models for the receptor in structure based drug design. *Current Pharmaceutical Design*, *9*(20), 1635–1648. doi:10.2174/1381612033454595 PMID:12871062

Teramoto, R., & Fukunishi, H. (2007a). Supervised consensus scoring for docking and virtual screening. *Journal of Chemical Information and Modeling*, *47*(2), 526–534. PMID:17295466

Teramoto, R., & Fukunishi, H. (2007b). Supervised scoring models with docked ligand conformations for structure-based virtual screening. *Journal of Chemical Information and Modeling*, *47*(5), 1858–1867. PMID:17685604

Terashi, G., Takeda-Shitaka, M., Kanou, K., Iwadate, M., Takaya, D., & Umeyama, H. (2007). The SKE-DOCK server and human teams based on a combined method of shape complementarity and free energy estimation. *Proteins*, *69*(4), 866–872. doi:10.1002/prot.21772 PMID:17853449

Terp, G. E., Johansen, B. N., Christensen, I. T., & Jørgensen, F. S. (2001). A new concept for multidimensional selection of ligand conformations (Multiselect) and multidimensional scoring (multiscore) of protein-ligand binding affinities. *Journal of Medicinal Chemistry*, *44*(14), 2333–2343. PMID:11428927

Terwilliger, T. C., Adams, P. D., Moriarty, N. W., & Cohn, J. D. (2006). Ligand identification using electron-density map correlations. *Acta Crystallographica. Section D, Biological Crystallography*, *63*(1), 101–107. doi:10.1107/S0907444906046233 PMID:17164532

Terwilliger, T. C., Klei, H., Adams, P. D., Moriarty, N. W., & Cohn, J. D. (2006). Automated ligand fitting by core-fragment fitting and extension into density. *Acta Crystallographica. Section D, Biological Crystallography*, *62*(Pt 8), 915–922. doi:10.1107/S0907444906017161 PMID:16855309

The UniProt Consortium. (2008). The Universal Protein Resource (UniProt). *Nucleic Acids Research*, *36*, D190–D195. PMID:18045787

Thomas, P. D., & Dill, K. A. (1996). Statistical potentials extracted from protein structures: How accurate are they? *Journal of Molecular Biology*, *257*(2), 457–469. PMID:8609636

Thompson, J. D., Gibson, T. J., & Higgins, D. G. (2002). Multiple sequence alignment using ClustalW and ClustalX. In Curr Protoc Bioinformatics. doi:10.1002/0471250953.bi0203s00

Thompson, J. D., Higgins, D. G., & Gibson, T. J. (1994). CLUSTAL W: Improving the sensitivity of progressive multiple sequence alignment through sequence weighting, position-specific gap penalties and weight matrix choice. *Nucleic Acids Research*, *22*(22), 4673–4680. doi:10.1093/nar/22.22.4673 PMID:7984417

Thomsen, R. (2003). Flexible ligand docking using evolutionary algorithms: Investigating the effects of variation operators and local search hybrids. *Bio Systems*, *72*(1-2), 57–73. doi:10.1016/S0303-2647(03)00135-7

Thomsen, R., & Christensen, M. H. (2006). MolDock: A new technique for high-accuracy molecular docking. *Journal of Medicinal Chemistry*, *49*(11), 3315–3321. doi:10.1021/jm051197e PMID:16722650

Tietze, S., & Apostolakis, J. (2007). GlamDock: Development and validation of a new docking tool on several thousand protein-ligand complexes. *Journal of Chemical Information and Modeling*, *47*(4), 1657–1672. doi:10.1021/ci7001236 PMID:17585857

Tintori, C., Magnani, M., Schenone, S., & Botta, M. (2009). Docking, 3D-QSAR studies and in silico ADME prediction on c-Src tyrosine kinase inhibitors. *European Journal of Medicinal Chemistry*, *44*(3), 990–1000. doi:10.1016/j.ejmech.2008.07.002 PMID:18722033

Tomašič, T., Hajšek, D., Švajger, U., Luzar, J., Obermajer, N., Petit-Haertlein, I., & Anderluh, M. et al. (2014). Monovalent mannose-based DC-SIGN antagonists: Targeting the hydrophobic groove of the receptor. *European Journal of Medicinal Chemistry*, *75*, 308–326. doi:10.1016/j.ejmech.2014.01.047

Tondel, K., Anderssen, E., & Drablos, F. (2006). Protein Alpha Shape (PAS) Dock: A new gaussian-based score function suitable for docking in homology modelled protein structures. *Journal of Computer-Aided Molecular Design*, *20*(3), 131–144. doi:10.1007/s10822-006-9041-7 PMID:16652207

Torchala, M., & Bates, P. A. (2014). Predicting the structure of protein-protein complexes using the SwarmDock Web Server. *Methods in Molecular Biology (Clifton, N.J.)*, *1137*, 181–197. doi:10.1007/978-1-4939-0366-5_13 PMID:24573482

Torchala, M., Moal, I. H., Chaleil, R. A., Fernandez-Recio, J., & Bates, P. A. (2013). SwarmDock: A server for flexible protein-protein docking. *Bioinformatics (Oxford, England)*, *29*(6), 807–809. doi:10.1093/bioinformatics/btt038 PMID:23343604

Totrov, M., & Abagyan, R. (2008). Flexible ligand docking to multiple receptor conformations: A practical alternative. *Current Opinion in Structural Biology*, *18*(2), 178–184. doi:10.1016/j.sbi.2008.01.004 PMID:18302984

Tovchigrechko, A., & Vakser, I. A. (2006). GRAMM-X public web server for protein-protein docking. *Nucleic Acids Res, 34*, W310-314. doi: 10.1093/nar/gkl206

Tovchigrechko, A., & Vakser, I. A. (2006). GRAMM-X public web server for protein-protein docking. *Nucleic Acids Research, 34*, 310–314.

Tripathi, S. K., Muttineni, R., & Singh, S. K. (2013). Extra precision docking, free energy calculation and molecular dynamics simulation studies of CDK2 inhibitors. *Journal of Theoretical Biology, 334*, 87–100. doi:10.1016/j.jtbi.2013.05.014 PMID:23727278

Tripathi, S., Kumar, A., Kumar, B. S., Negi, A. S., & Sharma, A. (2015). Structural investigations into the binding mode of novel neolignans Cmp10 and Cmp19 microtubule stabilizers by in silico molecular docking, molecular dynamics, and binding free energy calculations. *Journal of Biomolecular Structure & Dynamics*, 1–9. doi:10.1080/07391102.2015.1074941 PMID:26212016

Trosset, J. Y., & Scheraga, H. A. (1998). Reaching the global minimum in docking simulations: A Monte Carlo energy minimization approach using Bezier splines. *Proceedings of the National Academy of Sciences of the United States of America, 95*(14), 8011–8015. doi:10.1073/pnas.95.14.8011 PMID:9653131

Trott, O., & Olson, A. J. (2010). AutoDock Vina: Improving the speed and accuracy of docking with a new scoring function, efficient optimization, and multithreading. *Journal of Computational Chemistry, 31*, 455–461.

Tsai, C.-J., Kumar, S., Ma, B., & Nussinov, R. (1999). Folding funnels, binding funnels, and protein function. *Protein Science, 8*(6), 1181–1190. PMID:10386868

Tsai, T. Y., Chang, K. W., & Chen, C. Y. (2011). iScreen: World's first cloud-computing web server for virtual screening and de novo drug design based on TCM database@Taiwan. *Journal of Computer-Aided Molecular Design, 25*(6), 525–531. doi:10.1007/s10822-011-9438-9 PMID:21647737

Tseng, G. N., Sonawane, K. D., Korolkova, Y. V., Zhang, M., Liu, J., Grishin, E. V., & Guy, H. R. (2007). Probing the outer mouth structure of the HERG channel with peptide toxin footprinting and molecular modeling. *Biophysical Journal, 92*(10), 3524–3540. doi:10.1529/biophysj.106.097360 PMID:17293393

Tuccinardi, T. (2009). Docking-based virtual screening: Recent developments. *Combinatorial Chemistry & High Throughput Screening, 12*(3), 303–314. PMID:19275536

Tuccinardi, T., Botta, M., Giordano, A., & Martinelli, A. (2010). Protein kinases: Docking and homology modeling reliability. *Journal of Chemical Information and Modeling, 50*(8), 1432–1441. doi:10.1021/ci100161z PMID:20726600

Tuccinardi, T., Nuti, E., Ortore, G., Supuran, C. T., Rossello, A., & Martinelli, A. (2007). Analysis of human carbonic anhydrase II: Docking reliability and receptor-based 3D-QSAR study. *Journal of Chemical Information and Modeling, 47*(2), 515–525. doi:10.1021/ci600469w PMID:17295464

Tucker, H. M., Kihiko, M., Caldwell, J. N., Wright, S., Kawarabayashi, T., Price, D., & Estus, S. et al. (2000). The plasmin system is induced by and degrades amyloid-β aggregates. *The Journal of Neuroscience, 20*, 3937–3946. PMID:10818128

Tuffery, P., & Derreumaux, P. (2012). Flexibility and binding affinity in protein-ligand, protein-protein and multi-component protein interactions: Limitations of current computational approaches. *Journal of the Royal Society, Interface, 9*(66), 20–33. PMID:21993006

Tuszynska, I., Magnus, M., Jonak, K., Dawson, W., & Bujnicki, J. M. (2015). NPDock: A web server for protein-nucleic acid docking. *Nucleic Acids Research, 43*(W1), W425-430. doi:10.1093/nar/gkv493 PMID:25977296

Vagner, J., Qu, H., & Hruby, V. J. (2008). Peptidomimetics, a synthetic tool of drug discovery. *Current Opinion in Chemical Biology*, *12*(3), 292–296. doi:10.1016/j.cbpa.2008.03.009 PMID:18423417

Vajda, S., Hall, D. R., & Kozakov, D. (2013). Sampling and scoring: A marriage made in heaven. *Proteins*, *81*(11), 1874–1884. doi:10.1002/prot.24343 PMID:23775627

Vakser, I. (1995). Protein docking for low-resolution structures. *Protein Engineering*, *8*(4), 371–377. doi:10.1093/protein/8.4.371 PMID:7567922

Vakser, I. A. (2014). Protein-Protein Docking: From Interaction to Interactome. *Biophysical Journal*, *107*(8), 1785–1793. doi:10.1016/j.bpj.2014.08.033 PMID:25418159

van der Spoel, D., Lindahl, E., Hess, B., Groenhof, G., Mark, A. E., & Berendsen, H. J. C. (2005). GROMACS: Fast, Flexible and Free. *Journal of Computational Chemistry*, *26*(16), 1701–1718. doi:10.1002/jcc.20291 PMID:16211538

van Dijk, A. D., & Bonvin, A. M. (2006). Solvated docking: Introducing water into the modelling of biomolecular complexes. *Bioinformatics (Oxford, England)*, *22*(19), 2340–2347. doi:10.1093/bioinformatics/btl395 PMID:16899489

Van Dijk, M., & Bonvin, A. M. J. J. (2008). A protein-DNA docking benchmark. *Nucleic Acids Research*, *36*(14), e88. doi:10.1093/nar/gkn386 PMID:18583363

Van Dijk, M., van Dijk, A. D. J., Hsu, V., Boelens, R., & Bonvin, A. M. J. J. (2006). Information-driven protein-DNA docking using HADDOCK: It is a matter of flexibility. *Nucleic Acids Research*, *34*(11), 3317–3325. doi:10.1093/nar/gkl412 PMID:16820531

Van Dijk, M., Visscher, K. M., Kastritis, P. L., & Bonvin, A. M. J. J. (2013). Solvated protein-DNA docking using HADDOCK. *Journal of Biomolecular NMR*, *56*(1), 51–63. doi:10.1007/s10858-013-9734-x PMID:23625455

Van Drie, J. H. (2007). Computer-aided drug design: The next 20 years. *Journal of Computer-Aided Molecular Design*, *21*(10-11), 591–601. doi:10.1007/s10822-007-9142-y PMID:17989929

Van Lipzig, M. M. H., terLaak, A. M., Jongejan, A., Vermeulen, N. P. E., Wamelink, M., Geerke, D., & Meerman, J. H. N. (2004). Prediction of ligand binding affinity and orientation of xenoestrogens to the estrogen receptor by molecular dynamics simulations and the linear interaction energy method. *Journal of Medicinal Chemistry*, *47*(4), 1018–1030. doi:10.1021/jm0309607 PMID:14761204

Vanommeslaeghe, K., Hatcher, E., Acharya, C., Kundu, S., Zhong, S., Shim, J., Mackerell, A. D. (2009). CHARMM general force field: A force field for drug-like molecules compatible with the CHARMM all-atom additive biological force fields. *Journal of Computational Chemistry*. http://doi.org/<ALIGNMENT.qj></ALIGNMENT>10.1002/jcc.21367

Vasseur, R., Baud, S., Steffenel, L. A., Vigouroux, X., Martiny, L., Krajecki, M., & Dauchez, M. (2015). Inverse docking method for new proteins targets identification: A parallel approach. *Parallel Computing*, *42*, 48–59. doi:10.1016/j.parco.2014.09.008

Vass, M., Schmidt, E., Horti, F., & Keseru, G. M. (2014). Virtual fragment screening on GPCRS: A case study on dopamine D3 and histamine H4 receptors. *European Journal of Medicinal Chemistry*, *77*, 38–46. doi:10.1016/j.ejmech.2014.02.034 PMID:24607587

Velec, H. F., Gohlke, H., & Klebe, G. (2005). DrugScore(CSD)-knowledge-based scoring function derived from small molecule crystal data with superior recognition rate of near-native ligand poses and better affinity prediction. *Journal of Medicinal Chemistry*, *48*(20), 6296–6303. doi:10.1021/jm050436v PMID:16190756

Venclovas, C., Zemla, A., Fidelis, K., & Moult, J. (2003). Assessment of progress over the CASP experiments. *Proteins*, *53*(S6), 585–595. doi:10.1002/prot.10530 PMID:14579350

Venhorst, J., Núñez, S., Terpstra, J. W., & Kruse, C. G. (2008). Assessment of scaffold hopping efficiency by use of molecular interaction fingerprints. *Journal of Medicinal Chemistry*, *51*(11), 3222–3229. doi:10.1021/jm8001058 PMID:18447336

Venhorst, J., ter Laak, A. M., Commandeur, J. N. M., Funae, Y., Hiroi, T., & Vermeulen, N. P. E. (2003). Homology Modeling of Rat and Human Cytochrome P450 2D (CYP2D) Isoforms and Computational Rationalization of Experimental Ligand-Binding Specificities. *Journal of Medicinal Chemistry*, *46*(1), 74–86. doi:10.1021/jm0209578 PMID:12502361

Venkatachalam, C. M., Jiang, X., Oldfield, T., & Waldman, M. (2003). LigandFit: A novel method for the shape-directed rapid docking of ligands to protein active sites. *Journal of Molecular Graphics & Modelling*, *21*(4), 289–307. http://www.ncbi.nlm.nih.gov/pubmed/12479928 doi:10.1016/S1093-3263(02)00164-X PMID:12479928

Venkatraman, V., & Ritchie, D. W. (2012). Flexible protein docking refinement using pose-dependent normal mode analysis. *Proteins: Structure. Function and Bioinformatics*, *80*(9), 2262–2274. doi:10.1002/prot.24115

Venkatraman, V., Yang, Y. D., Sael, L., & Kihara, D. (2009). Protein-protein docking using region-based 3D Zernike descriptors. *BMC Bioinformatics*, *10*(1), 407. doi:10.1186/1471-2105-10-407 PMID:20003235

Verdonk, M. L., Cole, J. C., Hartshorn, M. J., Murray, C. W., & Taylor, R. D. (2003). Improved protein-ligand docking using GOLD. *Proteins*, *52*(4), 609–623. doi:10.1002/prot.10465 PMID:12910460

Veselovsky, A. V., & Ivanov, A. S. (2003). Strategy of computer-aided drug design. *Current Drug Targets. Infectious Disorders*, *3*(1), 33–40. doi:10.2174/1568005033342145 PMID:12570731

Vijayakrishnan, R. (2009). Structure-based drug design and modern medicine. *Journal of Postgraduate Medicine*, *55*(4), 301–304. doi:10.4103/0022-3859.58943 PMID:20083886

Vilar, S., Cozza, G., & Moro, S. (2008). Medicinal chemistry and the molecular operating environment (MOE): Application of QSAR and molecular docking to drug discovery. *Current Topics in Medicinal Chemistry*, *8*(18), 1555–1572. doi:10.2174/156802608786786624 PMID:19075767

Villoutreix, B. O., Eudes, R., & Miteva, M. A. (2009). Structure-based virtual ligand screening: Recent success stories. *Combinatorial Chemistry & High Throughput Screening*, *12*(10), 1000–1016. PMID:20025565

Villoutreix, B. O., Kuenemann, M., Poyet, J. L., Bruzzoni-Giovanelli, H., Labbé, C., Lagorce, D., & Miteva, M. A. et al. (2014). Drug-like protein-protein interaction modulators: Challenges and opportunities for drug discovery and chemical biology. *Molecular Informatics*, *33*(6-7), 414–437. doi:10.1002/minf.201400040 PMID:25254076

Villoutreix, B. O., Renault, N., Lagorce, D., Sperandio, O., Montes, M., & Miteva, M. A. (2007). Free resources to assist structure-based virtual ligand screening experiments. *Current Protein & Peptide Science*, *8*(4), 381–411. doi:10.2174/138920307781369391 PMID:17696871

Viricel, C., Ahmed, M., & Barakat, K. (2015). Human PD-1 Binds Differently To Its Human Ligands: A Comprehensive Modelling Study. *Journal of Molecular Graphics & Modelling*, *57*, 131–142. doi:10.1016/j.jmgm.2015.01.015 PMID:25723350

Viswanath, S., Ravikant, D. V., & Elber, R. (2014). DOCK/PIERR: Web server for structure prediction of protein-protein complexes. *Methods in Molecular Biology (Clifton, N.J.)*, *1137*, 199–207. doi:10.1007/978-1-4939-0366-5_14 PMID:24573483

Vitkup, D., Melamud, E., Moult, J., & Sander, C. (2001). Completeness in structural genomics. *Nature Structural Biology*, *8*(6), 559–566. doi:10.1038/88640 PMID:11373627

Vitorovic-Todorovic, M. D., Cvijetic, I. N., Juranic, I. O., & Drakulic, B. J. (2012). The 3D-QSAR study of 110 diverse, dual binding, acetylcholinesterase inhibitors based on alignment independent descriptors (GRIND-2). The effects of conformation on predictive power and interpretability of the models. *Journal of Molecular Graphics & Modelling*, *38*, 194–210. doi:10.1016/j.jmgm.2012.08.001 PMID:23073222

Vreven, T., Pierce, B. G., Hwang, H., & Weng, Z. (2009). Performance of ZDOCK in CAPRI rounds 20-26. *Proteomics*, *9*(12), 2175–2182. doi:10.1002/pmic.200800646 PMID:24123140

Vreven, T., Pierce, B. G., Hwang, H., & Weng, Z. (2013). Performance of ZDOCK in CAPRI rounds 20-26. *Proteins*, *81*(12), 2175–2182. doi:10.1002/prot.24432 PMID:24123140

Vriend, G. (1990). WHAT IF: a molecular modeling and drug design program. *J Mol Graph, 8*(1), 52-56.

Vuorinen, A., Engeli, R., Meyer, A., Bachmann, F., Griesser, U. J., Schuster, D., & Odermatt, A. (2014). Ligand-based pharmacophore modeling and virtual screening for the discovery of novel 17β-hydroxysteroid dehydrogenase 2 inhibitors. *Journal of Medicinal Chemistry*, *57*(14), 5995–6007. doi:10.1021/jm5004914 PMID:24960438

Vyas, V., Jain, A., Jain, A., & Gupta, A. (2008). Virtual screening: A fast tool for drug design. *Scientia Pharmaceutica*, *76*(3), 333–360. doi:10.3797/scipharm.0803-03

Wada, M., Kanamori, E., Nakamura, H., & Fukunishi, Y. (2011). Selection of *in-silico* drug screening results for G-protein-coupled receptors by using universal active probes. *Journal of Chemical Information and Modeling*, *51*(9), 2398–2407. doi:10.1021/ci200236x PMID:21848279

Waksman, G., & Sansom, C. (2005). Introduction : Proteomics and Protein – Protein Interactions : Biology, Chemistry, Bioinformatics, and Drug Design. In G. Waksman (Ed.), *Proteomics and Protein – Protein Interactions : Biology, Chemistry, Bioinformatics, and Drug Design* (pp. 1–18). New York: Springer. doi:10.1007/0-387-24532-4_1

Walensky, L. D., Kung, A. L., Escher, I., Malia, T. J., Barbuto, S., Wright, R. D., & Korsmeyer, S. J. et al. (2004). Activation of apoptosis in vivo by a hydrocarbon-stapled BH3 helix. *Science*, *305*(5689), 1466–1470. doi:10.1126/science.1099191 PMID:15353804

Wallace, A. C., Laskowski, R. A., & Thornton, J. M. (1995). LIGPLOT: A program to generate schematic diagrams of protein-ligand interactions. *Protein Engineering*, *8*(2), 127–134. doi:10.1093/protein/8.2.127 PMID:7630882

Wallner, B., & Elofsson, A. (2005). All are not equal: A benchmark of different homology modeling programs. *Protein Science*, *14*(5), 1315–1327. doi:10.1110/ps.041253405 PMID:15840834

Walters, W. P., Stahl, M. T., & Murcko, M. A. (1998). Virtual screening - an overview. *Drug Discovery Today*, *3*(4), 160–178. doi:10.1016/S1359-6446(97)01163-X

Wang, J. C., Chu, P. Y., Chen, C. M., & Lin, J. H. (2012). idTarget: a web server for identifying protein targets of small chemical molecules with robust scoring functions and a divide-and-conquer docking approach. *Nucleic Acids Res, 40*, W393-399.

Wang, J., Wang, C., Cao, J., Liu, X., Yao, Y., & Dai, Q. (2014). Prediction of protein structural classes for low-similarity sequences using reduced PSSM and position-based secondary structural features. *Gene, 554*(2), 241-248. doi: S0378-1119(14)01197-4

Wang, R., Liu, L., Lai, L., & Tang, Y. (1998). SCORE: A New Empirical Method for Estimating the Binding Affinity of a Protein-Ligand Complex. *Molecular Modeling Annual, 4*(12), 379-394. doi: 10.1007/s008940050096

Wang, C. H. U., Schueler-furman, O. R., & Baker, D. (2005). Improved side-chain modeling for protein – protein docking. *Protein Science*, *14*(5), 1328–1339. doi:10.1110/ps.041222905 PMID:15802647

Wang, F., Liu, D., Wang, H., Luo, C., Zheng, M., Liu, H., & Jiang, H. et al. (2011). Computational screening for active compounds targeting protein sequences: Methodology and experimental validation. *Journal of Chemical Information and Modeling*, *51*(11), 2821–2828. doi:10.1021/ci200264h PMID:21955088

Wang, J.-C., & Lin, J.-H. (2013). Scoring functions for prediction of protein-ligand interactions. *Current Pharmaceutical Design*, *19*(12), 2174–2182. PMID:23016847

Wang, J., & Hou, T. (2012). Develop and test a solvent accessible surface area-based model in conformational entropy calculations. *Journal of Chemical Information and Modeling*, *52*(5), 1199–1212. doi:10.1021/ci300064d

Wang, J., & Verkhivker, G. M. (2003). Energy landscape theory, funnels, specificity, and optimal criterion of biomolecular binding. *Physical Review Letters*, *90*(18), 188101. PMID:12786043

Wang, J., Wang, W., Huo, S., Lee, M., & Kollman, P. A. (2001). Solvation model based on weighted solvent accessible surface area. *The Journal of Physical Chemistry B*, *105*(21), 5055–5067.

Wang, J., Wolf, R. M., Caldwell, J. W., Kollman, P. A., & Case, D. A. (2004). Development and testing of a general amber force field. *Journal of Computational Chemistry*, *25*(9), 1157–1174. doi:10.1002/jcc.20035

Wang, J., Zheng, X., Yang, Y., Drueckhammer, D., Yang, W., Verkhivker, G., & Wang, E. (2007). Quantifying intrinsic specificity: A potential complement to affinity in drug screening. *Physical Review Letters*, *99*(19), 198101. PMID:18233118

Wang, R., Fang, X., Lu, Y., & Wang, S. (2004). The pdbbind database: Collection of binding affinities for protein-ligand complexes with known three-dimensional structures. *Journal of Medicinal Chemistry*, *47*(12), 2977–2980. PMID:15163179

Wang, R., Lai, L., & Wang, S. (2002). Further development and validation of empirical scoring functions for structure-based binding affinity prediction. *Journal of Computer-Aided Molecular Design*, *16*(1), 11–26. doi:10.1023/A:1016357811882

Wang, R., & Wang, S. (2001). How does consensus scoring work for virtual library screening? an idealized computer experiment. *Journal of Chemical Information and Computer Sciences*, *41*(5), 1422–1426. PMID:11604043

Wang, T., Wu, M.-B., Chen, Z.-J., Chen, H., Lin, J.-P., & Yang, L.-R. (2015). Fragment-Based Drug Discovery and Molecular Docking in Drug Design. *Current Pharmaceutical Biotechnology*, *16*(1), 11–25. doi:10.2174/13892010156 66141122204532

Wang, Y., Suzek, T., Zhang, J., Wang, J., He, S., Cheng, T., & Bryant, S. H. et al. (2014). PubChem BioAssay: Update. *Nucleic Acids Research*, *42*(1), D1075–D1082. doi:10.1093/nar/gkt978 PMID:24198245

Warshaviak, D., Golan, G., Borrelli, K. W., Zhu, K., & Kalid, O. (2014). Structure-based virtual screening approach for discovery of covalently bound ligands. *Journal of Chemical Information and Modeling*, *54*(7), 1941–1950. doi:10.1021/ci500175r

Wass, M. N., Fuentes, G., Pons, C., Pazos, F., & Valencia, A. (2011). Towards the prediction of protein interaction partners using physical docking. *Molecular Systems Biology*, *7*(469), 469. PMID:21326236

Weerapana, E., Wang, C., Simon, G. M., Richter, F., Khare, S., Dilon, M. B. D., & Cravatt, B. F. et al. (2010). Quantitative reactivity profiling predicts functional cysteines in proteomes. *Nature*, *10*(12), 790–795. doi:10.1038/nature09472

Weill, N., Valencia, C., Gioria, S., Villa, P., Hibert, M., & Rognan, D. (2011). Identification of Nonpeptide Oxytocin Receptor Ligands by Receptor-Ligand Fingerprint Similarity Search. *Molecular Informatics*, *30*(6-7), 521–526. doi:10.1002/minf.201100026

Weiner, S. J., Kollman, P. A., Nguyen, D. T., & Case, D. A. (1986). An all atom force field for simulations of proteins and nucleic acids. *Journal of Computational Chemistry*, *7*(2), 230–252. doi:10.1002/jcc.540070216

Weine, S. J., Kollman, P. A., Case, D. A., Singh, U. C., Ghio, C., Alagona, G., & Weiner, P. et al. (1984). New force field for molecular mechanical simulation of nucleic acids and proteins. *Journal of the American Chemical Society*, *106*(3), 765–784. doi:10.1021/ja00315a051

Weininger, D. (1988). SMILES, a chemical language and information system. 1. Introduction to methodology and encoding rules. *Journal of Chemical Information and Modeling*, *28*(1), 31–63. doi:10.1021/ci00057a005

Wei, Y., Li, J., Chen, Z., Wang, F., Huang, W., Hong, Z., & Lin, J. (2015). Multistage virtual screening and identification of novel HIV-1 protease inhibitors by integrating SVM, shape, pharmacophore and docking methods. *European Journal of Medicinal Chemistry*, *101*, 409–418. doi:10.1016/j.ejmech.2015.06.054 PMID:26185005

Welch, W., Ruppert, J., & Jain, A. N. (1996). Hammerhead: Fast, fully automated docking of flexible ligands to protein binding sites. *Chemistry & Biology*, *3*(6), 449–462. doi:10.1016/S1074-5521(96)90093-9 PMID:8807875

Westhead, D. R., Clark, D. E., & Murray, C. W. (1997). A comparison of heuristic search algorithms for molecular docking. *Journal of Computer-Aided Molecular Design*, *11*(3), 209–228. doi:10.1023/A:1007934310264 PMID:9263849

Wichapong, K., Lawson, M., Pianwanit, S., Kokpol, S., & Sippl, W. (2010). Postprocessing of Protein–Ligand Docking Poses Using Linear Response MM-PB/SA: Application to Wee1 Kinase Inhibitors. *Journal of Chemical Information and Modeling*, *50*(9), 1574–1588. doi:10.1021/ci1002153 PMID:20712342

Wiederstein, M., & Sippl, M. J. (2007). ProSA-web: interactive web service for the recognition of errors in three-dimensional structures of proteins. *Nucleic Acids Res*, *35*, W407-410.

Wiehe, K., Peterson, M. W., Pierce, B., Mintseris, J., & Weng, Z. (2008). Protein – Protein Docking : Overview and Performance Analysis. In M. Zaki & C. Bystroff (Eds.), Protein Structure Prediction (2nd ed.; vol. 413, pp. 283–314). Humana Press.

Wiehe, K., Pierce, B., Mintseris, J., Tong, W. W., Anderson, R., Chen, R., & Weng, Z. (2005). ZDOCK and RDOCK performance in CAPRI rounds 3, 4, and 5. *Proteins*, *60*(2), 207–213. doi:10.1002/prot.20559 PMID:15981263

Wiehe, K., Pierce, B., Tong, W. W., Hwang, H., Mintseris, J., & Weng, Z. (2007). The performance of ZDOCK and ZRANK in rounds 6-11 of CAPRI. *Proteins*, *69*(4), 719–725. doi:10.1002/prot.21747 PMID:17803212

Willet, P., & Barnard, J. M. (1998). Chemical similarity searching. *Journal of Chemical Information and Computer Sciences*, *38*(6), 983–996. doi:10.1021/ci9800211

Williams, P. A., Cosme, J., Ward, A., Angove, H. C., Matak Vinkovic, D., & Jhoti, H. (2003). Crystal structure of human cytochrome P450 2C9 with bound warfarin. *Nature*, *424*(6947), 464-468.

Wilson, G. L., & Lill, M. A. (2011). Integrating structure-based and ligand-based approaches for computational drug design. *Future Med Chem*, *3*(6), 735–750. doi:10.4155/fmc.11.18 PMID:21554079

Wishart, D. S., Knox, C., Guo, A. C., Shrivastava, S., Hassanali, M., Stothard, P., & Woolsey, J. et al. (2006). DrugBank: A comprehensive resource for in silico drug discovery and exploration. *Nucleic Acids Research*, *34*(90001), D668–D672. doi:10.1093/nar/gkj067 PMID:16381955

Wlodawer, A. (2002). Rational approaches to AIDS drug design through structural biology. *Annual Review of Medicine*, *53*(1), 595–614. doi:10.1146/annurev.med.53.052901.131947 PMID:11818491

Wodak, S. J., & Janin, J. (1978). Computer analysis of protein-protein interaction. *Journal of Molecular Biology*, *124*(2), 323–342. doi:10.1016/0022-2836(78)90302-9 PMID:712840

Wodak, S. J., & Méndez, R. (2004). Prediction of protein-protein interactions: The CAPRI experiment, its evaluation and implications. *Current Opinion in Structural Biology*, *14*(2), 242–249. doi:10.1016/j.sbi.2004.02.003 PMID:15093840

Wong, C. F. (2008). Flexible ligand–flexible protein docking in protein kinase systems. *Biochimica et Biophysica Acta*, *1784*(1), 244–251. doi:10.1016/j.bbapap.2007.10.005 PMID:17996742

Wu, G., Robertson, D. H., Brooks, C. L. III, & Vieth, M. (2003). Detailed analysis of grid-based molecular docking: A case study of CDOCKER-A CHARMm-based MD docking algorithm. *Journal of Computational Chemistry*, *24*(13), 1549–1562. doi:10.1002/jcc.10306 PMID:12925999

Wu, J., Hong, B., Takeda, T., & Guo, J. T. (2012). High performance transcription factor-DNA docking with GPU computing. *Proteome Science*, *10*(Suppl 1), S17. doi:10.1186/1477-5956-10-S1-S17 PMID:22759575

Xiang, Z., Soto, C. S., & Honig, B. (2002). Evaluating conformational free energies: The colony energy and its application to the problem of loop prediction. *Proceedings of the National Academy of Sciences of the United States of America*, *99*(11), 7432–7437. PMID:12032300

Xu, D., Jaroszewski, L., Li, Z., & Godzik, A. (2014). AIDA: ab initio domain assembly server. *Nucleic Acids Res*, *42*, W308-313.

Xu, B., Yang, Y., Liang, H., & Zhou, Y. (2009). An all-atom knowledge-based energy function for protein-DNA threading, docking decoy discrimination, and prediction of transcription-factor binding profiles. *Proteins*, *76*(3), 718–730. doi:10.1002/prot.22384 PMID:19274740

Xu, D., Xu, Y., & Uberbacher, E. C. (2000). Computational tools for protein modeling. *Current Protein & Peptide Science*, *1*(1), 1–21. doi:10.2174/1389203003381469 PMID:12369918

Xue, L. C., Jordan, R. A., El-Manzalawy, Y., Dobbs, D., & Honavar, V. (2014). DockRank: Ranking docked conformations using partner-specific sequence homology-based protein interface prediction. *Proteins*, *82*(2), 250–267. doi:10.1002/prot.24370 PMID:23873600

Xue, M., Zheng, M., Xiong, B., Li, Y., Jiang, H., & Shen, J. (2010). Knowledge-based scoring functions in drug design. 1. developing a target-specific method for kinase-ligand interactions. *Journal of Chemical Information and Modeling*, *50*(8), 1378–1386. PMID:20681607

Xu, M., & Lill, M. A. (2013). Induced fit docking, and the use of QM/MM methods in docking. *Drug Discovery Today. Technologies*, *10*(3), e411–e418. doi:10.1016/j.ddtec.2013.02.003 PMID:24050138

Xu, Y., Wang, Y., Meng, X. Y., Zhang, M., Jiang, M., Cui, M., & Tseng, G. N. (2013). Building KCNQ1/KCNE1 channel models and probing their interactions by molecular-dynamics simulations. *Biophysical Journal*, *105*(11), 2461–2473. doi:10.1016/j.bpj.2013.09.058 PMID:24314077

Xu, Y., Xu, D., & Uberbacher, E. C. (1998). An efficient computational method for globally optimal threading. *Journal of Computational Biology*, *5*(3), 597–614. doi:10.1089/cmb.1998.5.597 PMID:9773353

Yachdav, G., Kloppmann, E., Kajan, L., Hecht, M., Goldberg, T., Hamp, T., . . . Rost, B. (2014). PredictProtein--an open resource for online prediction of protein structural and functional features. *Nucleic Acids Res*, *42*, W337-343. doi: 10.1093/nar/gku366

Yakar, R., & Akten, E. D. (2014). Discovery of high affinity ligands for beta2-adrenergic receptor through pharmacophore-based high-throughput virtual screening and docking. *Journal of Molecular Graphics & Modelling*, *53*, 148–160. doi:10.1016/j.jmgm.2014.07.007 PMID:25137647

Yan, A., Nie, X., Wang, K., & Wang, M. (2013). Classification of Aurora kinase inhibitors by self-organizing map (SOM) and support vector machine (SVM). *European Journal of Medicinal Chemistry*, *61*, 73–83. doi:10.1016/j.ejmech.2012.06.037 PMID:22796044

Yang, C. Y., Sun, H., Chen, J., Nikolovska-Coleska, Z., & Wang, S. (2009). Importance of ligand reorganization free energy in protein-ligand binding-affinity prediction. *Journal of the American Chemical Society*, *131*(38), 13709–13721. doi:10.1021/ja9039373 PMID:19736924

Yang, J.-M., & Chen, C.-C. (2004). GEMDOCK: A generic evolutionary method for molecular docking. *Proteins*, *55*(2), 288–304. doi:10.1002/prot.20035 PMID:15048822

Yang, J., Yan, R., Roy, A., Xu, D., Poisson, J., & Zhang, Y. (2015). The I-TASSER Suite: Protein structure and function prediction. *Nature Methods*, *12*(1), 7–8. doi:10.1038/nmeth.3213 PMID:25549265

Yang, S., Chang, S., Chen, H., & Chen, Y. (2011). Identification of Potent EGFR Inhibitors from TCM Database@Taiwan. *PLoS Computational Biology*, *7*(10), e1002189. doi:10.1371/journal.pcbi.1002189 PMID:22022246

Yang, Y., Zhao, H., Wang, J., & Zhou, Y. (2014). SPOT-Seq-RNA: Predicting protein-RNA complex structure and RNA-binding function by fold recognition and binding affinity prediction. *Methods in Molecular Biology (Clifton, N.J.)*, *1137*, 119–130. doi:10.1007/978-1-4939-0366-5_9 PMID:24573478

Yan, Z., & Wang, J. (2012). Specificity quantification of biomolecular recognition and its implication for drug discovery. *Scientific Reports*, *2*, 309. PMID:22413060

Yan, Z., & Wang, J. (2015). Optimizing the affinity and specificity of ligand binding with the inclusion of solvation effect. *Proteins: Structure, Function, and Bioinformatics*, 83(9), 1632-1642.

Yan, Z., Zheng, X., Wang, E., & Wang, J. (2013). Thermodynamic and kinetic specificities of ligand binding. *Chemical Science*, *4*(6), 2387–2395.

Yi, H., Qiu, S., Cao, Z., Wu, Y., & Li, W. (2008). Molecular basis of inhibitory peptide maurotoxin recognizing Kv1.2 channel explored by ZDOCK and molecular dynamic simulations. *Proteins*, *70*(3), 844–854. doi:10.1002/prot.21706 PMID:17729277

Yin, K. J., Cirrito, J. R., Yan, P., Hu, X., Xiao, Q., Pan, X., & Turk, J. et al. (2006). Matrix metalloproteinases expressed by astrocytes mediate extracellular amyloid-beta peptide catabolism. *The Journal of Neuroscience*, *26*(43), 10939–10948. doi:10.1523/JNEUROSCI.2085-06.2006 PMID:17065436

Yin, S., Biedermannova, L., Vondrasek, J., & Dokholyan, N. V. (2008). MedusaScore: An accurate force field-based scoring function for virtual drug screening. *Journal of Chemical Information and Modeling*, *48*(8), 1656–1662. doi:10.1021/ci8001167 PMID:18672869

Yin, S., Biedermannova, L., Vondrasek, J., & Dokholyan, N. V. (2008). Medusascore: An accurate force field-based scoring function for virtual drug screening. *Journal of Chemical Information and Modeling*, *48*(8), 1656–1662. PMID:18672869

Yoo, C., Ahn, K., Park, J. E., Kim, M. J., & Jo, S. A. (2010). An aminopeptidase from *Streptomyces sp.* KK565 degrades beta amyloid monomers, oligomers and fibrils. *FEBS Letters*, *584*(19), 4157–4162. doi:10.1016/j.febslet.2010.08.048 PMID:20828565

Yue, S. Y. (1990). Distance-constrained molecular docking by simulated annealing. *Protein Engineering*, *4*(2), 177–184. doi:10.1093/protein/4.2.177

Yu, J., Zhou, Y., Tanaka, I., & Yao, M. (2010). Roll: A new algorithm for the detection of protein pockets and cavities with a rolling probe sphere. *Bioinformatics (Oxford, England)*, *26*(1), 46–52. doi:10.1093/bioinformatics/btp599 PMID:19846440

Yu, R., Wang, J., Wang, R., Lin, Y., Hu, Y., Wang, Y., & Lin, Z. et al. (2015). Combined pharmacophore modeling, 3D-QSAR, homology modeling and docking studies on CYP11B1 inhibitors. *Molecules (Basel, Switzerland)*, *20*(1), 1014–1030. doi:10.3390/molecules20011014 PMID:25584832

Yuriev, E., Agostino, M., & Ramsland, P. A. (2011). Challenges and advances in computational docking: 2009 in review. *Journal of Molecular Recognition*, *24*(2), 149–164. doi:10.1002/jmr.1077

Yuriev, E., Holien, J., & Ramsland, P. A. (2015). Improvements, trends, and new ideas in molecular docking: 2012-2013 in review. *Journal of Molecular Recognition*, *28*(10), 581–604. doi:10.1002/jmr.2471 PMID:25808539

Yuriev, E., & Ramsland, P. A. (2013). Latest developments in molecular docking: 2010–2011 in review. *Journal of Molecular Recognition*, *26*(5), 215–239. doi:10.1002/jmr.2266

Yusuf, D., Davis, A. M., Kleywegt, G. J., & Schmitt, S. (2008). An alternative method for the evaluation of docking performance: RSR vs RMSD. *Journal of Chemical Information and Modeling*, *48*(7), 1411–1422. PMID:18598022

Yu, X., Martinez, M., Gable, A. L., Fuller, J. C., Bruce, N. J., Richter, S., & Wade, R. C. (2015). webSDA: A web server to simulate macromolecular diffusional association. *Nucleic Acids Research*, *43*(W1), W220-224. doi:10.1093/nar/gkv335 PMID:25883142

Zacharias, M. (2003). Protein-protein docking with a reduced protein model accounting for side-chain flexibility. *Protein Science : A Publication of the Protein Society*, *12*(6), 1271–1282.

Zeng, Y., Cao, R., Zhang, T., Li, S., & Zhong, W. (2015). Design and synthesis of piperidine derivatives as novel human heat shock protein 70 inhibitors for the treatment of drug-resistant tumors. *European Journal of Medicinal Chemistry*, *97*, 19–31. doi:10.1016/j.ejmech.2015.04.043 PMID:25935385

Zhang, C., & Lai, L. (2011). SDOCK: A global protein-protein docking program using stepwise force-field potentials. *Journal of Computational Chemistry*, *32*(12), 2598–2612. doi:10.1002/jcc.21839 PMID:21618559

Zhang, C., Liu, S., & Zhou, Y. (2005). Docking prediction using biological information, ZDOCK sampling technique, and clustering guided by the DFIRE statistical energy function. *Proteins*, *60*(2), 314–318. doi:10.1002/prot.20576 PMID:15981255

Zhang, X., Song, Z., Qin, B., Zhang, X., Chen, L., Hu, Y., & Yuan, Z. (2013). Rupintrivir is a promising candidate for treating severe cases of enterovirus-71 infection: Evaluation of antiviral efficacy in a murine infection model. *Antiviral Research*, *97*(3), 264–269. doi:10.1016/j.antiviral.2012.12.029 PMID:23295352

Zhang, X., Wang, H., Li, Y., Cao, R., Zhong, W., Zheng, Z., & Li, S. et al. (2013). Novel substituted heteroaromatic piperazine and piperidine derivatives as inhibitors of human enterovirus 71 and coxsackievirus a16. *Molecules (Basel, Switzerland)*, *18*(5), 5059–5071. doi:10.3390/molecules18055059 PMID:23629759

Zhang, Y. (2007). Template-based modeling and free modeling by I-TASSER in CASP7. *Proteins*, *69*(S8Suppl 8), 108–117. doi:10.1002/prot.21702 PMID:17894355

Zhang, Y. (2008). I-TASSER server for protein 3D structure prediction. *BMC Bioinformatics*, *9*(1), 40. doi:10.1186/1471-2105-9-40 PMID:18215316

Zhang, Y., Zheng, N., Hao, P., Cao, Y., & Zhong, Y. (2005). A molecular docking model of SARS-CoV S1 protein in complex with its receptor, human ACE2. *Computational Biology and Chemistry*, *29*(3), 254–257. doi:10.1016/j.compbiolchem.2005.04.008 PMID:15979045

Zhang, Z., Schindler, C. E. M., Lange, O. F., & Zacharias, M. (2015). Application of Enhanced Sampling Monte Carlo Methods for High-Resolution Protein-Protein Docking in Rosetta. *PLoS ONE*, *10*(6), e0125941. doi:10.1371/journal.pone.0125941 PMID:26053419

Zhao, Y., & Sanner, M. F. (2007). FLIPDock: Docking flexible ligands into flexible receptors. *Proteins*, *68*(3), 726–737. doi:10.1002/prot.21423 PMID:17523154

Zhong, H., Tran, L. M., & Stang, J. L. (2009). Induced-fit docking studies of the active and inactive states of protein tyrosine kinases. *Journal of Molecular Graphics & Modelling*, *28*(4), 336–346. doi:10.1016/j.jmgm.2009.08.012 PMID:19767223

Zhou, H., & Skolnick, J. (2013). FINDSITE(comb): A threading/structure-based, proteomic-scale virtual ligand screening approach. *Journal of Chemical Information and Modeling*, *53*(1), 230–240. doi:10.1021/ci300510n PMID:23240691

Zhou, T., Huang, D., & Caflisch, A. (2010). Quantum Mechanical Methods for Drug Design. *Current Topics in Medicinal Chemistry*, *10*(1), 33–45. doi:10.2174/156802610790232242 PMID:19929831

Zhou, Y., & Johnson, M. E. (1999). Comparative molecular modeling analysis of-5-amidinoindole and benzamidine binding to thrombin and trypsin: Specific H-bond formation contributes to high 5-amidinoindole potency and selectivity for thrombin and factor Xa. *Journal of Molecular Recognition*, *12*(4), 235–241. doi:10.1002/(SICI)1099-1352(199907/08)12:4<235::AID-JMR460>3.0.CO;2-X PMID:10440994

Zhou, Z., & Madura, J. D. (2004). CoMFA 3D-QSAR analysis of HIV-1 RT nonnucleoside inhibitors, TIBO derivatives based on docking conformation and alignment. *Journal of Chemical Information and Computer Sciences*, *44*(6), 2167–2178. doi:10.1021/ci049893v PMID:15554687

Zhu, K., Borrelli, K. W., Greenwood, J. R., Day, T., Abel, R., Farid, R. S., & Harder, E. (2014). Docking covalent inhibitors: A parameter free approach to pose prediction and scoring. *Journal of Chemical Information and Modeling*, *54*(7), 1932–1940. doi:10.1021/ci500118s

Zilian, D., & Sotriffer, C. A. (2013). Sfcscore rf: A random forest-based scoring function for improved affinity prediction of protein–ligand complexes. *Journal of Chemical Information and Modeling*, *53*(8), 1923–1933. PMID:23705795

Zilian, D., & Sotriffer, C. A. (2013). SFCscore(RF): A random forest-based scoring function for improved affinity prediction of protein-ligand complexes. *Journal of Chemical Information and Modeling*, *53*(8), 1923–1933. doi:10.1021/ci400120b PMID:23705795

Zou, K., Yamaguchi, H., Akatsu, H., Sakamoto, T., Ko, M., Mizoguchi, K., & Kosaka, K. et al. (2007). Angiotensin converting enzyme converts Amyloid beta-protein1–42 (Aβ1–42) to Aβ1–40, and its inhibition enhances brain Abeta deposition. *The Journal of Neuroscience*, *27*, 8628–8635. doi:10.1523/JNEUROSCI.1549-07.2007 PMID:17687040

Zou, X., Sun, Y., & Kuntz, I. D. (1999). Inclusion of solvation in ligand binding free energy calculations using the generalized-born model. *Journal of the American Chemical Society*, *121*(35), 8033–8043.

Zsila, F., Bikadi, Z., Malik, D., Hari, P., Pechan, I., Berces, A., & Hazai, E. (2011). Evaluation of drug-human serum albumin binding interactions with support vector machine aided online automated docking. *Bioinformatics (Oxford, England)*, *27*(13), 1806–1813. doi:10.1093/bioinformatics/btr284 PMID:21593135

Zsoldos, Z., Reid, D., Simon, A., Sadjad, B. S., & Johnson, A. P. (2006). eHiTS: An innovative approach to the docking and scoring function problems. *Current Protein & Peptide Science*, *7*(5), 421–435. doi:10.2174/138920306778559412 PMID:17073694

Zwanzig, R. W. (1954). High-temperature equation of state by a perturbation method. I. Nonpolar gases. *The Journal of Chemical Physics*, *22*(8), 1420–1426. doi:10.1063/1.1740409

About the Contributors

Siavoush Dastmalchi graduated as Doctor of Pharmacy from Tabriz University of Medical Sciences (TUMS). Then he moved to Sydney where he received his PhD from the Faculty of Pharmacy at the University of Sydney in 2002. Since then he has worked as a full academic in the Medicinal Chemistry Department at the School of Pharmacy, TUMS, teaching medicinal chemistry, instrumental drug analysis and bioinformatics to graduate and postgraduate students. He is currently the Director of the Biotechnology Research Centre at TUMS where he leads his research team mainly with interests in molecular modelling, structural biology, and chemo-bioinformatics for their application to drug discovery.

Maryam Hamzeh-Mivehroud is an associate professor who graduated as Doctor of Pharmacy from Tabriz University of Medical Sciences in 2004 and received her PhD in Medicinal Chemistry from this university in 2011. Since then she has worked as a full academic member in the Medicinal Chemistry Department at the School of Pharmacy, and teaches medicinal chemistry at the undergraduate and postgraduate levels. Her main research interests are focused on QSAR and molecular modelling based on various description selection algorithms and machine learning methods. She is also professional in working with modelling and statistical softwares such as GROMACS, GOLD, AMBER, GOLPE, MOE, etc.

Babak Sokouti is an assistant professor and has over 17 years IT technical management and consulting experience, including managing and maintaining sophisticated network infrastructures. He has obtained Bachelor of Science in Electrical Engineering with a specialization in Control from Isfahan University of Technology, Isfahan, Iran; a Master of Science in Electrical Engineering with a specialization in Electronics (with background of biomedical engineering) from Tabriz Branch, Islamic Azad University, Tabriz, Iran; a Master of Science in Information Security with Distinction from Royal Holloway University of London, London, UK; and obtained PhD in Bioinformatics from Biotechnology Research Center, Tabriz University of Medical Sciences, Tabriz, Iran. In addition, he has obtained IT industry certifications including MCP, MCSA 2003, MCDBA 2000, MCSE 2003, and MCTS 2008. His research interests include cryptographic algorithms, information security, network security and protocols, image processing, protein structure prediction, and hybrid intelligent neural network systems based on genetic algorithms.

* * *

Pravin Ambure received his M.S. (Pharm.) in Pharmacoinformatics from the National Institute of Pharmaceutical Sciences, Mohali (INDIA) in 2012. In 2012, he was appointed as junior research fellow (JRF) under Department of Biotechnology (DBT), Government of India funding scheme in the Department of Pharmaceutical Technology, Jadavpur University. In continuation, currently he is pursuing PhD under Dr. Kunal Roy in the same department. The field of his research interest is Molecular Modeling. He has published about 8 research papers and 2 review articles in peer-reviewed journals and also contributed 2 book chapters.

Marko Anderluh studied Pharmacy at the University of Ljubljana, Faculty of Pharmacy (UNI-LJ), where he obtained his Master degree in 2000 and his Ph. D. in Pharmacy in 2004. During his PhD study, he visited the group of Prof. Dr. Athanassios Giannis at the University of Karlsruhe, Institute for Organic Chemistry (2002), where he worked on in vitro assay for integrin receptor modulators. He was a postdoctoral associate at the University of Milan, Chemistry Department in the group of Prof. Dr. Anna Bernardi (2007-2008), where he worked on synthesis and characterisation of glycomimetic compounds. In 2008 he became Assistant Professor, and in 2013 he was promoted to Associate Professor at the same University (UNI-LJ). His research interests involve the design, synthesis and biological evaluation of molecular probes and drug candidates. In chemical sense, he focuses his research primarily on the design and synthesis of glycomimetics and glycoconjugates.

Khaled Barakat received his PhD in biophysics from the University of Alberta in 2012. He then joined Professor Michael Houghton's Lab at UofA as a postdoctoral fellow in 2012. In 2014, he got his first academic appointment as a Research Assistant Professor at the school of Pharmacy at UofA. Dr. Barakat's research stands at the multidisciplinary interface of physics, biology and computer science. His major focus is on developing and applying state-of-the-art computational drug discovery tolls to discover new antiviral and immunotherapeutic small molecule drugs. Throughout his research career, Dr. Barakat has made great contributions in understanding the nature and biophysical processes underlying protein-drug interaction, protein-protein interactions, protein-DNA interactions, drug off-target interactions and predicting drug-mediated toxicity. He has recently discovered novel small molecule inhibitors for various targets including the HCV polymerase and the NS5A protein. His lab also designed small molecule inhibitors for the PD-1 immune checkpoint pathway. During his career, he received numerous awards including the CIHR and AIHS postdoctoral fellowships, the prestigious UofA dissertation award, the ACRI Studentship and many distinction awards throughout his undergraduate and graduate studies. Dr. Barakat is also the editor of a number of journals including the journal of Pharmaceutical Care & Health Systems, Austin Journal of Drug Discovery, Development and Delivery and the Journal of MOJ Bioequivalence & Bioavailability. He also serves as a guest reviewer for the journals of PLOS ONE, BMC Clinical Pathology and Molecular Modeling.

Sonika Bhatnagar completed her Ph.D. in Biophysics from A.I.I.M.S. and currently works as Associate Professor of Biotechnology. Her research interests lie in the application of Computational and Structural Biology tools for gaining insights into Cardiovascular Disease and Bacterial Stress response with a long term focus on Drug target selection, Drug design and Repurposing.

Simanti Bhattacharya is a bioinformatician undertaking researches in relation to diseases. She completed her Ph.D. research exploring different structural and functional aspects of mutations related to muscular dystrophies. She holds a Masters Degree from University of Calcutta in Bioinformatics and Biophysics. She has also worked for 3.5 years at IISER-Kolkata in the areas of protein biochemistry of actin binding proteins. Presently she is a Scientific Associate at the Drug-Re-purposing group of GVK Biosciences, Hyderabad.

Jure Borišek obtained his Pharmacist degree from the University of Ljubljana, Faculty of Pharmacy, in 2011. Currently, he is a PhD student at University of Ljubljana (Slovenia), program Biomedicine, under the supervision of Prof. Dr. Marjana Novič. His PhD thesis focuses on design of inhibitors of cathepsin K and autolysin with application of chemometrics and structure-based drug design.

Amit Das received his M.Sc. in Biophysics and Molecular Biology in 2009 from the University of Calcutta. During his M.Sc. he has worked as summer trainee at the AstraZencea India R&D Pvt. Ltd., performing protein purification, assay development and molecule screening. Post M.Sc. he has worked at IISER-Kolkata till June 2013 as a research fellow undertaking cellular and molecular biology research with respect to actin exoskeleton. Thereafter he switched to Bioinformatics during his Ph.D. and performed molecular modeling, docking and simulations studies. Scripting, web development, knowledge resource creations are few areas of his prime interest.

Maruti Jayram Dhanavade working as research student at Department of Microbiology, Shivaji University, Kolhapur. He has worked on DST-PURSE scheme as a research assistant. He has published eight research papers in international journals and presented his research findings in several national and international conferences.

Ali HajiEbrahimi was a regular contributor of programming, web development and robotic engineering as a great experience before and within first years of his university years. After his entrance to pharmacy faculty of Shiraz university of medical sciences at 2010, his approaches widely directed toward hiring computer modeling and bioinformatics in biological and biotechnological systems and eventually achieved the first place award of medicinal chemistry in 19th Iran pharmacy students seminar for his novel algorithm on "interaction fingerprints". He recently collaborated with the department of biotechnology to finish his final thesis of Pharm.D about assessing inter-atomic interactions in the percise evaluation of molecular simulations.

Hamidreza Ghafouri is studying pharmacy at Shiraz University of Medical Science, class 2016. He is interested in Molecular modeling and computational chemistry .He is also interested in combining of pharmaceutical sciences and computational modeling to propose new dosage forms. In addition he works on new methods on 3D-QSAR analysis. He could win the first place prize for "DOCKFACE: an easy to use application programming interface(API) for virtual ligand screening (VLS) using AUTODOCK software" at 18th Iranian Pharmacy Student Seminar was held in Tabriz. He hopes to be a medical researcher in higher levels.

Ruchika Goyal is pursuing her Doctorate degree in Pharmaceutical Sciences from the Department of Pharmaceutical Sciences, Guru Jambheshwar University of Science and Technology, Hisar, Haryana, India. She has worked for more than one year in Drug Regulatory Affairs department of Medicamen Biotech Limited, Bhiwadi (RJ). She has bagged the University Gold medal in M.Pharm (Pharmaceutical Chemistry) examination from Guru Jambheshwar University of Science and Technology, Hisar. She has qualified Graduate Pharmacy Aptitude Test with 95.4 percentile. She also published two research papers in impact factor international journals. Her research area includes computationally designed and directed synthesis and evaluation of therapeutically active novel chemical entities.

Akanksha Gupta is a faculty member in the Department of Biotechnology, IMS Engineering College, U.P. She is currently on study leave and pursuing her Ph. D. degree. Her research work is in the field of Mining and Modeling of drug targets in Atherosclerosis. She has previously completed M.Tech. (Bioinformatics) from Jamia Hamdard, New Delhi.

Horia Jalily Hasani studied as an undergraduate at the University of Pune in India. She is currently a Master's student in the Faculty of Pharmacy and Pharmaceutical Sciences at the University of Alberta. Her research is focused on the computational modeling of cardiac ion channels, cardiotoxicity assessment studies using novel computational techniques and application of sophisticated simulations to challenges in modeling ion channels.

Md. Imtaiyaz Hassan is currently serving as a group leader in the Jamia Millia Islamia, New Delhi. He mostly use X-ray crystallography method to determine protein structures in order to understand their biological function, and to provide biotechnology applications. Structural studies are combined with a range of biophysical techniques to interrogate the structure-function relationship of the target proteins. Structure-based drug design is a major focus, where he design and screen small molecules that contribute to both drug development and understanding biological functions in human body.

Sandeep Jain is working as Associate Professor in Pharmaceutical Sciences at the Department of Pharmaceutical Sciences, Guru Jambheshwar University of Science and Technology, Hisar, Haryana, India. He has done his Ph.D from Dr. H. S. Gaur, University Sagar, M.P. He has about 19 years of teaching and research experience. He has to his credits more than eighty research papers in national and international journals. Six students have been awarded Ph.D. under his guidance and two are pursuing.

Amandeep Kaur Kahlon has completed postgraduation in Toxicology and PhD in Biotechnology. Her research interests are host-pathogen interactions, structural biology, drug discovery, microbiology and bioinformatics. At present, she works as Research Associate at ICGEB, N. Delhi, India and on host-pathogen interaction studies to find peptide inhibitors targeting M. tuberculosis pathogen.

Akhil kumar has obtained his M.Tech. degree from Indian Institute of Technology (IIIT)-Allahabad, in 2010. Currently, he is pursuing his Ph.D. degree from CSIR-Central Institute of Medicinal and Aromatic Plant, Lucknow. His research interest includes drug discovery and interaction studies of Alzheimer's disease target proteins.

Ashwani Kumar is working as Assistant Professor in Pharmaceutical Sciences at the Department of Pharmaceutical Sciences, Guru Jambheshwar University of Science and Technology, Hisar, Haryana, India since 2006. He has taught at Hindu college of Pharmacy, Sonepat (HRY) for more than one year. He has to his credits more than twenty research papers in international journals of repute. He has guided several M. Pharm. students for their research projects. He is dynamically engaged in teaching and research as well as other professional activities. He has completed his Ph.D., M. Pharm. and B. Pharm. degrees from Department of Pharmaceutical Sciences, Guru Jambheshwar University of Science and Technology, Hisar, Haryana, India. He has been conferred with "Outstanding Reviewer Award, March, 2015" by European Journal of Medicinal Chemistry in recognition of the contributions made to the quality of the journal.

Pallavi Mohanty is currently pursuing her Ph.D. degree. Her research work lies in the area of Modeling and simulation of critical drug targets in Cardiac Hypertrophy. She completed her M.Sc. (Biotechnology) from Utkal University, Orissa. Subsequently, she has research experience of 3 years at GVK Biosciences, IIT(Delhi) and National Institute of Malaria Research, New Delhi.

Marjana Novič was born in Ljubljana, Slovenia, where she took studies in Physical Chemistry at the University of Ljubljana during 1974-79. She obtained Ph. D. in 1985 at the Faculty of Chemistry and Chemical Technology, University of Ljubljana with the thesis "Hierarchical Clustering and Recognition of Chemical Structures and Structural Fragments on the Basis of 13C NMR spectra. She started her career at the National Institute of Chemistry in Ljubljana, initially developing automated information systems in for infrared and NMR spectroscopy. Post-doctoral specialization was carried out at University of Lausanne, Switzerland (1986-87) in the field of automated pattern recognition in 2-D NMR spectra. In 1989 she has obtained "Boris Kidrič" national award for scientific achievements. Later she visited several research laboratories (in Tarragona and Cordoba, Spain, and in Hobart, Tasmania), while being employed in the Laboratory of Chemometrics at the National Institute of Chemistry, Ljubljana, Slovenia; currently she has a position of a head of the Laboratory. Her expertise includes the development of chemometrics methods, QSAR and ANN modelling, structural elucidation of transmembrane segments of membrane proteins, innovative merging of chemometrics methods with molecular modeling, which facilitates an effective drug design. She is also teaching chemometrics at the University of Ljubljana.

Mohsen Ranjbar was entered Pharm.D project of Shiraz University of Medical Science at 2010. He is a member of class 2016. His thesis is about synthesis of some new azole derivatives for antifungal effects. In addition to benchworks, he interests in computational chemistry and programming. Also, he has entered in some projects to develop some new methods for 3D QSAR. In other hand, he is deeply interested in pulmonary formulations and wants to be involved into related pharmaceutical projects.

Kunal Roy (http://sites.google.com/site/kunalroyindia/) is a Professor in the Department of Pharmaceutical Technology, Jadavpur University, Kolkata 700 032, India. Dr. Roy has been a Marie Curie International Incoming Fellow in the University of Manchester during 2013-15. He is the Editor-in-Chief of International Journal of Quantitative Structure-Property Relationships (IJQSPR) of IGI Global. He is also an Associate Editor of the Springer Journal Molecular Diversity and a member of the Editorial Advisory Board of European Journal of Medicinal Chemistry (Elsevier). The field of his research interest is QSAR and Molecular Modeling. Dr. Roy has published more than 240 research papers in refereed journals (current h-index 33).

Amirhossein Sakhteman (Pharm D, PhD) has been graduated from Tehran University of Medical Sciences by degree of PhD. He was one of the founders of Faculty of Pharmacy, Shahid sadoughi University of Medical Sciences, Yazd, Iran where he spent three years on teaching oraganic chemistry, medicinal chemsitry and drug design. He is now an assistant professor of medicinal chemistry at school of pharmacy, Shiraz University of Medical Sciences. His research topics are mainly focused on computer assisted drug design, molecular dynamic simulation and synthesis of small molecule compounds leading to publication of 25 papers and a book chapter (GPCRs) so far.

Ashok Sharma, PhD, Chief Scientist and Head, Biotechnology Division, CSIR-Central Institute of Medicinal and Aromatic Plants, is Professor of Biological Sciences in AcSIR (Academy of Scientific and Industrial Research), CSIR, and Adjunct Professor, JNU, New Delhi, for CIMAP-JNU PhD programme. He is also Coordinator of Bioinformatics Centre at CSIR-CIMAP, Lucknow, India. His research interest includes bioinformatics and computational biology of medicinal and aromatic plants, bioprospection of phytomolecules in infectious diseases and metabolic diseases for human health, metabolic pathway elucidation and modulation, analysis of regulatory elements, DNA barcoding of plants, microRNA and Phytoremediation. He has published more than 70 research and review papers, 10 books and book chapters.

Kailas D. Sonawane, presently working as Professor, Department of Biochemistry, Shivaji University, Kolhapur. He is also working as Head (I/c) Department of Microbiology, Shivaji University, Kolhapur since 2007. He is a founder member of one year Post Graduate Diploma in Bioinformatics course started in 2008. Prof. Sonawane did his PhD from National Chemical Laboratory, Pune and thereafter earned his post-doctoral fellowship from National Institutes of Health (NIH), National Cancer Institute (NCI), Bethesda, Maryland, USA. He has authored more than 47 national/international publications and serves as reviewer in the editorial board of different national/international journals. He has received research financial assistance from various funding agencies in India. He has delivered several invited talks in national/international conferences, workshops and seminars. He has worked as a member Board of Studies in Biotechnology and Bioinformatics of Dr. Babasaheb Ambedkar Marathawada University, Aurangabad and Ad-hoc board for Model College, Hingoli, S. R. T. M. University, Nanded.

Vijay Kumar Srivastav is presently working as Assistant Professor (Medicinal and Pharmaceutical Chemistry) in Department of Pharmacy, Shri Govindram Seksaria Institute of Technology and Science, Indore, MP, India. He has completed his M. Pharm (Medicinal and Pharmaceutical Chemistry) in 2009 and currently pursuing PhD under the supervision of Prof. Meena Tiwari, in the same Institute. His research interests are Drug-design, Synthetic chemistry and Pharmaceutical analysis. He has authored several research papers in national and international Journals.

Tjaša Tibaut obtained the master of biotechnology from the University of Ljubljana, Biotechnical Faculty, in 2014. Currentyl, she is a PhD student at University of Ljubljana (Slovenia), program Biomedicine, under the supervision of Prof. dr. Marjana Novič. Her PhD thesis focuses on design of inhibitors of autolysin and molecular modeling of bilitranlocase with application of chemometrics and structure-based drug design.

Meena Tiwari is presently working as Professor (Medicinal and Pharmaceutical Chemistry) in Department of Pharmacy, Shri Govindram Seksaria Institute of Technology and Science, Indore, MP, India. She obtained her PhD degree in Pharmacy in 1988 from Dr. H. S. Gour University, Sagar, MP, India. She has guided more than 50 M. Pharm and 3 PhD students. She is principal investigator of UGC (University Grant Commission) and AICTE (All India Council of Technical Education) funded research projects. She has published more than 40 research papers in national and international SCI Journals. Her research interest mainly focused on Drug design, Synthetic chemistry, and Polymer chemistry.

Tihomir Tomašič was born in 1981 in Novo Mesto, Slovenia. He studied pharmacy at the Faculty of Pharmacy, University of Ljubljana, Slovenia and obtained a Ph. D. in pharmacy in 2011. During his Ph. D. he performed a part of his research work in Prof. Dr. Gerhard Klebe's group at the Institut für Pharmazeutische Chemie, Philipps Universität Marburg, Germany. Currently, he is employed as an assistant professor at the Chair of Pharmaceutical Chemistry at the Faculty of Pharmacy, University of Ljubljana. His research work involves mainly the design, synthesis and computer-aided design of biologically active compounds, particularly with antibacterial and antiviral activity.

Shubhandra Tripathi hasobtained his M.Sc. degree from University of Allahabad in 2012. Currently, he is pursuing his Ph.D. degree from CSIR-Central Institute of Medicinal and Aromatic Plants, Lucknow. He is working on *in silico* molecular interaction studies of breast cancer targets. His research interest includes molecular modeling, molecular docking, molecular dynamics simulation and binding free energy estimation methods.

Jin Wang is currently a professor of Chemistry and Physics at Stony Brook University. Prof. Jin Wang got his Ph.D. in Astrophysics in University of Illinois at Urbana-Champaign in 1991. He then became a postdoctoral research associate on biophysics working on protein dynamics, protein folding and single molecules until 1996. He was a guest scientist at NIH from 1996 to 1997. He was a vice president and a senior analyst at Global Strategic Analytical Unit in Citibank (Citigroup) from 1997-2004. He joined the faculty of Stony Brook in 2004. Prof. Jin Wang has been working on biophysics of molecular and systems biology including protein folding, protein conformation dynamics, molecular recognition, cellular networks, differentiation and development, cancer, evolution and ecology. He has also been working on classical and quantum equilibrium and non-equilibrium statistical mechanics. He has developed and quantified the concept of intrinsic specificity and applied to ligand-protein interactions. He and Dr. Zhiqiang Yan developed a new scoring function called SPA for quantifying both affinity and specificity of ligand-protein interaction. He has also developed a multi-dimensional virtual screening strategy of selecting top candidates for drug discovery and design.

Zhiqiang Yan is currently working as associate professor in Changchun Institute of Applied Chemistry, Chinese Academy of Sciences. He received his Ph.D. from physical department of Nanjing University in 2010. He then worked as postdoctoral associate in the chemistry department of Stony Brook University from 2010 to 2013. His research interests include molecular modeling of bimolecular recognition, computer-aided drug design and discovery, as well as molecular dynamics simulations on the biomolecules.

Index

E

F

G

H

I

K

L

M

O

P

Q

R

S

T

V

www.ingramcontent.com/pod-product-compliance
Lightning Source LLC
LaVergne TN
LVHW060355300926
849162LV00006B/1273
* 9 7 8 1 5 2 2 5 0 1 1 5 2 *